The Congress was organized by the UK Scientific Committee on behalf of the International Advisory Committee and through the auspices of the Manchester Conference Centre at UMIST.

Congress and Exhibition Organizer
P B Kenway   Manchester Materials Science Centre, University of Manchester/UMIST, Grosvenor Street, Manchester M1 7HS, UK

UK Scientific Committee

Professor P J Duke	SERC Daresbury Laboratories
Professor G W Lorimer	University of Manchester
Professor T Mulvey	University of Aston
Dr I W Drummond	Kratos Analytical, Manchester
Dr G Love	University of Bath
Dr A G Michette	King's College, London
Dr M Stedman	NPL, Teddington

Editors
P B Kenway, P J Duke, G W Lorimer, T Mulvey, I W Drummond, G Love, A G Michette, M Stedman

International Advisory Committee

Professor J D Brown	University of Western Ontario, Canada
Professor O Brummer	Martin Luther University, Wittenburg, Germany
Professor R Castaing	Universite Paris Sud, Orsay, France
Professor T Ichinokawa	Waseda University, Tokyo, Japan
Professor S Jasienska	Academy of Mining and Metallurgy, Poland
Professor R E Ogilvie	MIT, Massachusetts, USA
Dr Kurt F J Heinrich	Center for Analytical Chemistry, MD, USA
Dr G Mollenstedt	Universitut Tubingen, Tubingen, Germany
Dr D B Wittry	University of Southern California, USA

X-Ray Optics and Microanalysis 1992

Proceedings of the Thirteenth International Congress, UMIST, Manchester, UK, 31 August–4 September 1992

Edited by P B Kenway, P J Duke, G W Lorimer, T Mulvey, I W Drummond, G Love, A G Michette and M Stedman

Institute of Physics Conference Series Number 130
Institute of Physics, Bristol and Philadelphia

CODEN IPHSAC 130 1–652 (1993)

British Library Cataloguing in Publication Data

A catalogue record for this book is available from the British Library

ISBN 0-7503-0255-0

Library of Congress Cataloging-in-Publication Data are available

Published by IOP Publishing Ltd, a company wholly owned by the Institute of Physics, London
Techno House, Redcliffe Way, Bristol BS1 6NX, England
US Editorial Office: IOP Publishing Inc., The Public Ledger Buildings,
Suite 1035, Independence Square, Philadelphia, PA 19106, USA

Printed in Great Britain by Bookcraft, Midsomer Norton, Nr. Bath, Avon

Preface

This, the XIII ICXOM, marked the return of Congress to the UK for the first time since Dr V E Cosslett organized the first meeting in Cambridge in 1956. Unfortunately, as Ellis Cosslett died in November 1990, he was unable to witness the event and it was decided to dedicate both Congress and the Proceedings to his memory and organize a symposium in his honour.

The Symposium began with a showing of part of a video of Ellis Cosslett being interviewed by Tom Mulvey in 1981 (Royal Microscopical Society Archive) in which Ellis recalled his early involvement in much of the early research in electron microscopy, X-ray microscopy and microanalysis. The video was followed by talks given by eight invited speakers—T Mulvey, W C Nixon, R W Horne, Audrey Glauert, J V P Long, P Duncumb, A Boyde and P W Hawkes—all of whom had long associations with Cosslett and ICXOM. They reviewed developments in electron and X-ray microscopy and microanalysis and in many cases related these to their own early investigations. It was apparent from these reviews that Ellis Cosslett played a crucial role in stimulating research, in motivating both students and staff, and perhaps most importantly in sensing what was achievable at the time.

Delegates were welcomed to Congress by Professor J D Brown, Chairman of the International Advisory Committee, and then delegates separated to join one of the two parallel sessions running throughout the week. One of these was on the theme of electron beam microscopy, microanalysis and complementary microscopies and the other on surface analysis and X-ray sources, optics and microscopy.

The standard for the EPMA sessions was set by S J B Reed's excellent paper in which both instrumental and theoretical developments were comprehensively reviewed. From this it was evident that much effort is being expended currently upon the analysis of thin surface films and this was emphasized subsequently in the papers presented by Bishop, Laurie and Auclair which described the progress being made, the levels of precision now achievable and the pitfalls involved in obtaining quantitative information.

A review of quantitative procedures for conventional EPMA by K F J Heinrich illustrated current performance but also highlighted deficiencies in existing databases and in the quality of some input parameters. The latter point was taken up by J D Brown with new measurements and calculations of $\varnothing(0)$ and by R E Ogilvie who outlined a new mass absorption equation. In a most interesting paper A P Mackenzie explained how an uncertain mass absorption coefficient could be treated as an additional unknown in the correction routine provided the measurements were performed at more than one voltage.

After several years there was a resurgence of interest in studying peak shifts and peak shape changes arising from the effects of chemical bonding. It was shown how such work could provide information on, for example, valency (Amthauer) and oxidation states (Kucha) and how peak deconvolution could be used for quantitative mapping (Critchell) and to aid specimen characterization (Love).

In an invited paper S J Pennycook described how high angular detectors can be used to obtain images from a STEM at atomic resolution which show strong compositional sensitivity and are easy to interpret. This was followed by an interesting application of STEM to quantify grain boundary segregation by J A S Ikeda. The sometimes overlooked problem of specimen damage by the small, high current density probes of a STEM was highlighted by P E Champness.

D B Williams described developments in the quantitative analysis of thin specimens in ATEM with particular reference to the technique of 'spectrum imaging' in EELS. This technique involves the acquisition of complete EELS spectra at each pixel followed by rapid complex data manipulation and compositional imaging.

The session on complementary microscopy was an innovation for ICXOM, recognizing the importance of the new scanning probe microscopies that have emerged during the last decade. H J Butt reviewed the surface imaging capabilities of STM and AFM, whilst M Stedman stressed the importance of traceable calibrations of these instruments to validate the dimensional measurements from the images. A novel means of investigating tip shape was presented in the paper by P Seidle. In the same session A Boyde presented an in-depth review of the recovery of 3D information from SEM images.

P W Hawkes's invited paper examined the fundamentals and various techniques of image processing. The following paper on image processing of scanning electron images of clays by Xiaoling Leng proved to be interesting, as were the papers by E van Cappellen and R Al-Saffar.

The surface science sessions were well endowed with outstanding invited speakers and a full update on much of the present thrust of work in the fields of XPS and AES was presented. The perennial problem of secondary electron background was enthusiastically approached by J A D Matthew under the intriguing title 'Never mind the baby, how about the bath water?'! Modern developments in electron spectroscopy detector technology were outlined by J Comer and we were lucky enough to catch the nomadic J A Venables on this side of the Atlantic to update us on his teams' projects at Sussex and Arizona. P Kruit gave us food for thought with the intriguing prospect of very high spatial resolution core photoelectron spectroscopy using electrons to create core vacancies. D W Turner showed some breathtaking slides taken from the Oxford University Photoelectron Spectromicroscope (PESM).

Contributions on ionization functions and backscattering were made by the ever ebullient J Cazaux and by A Assa'd. Another 'perpetual' area of investigation, namely depth profiling, received attention from G N Van Wyk and from S V Kuchaev whilst L Frank illustrated an elegant method of checking surface cleanliness.

P Morin discussed aspects of Auger images whilst K Robinson showed the value of magnetic lenses in imaging XPS and presented preliminary results on in-situ polymer lithography. The only SIMS contribution was on a novel SIMS/SNMS optical system presented by S M Smith. D Golijanin gave a thoroughly professional contribution on optical photoemission spectroscopy.

Perhaps the most lively discussions were stimulated by the presentations of P J de Lange and by A Wirth on the XPS of wood and automotive brake material respectively. We all know a lot about wood and car brakes!

The X-ray sources, optics and microscopy sessions were most enlightening and brought us up-to-date with the important developments in the use of non-synchrotron sources. A review of laser-plasma x-ray sources was given by F Bijkerk, and their specific use in giving the first non-synchrotron scanning X-ray micrographs was described by A G Michette. Other novel X-ray microscopes were described by J Theime, using a pinch plasma source, and R E Burge, using an XUV laser. The progress of X-ray optics to higher efficiencies and better resolutions was discussed by J Theime and D Morris. The potential of X-ray microscopy was demonstrated by the increasing range of applications in biological and materials problems as described in papers by C Jacobsen, J Theime, P C Cheng and J Cazaux.

The availability of X-ray sources based on synchrotron radiation (SR) produced from high energy electron storage rings is opening a new era in the application of both soft and hard X-rays. At the same time, the extremely high power of these sources is generating new challenges to the designers of X-ray optical equipment. The purpose of the synchrotron radiation session was to introduce delegates to the basic principles of these sources, and to describe how they are extending the application of X-rays into new areas. P Pattison described the new sources which are being constructed in Europe, the USA, and Japan, which will offer experimenters X-ray beams of unparalleled brightness and brilliance. Although the beams from these sources are at the design stage the amount of radiation is calculable to a high degree of accuracy, enabling collaborating groups of engineers and physicists to prepare designs and construct beam line optics which will handle the kW/cm^{-2} X-ray thermal loads which are anticipated.

L Vinze described one particular design being prepared at the European SR facility for X-ray fluorescence microscopy. In the soft to medium energy X-ray region, X-ray microscopy at high resolution is already a reality at existing SR sources as demonstrated by C J Buckley in his description of a variety of microscopies being applied to the imaging of calcium deposits in human cartilage. X-ray microprobes produced by using Fresnel zone plates are being used (Buckley) but progress towards using Bragg–Fresnel optics was described by A I Erko. A wide variety of techniques, which are being applied to the science of materials, were described by J V Smith. These techniques permit sensitivity levels down to parts per million using fluorescence from a $10\mu m$ beam and Laue or Bragg diffraction from crystals of $1-10\mu m$ dimensions. A specific application of time resolved X-ray diffraction to the study of semi-crystalline polymers was described by W Bras. D P Woodruff demonstrated that these new directions of X-ray experimentation are not confined to energetic X-rays. The use of magnetic undulators installed in electron storage rings is transforming the study of surfaces and near-surface regions of solid materials. Spectroscopy and diffraction, stimulated by intense beams of far VUV and soft X-rays, give high spectral resolution and make possible time resolved studies of absorbed chemical species.

The SR session concluded with a visit to the synchrotron radiation source at the nearby SERC Daresbury Laboratories. This gave delegates the opportunity to see at first hand the actual operation of an SR source and to talk with scientists developing and using the advanced equipment which is available for the users of such a facility.

A number of companies supported the associated trade exhibition and the organizer thanks them for their participation particularly in view of the present economic parsimony.

On the social side the provision of coffee, lunch, tea and dinner for the delegates throughout Congress promoted a convivial atmosphere in which much discussion took place. The Tuesday evening reception at the Museum of Science and Industry amongst the working exhibits of steam engines certainly revived a few nostalgic memories and in the Air and Space Gallery afterwards the chance to clamber up into a Halifax long-range search and rescue plane and 'pilot' it proved irresistible. On the Thursday evening delegates wined and dined sumptuously at the banquet held in a local Chinese restaurant.

P B Kenway
P J Duke
G W Lorimer
T Mulvey
I W Drummond
G Love
A G Michette
M Stedman

Acknowledgments

The organizer acknowledges and thanks the following institutions and companies for their generous financial sponsorship which enabled us to offer many young scientists burseries to attend Congress and to organize the Cosslett Symposium.

> The Royal Society
> The Institute of Physics
> The Royal Microscopical Society
> Imperial Chemical Industries
> BP International Limited
> Weir Materials Limited
> Unilever plc

The organizer also thanks the Royal Microscopical Society for the loan of the Cosslett interview video from their archives.

Thanks also to the UK Scientific Committee members for their effort and support, to the UMIST Conference Centre and in particular Helen McGlashan for all her help, to Joan Colclough for typing the hundreds of letters to delegates, and to my colleagues Ian Brough and Graham Cliff who shouldered some of my workload to allow me time to organize the Congress.

P B Kenway

The Congress and these Proceedings are dedicated to
Dr V E Cosslett FRS

V E Cosslett at the Annual Dinner of the Institute of Physics in May 1971, when he shared the Duddell Medal with K C A Smith for outstanding contributions to the design and construction of High Voltage Electron Microscopes. Looking at the Medal are Dr Audrey Glauert, President, and Dr Gerard Turner, Vice President, of the Royal Microscopical Society, whose Electron Microscopy Section was formed in Ellis Cosslett's seminal Presidency of the RMS.

Dr V E Cosslett FRS, 16 June 1908–21 November 1990

In 1946 Dr V E Cosslett (Ellis) founded the Electron Optics Section of the Cavendish Laboratory in the Department of Physics at Cambridge University. The laboratory attracted researchers from all over the world to carry out seminal research in X-ray optics and microanalysis.

It was characteristic of Ellis Cosslett that he took a close interest in all disciplines that could benefit from X-ray and electron microscopy. He took enormous trouble to assimilate the growing research literature in these areas and to get to know all the researchers, young and old, who were engaged in a most diverse range of disciplines worldwide. Thus he was able to invite some of the most promising researchers in the world to come and join his Group in Cambridge. Ellis also offered his microscope facilities and expertise to researchers in other Departments within Cambridge University. Many of these Departments eventually acquired their own instruments and soon there was a substantial growth in electron microscopy in Cambridge with strong world-wide links.

The idea of an International Congress on X-ray Optics and Microanalysis (ICXOM) was the brain-child of Ellis and he organized the first Congress in Cambridge in 1956. It was an outstanding success and many lifelong friendships were formed at the Congress. It was agreed to hold a similar Congress every three years and subsequent Congresses have been held in eleven different sites around the world.

At the XII ICXOM 89 held in Cracow it was planned that Ellis should preside at XIII ICXOM 92 to be held in the UK. It was not to be; Ellis died on 21 November 1990 at the age of 82. It was decided that XIII ICXOM 92 should be dedicated to his memory and that a special symposium be held in his honour.

On 21 November 1981, at the age of 73, Ellis was interviewed by Tom Mulvey and this was recorded on video tape for the Royal Microscopical Society archives. Part of this video was shown by Tom Mulvey as the opening to the Cosslett Symposium. During the remainder of the symposium friends, colleagues, and ex-research students reviewed developments in X-ray optics and microanalysis in the light of the inspiration and stimulus that Ellis had provided. From these talks it was clear that in the 1950's and 60's great ingenuity was required in building equipment and in interpreting the results. Recently, the advent of bright X-ray sources and the impact of computers has given renewed impetus to all the above disciplines and many of the ideas initiated by Ellis can now be realized.

Contents

Chapter 4: Complementary Microscopy, STM, AFM

Chapter 7: X-Ray Sources, Optics and Imaging

The Cosslett Symposium

The Cosslett Symposium: Chairman's introductory remarks
T Mulvey

X-ray projection microscopy and transmission electron microscopy
W Nixon

Early days in the electron microscope section of the Cavendish Laboratory
R W Horne

Ellis Cosslett: a physicist fascinated by biology
Audrey Glauert

Microanalysis
J V P Long

Scanning electron probe microanalysis
P Duncumb

In memory of V E Cosslett: from scanning with electrons to scanning with light in the investigation of mineralised tissues
A Boyde

The end of an era?
P W Hawkes

Inst. Phys. Conf. Ser. No 130: Chapter 1
Paper presented at Int. Congr. X-ray Optics and Microanalysis, Manchester, 1992

1

The Cosslett Symposium: Chairman's introductory remarks

T Mulvey

Department of Electronic Engineering and Applied Physics, Aston University, Birmingham B4 7ET

ABSTRACT: In this symposium, dedicated to Vernon Ellis Cosslett (VEC), principal founder of the ICXOM Congresses, eight speakers, with long associations with VEC and with ICXOM, review the present state of X-ray optics and microanalysis in the light of the early beginnings, many of which were stimulated by VEC and his group.

1. INTRODUCTION.

The first ICXOM Congress took place in 1956, in Cambridge, when Cosslett realised that there was a need for physicists, designers of equipment and those who were using X-ray microscopes and electron probe analysers to come together every three years to exchange experiences, learn what was going on in the subject as a whole, and gain new ideas for further research. The following speakers will be mainly concerned by events following that first Congress. I would like to set the scene to indicate how it came about that Cosslett acquired the broad inter-disciplinary skills that he undoubtedly possessed, together with the persuasive powers of attracting gifted research students and post doctoral researchers to the Cavendish to work on the theory, practice and instrumentation of X-ray optics and electron probe microanalysis.

V E Cosslett was born on June 16, 1908 and came to the Cavendish Laboratory in 1947 at the age of 39, starting as a "mature post-doctoral researcher" holding an ICI Research Fellowship, tenable for three years. He stayed there for over 30 years becoming a leading international figure in research into X-ray projection microscopy, electron probe analysis and electron microscopy. He began his research career at the very beginning of the invention of the electron microscope and continued until his death on 21 November 1990 to take a detailed interest in these fields and the people working in them. The following is a condensed and edited version of a hitherto unpublished interview with Ellis Cosslett, recorded in 1981, by courtesy of the Royal Microscopical Society as part of its archives. In this interview, I put a few questions to him:

2. AN INTERVIEW WITH ELLIS COSSLETT

TM. I think, Ellis, that we are now, in 1981, in a trade depression; you also started your career in not very auspicious circumstances, namely the depression of 1931. Can you recollect how you came into this field?
VEC. Yes, it is difficult to remember the precise order of happenings, but I have some definite remembrance of when I first heard of the electron microscope. It was way back in 1931. At that time I was a research student at Bristol University, where I previously graduated and one day Professor

Tyndall asked me if I would give a talk in the weekly colloquium on a new instrument which he had just read about, in the German literature, called the electron microscope. He said "You've got some German, Cosslett, would you like to mug it up and tell us about it", which I did; it must have been about the end of 1931 or the beginning of 1932.

TM. And you were an ICI Research Fellow?

VEC. No. At that time I was an H H Wills Memorial Fellow. Bristol was very much supported by the Wills family and we were working in the Wills Physical Laboratory. And so I mugged it up and gave a talk and I became very interested in it. I had begun to do some electron diffraction in London, in University College London, the previous year.

TM. That was the Heroic Age, where safety regulations were not foremost in your minds when you were adjusting the high voltage equipment?

VEC (laughing). That's very true. We had what we regarded then as very high voltage, of the order of 60 to 70 kilovolts, inside a wire cage; it had chicken netting around it to prevent you from getting into it, and I remember very well, once upon a time, someone pointing up at the gun and getting a spark off it. He just said "I must stand further back!". But those were the days, you are quite right. And it is quite true with the x-ray apparatus; it was also very ill-shielded.

TM. And that was also in the Cavendish tradition. I think that George Crowe lost a finger due to handling radioactive materials?

VEC. That is right. Yes, we may come back to Crowe presently because he was Rutherford's personal assistant; he used to handle all the radioactive materials. It was all very crude. We had no lead gloves etc. Anyway he later was in charge of of the electron microscope at the Cavendish, when I first came there, later on.

TM. I think that in those days, it was every man for himself in research; no one was coming to help you to assemble equipment or mend vacuum systems.

VEC. Oh, that's not quite true, Tom. There were a certain number of assistants in the Cavendish, for instance, or in Bristol. Let's go back to Bristol. Mr Zen, the Head Assistant, had a first-class glassblower, a man called Burrows; you must remember in those days technology was very primitive, you didn't go and buy a vacuum pump; you asked the glassblower to blow you a mercury pump. And there were very few people able to do that. You would have a chance to get some help, but you had to look for all the leaks and troubles and so on.

TM. And while you were doing this, I think Knoll and Ruska were doing similar hair-raising experiments in Berlin with their electron microscope. Would you like to say something about that?

VEC. The first electron microscope was built by Ernst Ruska and his supervisor Max Knoll, who later went into television so that the whole of the further development of the subject in the Technological University of Berlin, was carried on by Ruska and his collaborators.

TM. Would you think that it would be fair to say that although Knoll was a brilliant scientist, he hadn't a good feeling for the development of electron microscopy?

VEC. I think that's almost certainly true; in some sense it is true of Ruska himself. One has to say that the impetus for the further development, the negotiations with manufacturers and so on, was carried out by von Borries from the same lab, from the same room, in fact with Ruska. They were great friends. Later on von Borries married the sister of Ernst Ruska.

TM. And also, I think, in Berlin at that time a surprising number of people became interested in electron optics and electron microscopy; for instance the AEG Gesellschaft?

VEC. Yes it is a bit difficult to unravel at this date whether it all came from a common source, whether people picked up ideas from Ruska or vice versa. I think a lot of it was independent. It happened that at the AEG,

Brueche and Scherzer went in the direction of electrostatic lenses. There was another independent group under von Ardenne.

TM. He is a very interesting character. I first met von Ardenne´s work in the public library in Stockport around 1935 when I was a schoolboy.

VEC. Really!

TM. He wrote many excellent books on television and was an ingenious inventor and manufacturer. He was a source of inspiration to you later, I think, for the X-ray work?

VEC. Yes, indeed! He was a great promoter of his own ideas but from his publications you could never be quite sure whether he had made the thing work. I was lucky; I saw the last issue of Naturwissenschaften to come to the UK before the war. It had one of his ideas about X-rays which we followed up.

TM. In 1935 you went to Faraday House?

VEC. Yes, I had run out of my time at Bristol as a Wills Research Fellow. I tried all sorts of places, as the young men are also doing in 1981-82, I suppose, and I got nothing in the way of an academic place, till an advertisement appeared from Faraday House, which described itself as an "Engineering College for the sons of Gentlemen". It was set up in the 1880´s before Electrical Engineering was respectable at universities; the Engineering industry set it up on their own account and funded it themselves. They called it Faraday House, understandably. They had a grand man in charge, one Alexander Robinson, a man of some eminence, he was an FRS; he was running the thing very well at a level we would now call HNC, Higher National Certificate, Higher National diploma level, not quite high enough to be recognised by the Universities.

TM. And what princely salary was offered?

VEC. I think my salary was of the order of £500 or £600 a year, but the load was not very heavy; the level was very low; I was teaching elementary physics and fairly elementary chemistry, which gave me time.

TM. Did that give you the idea of doing something at Birkbeck College, part-time?

VEC. Well, it gave me the idea of finding some place to follow up my own interests, as it was unheard of in Faraday House for anybody to do research. And I learned, that at Birkbeck there was Professor Blackett running the Physics Department. I thought I might help with building electron lenses for focussing very high energy electrons, beta particles, from radioactive sources.

TM. So he took you on. I think that Blackett had been a Naval officer. He was a very forceful personality but he knew how to organise research. Could you say something about his influence?

VEC. He had been through the Cavendish, so he had the Cavendish tradition behind him. He left the Cavendish because he couldn´t get on too well with Rutherford; they were both very forceful characters, so he took the chair at Birkbeck. And the set-up was in a basement. You did research setting up electron lenses in basements, and I say "we" because I was working with an Indian student called Sen Gupta. Later on, I worked in the basement of the Cavendish. But Birkbeck was a nice place, in the middle of London. They certainly welcomed me. I must say I learnt a great deal from the attitude to research - the positive attitude of Blackett. He always assumed that there would be an answer; he was not one of these pessimists who emphasise the difficulties. "Go away and think about it" was his attitude.

TM. And also, this idea of doing a trial calculation or a trial experiment was also a characteristic?

VEC. That was partly a Cavendish characteristic which fitted him very well. Doing things on the back of an envelope as one says to see whether the thing was likely to come out and what the difficulties might be and what the limitations might be. He was very good at that.

TM. And that was very good experience for you when you had to undertake very large projects.

VEC. Yes.

TM. By 1938, things were looking a little bit better, but unfortunately the war intervened and just when you had to apply for another post......

VEC. -If I may interrupt- it happened that Blackett had been promoted to the Chair at Manchester and Bernal, who was a bright young crystallographer in Cambridge, again who did not get on with the authorities, took the Chair at Birkbeck. The apparatus that I had been working on with Blackett went with him to Manchester, and so I had to work out with Bernal what help I might be to him in London. There was no possibility of going to Manchester. I stayed on at the job at Faraday House, but I said to Bernal "Look, we ought to be able to improve X-ray tubes. After all, they have electron beams in them; if we focus the electron beams properly, we ought to be able to get very high intensities; let´s have a go at that. And how to do it? There happened to be in London University a research fellowship called the Keddy Fletcher Award, so I put in for that in 1939.

TM. Now, you never knew if you got it because the war intervened?

VEC. Yes, I knew I got it; it was to begin on the first of September 1939, almost the sme day as the war started (laughter from VEC).

TM. So that was another set-back for the plans. What happened then?

VEC. Well, we didn´t win the war straightaway, nor were we blown up. Nevertheless, there were evacuation plans for scientific research groups. Quite a number went to Cambridge or Oxford or Bristol. Bernal´s group were transported to Oxford because Dorothy Hodgkin, who was one of the leading crystallographers, had a group there, so his group went there. I tagged along; at first I didn´t think I could find laboratory space there but I was told that there was in fact the Old Electrical Laboratory, beside the Clarendon, which was the modern one. Lindemann, Lord Cherwell as he later became, was in charge of the Clarendon. The Electrical Laboratory was under Professor C S E Townsend. I was accepted in the laboratory because they had plenty of space and a good workshop. Townsend was quite interested to have someone do something new.

TM. Now, when did you actually see your first electron microscope, because that is always a crucial point in the development of the subject?

VEC. True! Yes, I had often read about it and written about it here and there, at the beginning of the war. I´ve quite forgotten how I learned that there was such an instrument in the National Physical Laboratory. It was commissioned, partly designed, by Professor L C Martin, who was Professor of Technical Optics, I think they called it, or Applied Optics, at Imperial College, London. I do not know how he got in contact with Metropolitan Vickers (M-V). But the Metropolitan Vickers experts were, at that time, Parnum and Whelpton. Between them and Martin, they developed this electron microscope. The M-V Research Laboratories in Manchester (later AEI Ltd.) made it. It was brought out with a unique design. It followed of course the electron lenses of Ruska, but near the gun, there was a curious castle-like affair, looking like the bridge of a ship. That was an idea of Martin, who was very doubtful about the electron microscope, so I understand, so he had this arrangement put in: a specimen chamber in which the specimen was on a large disc so that it could be observed in an optical microscope, situated to the left of the electron gun. When they had sorted out an interesting particle to look at, it was wheeled round in the vacuum to be looked at under the electron beam. So it was a curious form of specimen chamber to give an immediate cross-check on what you had hoped to see in the electron microscope.

TM. And I think the Royal Society were involved here; didn´t they give a grant to Professor Martin to have this constructed?

VEC. Yes, they helped him out, because it would be too expensive for those days. I don't think I saw it in operation. The man who was in charge of it at that time was G P Preston, who was later Professor in Dundee. They did a few investigations on dust particles; two short notes on dust particle sizing but nothing I remember substantial, partly because of the war. And there it sat, as you know, until it was taken back to Metropolitan Vickers, as the basis for later models, which they constructed and marketed after the war.

TM. Which was the first microscope that actually came under your control? Was it the RCA model?

VEC. Yes, the first one that I had any real responsibily for and the first one that I laid hands on was the RCA model in Cambridge. This is one of half a dozen which were brought over to this country in the great Lend-Lease Agreement with the United States, after the States came into the war. These RCA electron microscopes were a different design from anything we had seen so far. It was a unitary design; all the electronics were in a rack at the back of the column and at the top is the high voltage tank that produced up to 60 kilovolts.

TM. That is when you were actually in the Cavendish?

VEC. No, we have go a little futher back than that, Tom. They came to this country, I think, at the end of 1942 and there was a Scientific Committee of the War Cabinet, who decided where they should go. The man primarily responsible was Sir Charles Darwin, grandson of the original Charles Darwin; he was at that time Head of the National Physical laboratory and knew about electron microscopes. G P Thomson, who was one of the scientific advisors for our Government in Washington, also knew about them. They brought six TEMs in and decided to put one in the Wool Research Laboratory in Leeds under Astbury, one in the Cotton Research Laboratory Institute in Didsbury, Manchester, one in the Agricultural Institute, one in the Medical Institute, and the Cavendish got one, before I actually got there.

TM. When the war was over, you then had to think about electron microscopy and what your next move would be? You were thinking of applying for another Fellowship?

VEC. Well, I had to apply for something else, because although I'd been able to carry on a bit of research at Oxford it was not likely that I could stay there. We had set up an electron lens and found the troubles with it worse than I expected. One doesn't realise how bad spherical aberration is until one tries to get rid of it. I tried for several jobs including a Fellowship at Cambridge at the beginning of 1946.

TM. Now, how did you get to the Cavendish, eventually?

VEC. As I say, I applied for one of the ICI Fellowships, which had been devised by ICI after the war, specifically for the purpose of helping "mature" researchers, I was nearly 40, to get back into academic life. Cambridge had four or five, I think. So I applied; it seemed just the very thing. I applied for other jobs as well. After some time, this must have been about June 1946, I had heard nothing at all, so I wrote to Bragg and said that I had had an offer from elsewhere for an Assistant Lectureship and when was I going to hear. He wrote back at once to say "We have given the Awards; they've lost your papers but come for an interview"! Fortunately, by that time, I had been lecturing on electron optics and had written up my lectures in the form of a book, the proofs of which had just come from the Oxford University Press. So off I hared to Cambridge. Bragg and his Reader, Ratcliffe interviewed me and somehow they conjured up another Fellowship!

TM. And your books were the beginning of what one might call your apostolic mission in electron microscopy?

VEC. (laughing) Well, they helped me to sort out my ideas as to what was needed to establish the subject, as it turned out, in the Cavendish. But I was not yet in the Cavendish; all this came from Oxford.

TM. By the way, on your way to the interview, did you take the proofs with you?

VEC. Yes, the "Introduction to Electron Optics". And that suited Bragg in the Cavendish because there, we were to be concerned with the RCA instrument and later with the Siemens Microscope. Furthermore, they wanted someone to help keep the system running, to set up a Centre, if you like, as we would now say.

TM. Now, I think you have always been good, if I may say so, at making a case for something to be done, looking ahead and making practical proposals even though there are difficulties. I think you made an ambitious plan for future electron microscopy in the Cavendish? Can you remind us what you had in mind?

VEC. Yes, then as now , one had to go outside the laboratory to get the money that really mattered. The Cavendish was still in the string and sealing wax era; it was not easy to get big money; in fact neither the laboratory nor the University had big money. They had the electron microscopes, but for any further development, one had to go to the Research Council, which was then called the Department of Scientific Research (DSIR). And you´re quite right, Tom, I wrote them what I now, with hindsight, see as an extraordinarily ambitious programme, which has taken 20 or 30 years to work out, but then not surprising in view of all the difficulties, and so DSIR came up with a certain amount of support. Fortunately for me, within that plan, there was a proposal for a high resolution, high voltage microscope and as a long-term aim as well as the immediate development of technology. And money of course to aid research, i.e. help and assistance.

TM. You were also interested, I think, in X-ray projection microscopy and the possibility of biological application with X-rays.

VEC. That is to go back to von Ardenne´s idea. Yes, I took that up already at Oxford and I had one research student there. And that is another thing for which we asked DSIR´s help. That funding came with me and a research assistant to the Cavendish. And from there we went on, especially with Bill Nixon´s green fingers to make the system work for X-ray imaging and as you say, primarily for biological material, which you could have outside the vacuum, whereas in the electron microscope you normally have to have the specimens dried and inside the vacuum.

TM. Now at the same time, I think, you were able to acquire a "bathtub" electron microscope, the Siemens EM100.

VEC. Yes, that was one of 40 or 50 that Siemens made; they began to make them just before the war. In fact when I was with Bernal, he was about to raise funds to buy one of these models, which came out at the time of the one that Metrovick made. The column is like any other column, but at the top is this huge "bath tub"; it was almost as large as a bath tub, and inside was all the high voltage gadgetry including the batteries, which were used for powering the cathode. It came to the Cavendish in October 1942, as booty of war from Krupps Armaments in Essen. Several Siemens Microscopes were brought over and we got this one. Bob Horne, who came to me at that time helped us enormously at that time, getting it working with the help of George Crowe.

TM. Now, talking of Bob Horne and many of your other associates, I think you were very fortunate in being able to attract such high calibre people. They all seemed to blossom out by being members of the EM Section?

VEC. Well, by being given opportunities, I think and having a nose for research. Horne, for instance, didn´t have a degree. Partly because of the war, he never went to University; he was in the RAF, I think. But he had a keen interest in research so we were very glad to have him; he knew a lot of electronics which most of us didn´t. Yes, you´re quite right Tom, it was partly also this favourable atmosphere in the Cavendish.

TM. Is it an apocryphal story, that since Cambridge doesn't recognise the degrees of any other University that you were able to take on students without degrees, Ray Dolby for example?

VEC. Dolby came later on with a degree, it is true from Stanford, California. He registered in the ordinary way as a research student so by that time, which was a few years after the time we were speaking of, about the year '62 or '63; but on the other hand the regulations even then were not as tight as they are now; one still had to have quite a good degree from a recognised University, which Stanford was, of course, or Harvard or Berkeley, but not anybody would get in. But it's interesting that in other places as well as Cambridge, if you read the small print of the regulations, you will find somewhere "in normal circumstances" or some such phrase and that is how we got him in. Later on he took a Cambridge doctorate. I know of similar cases in Oxford; they also had the same small print. The Professor of Geology there, I discovered, had come in not quite as a lab boy, but more or less as a technician. If you know somebody it can help you to find a way through. And this is not unique with us.

TM. Now I think that while you were at Cambridge, or when you first started at Cambridge, you regarded yourself, I think, as a craftsman, concerned to seek the practice of the art, and also to make sure that apprentices to the art were properly trained. Was that always a guiding principle?

VEC. Yes, I think that this comes from my family atmosphere. My father was a craftsman, a cabinet maker originally, before he set up on his own and I saw the procedure, you know, of apprentices coming, and people doing the job as craftsmen and doing the job for the job, not necessarily for what they were paid for it. And I think this helped to form my attitude to the subject, looking into what would be necessary because in those days there was no training school; now we have courses through the Royal Microscopical Society and there are other training courses; some Technical Colleges run them. In those days there was nothing available and so we set out to run some, not entirely on our own planning. For instance, when we arranged summer schools in the subject, we had in mind the example of the X-ray diffraction summer schools which Bragg had started some while before.

TM. I think there were also some early conferences.

VEC. There was the one held before I got to Cambridge, when we were still in Oxford. These grew out of the self-help groups that were sponsored by the NPL. There were three more or less closed meetings held in the 1940's at the National Physical Laboratory, the Royal Society and the National Institute for Medical Research in previous years; this all came out of the installation of the RCA microscopes. After that we had a meeting in Manchester, which was the fourth of such meetings and there it was decided that we needed to set up some sort of organisation to carry things on and to find finance to organise meetings, to invite speakers. This Oxford meeting was the first one that really was international; we did have some foreign visitors at the Manchester meeting, which was held in January 1946, for instance the Dutch expert Le Poole was there, but this time we got some of the Germans, we also got Dupouy from Toulouse. Many experts came to Oxford at that time. I happened to be the Secretary because I was there and that is, I think, how I came to be Secretary of the Institute of Physics Electron Microscope Group which then was formed, again on the direct model of the X-ray Crystallography Group, which already existed under the aegis of the Institute of Physics. There had been long discussions in Manchester already, that whatever organisation that might be set up should be under the Royal Microscopical Society or under the Institute of Physics. The Committee was set up in Manchester, I wasn't a member of it, but Dr Haine was and they held a ballot among the users of the microscope, perhaps 40 or 50, not 40 or 50 microscopes, and they said: ok we'd better go to the

Institute of Physics; they had a better office, they had other such groups already. And so this was the first EM Group; under the Institute of Physics. They helped us. It was the first conference also at which we had a machine specially brought over. The Dutch brought over the first Philips transmission electron microscope.

TM. Yes, and if I remember rightly, the shaking it received on the way prevented it operating. Was that so?

VEC. That was a very sad outcome. They chartered a plane to fly to what was going to be Croydon Airport; it was brought by road to Oxford; and of course the thing wouldn't work. After the meeting was over, they found that an aperture had fallen down the column and blocked it.

TM. Now can you recollect how the Royal Microscopical Society came to be interested in electron microscopy, because you took a big interest, I think, in reorienting them?

VEC. Indeed, but I was not the only one. They were, from the last century, interested in "brass and glass" optics, looking at diatoms and animalculae and so on, describing the different species and so on. They were mostly amateurs, with just a few people concerned with the optical industry, but there were parsons and visitors and so on and they were brought up with a bump against this new instrument. And so it took some time for them to understand and to accept that here was a much more powerful and informative microscope and by no means an amateur instrument. So those who had made some approaches to the RMS were not entirely welcome at first. But by 1946 there was enough interest for us users to hold a ballot as mentioned above. This fell out in favour of the Institute of Physics. But we continued on very friendly terms and it was only a few years before joint meetings took place between the EM Group of the IOP and the RMS. When Drummond wrote his Handbook on specimen preparation techniques, for example, the RMS kindly agreed to publish it as a special issue of the Journal of the RMS.

TM. I would like to mention your wife Anna. She was greatly admired as a person and was a microscopist in her own right. She helped you by getting a good family atmosphere going in the Cavendish EM Group?

VEC. Yes, indeed. She was a physicist from Vienna and came to the UK in 1938 as a political refugee. She found a job as a cook and then she went, oddly enough, to work in Bristol for Garner, my old Professor. That is how we met. Later she took up research work in optical microscopy in the EM Section of the Cavendish. As you say, having such a warm-hearted person in the Section and in the Cavendish was very useful; research students tend to have all sorts of problems!

This concludes the part of the interview dealing with Cosslett's careeer before joining the Cavendish Laboratory. Subsequent contributors will deal largely with events after 1947.

3. ACKNOWLEGMENTS

The author and the ICXOM Committee is grateful to the Royal Microscopical Society for permission to publish this material, based on the Cosslett Archive interview. We also wish to thank the Royal Microscopical Society and the Institute of Physics for generous financial support of this Symposium, in recognition of the seminal part that Ellis Cosslett played in the founding of the Electron Microscopy Group of the Institute of Physics and later, the Electron Microscopy Section of the Royal Microscopical Society.

Inst. Phys. Conf. Ser. No 130: Chapter 1
Paper presented at Int. Congr. X-ray Optics and Microanalysis, Manchester, 1992

9

X-ray projection microscopy and transmission electron microscopy

William Nixon

Peterhouse, Cambridge University, England

ABSTRACT: Before joining Ellis Cosslett's EM Group in October 1949, coming from Canada, I had chosen to work on the x-ray projection microscope. The new instrument was constructed, tested and improved and a Ph.D. Degree awarded in June 1952. Many visitors came to see the instrument and the results. The universal comment was "How do you get such beautiful images from such a heap of junk?" The High Voltage Transmission Electron Microscope was a Cavendish project of the 1960's with KCA Smith supervising. The High Resolution Electron Microscope of the 1970's was a joint Engineering and Physics Departments project with Cosslett the Physics Principal Investigator until his retirement in 1975 and Nixon the Engineering Principal Investigator throughout.

1. INTRODUCTION

This contribution to the Cosslett Symposium is written from a personal viewpoint. We first met in October 1949 in Cambridge when I commenced research in the Electron Microscopy Group of the Cavendish Laboratory. We had very different backgrounds prior to our meeting and these earlier experiences have a bearing on our work from October 1949 onwards.

I was 12 years old at the outbreak of the Second World War in September 1939, with my home in Toronto, Canada. I entered Queen's University, Kingston, Ontario, Canada at the age of 17, in September 1944, to study Physics and Mathematics. On my eighteenth birthday I interrupted my University studies and joined the British Royal Navy Fleet Air Arm as a Pilot Cadet and travelled to England. Flying training occurred over Salisbury Plain, the Malvern Hills and Stonehenge with all the excitement of solo flying in a Tiger Moth. I also attended a two week course in Balliol College, Oxford, which turned my attention to the possibility of graduate research in England.

At the end of the war, I took up my interrupted University course again in September 1945 and with accelerated teaching graduated (First Class) in May 1948 (and married in June 1948). It was decided that research in Physics at the Cavendish Laboratory in Cambridge would be a suitable next step but that a one year Master's Degree in Physics with a written dissertation would be good preparation. I stayed on for this further year while negotiations were carried out with the Cavendish and with Trinity College. In addition, full funding was raised in Canada to cover the three years from 1949 to 1952 in Cambridge for all fees and living and travel costs. My wife and I travelled from Canada to Cambridge to arrive at the beginning of October, 1949, to start as a Research Student in the Cavendish Laboratory and as a Graduate Student of Trinity College, while already married, a war veteran of flying in the Royal Navy Fleet Air Arm in England, with a Canadian BA and MA and 22 years old.

The background of Ellis Cosslett is fairly well known. Tom Mulvey has introduced this Symposium in the previous paper with an outline of Cosslett's research career and I will only record a few additional facts. Ellis Cosslett was 19 years older than I and he was 41 years of age when we met in October 1949. He had received his Ph.D. from Bristol University in 1932, following work in Berlin. He was fluent in German with contacts in Germany and Austria. He had married in 1936; the marriage was dissolved in 1939. He married Anna in 1940. He spent the war years in teaching and research in Oxford and moved to the Cavendish Laboratory, Cambridge, as an ICI Fellow in 1946. He became a University Lecturer in Physics in Cambridge in October 1949 at the same time as that of our first meeting.

2. X-RAY PROJECTION MICROSCOPY IN THE 1950'S

The University of Toronto had a reputation in both low-temperature studies at liquid helium temperatures and in transmission electron microscopy, due to links with Britain and Europe through the movement in both directions of Faculty members. I considered both areas and chose electron microscopy – the fact that I brought all my personal funding with me, meant that I could also choose within electron microscopy. Ellis Cosslett also offered the use of electrostatic lenses for an attempt at correcting spherical aberration using Scherzer's methods. I declined that topic which was later taken up by Jack Burfoot.

The x-ray projection microscope was included in that Bible for work at the time by Zworykin, Morton, Ramberg, Hillier and Vance, published in 1945(1). I had purchased this book in Canada and read it through as background for work in Cambridge. Hillier was one of the many research students at the University of Toronto in the 1930's and early 1940's working with Professor Burton, who had visited Ruska in Germany in the early 1930's. Burton started a big electron microscopy activity with Hillier, Hall, Ellis and several others who made their careers in the field. Hillier had joined the RCA Group and made the EMB, one version of which had been supplied to the Cavendish Laboratory under the Lend/Lease scheme to help the war effort – maybe because it contained an electron gun! That instrument and an old 1939 Siemens "bathtub" electron microscope were the only two instruments in Cosslett's group when I arrived.

The interest in x-ray microscopy had a short period in Oxford with Taylor but there were no publications, almost no equipment and no work had been done by anyone for at least a year before my arrival. Coming from Toronto I was surprised at how small the effort was in Cambridge. However, I quickly convinced myself that there was a great deal that could be done and rapidly got on with the job. As noted above I brought all my necessary financial resources with me – but there was one important condition. I would only get my Canadian Scholarship money for the second year by submitting a fully written up report on the first year's work, and the same rule for the money for the third year only after a second year report. This necessity meant that I kept detailed notes and laboratory books from the very first day and produced full reports as required(2). With a report also from the first two terms of my third year I could write my Ph.D. Dissertation in the Easter vacation of the third year by using these three reports, submit the bound volume by the first week of the Easter Term, be examined and take the Ph.D. Degree in the Senate House, all within less than two years and nine months of starting in October 1949.

Ellis was my University appointed Internal Examiner and Denis Gabor of Imperial College was my External Examiner, each for both the written report on the written dissertation and for the joint oral examination. Gabor immediately put me at my ease, as he was very satisfied with the work. He stressed that I had made a much bigger advance than I had claimed as there had been almost no previous real research results. This left time to discuss holography in all its aspects – Gabor's first thoughts, early ideas, electron and light combinations, x-ray methods and much else – all in 1952, 40 years ago!

Ellis was pleased that the examination had gone so well with such a distinguished external examiner. Afterwards, we discussed the future work that could be undertaken with my equipment, both by Norman Dyson who started in October 1952 and in the light of Castaing's work in Paris on a static electron probe system using electrostatic lenses for x-ray microanalysis. I compared the current density for the same probe size, say one micrometre, available in Castaing's instrument and my x-ray microscope with magnetic lenses. My magnetic lens system could produce some thousand times more current – one nanoampere with electrostatic lenses could be raised to one microampere with magnetic lenses. This electron optical experience and expertise was continued in all the subsequent instruments in the Cavendish Laboratory and by myself in the Engineering Department of Cambridge University after being appointed there as Assistant Director of Research from October 1959, ten years after arriving in Cambridge.

Having myself supervised 26 Research Students in Cambridge University I have seen the following phenomena at close hand, which I experienced and worked out for myself when Ellis Cosslett's first Research Student in Cambridge. In the First Year of Research, the student thinks that his Supervisor knows everything because if any topic is raised or new proposal suggested, then the Supervisor already can quote a publication giving all the details of work already done. In the Second Year of Research, the student thinks that his Supervisor knows nothing because by that time the equipment is operating, new results are being produced by that equipment and the student is the first in the world to know. Therefore the Supervisor is, at best, only the second! In the Third Year of Research, the student then realises that the Supervisor does not know everything but the Supervisor also knows much more than nothing and that indeed there is far more work to do than the Student will be able to finish and write up in his dissertation given the time constraints.

Ellis Cosslett fitted well into all three categories; with his world-wide contacts he always had a good reference or bit of relevant news to help out. By this time Ellis did not do any experimental work himself and so all the instrumental design, construction, testing and interpretation on the x-ray projection microscope was done by myself. We would discuss the results in detail and ensure that full and prompt publication occurred, both in subscription Journals and by attending conferences. He spent a great deal of time, even in the early days, on the organisation of Electron Microscopy on a national and then international basis with a legacy that persists to-day.

After graduating in 1952 I worked for about one year with the American General Electric Company in Schenectady, New York State, building an x-ray microscope. We made both magnetic lens and electrostatic lens versions and the latter was marketed as the company had already produced a small electron microscope from the work of Simon Ramo. In the meantime, a copy of my x-ray microscope, with some slight mechanical changes for ease of use, was being constructed in the Cavendish Laboratory and shipped to Stanford University, California, by the summer of 1953. The agreement included an extended visit to Stanford by myself to commission the instrument and instruct those who would use it from then on in the Department of Physics. The two people involved were Albert Baez and Paul Kirkpatrick who had been investigating x-ray imaging and magnification with grazing incidence mirrors. The x-ray projection microscope could provide a suitable illuminating source for such a system.

By this time other groups were interested in acquiring an x-ray microscope and I returned to the Cavendish Laboratory in October 1953, by which time Peter Duncumb had just started as a research student. Further improvements were made and other instruments built, some in association with Raymond Ely. In the autumn of 1955 I was sponsored by a National Science Foundation Fellowship (U.S.A.) to visit Ralph Wyckoff at the National Institutes of Health, Bethesda, Maryland; Albert Baez at the University of Redlands, California; and Paul Kirkpatrick at Stanford University, California. By this time we had supplied both Ralph Wyckoff and Albert Baez with x-ray microscopes and the one at Stanford had been used for two years(3).

The first X-Ray Microscopy and Microanalysis International Conference was organised for Cambridge in the summer of 1956; I returned to Cambridge for that meeting with publications from the work abroad. I had been elected to an Associated Electrical Industries Research Fellowship at the Cavendish Laboratory for three years from October 1956 and helped in the expansion of the EM Group to a very much larger size than the few members of October 1949.

A natural break came in October 1959 when three of us moved on at the same time. I moved to the Cambridge University Engineering Department as Assistant Director of Research to Charles Oatley; Peter Duncumb and Jim Long also left Ellis Cosslett's group at that time. Their papers follow later in these proceedings and give their own version of their time in the Cavendish. We all agreed that it was the end of an exciting era but also the beginning of new fields for the three of us and the next chapter for Ellis Cosslett.

Shortly afterwards, in 1960, Cambridge University Press published the book "X-RAY MICROSCOPY" by VE Cosslett and WC Nixon. This book summarised ten years of work in Cambridge and in other laboratories and was the first in a field that is still active to-day(4).

3. THE HIGH VOLTAGE ELECTRON MICROSCOPE OF THE 1960'S

The High Voltage Electron Microscope was supported by the Paul Instrument Fund of the Royal Society and with University Funds. The Paul Fund was started with the aim of aiding the development of instruments and one example given was the electron microscope of Ruska in Berlin in the early 1930's which needed funds for development. The grants were raised by Ellis Cosslett and the design started in 1961/1962. KCA Smith had returned to Cambridge from Montreal in 1960 having installed the scanning electron microscope that he had made in the Cambridge University Engineering Department during 1956 – 1958 and then used in the Pulp and Paper Research Institute, Montreal, from 1958 to 1960. Ken Smith became the leading design person with Kevin Considine, also full time, with Julian Davey (part time), Norman Betts, Cavendish Electronics Workshop (part time) and later Peter Ward and Roger Camps continued development and operation of the microscope following the main construction phase.

The manufacturing was carried out for some fine mechanisms in the Cavendish Electron Microscope Section Workshops under Ron Pryor. The Cavendish Electronics Workshops produced all of the electronic equipment. The Cambridge University Engineering Department produced the magnetic lens pole pieces and some other fine mechanical parts as well as all the heavy sheet metal work, welding and other similar activities. Tube Investments at Hinxton Hall manufactured the lens casings and Haefly supplied the 750 kV high voltage generator.

The first images were produced in 1965/1966 with an early publication by KCA Smith, K Considine and VE Cosslett in the Proceedings of the Sixth International Congress for Electron Microscopy, Kyoto, Japan, 1966(5).

The microscope was designed for and fulfilled the following criteria.

1) A working voltage range from 75 kV to 750 kV, or a factor of ten, with the lower voltage to overlap with the usual 100 kV of normal microscopes. The voltage could be changed rapidly and sequences of micrographs could be taken at the various voltages.

2) The column could be separated easily and the specimen chamber and the lenses could be removed with the aid of a special fork lift truck for rapid alteration of the microscope from one use to another.

3) The effects of mechanical vibration and stray magnetic fields were reduced to a level consistent with the expected resolution predicted by the electron optical conditions.

4) The electronic stabilities reached were as follows: high voltage drift stability less than 4 parts per million; high voltage drift and ripple stability combined over a three minute period less than 8 parts per million; lens current stability less than 5 parts per million.

5) The controls were designed to be similar to a conventional instrument for ease of use in all modes such as bright-field, dark-field and diffraction conditions.

The lessons learned from the 750 kV electron microscope were transferred to the development of the AEI EM7 instrument of the late 1960's. Both Ellis Cosslett and Ken Smith were consultants to AEI during development and later were members of the Users' Committee chaired by Sir Peter Hirsch.

The Duddell Medal of the Institute of Physics was awarded to Ellis Cosslett and Ken Smith in 1971 for their work on the High Voltage Electron Microscope, with the presentation by Sir James Menter who had been a Paul Fund Advisor on the original application some ten years earlier.

4. THE HIGH RESOLUTION ELECTRON MICROSCOPE OF THE 1970'S

The success of all the electron optical research in both the Engineering Department and the Cavendish Laboratory led the SERC to invite the two Departments of Cambridge University to submit a joint project for the 1970's. Several projects were considered and the final agreement was to attempt atomic resolution at 600 kV. The first funding was received in 1972, after detailed plans and proposals had been made. Operation commenced in 1977 and the microscope facility was officially opened in 1979(6). After a very fruitful life of some 13 years the microscope was moved earlier in 1992 to the care of the Science Museum. The x-ray projection microscope described above as my Ph.D. research project is also on view at the Science Museum as well as in the Museum of the Cavendish Laboratory. The x-ray microscope of 1952 was the original progenitor of all the electron optical instruments with magnetic electron lenses devised in Cambridge ever since.

The 600 kV high voltage supply was purchased from Haefly, Basel, Switzerland, as they had already supplied high voltage units for microscopy to a high standard. The supply was carefully chosen by the manufacturers from their production line to show very high stability and very low ripple. The supply was tested and evaluated by John Catto, sent out to the factory by the Engineering Department, and he further improved it by a factor of 10 when the high voltage supply was assembled on site in Cambridge.

As a joint project the microscope could have been sited anywhere in Cambridge where room could be found. As much of the construction was done in the workshops of the Engineering Department a site nearby was considered but insufficient space could be made available. The Cavendish Laboratory was also very short of space but a move to new buildings in west Cambridge was going to occur with completion by 1974. Ellis Cosslett's group was not to move to West Cambridge, in part due to his coming retirement in 1975.

The space requirements were large for several reasons. We wished to use an indirectly heated lanthanum hexaboride electron gun, as pioneered by Alec Broers at IBM Yorktown Heights in the late 1960's after receiving his Ph.D. from the Cambridge University Engineering Department. The design used was similar to that of Ahmed and Nixon, produced in Cambridge after Haroon Ahmed had been on sabbatical leave in Alec Broers group at IBM. This gun used air insulation at 600 kV so that the electron source and illuminating system could be optimised without having to gain access through an oil-sealed container. As a result the electron source, electron accelerator, mid-section for joining to the electron optical column, the viewing chamber and the photographic and electronic recording and display units altogether occupied two high ceilinged rooms one above the other in part of the Old Cavendish. In addition, the special direct current low voltage supplies for the electron lenses had banks of electronics in distant rooms so that all mains frequency was kept far from the column, even with extensive mu-metal shielding. The lenses carried approximately 10 amperes, controllable to less than one microampere or one part in 10 million. This level of performance was achieved on all of the components used in the high resolution electron microscope.

The magnetic electron lenses were considered for design and construction completely in Cambridge. It was decided to save time by using a set of AEI EM7 high voltage microscope lenses with respect to the outer parts and reconstruct the pole pieces and other parts to be capable of high resolution. The electron optical design for lenses and alignment and deflecting systems was based on variants of the computer programs devised by Eric Munro of the Engineering Department. One of Eric Munro's early papers was presented at a European Electron Microscopy Conference here in UMIST, Manchester, in 1972, 20 years ago! The design was done by John Cleaver of the Engineering Department who had been one of KCA Smith's research students, as had John Catto. Ken Smith had been the engineer in charge of the high voltage microscope as above and had been a teaching staff member of the Engineering Department since 1965. Ken was an essential member of the engineering team who created the high resolution electron microscope for a variety of departments and disciplines including the physicists of the Cavendish Laboratory.

The mechanical design and construction of the anti-vibration system and the mechanical assembly of the microscope including the accelerator and electron source unit were the domain of Arthur Timbs, Senior Design Engineer of the Engineering Department. Early discussions were helped by Philip Turner, a senior colleague. The air suspension system used three compressed air cylinders on a raised tripod attached to the walls of the room. This supported 7 tons with a natural frequency well below one cycle per second. The control system maintained this performance down to the level where the only input of vibration was due to the sound waves if anyone in the room became too excited at seeing the remarkable results!

An equally essential part of the whole system was the performance of the electronic control units, both designed and supervised in building by John Catto. These used a pulse code modulation command system to send signals from the control console through black coated fibre optic cables up to the 600 kV electron source. Sensors and detectors then sent messages in the same way down to the control console to display the parameters such as voltage, current, position, battery state and other facts. A compressed air line of plastic tubing was also taken up to 600 kV and the electronic signals controlled the compressed air valves for vacuum regulation and other uses.

Ken Smith and John Catto also produced the electronic television pick-up system that was one of the first to record dynamic motion at the atomic level by using specimens of cadmium telluride provided by Bob Sinclair of Stanford University and who was present when the first recordings were made. Results from the 600 kV high resolution microscope were printed on the front cover of three issues of Nature; atomic resolution of gold, molecular images and four frames from a video tape of moving cadmium telluride lattice defects.

Throughout all this Ellis Cosslett's own group made excellent contributions through the skills of Ron Pryor and the members of the local workshop and photographic assistance by Ken Harvey; the operation and maintenance of the microscope by Roger Camps and the first full time operator David Smith. Many visitors came with specimens including John Fryer of the Chemistry Department of Glasgow University. In most cases the images would be computer processed by Owen Saxton for revealing all the details. In February 1992, John Fryer et al published in Nature a full computer analysis of amplitude and phase information from images taken some time ago by the 600 kV high resolution microscope. These results showed that the nominal 1.7 Angstrom resolution achieved early on by the microscope was indeed considerably better when the true information content was revealed. In other words, the engineering quality was there all along, waiting to be revealed when the image processing had advanced sufficiently far.

5. OTHER TOPICS

Ellis Cosslett contributed to the scientific work described above and in the other papers in this Symposium. He also had a keen interest in many topics that were not scientific and a few examples are as follows.

Ellis was well known for his far-left political views and applied these views to his relationship with the University. He was a key figure for a certain time in the Association of University Teachers who as a group have not been very influential in the Cambridge context.

He also had strong views on the role of the Colleges within Cambridge University. These views were pressed by him privately but also in public in a particular Cambridge way. "Discussions" are held at 2 pm on Tuesday afternoons in the Senate House at the heart of Cambridge University. The Topic for Discussion will be a Report that has been prepared by the Central Offices and which is of concern to the whole University. On one such occasion the topic was "The Colleges" with the implication that they should be somehow different than they were at that time. Ellis Cosslett spoke very vividly against the Colleges as unnecessary and an anachronism in the prestigious science based University with all the power in the Faculties. Such a Discussion will have a small number of people attending but the full text of each contribution is published a week later in "The Cambridge University Reporter" which is "Published by Authority" as printed on the cover page.

Some time passed and then Ellis Cosslett was offered a Fellowship at Corpus Christi College, in 1963. He went through agonies on whether he should accept, and therefore join the other side, or refuse and stay outside what is the mainstream of influence in Cambridge. I had been elected a Fellow of Peterhouse in 1962, the previous year, and was 19 years his junior. However, Ellis finally accepted but softened the blow by choosing to be interested in the new Graduate Student part of Corpus called Leckhampton House on Grange Road. Indeed, much later on he served for a period as Warden of Leckhampton. This ability to look and act in both directions at once, as shown by this example, could be seen in other facets of Ellis Cosslett's very interesting life.

In my first sentence I said that this contribution was written from a personal viewpoint. A final personal event will conclude this paper. My Canadian wife died in 1986 after 38 very happy married years. Both my wife and I and our two children as toddlers knew Glenys Hutchinson who worked as a Research Assistant in Ellis Cosslett's Group in the Cavendish around 1959/1960. Glenys married Alan Baker, a Research Student in Metallurgy, and lived for four years near Pittsburgh while Alan was at US Steel and then at Leeds University. They also had two children. In 1987 Glenys returned to Cambridge to a new position in the President's Lodge of Queens' College. We met again during the winter and in particular at a function in July 1988. Peter Duncumb and I organised a Dinner in Peterhouse to celebrate Ellis's eightieth birthday in June 1988. Some twenty people were invited, all of whom had worked in Ellis's Group 25 to 30 years earlier. From then on fate took over and Glenys and I were married in Queens' College Chapel a year later in July 1989.

Many memories of Ellis Cosslett remain and it was a privilege to be part of the activity from the very early days up to and beyond his eightieth birthday.

REFERENCES

1)   Zworykin, V.K., Morton, G.A., Rambert, E.G., Hillier, J. and Vance, A.W. "Electron Optics and the Electron Microscope", Wiley and Sons, New York, 1945, pages 112, 127, 268, 631 and 684.

2)   Cosslett, V.E. and Nixon, W.C. "X-ray Shadow Microscope", Nature, Vol. 168, p. 4262, 1951.

3)   Nixon, W.C. "High–resolution x–ray projection microscopy" Proceedings of the Royal Society, Vol. A 232, pages 475–485, 1955.

4)   Cosslett, V.E. and Nixon, W.C. "X–ray Microscopy",

Cambridge University Press, 1960. Pages xiv and 406 pages, 15 Chapters and Appendix, 32 full page photographic plates, many diagrams, 23 pages of references and an index.

5)   Smith, K.C.A., Considine, K. and Cosslett, V.E. "A New 750 kV Electron Microscope" Proceedings of the Sixth International Congress for Electron Microscopy, Kyoto, Japan, 1966, page 99.

6)   Cosslett, V.E., Camps, R.A., Saxton, W.O. and Smith, D.J. (Cavendish Laboratory, Cambridge University) and
            Nixon, W.C., Ahmed, H., Catto, C.J.D., Cleaver, J.R.A.,
            Smith, K.C.A., Timbs, A.E., Turner, P.W. and Ross, P.M.
(Engineering Department, Cambridge University)

        "Atomic resolution with a 600–kV electron microscope"
        Nature, Vol. 281, No. 5726, pp.49–51, September 6, 1979.

Inst. Phys. Conf. Ser. No 130: Chapter 1
Paper presented at Int. Congr. X-ray Optics and Microanalysis, Manchester, 1992

17

Early days in the electron microscope section of the Cavendish Laboratory

Robert W Horne
School of Biological Sciences, University of East Anglia, Norwich NR4 7TJ UK

ABSTRACT: The early years of electron microscopy in the Cavendish
Laboratory following Ellis Cosslett's move to Cambridge in the late 1940's
are described.   Two transmission electron microscopes in the form of an
RCA EMB and Siemens instrument were installed in the Old Cavendish
building between 1942 and 1946.   These electron microscopes formed the
basis for Cosslett's laboratory, which developed from modest beginnings to
become one of the leading centres for electron microscopy, electron physics,
theoretical studies and microanalysis.
Some early applications of the EM are summarised, together with two
important contributions made to metal physics and biology.

It was not my intention to work on electron microscopes following my service
in the RAF from 1941 to 1946.   For about three years during the war I was
attached to a joint high energy physics project developed by the Americans at
Harvard and MIT.   This complex system was under the direction of W W
Salisbury, who suggested at the end of hostilities that I should contact the
Cavendish Laboratory in Cambridge, as several high energy accelerators were
installed there.   However, the future use and location of these machines after
1946 were under discussion, with research and technical appointments being
shelved.

It was during my employment in Cambridge at the laboratory for Physical
Chemistry of Rubbing Solids, under J R Whitehead, that I heard of the proposal
to appoint a research assistant for electron microscopy.   Whitehead suggested
to Ellis Cosslett that my experience with high energy generators, vacuum
systems, electronics, etc. might be useful in running electron microscopes.

Most research workers of today engaged in electron microscopy or microanalysis
will find that the modern equipment and facilities are located in a laboratory
with relatively pleasant surroundings.   However, this was not the case in 1947
when I joined the Cavendish Laboratory to work in Ellis Cosslett's laboratory.

There were two electron microscopes located in separate rooms at the end of the semi-basement area of the Old Cavendish building.   This part of the laboratory was often referred to as the 'old dungeons', with bare walls of grey stone and bricks.   There were a few high bay windows overlooking Free School Lane.   Electric light bulbs hung from the ceiling by very old maroon-coloured wires.   The dungeon-like atmosphere was enhanced by a dark stained wood block floor, looking as though it had only occasionally been cleaned since Ernest Rutherford's time.

The RCA EMB electron microscope (allocated to the Cavendish about 1942 through the Lease-Lend Agreement), was installed in one of the end rooms. This instrument was under the care of George Crowe, who was formerly Rutherford's personal assistant.   In the same room as the RCA microscope were bits and pieces of old apparatus, also a derelict desk containing some old exercise books and faded papers.   These papers, George Crowe informed me, were some of J J Thompson's notes!

The second microscope was a very early Siemens instrument which had been removed from Germany by the Admiralty as part of the war reparations and then sent to the Cavendish Laboratory.   This was installed in an adjacent room to the RCA EMB.   The base of the Siemens, control table, large mercury diffusion pump, column support block, etc. had been assembled, but the lens units, insulator and gun, filament battery (3 two volt wet batteries in a metal box) precariously perched on a porcelain insulator, were arranged on a table covered with brown paper.   None of the wiring had been attempted, as the only information about connections and cables was limited to an old diagram on a folded sheet about the size of a local town map.   The symbols were, of course, in German.

The wiring of the Siemens microscope, high tension tank, magnetic stabilizers and lens supplies (the only suitable source of DC was a battery bank located in the building at least 50 metres away from the microscope!), took several weeks to complete.   To our surprise, the first 'switching on' of the EHT generator met with success;   when the filament was switched on, the instrument produced a spot on the final screen!   With Ellis Cosslett's advice and guidance we produced an image of a diatom, following a few days of rough alignment and adjustment.   Fig. 1.

From these modest beginnings, Cosslett's electron microscope laboratory started to attract attention and interest from other departments within the University of Cambridge.   First from the metal physics people in the Cavendish and then from research students from Professor Austen's adjacent Metallurgy Department.

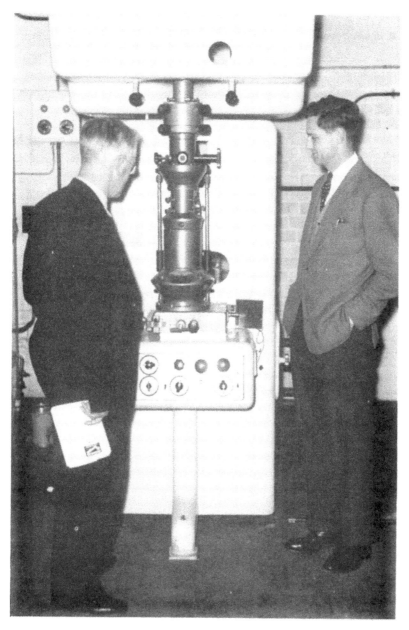

Fig. 1   Vernon Ellis Cosslett (left) and Robert Horne viewing the Siemens
microscope which formed one of the two original instruments
installed in the Cavendish Laboratory between 1941 and 1946.

Arthur Brown began studies of metal specimens which were subjected to strain with the aid of plastic or aluminium replicas.   The examination of polished metal specimens by replica methods gathered momentum, particularly with the development and applications of shadow-casting techniques.   Jack Nutting, both as a research student and post-doctorate researcher from Metallurgy, was also one of the first users of the EM facilities in the Cavendish Laboratory.

There was a gradual increase in the number of people studying non-biological specimens, which resulted in the EM laboratory being placed on a basis whereby there was a small charge for the facilities and photographic materials, etc. Several of the metallurgists were trained to use the Siemens instrument, where 'driving licences' were issued by Cosslett to those he felt were competent to use the instrument.   However, the alignment of the Siemens was a strange mixture of skill and luck, also the changing of the filament could only be undertaken by one or two people in the EM laboratory, as it meant leaning into the large bathtub-like protecting canopy to remove the electron gun.   Both the EHT and gun were discharged by an electrically driven rod, which was known to stick in the halfway position from time to time.   If there wasn't a large spark and sound like a rifle report when the EHT was switched off, the operator knew the rod had jammed and the gun was still at 80 kV!

By 1950 there were about ten to twelve regular non-biological users of the Siemens instrument.   The RCA microscope produced relatively low beam intensity for metallurgical specimens until a new biased gun was fitted in 1952. It was, therefore, more suitable for the examination of biological specimens.

The biologists were at first hesitant to make use of these two 'super microscopes', as looking at the incinerated remains of their specimens was not very attractive.   One of the first biological studies carried out in the Cavendish using both the Siemens and RCA microscopes was concerned with the structure of small plant viruses.   Cosslett and Markham (1948).   This was followed by other investigations in the field of microbiology, where whole mounts were formed by placing droplets from liquid suspensions containing bacteria, viruses or rickettsia directly onto plastic filmed grids, then shadowed in a primitive evaporator (see Brieger and Cosslett 1949; Brieger et al 1951; Salton and Horne 1951; Salton, Horne and Cosslett 1951).

From these initial successful applications to biological particles, the demand from other biologists began to increase.   Ellis Cosslett was keen to encourage a wider collaboration between his laboratory and the departments in the University engaged in biological research.   It should be stressed that, at the time, it was only possible to prepare whole mounts as specimens. Facilities were not yet available for thin sectioning of specimens. Nevertheless, the early results attracted more people, with special reference to several biochemists mentioned below who wished to study morphological changes resulting from biochemical treatment. The separation of bacterial cell walls by Milton Salton working in the Department of Colloid Science to provide pure suspensions of this bacterial component, was an important step in microbiology, where the purity and concentration of the isolated walls could be monitored by electron microscopy (see Salton and Horne 1951). The chemical structure of yeast cell walls by Northcote and Horne (1952) and the intracellular components of skeletal muscle by Perry and Horne (1952) were other examples of published work which began in the early days.

The output of papers in the form of collaborative studies between the EM laboratory and the Departments of Biochemistry, Pathology, Zoology, Strangeways Laboratory, Veterinary School and others, allowed additional funding to be allocated from the University authorities for electron microscopy. The dungeons took on a new look! The walls were decorated, new lighting was fitted and experimental apparatus was built in the EM workshop; this formed a vital part of the laboratory in the early fifties. Moreover, the large arched semi-basement floor area now became the Electron Microscope Section of the Cavendish Laboratory. The archways formed convenient divisions allowing the large space to be divided up into smaller rooms to house new preparation areas, a secretarial office, a workshop, etc. Research students were then engaged to build apparatus and design experiments, including a reflecting electron microscope by Derek Jones (see other contributions to this symposium). The primitive facilities were enhanced by more advanced equipment following the appointment of Ken Harvey as a professional photographer. The technical staff was also increased, including Ron Pryor who joined the group to cope with the demand for more apparatus and equipment for the EM collaborative work. The training of suitable people to use the machines independently, both in the biological and non-biological fields, became an essential part of the activities in the Cavendish for those engaged in long-term applications of electron microscopy.

By 1953 thousands of electron micrographs had been recorded on the two microscopes. The task of keeping the original instruments and their subsequent modifications to high pitch was becoming increasingly difficult.

The decision by Ellis Cosslett in 1954 to install the first Siemens Elmiskop I in the Cavendish Laboratory was to produce far reaching results in the fields of metal physics and biology.

It was mentioned earlier that originally metallurgists had to rely on replica methods to study their specimens, although Heidenreich in the late 1940's was engaged in producing electrolytically thinned specimens from metals. During the 1950's new methods were being developed to provide thin foils of certain metals. The aim of this work was to produce a specimen thin enough to allow penetration by the electron beam.

Thin foil specimens were also being prepared in the Cavendish, by beating and etching, at about the time the Siemens Elmiskop I instrument was moved from the International Congress on Electron Microscopy held in late 1954 to Cosslett's EM Section. The final checking of the new Elmiskop was carried out by Ernst Ruska in 1955.

During 1955 I was approached by Peter Hirsch and a research student, Mike Whelan, working in the Cavendish Metal Physics Section, to discuss the possibilities of using the new Elmiskop for examining the thin foils of aluminium. I recall that, at the time, Cosslett had grave doubts about whether these thin foil specimens would yield much information, especially as it seemed unlikely to achieve effective beam penetration at 80 to 100 kV. However, it was decided that some aluminium foils should be prepared and examined using the double condenser lens arrangement fitted to the new Elmiskop.

These experiments coincided with my return from Berlin, where I had spent a second visit working in the Siemenstadt factory on the detailed assembly and operation of the new Elmiskop I. The Siemens Elmiskop was far in advance of other production electron microscopes available at the time, consequently the operation, alignment and use of the first double condenser lens system to be fitted to a production instrument, required special instruction. The double condenser lens design and its aperture system produced, in my view, one of the most important effects observed by transmission electron microscopy of thin metal specimens.

The low magnification images from the thin aluminium foils showed a structure of subgrains and boundaries of about 1 μm or more across. At higher magnification it was possible to see distributions of dislocations and high contrast networks of lines or dots spaced at about 10 nm apart. Although there was reasonable contrast on the final screen of the microscope, the brightness of the image was relatively low. Increasing the beam current produced a small improvement. I then decided to increase the size of the second condenser lens aperture by selecting one of the three apertures on the mechanical slide. By mistake I withdrew the entire aperture slide. This, coupled with the high setting of the beam current, produced dramatic effects. The electron-dense dots and lines began to move, coupled with striking contrast changes and the appearance of 'tramline' patterns. By reducing the beam current the process was slowed down.

These observations were repeated on another aluminium foil showing the same dynamic effects from the dots, lines and contrast changes. Peter Hirsch began to pace up and down the EM room, first in silence and then repeating over and over again that he did not believe what was happening, as these dynamic changes <u>must</u> be the movement of dislocations in the specimen.

We quickly obtained a secondhand Kodak Ciné Special camera to record the movement of the dislocations seen on the viewing screen, first by filming the dots and lines in a static state and then their movements following the changes in illumination. Ciné films were made from images obtained from aluminium and stainless steel foils. Those interested in the historical aspects of electron microscopy are referred to the original publication of Hirsch, Horne and Whelan (1956). Thus began an application of the transmission electron microscope to an area of metal physics which subsequently became a vast subject within its own right.

At about the same time the work on the thin foils began. Sydney Brenner had joined the MRC Unit for Molecular Biology located in a small hut outside the Austin Wing of the Cavendish Laboratory. His research programme was to isolate, purify and analyse the structure of the T2 bacterial virus. The biophysical and biochemical procedures required some means of monitoring morphologically the various stages of preparation, purity and concentration of the final fractions of bacteriophage protein components. The electron microscope was obviously considered to be an ideal technique, but the number of protein fractions being prepared required a specimen preparative method which was both rapid and reliable, also free of serious artefacts. The preparation and processing of biological material by thin sectioning was beginning to make progress, but it was a relatively lengthy process. Moreover, very few results had been achieved at the time with particles as small as viruses. The method of air-drying as well as freeze-drying viruses from liquid suspensions, which were subsequently shadowed, was attempted but the disruption of the protein components, coupled with the background granulation from the shadowing material, made interpretation of the images very difficult.   One of the disadvantages with the new Elmiskop was that the high resolution and range of magnification revealed with considerable clarity the defects in the specimen preparation of biological materials!

Attempts were made to use positive staining techniques, but the low pH of such heavy metal stains as phosphotungstic acid (PTA) produced disastrous results by disrupting the phage protein to form unrecognisable structures when seen at high resolution.   It was then found that by adjusting the PTA to a neutral pH the virus and its components remained stable.   When examined in the Elmiskop, the images of the T2 phages showed a remarkable state of contrast and preservation.   The normal electron dense specimen after acid PTA staining was completely reversed in the case of the neutral phosphotungstate. It appeared that the phosphotungstate had formed an electron dense 'glass' around the specimen which provided a remarkable state of preservation of the virus protein.   This technique allowed high magnification images of the T2 phages and their components to be recorded showing considerable detail, with very low random background effects.   This simple and rapid preparative method was called 'negative staining'.   The technique was subsequently applied to a wide range of animal, plant and bacterial viruses as a result of extensive collaboration between Cosslett's EM Section and other departments within the University of Cambridge.   Between 1958 and 1962 some twenty papers were published in international journals dealing with the application of negative staining to viruses studied in the Cavendish Laboratory.   Figs. 2 and 3.

Negative staining continues to be used as a reliable technique for specimen preparation of a wide range of biological materials where high resolution is required.   Its application to diagnostic virology is regarded as being of some importance in relation to rapid identification of plant and animal infectious diseases.

The development and background to the method of negative staining is of great interest and the reader is referred to a recent historical article by Horne (1992).

I have attempted to recreate the atmosphere of the research during the early days in the Electron Microscope Section of the Cavendish Laboratory.   Ellis Cosslett was always helpful and positive in supporting the collaborative work which brought so many people to work with us in the Cavendish.   The EM Section grew rapidly from about 1953 onwards, with Ellis having to deal with an increasing level of teaching and laboratory administration.   If the research was going well he tended to leave people to develop their ideas and projects.   He

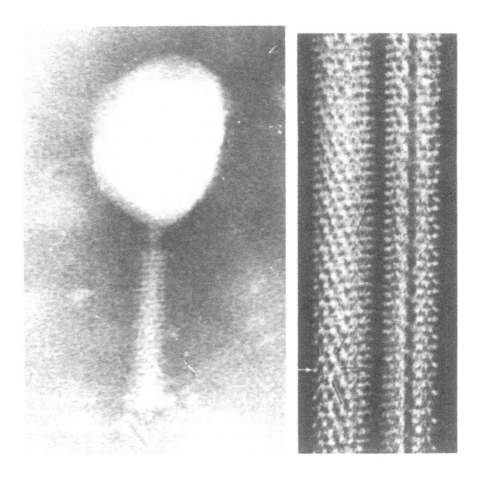

Fig. 2   Fig. 3

Fig. 2   The electron micrograph shows the first image of a negatively stained
bacteriophage, which was photographed on the 4th October 1957 in the Electron
Microscope Section of the Cavendish Laboratory using the Siemens Elmiskop I
instrument.   Magnification X 650,000   (R W Horne and S Brenner)

Fig. 3   High resolution electron micrograph of two rods of tobacco mosaic virus
embedded in negative stain.   The micrograph has been subjected to image
processing to enhance the repeating regular features of the protein molecules
(arrows) which are arranged in a single helix.   The regular periodicity of the
helical array seen along the edge of the rods is 2.3 nm
The central 4 nm diameter axial hole in the particle on the right has been
penetrated by electron dense stain.   (From R W Horne, J M Hobart and
R Markham   J gen Virol 1976   31 265 Cambs Univ Press)

was always available and prepared to devote time to discussions about the work and its progress, also he ensured that any distinguished visitor to the EM Section was introduced to all members of the Section.

The Electron Microscope Section provided a very friendly and welcoming atmosphere, where one had the feeling of belonging to a large family.   In addition, the collaborative encouragement generated by Ellis Cosslett from the very beginning, resulted in a considerable output of original publications within the University of Cambridge and associated laboratories.   It is interesting to look back and to note that I was personally involved in work which resulted in the publication of 69 papers during the period from 1951 to 1962.

REFERENCES

Cosslett V E and Markham R 1948 Nature Lond 161 250
Brieger E M and Cosslett V E 1949 Nature Lond 164 352
Brieger E M, Miles J A R, Cosslett V E and Horne R W 1951 Nature Lond
    168 96
Hirsch P B, Horne R W and Whelan M J 1956 Phil Mag 1   1
Horne R W 1992 Micron 22 321
Northcote D H and Horne R W 1952 Biochem J 51 232
Perry S V and Horne R W   1952 Biochim Biophys Acta 8 483
Salton M R J, Horne R W and Cosslett V E 1951 J Gen Microbiol 5 405
Salton M R J and Horne R W   1951 Biochim Biophys Acta 7 177

Inst. Phys. Conf. Ser. No 130: Chapter 1
Paper presented at Int. Congr. X-ray Optics and Microanalysis, Manchester, 1992   27

Ellis Cosslett: a physicist fascinated by biology

Audrey Glauert

Clare Hall, University of Cambridge, Herschel Road, Cambridge CB3 9AL

ABSTRACT: Ellis Cosslett was one the first physicists to have enthusiasm for the study of biological specimens by electron microscopy. Biologists from many departments and institutes in Cambridge were encouraged to come to the Cavendish Laboratory in the late 1940s and early 1950s to use the early Siemens microscope. In consequence the EM Section at the Cavendish became a fruitful meeting place and a number of the most important developments in specimen preparation techniques were made there. Subsequently, Ellis Cosslett provided similar encouragement to biologists during the development of the high voltage electron microscope.

1. EARLY BEGINNINGS

Ellis Cosslett's first degree was in physical chemistry, but later he became involved in electron diffraction and electron optics during his studies for a PhD. By a happy accident he was required by his Professor at Bristol to read the papers of Knoll and Ruska and to give a colloquium in 1932. In consequence he became very interested in the possibilities of electron microscopy and it seems that he was already fascinated by biology when he moved to Cambridge in 1946, since his application for an ICI Fellowship included the aim 'Construction of a shadowgraph type of (X-ray) microscope to allow the observation of biological specimens in air' (Cosslett 1981).

At the Cavendish Cosslett found the RCA EMB electron microscope (see Bob Horne's contribution) and this was the first EM that he actually laid hands on. This instrument had already attracted a great deal of interest among biologists from laboratories and research institutes in Cambridge. It enabled the detailed shape of micro-organisms to be seen for the first time, although very little internal structure was visible. Soon after Cosslett's arrival the early Siemens microscope (see photograph in Cosslett 1985) was installed and under the expert and patient guidance of Bob Horne it was to become the preferred instrument for biological users.

Ellis Cosslett now played a major role in encouraging biologists to use the electron microscope and he helped to overcome the strong opposition from certain eminent scientists, who were convinced that nothing of value would be seen because of the drastic effects of dehydration and damage by the electron beam. In fact, two specimen preparation methods were to be introduced in Cambridge in which the specimen is largely protected from the effects of irradiation. One of these was the development of a method of embedding specimens for ultramicrotomy in an epoxy resin (Araldite)(Glauert et al 1956; Glauert and Glauert 1958) which is relatively radiation-resistant, and the other was the technique of negative staining (see Bob Horne's contribution).

2. THE ELECTRON MICROSCOPY SECTION AT THE CAVENDISH

The establishment of the EM Section as a centre for biological research in Cambridge in the early 1950s was entirely due to Cosslett's enthusiasm and particularly to his appointment in 1949 of John Bradfield (the founder of the Cambridge Science Park) as a University Assistant in Research in Electron Microscopy (believed to be the first UK university appointment in the subject). Bradfield was a cytochemist and had held a Research Fellowship at Trinity College and a Commonwealth Fellowship in Chicago. He now became the 'resident' biologist at the Cavendish and provided a focus for all the many users who came from other laboratories. Bradfield's own interest was mainly in the structure of flagella and bacteria, but he also contributed to the very early observations on the structure of cell and nuclear membranes, and on mitochondria and the endoplasmic reticulum, and also was the first to show the connections between the brush border and mitochondria. In these studies he was ably assisted by a series of research students, including Ian Gibbons, Barbara Barnes, Robin Valentine and Françoise Hawkes.

3. A STIMULATING ENVIRONMENT FOR YOUNG SCIENTISTS

My own involvement with the EM Section started in 1950. I had obtained BSc and MSc degrees in physics at Bedford College in London and had then been an Assistant Lecturer in Physics at Royal Holloway College with Tolansky. My research was already concerned with biology since I had applied multiple-beam interferometry to the examination of thinly spread cells. I had hoped to continue these studies in Cambridge but was encouraged by Cosslett to consider changing to electron optics. Since no post was immediately available in the Cavendish, he advised me to apply for a vacant post as a Research Assistant to Dr Brieger at the Strangeways

Research Laboratory. Brieger was a medical scientist with interests in all
aspects of tuberculosis. He was already collaborating with Cosslett in the
examination of tubercle bacilli in the electron microscope (Brieger and
Cosslett 1949), and Cosslett suggested that I take this temporary post
with the expectation of soon moving to the Cavendish. In the event I
remained at the Strangeways for nearly 40 years! In some ways I had the
best of both worlds. The Strangeways is an independent research institute
with a particular emphasis on the application of tissue culture to
biological and medical problems and my colleagues there have enabled me to
gain an insight into the scientific criteria of biology, which are
considerably different from those of physics. At the same time I was
accepted as a full member of the EM Section in the Cavendish.

The EM Section I joined in 1950 was an ideal environment for stimulating
the development of biological studies. In addition to John Bradfield, the
staff included Bob Horne, who coaxed such good results out of the early
Siemens microscope and later taught us how to handle the Siemens
Elmiskop I, Elizabeth Green (now Crawford), who helped us with specimen
preparation and particularly with the delicate operation of making
support films, and Ken Harvey, a photographer of wide capabilities who
played such an essential role in the darkroom. There was an atmosphere
of great enthusiasm as we were often seeing the ultrastructure of a
particular specimen for the very first time. There was also great
friendliness and informality. I quickly had to listen to the radio to
learn Goon-speak so that I could understand what was going on! In all this
a key role was played by Anna Cosslett, who not only contributed to the
research of the Section through her studies with the ultra-violet and
interference microscopes, but also created a 'family' atmosphere through
her warmth and personal concern for the welfare of us all. She was the
inspiration for the annual garden parties held at the Cosslett home in
Long Road, at which members of the Section, together with their spouses
and children, would be introduced to the delights of Viennese cakes. The
setting was ideal since both Anna and Ellis had a deep interest in, and
love for, their garden. In addition, Ellis was less reserved in this
relaxed atmosphere and I, for one, became his personal friend.

The scientific achievements of the biologists working in the EM Section
are too numerous to describe here, although Bob Horne has high-lighted
some of them in his contribution. In parallel with the TEM studies,
pioneering results were also obtained by X-ray microscopy (see Bill
Nixon's contribution), by analytical methods (see contributions by

Peter Duncumb and Jim Long), and by ultra-violet and interference micro-scopy (by Anna Cosslett). Anna used an ultra-violet microscope of high resolution to make comparative studies of intact and sectioned bacteria by light and electron microscopy (Anna Cosslett 1960) and was able to distinguish dense cytoplasmic granules from DNA-containing regions of the cytoplasm. These observations helped to settle a matter of considerable controversy at the time. She was also the first to make quantitative measurements of the refractive index and thickness of sections. Using the interference microscope she compared the effect of the electron beam on four different embedding media (methacrylate, Araldite, Vestopal and Aquon) and concluded that Araldite was superior to the other media in its resistance to irradiation (Anna Cosslett 1960), much to my delight. These results still stand and are quoted today.

Gradually most of the biological users obtained their own instruments in the late 1950s and at one time Cambridge at the highest concentration of EMs in the UK. However, we did not lose contact with the EM Section, since Cosslett actively encouraged regular meetings in the Cavendish to discuss our latest results and publications in journals. He was always an enthusiastic member of the biological sessions and was able to contribute much from the close contacts he maintained with biologists in other countries.

4. THE NATIONAL AND INTERNATIONAL SCENE

Throughout his career Ellis Cosslett made a major contribution to the advancement of all aspects of electron microscopy and to the training of electron microscopists. He described his attitude as 'rather like that of a craftsman concerned to see the practice of his art perfected and apprentices to it properly trained' (Cosslett 1981). He organised a series of Summer Schools in the Cavendish in the early days and I was fortunate in being able to attend the one held in 1950 (see photograph in Cosslett 1985). Subsequently Cosslett was a frequent participant in the courses organised by Ugo Valdrè for the International School of Electron Microscopy at Erice in Sicily. Valdrè collaborated with Cosslett and was a regular visitor to Cambridge from Bologna. As a consequence he became well known in the EM Section and both Bob Horne and myself were asked to contribute to the courses in Erice, a delightful centre for the purpose. The courses were attended by students from many different countries and were excellent opportunities for fruitful discussions between physicists and biologists. In addition, the welcoming atmosphere provided by Ugo Valdrè and his family, and by the course organisers Ivonne Pasquali Ronchetti and Umberto Muscatello, led to the formation of many international friendships by Ellis Cosslett and other participants.

Members of the EM Section at the Cavendish also benefited greatly from Cosslett's involvement with both the EM Group of the Institute of Physics (IOP) and the EM Section of the Royal Microscopical Society (RMS). In the 1940s the RMS did not consider the electron microscope to be a 'proper' microscope and biologists found their first home in the EM Group of the IOP which was formed in 1946 (Turner 1989). This was satisfactory in the early days but gradually the IOP became unhappy about the increasing number of biological sessions included in the meetings of the Group. It was made clear that we were unwelcome and I resigned from membership of the Committee. Cosslett now played a vital role in changing the views of the RMS. He was elected President of the RMS in 1961 and during his tenure of office (1961-1963) a number of symposia on EM were held, leading to the full acceptance of the electron and X-ray microscopes by the Society. Most notable was their inclusion in the programme for the first American Meeting of the Society, which was held in Bethesda, Maryland, in April 1963 (see Journal of the Royal Microscopical Society, Volume 83, parts 1 and 2, 1964), and which celebrated the tercentenary of 'The Microscope in Living Biology'. Messages were received from the Queen and from President John F Kennedy, and Cosslett, as President of the RMS, made the Opening and Closing Remarks and was already looking forward to the possible achievements of the high voltage electron microscope in biology.

The culmination of Cosslett's advocacy for electron microscopy within the RMS was the formation of the EM Section in 1965. I was elected the first Chairman of the Section, and later was to follow Cosslett as President of the RMS (1970-1971), becoming the first woman to hold the position since the founding of the Microscopical Society of London in 1839 and the award of a Royal Charter to the Royal Microscopical Society in 1866. The EM Section of the RMS is now the major meeting point for biological electron microscopists in the UK.

5. THE FINAL PHASE. HIGH VOLTAGE AND HIGH RESOLUTION ELECTRON MICROSCOPY

The final contribution that Ellis Cosslett was to make to the study of biological specimens was his eagerness in developing the high voltage electron microscope. Physicists had long been fascinated by the possibility of viewing living cells in the EM and Cosslett was one of the few who had sufficient understanding of biological specimens to know that this would be impossible as a consequence of radiation damage (Cosslett 1964).

A lethal dose of electrons is required to record a single image. However, the hope of examining hydrated biological specimens, for which water is such an important structural component (Cosslett 1974; Glauert 1974), has been realised. This was one of the aims behind the development of the 750 kV high voltage microscope at the Cavendish, which became the prototype for the AEI EM7, and for Cosslett's involvement with the 600 kV high resolution microscope. The completion and testing of this microscope coincided with the final closing of the EM Section in 1979 after over 30 years of pioneering contributions to biological electron microscopy.

Ellis Cosslett's fascination with biology led him to make extensive contacts with biologists and to take a great interest in their observations. In consequence he was one of the few physicists who was able to understand the requirements of biologists and to ensure that these requirements were met. All biologists are in his debt and he has left us a very rich inheritance.

6. REFERENCES

Brieger E M and Cosslett V E 1949 Preparation of bacteria for electron microscopy Nature 164, 352

Cosslett A 1960 Some applications of the ultra-violet and interference microscopes in electron microscopy J Roy Microsc Soc 79, 263-271

Cosslett V E 1964 Closing remarks of the President J Roy Microsc. Soc. 83, 10-11

Cosslett V E 1974 Current developments in high voltage electron microscopy J Microscopy 100, 233-246

Cosslett V E 1981 The development of electron microscopy and related techniques at the Cavendish Laboratory, Cambridge, 1946-79. Part 1 1946-60 Contemp Phys 22, 3-36

Cosslett V E 1985 Random recollections of the early days. In: The Beginnings of Electron Microscopy, ed P W Hawkes (London: Academic Press) pp 23-61

Glauert A M 1974 The high voltage electron microscope in biology J Cell Biol 63, 717-748

Glauert A M and Glauert R H 1958 Araldite as an embedding medium for electron microscopy J Biophys Biochem Cytol 4, 191-194

Glauert A M, Rogers G E and Glauert R H 1956 A new embedding medium for electron microscopy Nature 178, 803

Turner G L'E 1989 God Bless the Microscope! A History of the Royal Microscopical Society over 150 Years (Oxford: Royal Microscopical Society)

Inst. Phys. Conf. Ser. No 130: Chapter 1
Paper presented at Int. Congr. X-ray Optics and Microanalysis, Manchester, 1992

33

Microanalysis

J.V.P. Long

Department of Earth Sciences, Bullard Laboratories,

Madingley Rise, Madingley Road, Cambridge CB3 OEZ.

ABSTRACT: The Electron Microscope Section in the Cavendish, under V.E.Cosslett, entered the field of microanalysis both by way of the point-projection Cosslett-Nixon microscope which offered possibilities of performing differential X-ray absorption analysis, and also by virtue of the obvious similarities between the technology of the X-ray microscope and Castaing's Sonde Electronique. This paper attempts to correlate some of the developments in instrumentation and physical theory made by the Cambridge group and to show how Ellis Cosslett's wide-ranging view of the subject actively promoted significant advances, not only in microanalysis but in many fields dependent upon it.

1. INTRODUCTION

In attempting to set the contribution of the Cavendish to microanalysis into perspective it is appropriate, at a distance of forty years, to recall the state of development of the relevant science and technology when work in Ellis Cosslett's group was gathering momentum. Many techniques for investigating the solid state, which today are established fields of research, some with their own specialist journals and certainly all with individual acronyms, were in the post-war decade either as yet unidentified or, at best, still at the stage of being interesting physical phenomena whose application remained to be developed. Microscopy, whether with light, electrons or X-rays, was primarily concerned with imaging, with only the first dawn of the subsequent developments in localised analysis upon the horizon. There were already important exceptions to this generalization, notably in the work of Engström (1946,1957) and his school at the Karolinska Institute in Stockholm where densitometry of contact microradiographs was developed both for the determination of tissue dry-weight, and by using differential absorption across characteristic absorption jumps, for the measurement of specific elements such as calcium (Lindström, 1955, 1957).

The outstanding portent of the future was, of course, Raymond Castaing's 'Sonde Electronique' described in his doctoral dissertation to the University of Paris (Castaing, 1951). However, it was perhaps inevitable that in the early years, many of those concerned with the physics of microanalysis did not appreciate the almost infinite range of analytical problems awaiting solution in the sciences concerned with materials, geology and biology, and conversely, that workers in those disciplines were, in general, less aware of the potential value of physical techniques than would be their present-day successors. Indeed, one of the remarkable features of the 1954 London E.M. Conference was the scant interest displayed by the assembled electron microscopists in the account of recent work presented by Castaing. After all, the transmission microscope of the time could resolve some 10 Å: real imagination was required to envisage any application for a technique that could, at best, achieve 1μm.!

Against this general background, the Electron-Microscope Section, and Ellis Cosslett in particular, can now be seen to have played an important role in bringing together the problems and the physical methods of solving them. The basic technology of the electron microprobe was identical with that of the X-ray microscope developed independently in the group, so that it was a natural step to encompass X-ray emission techniques along with those based on absorption. The long-standing tradition of the Cavendish in the field of instrument development and the association with the Cambridge Instrument Company were also to play an important part in the subsequent rapid emergence of commercial instruments based on work in the group.

2. ANALYSIS WITH THE COSSLETT-NIXON X-RAY SOURCE.

As a new recruit to the group in the Spring of 1954, I experienced the cultural shock of a transfer from the comparative affluence and security the Scientific Civil Service, to a somewhat impoverished but dynamic environment where exciting new developments occurred, it seemed, on a time-scale measured in days rather than the accustomed months. My primary task, to set up techniques for quantitative spectrometry low-intensity beams of relatively soft X-rays, was initially directed towards differential absorption measurements for the study of mineralisation in biological tissue - the projection-microscopy equivalent of the contact techniques of Lindström in Stockholm.

Early attempts to use proportional counters to select suitable energy bands from the continuum transmitted through small areas of the specimen proved abortive, partly due to the limited energy resolution of the detector, partly to the relative crudity of the single-channel recording technique, and not least to a failure to realize what might be achieved with appropriate techniques of spectrum deconvolution. Resort to Bragg crystal spectrometers, unfettered at the time by any knowledge of the work of Johann and Johansson two decades earlier, led to the development of the simple but effective "semi-focussing X-ray spectrometer" (Long and Cosslett, 1957). This device was peculiarly suited to our needs and served well in a number of instruments including the Cambridge Instruments 'Microscan I', the commercial development of the Duncumb-

Melford (1960) design. In connection with these early essays into Bragg spectrometry it is a pleasure to record the many occasions when we drew upon the experience and advice of E.F. Priestley of the Royal Armaments Research Establishment at Woolwich in the treatment of monochromators (Priestley, 1959).

The point-projection absorption technique (Long,1958) was effective, if slow, and a long collaboration with Hans Röckert of the Department of Histology in Göteborg led to the publication of his doctoral thesis on the distribution of calcium in monkey teeth (Röckert 1958). The close relationship of the physical sciences in Cambridge (which was partly instrumental in bringing about the fruitful collaboration of David Melford of the Tube Investments Research Laboratories with Peter Duncumb) led also to a link with Mineralogy and Petrology, specifically with Dr. Stuart Agrell but also with Desmond McConnell among whose interests was the formation of hydrated calcium silicates in natural conditions closely approximating to those obtaining in setting cement. This thermally-sensitive material was not well suited to examination in the electron probe but appeared to be a good

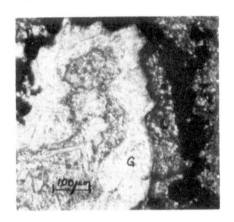

Fig.1. Optical micrograph showing part of a cavity in which larnite, L, (β-Ca_2SiO_4) is hydrated in a topochemical reaction to produce a hydrated gel (G). Measurements of mass thickness of Ca show that the gel is Ca-rich equivalent to $CaO:SiO_2 = 1.5$ (Long and McConnell 1959) .

candidate for the absorption technique and a study of 10μm diameter areas in the relatively coarse-textured natural material (Fig.1) yielded useful data on the calcium concentration of the hydrogel (Long and McConnell, 1959). It is an interesting commentary on the advances of the intervening years to record the results of the same type of measurement made recently by Morrison and colleagues (1990) using synchrotron radiation, the resolution now being such that the texture and calcium distribution among filamentous growths in commercial cement may be studied in detail within an area equal to that within which single concentration measurements were originally performed (Fig.2).

The high specific loading (~100kW /mm^2) of the Cosslett-Nixon tube was also exploited to produce fine, collimated beams to excite fluorescence in thin specimens, the secondary radiation being observed in transmission. Detection of potassium in single nerve cells was achieved with a very simple experimental arrangement (Long and Röckert, 1963) and an

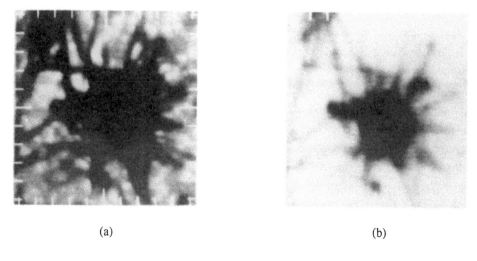

(a)   (b)

Fig. 2. Images taken with the scanning transmission X-ray microscope using synchrotron radiation showing filamentous growths from a single cement grain after 48 hours hydration in a 5:1 mixture of water and cement. a) image taken at 353eV where Ca absorbs strongly, b) Image at 355eV, above Ca L-edge resonance (Morrison et al. 1990). Reproduced by courtesy of Dr. Graeme Morrison.

extrapolation of these results suggested that sensitivities of the order of 10^{-10} to 10^{-11} g. would be possible in areas of the order of 10-20μm diameter. The practical limitation was the low intensity of the signal and just as with the differential absorption method, a re-emergence of the technique had to await the development of synchrotron radiation sources and solid-state detectors which together now give mass sensitivities in the region of 10^{-15}g.

3. ANALYSIS WITH THE ELECTRON PROBE

Thus it was inevitable that the electron microprobe, with it's versatility and ability to measure and map the distribution of a wide range of elements, should emerge as the dominant technique. It was not, however, without its deficiencies, notably in the analysis of light elements below Z=11 (Na) where rapidly-falling spectrometer efficiency, rising absortion corrections and uncertainties as to the applicability of accepted formulae for X-ray generation efficiency, all conspired to deter serious application.

It was left to a new research student, Raymond Dolby, to point out that the high collection efficiency of the proportional counter could be exploited in the soft X-ray region, and used to offset it's poor energy discrimination (Fig.3). Cosslett (1960) himself made an important

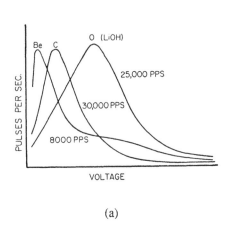

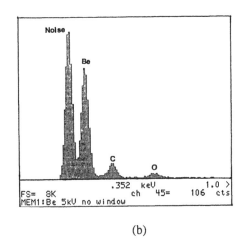

(a)   (b)

Fig. 3. K spectra of Be, C and O, a) recorded with a gas proportional counter (Dolby 1961) b) recorded with a modern Si-Li detector with very low-noise FET (Nashashibi and White 1990). Reproduced by courtesy of Oxford Instruments.

contribution by demonstrating that the reputedly very low generation efficiency of the characteristic radiation of elements such as carbon was in fact only real when expressed in terms of energy conversion: because the photon energies were very small, the quantum efficiency was relatively high, with that for carbon K radiation no worse than 5% of the value for copper at the same overvoltage. With high count rates available at probe currents of only a few nanoamps, an important component of the success of Dolby's network analysis technique (Dolby and Cosslett, 1960) for the on-line deconvolution of proportional counter spectra was the full analysis of the behaviour of these detectors at count rates in excess of 10^6/sec. (Dolby, 1961). By 1960, he had demonstrated scanning images taken with the K radiation of Be at 110 Å or 109 eV., this work being continued by Ian Wardell (Wardell 1965, Wardell and Cosslett, 1966).

It is impossible here to summarise the full development of non-dispersive (or energy-dispersive) X-ray spectrometry, but in describing the Cambridge scene, it is relevant to note that some twelve years later a very significant step forward in quantitative microprobe analysis resulted from work carried out in the Department of Mineralogy and Petrology by another research student, Peter Statham. We were exceptionally fortunate at that time in establishing contact with K. Kandiah and his collegues of A.E.R.E., Harwell, whose research in γ-ray spectrometry and the design of pulse-processing circuitry (Kandiah 1971, Kandiah, Smith and White 1975) with accurately predictable behaviour at high counting rates, presented opportunities for the recording and rigorous deconvolution of the spectra from lithium-drifted silicon detectors that were not available in less well-constrained systems. A prototype of this equipment was installed at Cambridge on a small microprobe constructed originally by Stephen Reed and the author in 1961 and with Statham's (1975, 1976) software, remained in almost continuous operation until 1987.

During its lifetime, this equipment performed over 200,000 complete mineral analyses, mostly for 10 or more elements, the data appearing in more than 200 papers published by members of the Department and by visiting scientists. On the rare occasions when the pulse-processor failed, it was placed in the boot of a car, driven to Harwell, restored to health by its designers, and was normally back in action within a day, the occasion having provided, as an incidental bonus, the opportunity for discussion between the two groups. It is also satisfying to record that this very informal collaboration played its part in the successful commercial development of energy-dispersive X-ray spectrometry by Link Systems Ltd., now the Microanalysis Group of Oxford Analytical Instruments. The scientific and commercial cost-effectiveness of this operation, and many others like it, which were unimpeded by the benefits of "effort returns" and other trappings of modern business efficiency now being imposed upon the scientific community, is something upon which our accountant masters may care to ponder.

4. THE SCIENCE

It is relatively easy to set down the contribution of Cosslett's group to the design of analytical instruments; much less so, to pursue fully the science derived from the application of those instruments. In addition, we need to recognise the advances made in the physics of X-ray production, for while the basic mechanisms were qualitatively well understood, there remained serious gaps in the quantitative knowledge required for relating observed characteristic intensities excited in compound targets to those in pure elements. N.A. Dyson (1956) had already completed a study of the intensity and angular distribution of the continuum; so in turn, Martin Green (1963, a, b) was given the task of making new experimental measurements of the efficiency of characteristic X-ray production and target absorption. Noel Thomas studied electron scattering (Cosslett and Thomas, 1964a,b 1965), and later, H. Bishop (1966a) made measurements of electron backscattering coefficients, work that was extended to low energies by Darlington and Cosslett (1972).

Until 1962, the process of electron retardation and X-ray production in the target was described by models of the average behaviour of many electrons. Martin Green (1963c), using the new EDSAC II computer constructed in the University Mathematical Laboratory, was the first to employ Monte-Carlo techniques in which the intensity of the emergent X-rays was obtained by summing the contributions of a large number of electrons whose individual paths and energy losses were modelled in discrete steps by a computer programme. The results of these calculations could be compared with experimental data and used to explore the effect of specific parameters; for example, Hugh Bishop (1966b) who extended Green's work was able to predict very accurately the variation of backscattering coefficient with atomic number, and the technique was employed by many subsequent workers.

It would be misleading to represent these studies as the main stream of the intense activity of the early 60's in evaluating correction procedures for microanalysis: the work of many individuals in the U.K., France, the U.S.A. and Japan are to be found in the proceedings of the 1962 and 1965 conferences in the present series and are also reviewed by Reed (1975). The Cambridge programme, extending over 20 years, was however particularly distinguished by a strong interaction of theory and measurement. A large body of new data was produced with equipment designed and assembled by individual students.

The early application of microprobe analysis in Cambridge and the U.K. was greatly influenced by Duncumb's scanning technique: the study of the microsegregation of copper and tin at the surface of mild steel, started in the Cavendish and pursued at the Tube-Investments Research Laboratories (Melford, 1960) illustrates very well the advantages of a two-dimensional presentation of the element distribution. Cosslett and Switsur (1963) also demonstrated the value of scanning images to reveal the distribution of metallic elements in biological tissue, while Stuart Agrell and the author (1960) applied the technique to the study of mineral intergrowths. There is no doubt too, that the Cambridge Instruments Microscan I owed much of its success to this facility. Scanning was also a standard feature of the Metropolitan Vickers microprobe (Page and Openshaw, 1960) which originated in the work of Tom Mulvey (1960). In France, on the other hand, the fuller development of the static probe technique in conjunction with fully-focussing spectrometers is reflected in experiments yielding more quantitative data, for example, the analysis of intermetallic diffusion couples.

The development of electron-probe analysis in the Department of Mineralogy and Petrology, where a collaboration between the author and the Cambridge Instrument Company led to the production of the "Geoscan" in 1965, was influenced by the fact that rocks, unlike metals, can be crushed and separated into their component minerals which, in turn, may be subjected to bulk analysis. As a result, petrologists had for decades been able to correlate distinctive optical properties with chemical composition. The probe provided the means of analysing minerals in situ and of studying variations in zoned minerals. The primary need, therefore, was for accurate multi-element quantitative analysis and this theme, in addition to application to specific problems in petrology, dominated the research effort for many years. Stephen Reed, who joined as a research student in 1961, has made important contributions in this area, while Rex Sweatman, a mature research student, 3 years older than his supervisor, performed an invaluable service in the rigorous experimental evaluation of matrix correction procedures (Sweatman and Long, 1969).

The analysis of biological material presented it s own problems: owing to the very variable density of the matrix, it was difficult to express the measured intensity of a characteristic line in terms of a meaningful concentration or mass thickness. Ted Hall, working first in the E.M. Section, and later in the Department of Zoology, solved this problem by recording also the intensity of the continuum and using this as a measure of the total mass of material in the excited volume (Hall and Gupta 1983).

5. EXPERIMENTAL EQUIPMENT

It is impossible to describe adequately the work of Cosslett's group in microanalysis without reference to the development of experimental equipment, some part of which has been covered by other contributors to this Symposium. In addition to the microanalysers constructed by Duncumb, Dolby, Switsur and the author, there were the absorption and fluorescence spectrometers and the special rigs set up by Green, Thomas and Bishop for the study of X-ray production and electron scattering, together with their supporting electronics.

The apparatus itself bore the unmistakable hallmarks of the Cavendish: sealing wax had had it's day by the 1950's but brass and soft solder were still in the ascendant. Such was the financial stringency and the need for economy, that durable second-hand components were much prized; my first microprobe used a lens inherited, I think, from Dyson's equipment and subsequently remobilised in the mineralogical instrument where it was in use until 1987. Pirani gauge heads were constructed by a simple glass-blowing operation on 15-watt lamp bulbs; the need for potentiometric chart recorders was met by a combined design effort by Peter Duncumb and myself, the result being sufficiently successful for a total of 18 to be built by the main workshop for use in other groups. I still possess the worm-and-wheel mechanism of a gunsight from the 1914-18 War that was adapted for the θ-drive of a Bragg spectrometer.

This continual evolution was only possible as a result of the assistance received from the Cavendish main workshop, the student's workshop (research students were expected to become proficient in the mechanical arts) and above all, from the E.M. Section workshop. Here, every new challenge was greeted by Ron Pryor and his collegues with a fund of experience and an enthusiasm which is a pleasure to recall.

6. CONCLUSION

This very selective account of the development of microanalysis is inevitably biassed towards the work of the E.M. Section and its siblings in the Cambridge area. It has been in no sense the intention to ignore the extensive work in other laboratories but rather to present a picture of the wide-ranging research that Ellis Cosslett initiated. Central to his own interests was the underlying physics, as evinced by the long line of experimental studies on X-ray production and electron scattering in which he was certainly more directly involved than in microanalytical technique. An equally strong driving force, however, was his desire to see the application of the physics, particularly in the biological field. However, here his control was very flexible and for example, when I myself found it difficult to enthuse about the analysis of biological material and turned to minerals, I received nothing but encouragement. Nevertheless, the objective was not abandoned: Roy Switsur was steered towards biology and Ted Hall, already experienced in the field, was brought into the group. In some ways it was a hard school, with the Cavendish sink-or-swim approach to new research students much in evidence: "...you might like to look at this problem; go away and think about it and meanwhile, build yourself an electron probe..". It says much for the confidence that V.E.C.- and Bill Nixon- engendered within the group that no-one drowned.

7. REFERENCES

Agrell S O and Long J V P 1960 In Engström, Cosslett and Patee pp.391-400

Bishop H E 1966a In Castaing et al. pp.153-

Bishop H E 1966b ibid. pp.112-

Castaing R 1951 Ph.D. Thesis, University of Paris

Castaing R, Deschamps P and Philibert J (Eds.) 1966 Optique des Rayons X et Microanalyse (Paris: Herman)

Cosslett V E 1960 In Engström et al. pp.346-50

Cosslett V E, Engström A and Pattee H H (Eds.) 1957 X-ray Microscopy Microradiography (New York: Academic Press)

Cosslett V E and Switsur V R 1963 In Pattee et al. pp.507-12

Cosslett V E and Thomas R N 1964a Brit. J. Appl. Phys. 15 883-907

Cosslett V E and Thomas R N 1964b Ibid. 15 1283-1300

Cosslett V E and Thomas R N 1965 Ibid. 16 779-96

Darlington E F and Cosslett V E 1972 J. Phys. D 5 1969-81

Dolby R M 1961 Ph.D Thesis, Univ. of Cambridge

Dolby R M and Cosslett V E 1960 In Engström et al. pp.351-7

Duncumb P and Melford D A 1960 In Engström et al. pp.358-64

Dyson N A 1956 Ph.D. Thesis, Univ. of Cambridge

Engström A 1946 Acta Radiologica Suppl. 63

Engström A 1957 In Cosslett Engström and Pattee pp.24-33

Engström A, Cosslett V E and Patee H H (Eds.) 1960 X-ray Microscopy and Microanaly: (Amsterdam:Elsevier)

Green M 1963a In Pattee et al. pp.185-92

Green M 1963b Ibid. pp.361-77

Green M 1963c Proc. Phys. Soc. 83 204-15

Hall T A and Gupta B L 1983 Quarterly Review of Biophysics 16 3 279

Kandiah K 1971 Nucl. Instrum. and Meth. 95 289-300

Kandiah K, Smith J. and White G. (1975) In Proc 2nd ISPRA Nucl. Elec. Symp (EUR 5370 (Luxembourg: Commission of Eur. Comm. Directorate.) pp 153-60

Lindström B 1955 Acta Radiologica Suppl. 125

Lindström B 1957 In Cosslett et al. pp.443-7

Long J V P 1958 J. Sci. Instrum. 35 323-9

Long J V P and Cosslett V E 1957 In Cosslett et al. pp.435-42

Long J V P and McConnell J D C 1959 Mineral. Mag. 32 117-27

Long J V P and Röckert H O E 1963 In Pattee et al. pp.513-21

Melford D A 1960 In Engström et al. pp.407-15

Morrison G R, Beswitherick J T, Browne M T, Burge R E, Cave R C, Charalambous P S, Duke J, Foster G F, Hare A R, Michette A G, Morris D, Potts A W and Taguchi T 1990 E Shinohara et al. Japan.Sci. Soc. Press Tokyo (Springer-Verlag:Berlin)

Mulvey T 1960 In Engström et al. pp.372-8

Nashashibi T and White G 1990 I.E.E.E. Trans on Nucl. Sci. 37 452-6

Page R S and Openshaw I K 1960 In Enström et al. pp.385-90

Pattee H H, Cosslett V E and Engström A (Eds.) 1963 X-ray Optics and X-ray Microanalys (New York: Academic Press)

Priestley E F 1959 Brit. J. Appl. Phys. 10 141-2

Reed S J B 1975 Electron Microprobe Analysis (Camb. Univ. Press)

Statham P J 1975 Ph.D. Thesis, Univ. of Cambridge.

Statham P J 1976 X-Ray Spectrometry 16-28

Sweatman T R and Long J V P 1969 J. Petrol 10 332-79

Wardell I R M (1975) Ph.D. Thesis, Univ. of Cambridge

Wardell I R M and Cosslett V E (1976) In McKinley T D, Heinrich K F J and Wittry D B (Eds The Electron Microprobe (N. York: Wiley) p.23

Inst. Phys. Conf. Ser. No 130: Chapter 1
Paper presented at Int. Congr. X-ray Optics and Microanalysis, Manchester, 1992

Scanning electron probe microanalysis

P. Duncumb

University of Cambridge

ABSTRACT: The development of the scanning electron microprobe analyser forms an interesting case study on the transformation of a scientific idea into an established physical technique. Ellis Cosslett initiated the work in 1953 in the belief that it would find widespread application in other disciplines and later in the decade this hope was realised. He was quick to grasp the opportunity for commercial exploitation and, as a result, the technique became widely available to users in the 1960's. I shall attempt to follow this development and to comment briefly on how the technique is employed today.

1. ORIGINS

I joined the Cavendish Laboratory as a new research student in 1953, to find a lively group - not lavishly equipped, but with access to an excellent workshop and stores, and full of ideas. For the six years that I was in the Electron Microscope Section, I was supported by standard grants from DSIR and there was no other special funding, so we were encouraged to build our own equipment wherever possible. There were two electron microscopes of commercial origin and two X-ray microscopes, one of which had been the result of Bill Nixon's PhD work (see Cosslett and Nixon 1960). As a new research student my first task was to learn some of the experimental techniques by measuring the X-ray intensity produced by this microscope using a gas proportional counter. I benefited greatly from the experience of Nixon and others on the X-ray microscope and, later, of Jim Long on X-ray counting techniques.

As part of this initiation I immersed myself in the literature, of which the thesis by Castaing (1952) on the electron microprobe was clearly obligatory reading, and has indeed been recognised as the foundation stone of practical electron microprobe analysis ever since. For a new research student the first reaction was: "It's all been done !" but fortunately I could find no mention of scanning. Cosslett, therefore, suggested rebuilding one of the electron microscopes as a scanning version of Castaing's instrument, using magnetic instead of electrostatic lenses to give sufficient intensity for X-ray image formation.

At an early stage I also visited Dennis McMullan and Ken Smith in the Engineering Laboratory, who were developing the first scanning electron microscopes under Charles Oatley (see Smith and Oatley 1955), and there I learned much of scanning techniques. The demarcation between our respective activities was never a problem, since X-rays were deemed to be the province of physicists, while engineers were the professionals in electronics design.

The origins of the scanning microanalyser were thus threefold: the X-ray projection microscope (using magnetic lenses): Castaing's work on the microprobe (providing the physical basis), and the scanning and display techniques developed for the SEM. A genealogy, including some of the developments to be referred to later, is given in Figure 1.

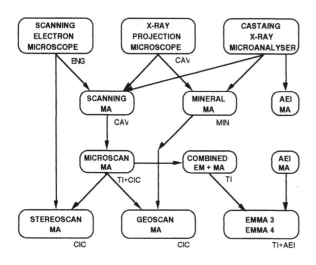

It was characteristic of the whole period that there was a great deal of interaction with other laboratories, not only in Cambridge but elsewhere. Cosslett's role in all this was central. His style was to interfere little with the work itself but to support it with a stream of ideas and contacts from the world outside. This is exemplified in Cosslett's own history of the EM Section (1981), covering the period 1946 - 79. It was a stimulating environent; good fortune played a major part in the way the work evolved, but the conditions were right for lucky events to occur.

Fig. 1. The genealogy of electron microprobe analysers in the UK from 1952 - 72, involving the Cavendish, Engineering and Mineralogy Laboratories at Cambridge University, and the Cambridge Instrument Company, Tube Investments and Associated Electrical Industries

More by luck than judgment, the science of the instrument design was influenced by the needs and experience of the user, thus aiding the evolution of the next generation of instrument. This feedback is hard to achieve, even in the organised environment of industry, and it is a tribute to the effectiveness of informal relations that this was able to occur.

2. EARLY DEVELOPMENT

In April 1954 I set to work modifying the RCA-EMB electron microscope, which had come to this country with lease-lend during the war. The instrument had served time as a transmission electron microscope in Cosslett's group. and had proved reasonably reliable despite a formidable array of vacuum tubes in the power supply. Later in the year Bob Horne returned to the Group, after a period as a service engineer with RCA, and knew the foibles of the instrument well. I often consulted him.

Also at this time, Jim Long joined the Group, bringing with him a great deal of practical experience on counting techniques, and one of the first visits we made together was to the Royal Institution in London where Uli Arndt had just invented the gas flow proportional counter for his work on X-ray diffraction (Arndt et al 1954). Long immediately recognised the potential for using this detector in the microprobe, first as an energy dispersive spectrometer and soon after as part of a simple crystal spectrometer. Later he built a scanning microprobe instrument of his own, particularly suited to mineralogical applications (Agrell and Long 1959), and subsequently moved to the Department of Mineralogy to pursue the work further.

Another major influence on the design of this type of equipment was the pioneering electron-optical work at the AEI Laboratory at Aldermaston Court. The work of Haine and Einstein (1952) on electron guns and that of Liebmann and Grad (1951) on magnetic lenses provided comprehensive design information which was not surpassed until finite element methods became practicable in the 1970's. The scanning microanalyser would not have come into existence when it did without these developments on electron optics and X-ray counting techniques. They also led to the parallel development of a microprobe instrument in the AEI laboratory under Tom Mulvey (1960), using the magnetic pinhole lens which he pioneered, and this instrument also became available commercially in the early 1960's.

In 1958 Mulvey and I were both invited to the US to take part in the first meeting there on microprobe analysis. This was arranged by Verne Birks of the Naval Research Laboratory and was attended by 30 - 40 people representing almost all the laboratories active in microprobe analysis at the time. We counted ten operational microprobe instruments in the world, of which the only two with scanning capability were in Cosslett's laboratory. After the meeting we felt we knew all that was going on in the subject, but this situation rapidly changed in the 1960's as more and more instruments became available.

Figure 2 shows the modified RCA microscope used in my own instrument, with the lower half of the column removed and an objective lens in place for the analysis of solid samples. X-rays were collected through the lens gap into a "semi-focusing" crystal spectrometer with gas flow proportional counter visible on the right of the column. The display tubes on the left showed the scanned image in terms of the selected X-radiation (now called element mapping), and also an image produced by high energy backscattered electrons for probe positioning. The instrument was completed in this form in

Fig 2. RCA-EMB electron microscope modified as a scanning microprobe analyser

1956. Thus it was possible to tune into a selected element, show its distribution across the surface and position the probe for spot analysis. What was lacking was an inbuilt optical microscope and a fully focusing crystal spectrometer, as provided in Castaing's static probe instrument, and these differences were the subject of much discussion among users for years to come. It was not until much later that a way was found to combine all these features in one instrument, and by then the silicon detector had largely replaced the crystal spectrometer, at least for imaging and qualitative analysis.

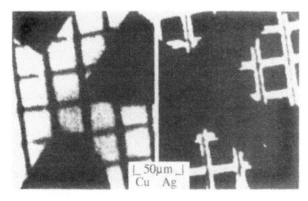

50μm
Cu Ag

Fig. 3. Early X-ray images: copper and silver grids imaged by CuK (left) and AgL emission (right)

Reverting to the RCA instrument, Figure 3 shows one of the first scanning images, presented at the first ICXOM in 1956 (Duncumb and Cosslett 1957). It consists of overlapping copper and silver grids, from which the K and L radiations respectively were collected, after passing through the sample, by a gas proportional counter. These images were rephotographed through red and green filters and superimposed to give a colour map - something which is now much more easily done in the computer. Subsequently, the technique for colour mapping was exploited much more thoroughly by Yakowitz and Heinrich (1969) at the National Bureau of Standards in Washington, and an up to date account of the technique has been given by Newbury (1992) of the same laboratory.

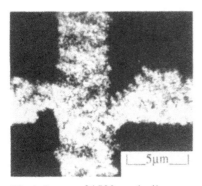

Fig.4. Image of 1500 mesh silver grid at 6kV (bar width 3 μm) formed by total X-ray emission

Later in 1956 the crystal spectrometer was brought into operation and afforded much better discrimination between neighbouring elements in the periodic table. However the proportional counter always provided better collection efficiency and hence was to be preferred when the X-ray intensity was low - for example when the sample was thin or the beam voltage reduced. By the time of the second ICXOM in 1959, both avenues had been explored, and Figure 4 shows a silver grid, imaged with a proportional counter at a beam voltage of 6 kV. The bar width is 3μm and, although some quantum noise is visible, the image shows a resolution of around 0.3 μm.

A similar resolution was also obtained on thin film samples of metal carbide particles, transparent to the electron beam (Duncumb 1960). The field of view was located first in a separate electron microscope and then transported to the RCA instrument. Needless to say this was a tedious process and the idea grew of building a combined electron microscope and microanalyser - a technique now known as analytical electron microscopy

By the late 1950's, the RCA instrument had served its purpose. It had demonstrated the fundamental limitations of X-ray scanning microanalysis, spawned several ideas which were to bear fruit in subsequent years, and pointed up the need for a more "user-friendly" version for routine microanalysis. By a stroke of good fortune, the way forward on this latter requirement was not long delayed.

3. APPLICATION AND COMMERCIAL MANUFACTURE

With Cosslett's encouragement, both Jim Long and myself were looking for applications to prove the value of microprobe analysis. In Long's case this led to a concentration on studies of mineralogical and meteoritic samples, and I looked at a number of others, including metallurgical, biological and even archaeological samples. From the beginning Cosslett had had in mind the possible application to the study of lung disease, and the instrument had no difficulty in detecting tin particles in lung sections from victims of stannosis. But it was not until Ted Hall joined the Group in the early 1960's with a particular interest in the biological applications that this field moved ahead (see Hall and Gupta 1983).

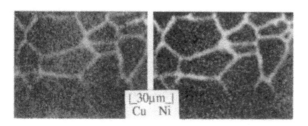

Fig.5. X-ray images showing copper and nickel segregation in grain boundaries in steel (after Melford)

Meanwhile I had concentrated on metallurgical samples and in August 1957 received a visit from David Melford, newly recruited to Tube Investments Research Laboratories at Hinxton Hall. He brought with him some samples of steel billet showing surface cracking and had come at the suggestion of Jim Menter there, who had followed the development of the instrument from the outset. Time was short, but I agreed to spend one day looking at the steel samples, and it proved to be a day well spent. We showed that the cause of the cracking was due to segregation of a phase containing copper, nickel and tin at the grain boundaries near the surface, which remained molten as the steel was deformed. An example of this segregation is illustrated in Figure 5 (Melford 1963). As a result of this and subsequent work, Melford was able to recommend a change in steel making practice to overcome the problem. It is not often that an experiment at the micron scale impacts so directly on the efficiency of a steel works but there was no doubt about the cost saving from scrap reduction in this instance. Menter quickly decided that TI should build an instrument of their own, and I was invited to assist Melford in the design and subsequently to join the company.

While the TI instrument was under construction in 1958, it was drawn to the attention of Cambridge Instrument Company, who were looking for a new strategic product line for their company. Under Menter's guidance a three-cornered arrangement was readily arrived at between the Cavendish, TI and CIC. As a result, the "Cambridge Microscan", which followed closely the Hinxton design, was announced at ICXOM in 1959 (Duncumb and Melford 1960). Nearly 80 were built in the early 1960's, to be replaced by the "Geoscan", which had derived from Long's work in the Department of Mineralogy. In the meantime the French Cameca instrument was becoming widespread in Europe, the AEI instrument was launched in this country, and there were several contenders in the US. If the 1950's saw the birth of the species, the 1960's saw the proliferation , and it was not until the 70's and 80's that evolutionary forces brought about a degree of convergence in the design of microprobe instruments.

Soon after work started with TI, Cosslett recruited into his group a research student Ray Dolby, later of audio fame, who set out to detect the long wavelength X-radiation from the light elements - notably carbon K radiation at 44 Å. In a series of careful experiments Dolby became the first person to count carbon K quanta, using a gas proportional counter, and later invented an electronic "matrix" method for unscrambling the overlapping peaks from light elements. In this way he was able to demonstrate separate scanning images of beryllium, carbon and oxygen present in a single sample (Dolby 1963), and the equipment was taken up by CIC and supplied with the first production instrument. Other forms of wavelength and energy dispersion soon became available, but it is interesting that the modern equivalent of the matrix method is to be found in computer image processing today

4. COMBINED ELECTRON MICROSCOPY AND MICROANALYSIS (EMMA)

It was in Cosslett's group that the idea emerged for analysing thin film samples in the electron microscope, and this was reinforced at Hinxton by the need to study carbide precipitation in stainless steel. It was not sufficient to add an X-ray spectrometer to the column of a conventional electron microscope, though some EM manufacturers attempted to do so, since the beam illuminating the sample was too broad for sub-micron resolution to be achieved. Consequently the decision was taken at Hinxton to build a prototype instrument from scratch, which became known as EMMA. The first results emerged in 1962 (Duncumb 1963), using a proportional counter detector, and by 1964 it was fitted with twin focussing spectrometers, enabling it to fulfil its original remit (Duncumb 1966).

Figure 6 shows an early application of the instrument using a proportional counter detector to study the variation of X-ray intensity over the surface of a single-crystal gold film. It had been predicted by Howie and

Fig. 6 Enhanced X-ray emission at extinction contours in single crystal gold film, observed in EMMA in 1962

Whelan, working in the Cavendish, that this intensity should vary with crystal orientation, and we devised an experiment in EMMA to test this prediction (see Duncumb 1962). Because the film was only 200 Å in thickness, count rates were low, but the enhancement of X-ray intensity at extinction contours in the transmission image is clearly visible. As much as a 30% variation was measurable, even in this early experiment, which raised a warning about the accuracy obtainable in the quantitative analysis of crystalline samples. Today the effect is well understood, and is harnessed to provide better information about the orientation and composition of such samples.

In 1968 the paths of AEI and TI crossed again and by 1970 a completely new instrument had been designed jointly between the two companies (Cooke and Duncumb 1969), which was sold by AEI as EMMA-4. It was rechristened the "Analytical Electron Microscope" and offered twin focussing spectrometers combined with a miniature probe-forming lens, giving a beam size down to 0.1μm. Whilst the instrument was equipped with the full diffraction capability to meet metallurgical needs, it also found good use in biological laboratories, and some of its applications are described by Hall and Gupta (1983). In general, raster scanning is less useful in analytical electron microscopy, because of the low intensities, but manual deflection of the probe is essential, to allow the beam to be steered in relation to the TEM image. The technique has been very fully studied by Cliff and Lorimer, who had one of the first EMMA4 instruments from AEI, and who now foresee that with modern equipment it may even be possible to detect the X-ray emission from single atoms (Cliff and Lorimer 1992).

5. EVOLUTION OF SCANNING MICROPROBE ANALYSIS

One further influence on the subject to emerge from Cosslett's group was the work of a series of research students in the early 1960's on the physics underlying the interaction of electrons with solid targets and the generation of X-rays. This was described, among other places, at the ICXOM meetings in 1962 and 1965, and led to a greatly improved understanding of the relationship between target composition and measured X-ray intensity. Partly as a result of this work, there now exists a sound basis for quantitative analysis, even for the light elements down to boron or carbon; however the subject is a major one and is outside the scope of this paper.

The 1960's also saw some major advances in technique, which have changed the face of scanning electron microprobe analysis and analytical electron microscopy. The advent of UHV techniques made the field emission gun practicable; the lithium-drifted silicon X-ray detector reduced (though did not remove) the need for mechanically complex crystal spectrometers; and developments in computing changed the way instruments were controlled and the data processed. These advances have resulted in broadly three classes of instrument for microprobe analysis available today:

- the thoroughbred microanalyser, which is now a highly automated and stable instrument, allowing multiple crystal spectrometers or an energy dispersive analyser to be programmed to analyse a series of samples with a high degree of accuracy.

- the scanning electron microscope, which is often equipped with a silicon detector for rapid (though generally less accurate) analysis, and which may have a field emission gun for optimum probe resolution.

- instruments for thin film analysis, such as the AEM or STEM, where X-ray intensities are low, requiring a short-focus probe-forming lens to be combined with a wide-angle X-ray detector.

In general these instruments have become easier to use - a tendency driven by the increasing need for one person to access several techniques of materials characterisation in parallel, and facilitated by the dramatic advances in the software interface. At the same time the enhanced degree of control offered by such software has allowed the skilled user to extract information about his sample which was hitherto unthinkable. This tendency will undoubtedly extend in the future, as more and more intelligence is built in to attune the instrument to the real needs of the user.

6. CONCLUSIONS

Ellis Cosslett saw all these developments taking place, and had the satisfaction of laying the foundations of much of what is in use today. Scientific curiosity alone was not a sufficient motivation for him; he had an eye also for the wider applications in disciplines beyond the physics of the technique itself. This ability to conceive the transition from science to application was accompanied by a "management style" well suited to technique development. In the first place, he was good at identifying potential project areas for his students - coherent but open-ended and with plenty of scope for originality. Often he would suggest two or three areas to choose from.

During the work itself, he did not interfere, but supported the student by occasional discussion and encouraged him or her to link up with others of similar interests outside. In these respects he was ably supported by Bill Nixon - together organising the first ICXOM in 1956 and writing the standard work on X-ray microscopy (Cosslett and Nixon 1960).

Finally, with his wife Anna, who also worked in the Group, he maintained a pleasant and lively social atmosphere - a model for a research group anywhere. All this taught me much about the research environment that was useful when, in 1959, I moved into industry. Six years in the Electron Microscope Section had provided more than a training in research; it had proved to be a training in life itself.

REFERENCES

Agrell S O and Long J V P 1960 Proc 2nd ICXOM 1959(Elsevier) pp 391-400

Arndt U W, Coates W A and Crathorn A R 1954 Proc. Phys. Soc. B 67 357

Castaing R 1952 PhD Thesis, 1951(Paris University)

Cliff G and Lorimer G W 1992 Proc EMSA/MAS Mtg 1992(San Francisco Press) pp

Cooke C J and Duncumb P 1969 Proc 5th ICXOM 1968(Springer) pp 245-247

Cosslett V E 1981 Contemp. Phys. 22 3

Cosslett V E and Nixon W C 1960 X-ray Microscopy (Cambridge University Press)

Dolby R M 1963 Proc 3rd ICXOM 1962(Academic) pp 483-493

Duncumb P and Cosslett V E 1957 Proc 1st ICXOM 1956 (Academic) pp 374-380

Duncumb P 1960 Proc 4th Int. Conf on Electron Microscopy 1958(Springer) p 267

Duncumb P and Melford D A 1960 Proc 2nd ICXOM 1959(Elsevier) pp 358-364

Duncumb P 1962 Phil. Mag. 7(84) 2101

Duncumb P 1963 Proc 3rd ICXOM 1962(Academic) pp 431-439

Duncumb P 1966 Proc. Symp. Electrochem. Society 1964(Wiley) pp 490-499

Haine M E and Einstein P A 1952 Brit. J. Appl. Phys. 3 40

Hall T A and Gupta B L 1983 Quarterly Review of Biophysics 16(3) 279

Liebmann G and Grad E M 1951 Proc. Phys. Soc. B 64 956

Melford D A 1963 Proc 3rd ICXOM 1962(Academic) pp 577-589

Mulvey T Proc 2nd ICXOM 1959(Elsevier) pp 372-378

Newbury D E 1992 Microscopy 22 11

Smith K C A and Oatley C W 1955 Brit. J. Appl. Phys. 6 391

Yakowitz H and Heinrich K F J 1969 J. Res. Nat. Bur. Stand. A Phys. Chem. 73A 113

Inst. Phys. Conf. Ser. No 130: Chapter 1
Paper presented at Int. Congr. X-ray Optics and Microanalysis, Manchester, 1992

In memory of V E Cosslett: from scanning with electrons to scanning with light in the investigation of mineralised tissues

Alan Boyde.

Dept of Anatomy and Developmental Biology, University College London, London WC1E 6BT.

ABSTRACT. In this text, I trace some of the threads of influence which V.E. Cosslett and his team and chain of contacts had on an outsider fortunate to meet him at the outset of a research career.

The investigations which I have been concerned with and which address the interrelationships between the structure, function and development of skeletal and dental tissues have used many different methods of microscopy: they owe a great deal to the opportunities and the encouragement provided by Dr. V. Ellis Cosslett (VEC) and the colleagues that I met in a chain reaction starting with him. As a young graduate, I entered the Anatomy Department of the London Hospital Medical College as a Junior Demonstrator, to be baptised into microscopy by Ronald William Fearnhead. Ron instructed me in the arts of electron microscopy and diffraction, soft x-ray microscopy and quantitative interference methods in light microscopy, the latter being much in vogue at the time under the influence of R. Barer - with whom VEC was later to edit an influential series of reviews covering many different methods of microscopy. Ron Fearnhead had a particular enthusiasm for x-ray microscopy and the Cosslett and Nixon (1960) volume was the bible when it appeared.

Ron Fearnhead had learnt about the Duncumb and Cosslett microanalyser at the Cavendish Laboratory, and then, as now, was not bashful about asking to have a go with what was then, in 1959, a unique facility. Fortunately, neither was VEC averse to discovering new areas of application for the emerging microscopies, and it turned out that he would put aside a substantial amount of time for us to obtain access to and really work with the method. This generosity of spirit was matched in V. Roy Switsur who ran and maintained the system(s) and actually did the work. It is most probable that I would afterwards have been quite reticent in proposing to show to other inventors of new microscopic methods what they were really invented for (i.e. to make some new contribution to the study of hard tissue structures!) if we had not met with that initial welcome from VEC:- the opposite of "once bitten, twice shy".

Thus in 1959 I began to travel regularly to Free School Lane, taking with me the best test specimens that we could invent in Whitechapel. Through Roy Switsur, I learnt that cruising in an old wooden boat without an engine might not have been such a bad training after all for working with prototype, ancient electron probes. In both cases, one seemed to spend a great deal of time going backwards, only rewarded occasionally by forward rushes when the wind and tide suited. Thus patience and resolve in scientific endeavour were also learnt through the good offices of VEC.

I very well remember the coffee and tea breaks in VEC's laboratory, through which I learnt the importance of the associated activities in a largish group. Apart from the fact that one was always greeted by VEC and that he always enquired keenly about the progress that we were making, I met such an incredible group of individuals who were all working together, sharing thoughts and goods in what seemed to me, as an outsider, to be an almost Utopian atmosphere. If ever they were dispirited, they certainly never showed it. I was also impressed that there was no class distinction, in the sense that graduate students, technicians, research assistants and faculty were all jammed in together, united in their commonality of interest. What happened there and then was right. Because of the timing of the 1960 Electron Microscopy Congress in Delft, I made good friends with several of the Cavendish PhD students: some of these brave souls sailed there in my leaky vessel: wiser ones just came and slept on board. The wisest just came and looked and went away again.

The analytical results that we obtained with the x-ray microanalyser in Cambridge are a matter of the record (Boyde et al 1961). VEC was well pleased with our efforts and cited them in a current review (Cosslett 1960). It was clear that further work with electron beam excited x-ray emission microanalysis of teeth would be limited by real physical problems. First, we were limited, by the nature of the material which we were studying, to work with bulk samples. Second, the electron beam penetrated and excited the characteristic radiation within too large a volume, and we did not know what that volume was, nor how it varied with the mineral packing density (Switsur and Boyde 1963; Boyde and Switsur 1963). Thus we were limited to making qualitative statements: for example that certain regions, such as the peritubular zones in dentine were more highly mineralised than closely neighbouring regions, the intertubular dentine. However, this information, although important, could not only already be deduced from the reflected electron image, but at higher resolution and with less noise, i.e. greater certainty. From the outset, therefore, I was convinced that the BSE SEM image represented the better way forward (and it is still one of our main enthusiasms at the present day). This meant looking away from the offspring of the recent Duncumb and Cosslett marriage of the scanning electron beam technology developed under Charles Oatley at Engineering Laboratory in Cambridge with the static probe analyser of Castaing. However, VEC was pleased with the results that ensued after Roy Switsur was kind enough to introduce me to Garry Stewart and Scroope House, Trumpington Street. Garry was Oatley's current SEM research student, very friendly and instantly helpful, and our joint work with his much higher resolution SEM began the day that we met. It therefore happened that I learnt within hours (1) about the very important contribution of topography in reflected (i.e. BSE) electron imaging, (2) that I did not know how to polish samples (to minimise the topographic contrast) for work in the better resolution range, and (3) that the hardest known biological tissue, dental enamel, underwent plastic flow under load (Stewart and Boyde, 1962). From then on, I was yearning for access to SEM for imaging surface topographies. Garry was whisked away to the Cambridge Instrument Company, where he worked with Mike Snelling on the development of the first commercial SEM as such (the Stereoscan), although the Instrument Co were already well ahead with manufacturing commercial scanning electron probe x-ray microanalyzers. The Stereoscan employed the then rather freshly invented Everhart-Thornley, biassed scintillator secondary electron detector. In retrospect, we can see that this meant that the newly developing field lost the advantages of the more interpretable and quantifiable BSE image. Via Garry, some access was available to the prototype Stereoscan and we were able to investigate aspects of the development of mammalian dental enamel (Boyde and Stewart, 1963). I also reported on these findings to the first Tooth Enamel Symposium in 1964 (Boyde 1965), where I met Hans-Jürgen Höhling. Back at the Cavendish, Ted Hall took over the mantle of the electron probe from Roy Switsur, and Hans went to work with Ted (Höhling et al 1967) before setting up his own probe in Münster, to which end he also recruited W.A. Patrick Nicholson from the Cavendish (Nicholson et al 1975). Having spent much more time with Cosslett and having met him at a later stage in his career, Hans was, if I may say so, really deeply impressed and we have often discussed his time there and his deep admiration for VEC. As I collaborated on and off with Hans over the years, this became another link in the chain of contacts leading back to VEC. Via Hans, I came to know Gerhard Pfefferkorn, who arranged for me to use the Stereoscan in Münster in 1966, an invitation which he extended to many other people in the process of establishing and popularising scanning electron microscopy. Both Gerhard Pfefferkorn and Hans Höhling collaborated in different ways with Ludwig Reimer (Höhling et al 1971), another man whom I came to respect for his energy, insight, physical intuition and the friendliness which he engendered in his group of graduate students. Later, he and I became founding editors of Scanning^{TM}, The Journal of Scanning Microscopy.

To overcome the frustration of not being able to study our SEM samples in an SEM-less (1963-1966) interim, I had to make do by removing a surface coating for study by TEM. Here, the grey uniformity of the carbon replica image proved to be more akin to the random dot patterns of Julesz than the intuitively obvious, easily interpretable (or so most think, even until today) SEM image. However, the 3-D shape of the replica could be recovered by stereo-imaging and stereo-photogrammetric measurement (Boyde 1966). Here again, in an indirect fashion, I ran into the hand of VEC. Through his influence, with Bill Nixon, in the development and popularisation of x-ray microscopy and stereoscopic imaging, he had persuaded Raymond Ely to develop a commercial x-ray microscope (or at least not dissuaded him from doing so). Raymond Ely, yet another outstanding gentleman, had established the M.R. Research Trust with his own funds and I worked at the laboratories in his house in Guildford with medium resolution projection x-ray imaging and systems for stereoscopic imaging and rotational microtomography. To aid in the analysis of these images and also to be able to start in the stereophotogrammetry of TEM replicas, a Hilger and Watts Stereometer, and also a Stereosketch, were purchased and donated by the M.R. Research Trust. Under no circumstances would such trust have been bestowed in me if it were not for the Cosslett connection.

When I was eventually successful in obtaining a grant from the SRC to purchase the Stereoscan (which also saw my removal in 1967 to JZ Young's Dept at UCL), we were therefore already experienced in both modelling (Stereosketch) and measuring surface topography from stereo-pair SEM images. This enabling technology allowed us to think of solving problems which would otherwise have been out of court. In particular, it led to the proposal to be able to provide a measure of the functional activity of single osteoclasts, the cells that resorb bone, by measuring the volumes of the pits that they excavate (Boyde 1968; Boyde and Jones 1992). This preoccupation in turn led to the development of faster, more user friendly solutions to 3-D measurement via confocal scanning optical microscopy.

Having received an early introduction to SEM via VEC, I was consequently an early SEM user, and if one person is a user and another is not, the latter may consider the former to be an expert, proving that such a term is only relative. Under these circumstances, I was invited to Czechoslovakia in 1970 to give a series of lectures about SEM and thus came to meet Mojmir Petran and Milan Hadravsky in Pilsen, where their real-time (meaning faster than video rate) direct-view confocal microscope, or "Tandem Scanning reflected light Microscope" (TSM) as these joint inventors described it, was demonstrated to me. On a series of later visits there, I became further interested in the potential of confocal microscopy, which eventually led to our proving its great merit in hard tissue research (Boyde et al 1983), which in turn convinced the MRC to finance the construction of a new instrument in Pilsen, delivered to UCL in 1983.

Confocal scanning optical microscopy proved to be a great adjunct to scanning electron microscopy, both because we could look below electron opaque surfaces - which could also be alive and wet - and because we could find the height of the surface with excellent precision due to the sharp peak in signal intensity which is a feature of confocality. Thus we eventually had the means to displace SEM in surveying surfaces at submicrometre resolution (Cox and Sheppard 1983), and although it took some time to develop, we could both do most of the work which had previously been within the realm of SEM stereophotogrammetry (Boyde and Jones 1992) and also found new means for 3-D imaging within semi-transparent materials (Boyde 1985).

Since I really only met VEC when I was a youngster, and he was already at the head of a remarkable team, I always held him in awe. Confidentiality in the reviewing process for papers submitted to journals and for applications to the granting authorities rarely permits one to know how those who know you judge you in private. I was very pleased to discover - and I cannot now remember how - that VEC had accepted a note which we wrote for Nature in which we described the use of the glass light microscope slide as part of the signal detection chain in an STEM arrangement in a standard Stereoscan SEM. The cathodoluminescence (CL) of the glass underlying a conventional histological section, dewaxed and coated, was reduced in accordance with the mass intercepting the beam in the section (Boyde and Reid 1983). I had always yearned for the ability to use CL in SEMs, since direct LM viewing of the CL of the spot under electron bombardment had been provided in both the Duncumb-Cosslett apparatus and in the Cambridge Instruments Microscan.

All the really important steps in the developments of microscopical technique which I have seen were done with little or no money, and certainly with nothing smacking of modern grant funding. The hard work was done first. We exchanged gifts of friendly assistance and time and access to instrumentation, behaviour which I learnt from a number of nature's real gentlemen, with Ellis Cosslett at the top of the list.

Ackowledgements. Ron Fearnhead allowed me to see that the greatest enjoyment in microscopy is to be in at the exploration and the evolution of a new technique [to which the counterpart is that, if a method actually takes off, one inevitably has the oldest, most inefficient and unreliable equipment in the field]. I have Ron alone to thank for putting me in touch with Ellis Cosslett.

References
Boyde A 1964 Tooth Enamel eds MV Stack, RW Fearnhead (Bristol: Wright) pp 163-7 & 192-4
Boyde A 1966 Adv. Fluorine Res. & Dental Caries Prevention 4 137
Boyde A 1968 Beitr. Elektronenmikroskop. Direktabb. Oberfl. (Münster) 1 97
Boyde A 1985 Science 230 1270-2
Boyde A and Jones SJ 1992 Photogrammetric Record 14 59-84
Boyde A, Petran M and Hadravsky M 1983 J. Microsc. 132 1-7
Boyde A and Reid SA 1983 Nature 302 522-3
Boyde A and Stewart ADG 1963 Nature 198 1102
Boyde A and Switsur VR 1963 X-ray Optics and X-ray Microanalysis ed HH Pattee, VE Cosslett and A Engstrom (New York: Academic Press) pp 499-506
Boyde A, Switsur VR and Fearnhead RW 1961 J. Ultrastruct. Res. 5 201
Cosslett VE 1960 Brit. J. Radiol. 1-20
Cosslett VE and Nixon WC 1960 X-ray Microscopy (Cambridge: Cambridge University Press)
Cox IJ and Sheppard CJR 1983 Image & Vision Computing 1 52-63
Höhling HJ, Hall TA and Boyde A 1967 Naturwissenschaften 54 617
Höhling HJ, Scholz F, Boyde A, Heiner HG and Reimer L 1971 Z.Zellf. 117 381
Nicholson WAP, Ashton BA, Höhling HJ and Boyde A 1975 Calcium Metabolism, Bone and Metabolic Bone Diseases eds F Kuhlencordt F, HP Kruse HP (Berlin: Springer-Verlag) pp 181-3
Stewart ADG and Boyde A 1962 Nature 196 81
Switsur VR and Boyde A 1963 X-ray Optics and X-ray Microanalysis ed HH Pattee, VE Cosslett, A Engstrom (New York: Academic Press) pp 495-7

Inst. Phys. Conf. Ser. No 130: Chapter 1
Paper presented at Int. Congr. X-ray Optics and Microanalysis, Manchester, 1992

59

The end of an era?

P W Hawkes

CEMES - LOE / CNRS, BP 4347, F-31055 Toulouse Cedex (France)

ABSTRACT : The Electron Microscopy Section of the Cavendish Laboratory, Cambridge, contributed to many of the developments in electron optics during its thirty years of existence. We set these contributions in the context of research elsewhere in that period.

1. ETERNAL SUNSET

Electron optics has been repeatedly if not continuously accused of being moribund. Generations of students of the subject have been encouraged to find a new theme to work on, on the grounds that the mine is worked out. No less repeatedly, those same students have proved the pessimists wrong and, although electron optics is not immune from the consequences of the research slump of today, I do not believe that it is worse affected than any of the many other less-than-fashionable branches of physics. In the remainder of this paper, the contributions of the numerous students of electron optics who worked in V.E. Cosslett's research group will be recalled and some attempt made to assess their place in a wider research setting.

2. THEMES

Cosslett himself published papers and a book on electron optics (Cosslett, 1940, 1945, 1946), though in later year he concentrated on other aspects of electron microscope imagery, notably scattering and radiation damage. The first and perhaps the most important contribution to pure electron optics by a member of his group was made by P.A. Sturrock, whose formation as a pure mathematician enabled him (Sturrock, 1951a) to improve on the basic Hamiltonian formation set out in detail by Glaser in the 1930s and later in his two great treatises (Glaser, 1952, 1956). Sturrock applied the theory to the study of parasitic aberrations (Sturrock, 1951b) as well as to the intrinsic aberrations of electron lenses and in the later work published in his book of 1955, he also studied the time-dependent fields of accelerator optics. In addition, he devoted a long paper (Sturrock, 1952) to the optics of general systems, with arbitrarily curved axes, a tour de force of only limited practical applicability. Sturrock's contribution is all the more remarkable that Glaser's work was not available to him in book form before 1952 and the only connected accounts of the subject were Cosslett's book and those of Picht (1939) and Zworykin *et al.* (1945). Although Sturrock's work and of course Glaser's were extensively used during the decades that followed, the fundamentals of electron optics were not further extended or developed in western Europe until the late 1960s, when H. Rose produced a new formulation. There were, however, developments of considerable interest in Russia during the 1950s, little appreciated elsewhere at a time when systematic translation of Russian-language journals did not exist or had only just begun to appear. The work of Vandakurov (1955, 1956, 1957), building on that of Grinberg (Grünberg) (1942, 1943a,b) and of Kas'yankov (1956) remains imperfectly known to this day.

The next theoretical contribution that I am aware of was concerned with one of the great themes of the 1950s and 1960s, namely, the correction of spherical and chromatic aberration by one or other of the means proposed by Scherzer (1947), author of one of the most unwelcome "theorems" ever derived (Scherzer, 1936). Throughout those decades, efforts were being made in Tübingen, Paris, Darmstadt, London, Leningrad, Cambridge, the AEI Research Laboratories and doubtless elsewhere to put Scherzer's ideas into practice. Despite the ingenuity of those who attempted to introduce controlled space-charge distributions or to use lenses with electrodes on the axis, departure from rotational symmetry had the largest following. The severity of the effect of misalignment in electron optical systems was well-known and the problem would clearly be even more acute if the number of electrodes or polepieces was multiplied, as it would have to be if quadrupoles and octopoles joined or supplanted the round objective lens. J.C. Burfoot (1953) therefore sought the corrector system with the smallest number of electrodes. Although the detailed features of his design are now known to be unnecessarily fussy, his general conclusions were correct and have influenced subsequent thinking about quadrupole-octopole corrector design. At the same period, work on the design of such systems was in course in Tübingen (Seeliger, 1948/9, 1949, 1951, 1953, Möllenstedt, 1956) and at the AEI Research Laboratories in Aldermaston and later Rugby, where Archard (1954, 1955) was exploring possible configurations with a view to the correction of a commercial microscope, an ambition that remained unrealized, at AEI and elsewhere. V.E. Cosslett was, however, much encouraged by Burfoot's findings despite the conclusion that such a system would have to be built with a precision beyond that of the machine-shops of the time. In 1959, he suggested that I reconsider the whole question and, in particular, the theory of the aberration coefficients of quadrupole systems. Expressions for these coefficients had been derived by Melkich (1947) during the war years, using variation of parameters to solve the equations of motion including aberration terms. This approach, although of course perfectly correct, has the twin disadvantages of failing to reveal relations between apparently unconnected coefficients and of yielding untidy-looking formulae. I therefore recalculated the expressions for all the aberration coefficients of systems including round lenses, quadrupoles and octopoles (Hawkes 1965a,b) and demonstrated, after very laborious and intricate partial integration, that they agreed with those of Melkich. I considered only geometrical aberrations, thereby missing the opportunity of showing that mixed electrostatic-magnetic quadrupoles can be free of chromatic aberration, discovered by Kel'man and Yavor (1961) and later by Septier (1963) for a particular field model ; I did however later show it to be true in general (Hawkes, 1964, 1965c).

The 1960s were the heyday of electron optical research in Cosslett's group, in the number of research projects at least. During those years, M.G.R. Thomson (1966, 1967) used the computing facilities then available to launch numerical electron optics : primitive though that EDSAC computer may seem today, it made it possible to examine problems beyond the capacity of the analogue devices that had been used previously - rubber sheets and resistance networks.

D.F. Hardy studied both theoretically and experimentally the optical properties of mixed quadrupoles (1967). J.M. Deltrap built and tested a quadrupole-octopole corrector and showed that the principle on which it was based was correct (Deltrap and Cosslett, 1962 ; Deltrap 1964a,b). Later, H.E.R. Preston showed that the potential functions for quadrupoles of lower symmetry could be calculated numerically (with the next generation of computer) and so too could their optical properties, therefore (Preston 1968, 1969, 1970, 1972).

Quadrupoles and octopoles were not the only type of corrector considered. Deltrap in his Ph.D. thesis also considered space charge, only to dismiss it as impractical. In the same decade, N.C. Vaidya came to the Laboratory as a post-doctoral visitor, with extensive experience of microwave techniques. One of the possible ways of correcting spherical aberration does indeed involve exciting a three-electrode "electrostatic" lens at high frequency, the idea being that the electrons travelling far from the axis, which would normally be focused too close to the lens thanks to the spherical aberration, experience a weaker field that the electrons close to the axis, which have arrived sooner, and all will hence be focused correctly. The phase of the a.c. signal when the electrons arrive is thus a vital parameter and the latter must therefore arrive in tight bunches. N.C. Vaidya built and tested a system for bunching and focusing the electrons (Vaidya and Hawkes, 1970) and continued with this work after his return to India (1972, 1975 a,b ; Vaidya and Garg, 1974). His project was taken up and considerably advanced by L.C. Oldfield, who built a much perfected system and made extensive measurements with it ; he also developed the theory and successfully computed the paraxial properties of such devices (Oldfield 1971, 1973a, b, 1974).

Another topic that attracted research in Cosslett's group was computer-aided design of electron optical systems. The best remembered research on this theme is that of Eric Munro, which was performed in the University Engineering Dept, not in the Cavendish, but I and shortly after, M.E.C. Maclachlan worked on the structure of aberration coefficients and ways of exploiting this to design complex systems in a systematic way (Hawkes 1968, 1970, 1973 ; Maclachlan, 1971 ; Maclachlan and Hawkes, 1970). During the same period, Hajime Ohiwa, the inventor of the Moving Objective Lens (MOL), which has become of great importance for electron microlithography, spent some time in Cosslett's group, working on the use of computer algebra for analysing deflection aberrations (Ohiwa, 1977).

Before turning to a rather different branch of 'sixties electron optics, we mention the work of D.W. Swift, who took up the idea promulgated by Hibi (1956) of replacing the primitive hairpin filament in triode guns by a pointed filament. Drechsler, Cosslett and Nixon had drawn attention to this improvement to the traditional triode-gun emitter at the Berlin Conference in 1958 and Swift published two experimental studies on such guns (Swift and Nixon 1960, 1962). There was little that could be achieved on the theoretical front in gun optics until sufficient computing power became available and these experimental explorations were hence indispensable.

The 1960s also witnessed a major development, the repercussions of which are still with us. In 1956, Lenz and Scheffels had considered the effect of defocus on the wave aberration, that is, the distance between the true wave surface in an optical system and the ideal spherical surface that would produce a point image of a point object. In 1959, the first edition of Born and Wolf's *Principles of Optics* appeared and, directly or indirectly, the notions of transfer functions and Fourier optics in general, introduced long before by a little-known genius, Pierre-Marie Duffieux (1946) and disseminated by E.H. Linfoot (1955, 1960, 1964), became widely known. They were introduced into electron optics by K.-J. Hanszen (Hanszen *et al.*, 1964 ;

Hanszen and Morgenstern, 1965 ; Hanszen, 1966) whereupon F. Thon (1965, 1966) set out to verify the correctness of the theory. He extended the reasoning of Lenz and Scheffels to include spherical aberration and demonstrated brilliantly that transfer theory does describe electron image formation for a certain class of objects. All the earlier thinking of the theoreticians was upset : like cholesterol today, there seemed to be '"good" spherical aberration and "bad" spherical aberration. The good helped to convert invisible phase variations into visible amplitude patterns while the bad cut off the range of size of structural detail transmitted along an electron microscope (in an indirect way, since the cut-off is in Fourier space, not direct space). The conclusion was inescapable : it should be possible to correct the image afflicted by aberrations, after recording it, by some kind of filtering operation, provided the effects of the zeros in the transfer function could be palliated. Moreover, the climate was just ripe for such attempts - the shortcomings of the earliest American satellite pictures had led to the birth of a new subject, digital image processing, a first textbook by A. Rosenfeld appeared in 1969 shortly followed by that of Andrews (1970). And in the Cambridge Laboratory of Molecular Biology, the first three-dimensional reconstructions from electron micrographs were being made. With Cosslett's enthusiastic support, I applied to the S.R.C. for funds to purchase an advanced computer (The DEC PDP-8 initially with 4k of memory) and a Joyce-Loebl Scandig microdensitometer, to which an Optronics Filmwriter was shortly added, and an entirely new research activity began in the Electron Microscopy Section : digital electron image processing. Progress was rapid, thanks largely to R.W. Gerchberg, W.O. Saxton and M. Horner. The first two found a computationally acceptable way of solving the phase problem, that is, of extracting the phase and amplitude of a complex signal from recordings of the intensity (Gerchberg and Saxton, 1972, 1973). Martyn Horner was largely responsible for devising an image-handling language for microcomputers, SEMPER (Horner, 1975) while W.O. Saxton produced IMPROC for a mainframe (Saxton, 1974). Image processing continued to be a major activity during the remaining years of Cosslett's direction of the electron microscopy section and beyond and Owen Saxton's book (1978) soon became the standard reference. In particular, J. Frank came to Cambridge from W. Hoppe's laboratory in Munich and laid the groundwork of his studies of correlation techniques (Frank, 1980). He also showed that the effect of partial spatial coherence on the coherent transfer functions can be represented by an envelope function (Frank, 1973) and introduced a useful method for estimating resolution (Frank, 1975). With Layla Al-Ali, he examined the cross-correlation between images recorded at different defocus and signal-to-noise ratio (Frank and Al-Ali, 1975 a, b ; Al-Ali 1975, 1976).

It is not easy to situate in a few words the place of this image processing activity relative to research elsewhere, for the latter was of many different kinds, theoretical as well as experimental. Moreover, even the vocabulary is confusing in that many practitioners of the filtering that we mentioned earlier preferred the language of holography, for the electrons that traverse a specimen in a microscope unscattered can be regarded as a reference beam, which subsequently interferes with the electrons that are scattered : this is indeed the principle of in-line hologram formation.

The following comparisons are thus inevitably incomplete. The phase problem was studied in the laboratory of R.E. Burge in Queen Elizabeth College, London (notably by D.L. Misell) in that of H.A. Ferwerda in Groningen and by J. Gassmann in Munich. Interest in transfer theory and inverse filtering (or holographic deconvolution) was of course widespread : in Braunschweig, notably K.-J. Hanszen, L. Trepte, R. Lauer, G. Ade ; in Tübingen where F. Lenz did much to propagate the theory ; in Munich, led by W. Hoppe ; in Buffalo where G.W.

Stroke and collaborators, including Willasch of Siemens, attempted optical inverse filtering ; in Berlin, where P. Schiske designed a generalized Wiener filter ; by O. Kübler, M. Hahn and J. Seredynski.

3. RETROSPECT AND PROSPECTS

> The curtain falls ; the play is done.
> The ladies and the gents have gone.
> And did they like it ? Well, some paws,
> I think, beat out some gloved applause.
>
> ...

Two later lines in Heine's poem tell us that "But now the building has gone dumb : / light and delight give way to gloom", which happily is not true, for at least a part of the space in the Old Cavendish occupied by Cosslett's research group is still devoted to electron microscopy and electron image processing. Even so, the jargon of the environment is all too applicable to practitioners of electron optics : an endangered species, they. Increased computational sophistication is at once exciting and potentially lethal : powerful, flexible, easily accessible program suites could render the solution of (almost) any problem straightforward. There are those who think that the arrival of new types of high-resolution microscopy weakens the position of the electron microscope itself (e.g. Howie, 1989) though one could argue that they are complementary.

It is inevitable that some branches of electron optics should be in decline, where seemingly little remains to be done. Thus, expressions for most of the aberration coefficients that we are likely to need are available ; only the electron mirror situation remains unsatisfactory. And yet, recent work by Lenc (1992) draws attention to the fact that a little-used form of the expression for the spherical aberration coefficient of round magnetic lenses has secrets to reveal. Presumably, the present lively activity in numerical electron optics will calm down once portable well-documented program suites become commonplace. Meanwhile, other branches of the subject show no signs of having reached their zenith : the multifarious aspects of electron image processing are by no means all well understood and since there are branches of image processing that are currently in vigorous growth, the same can be expected to be true of their electron counterparts. True, these are mostly concerned with image analysis, since funding is easier to obtain for robot vision than for electron-microscope-aided human vision but there is much to be done to improve, extend and above all, increase the reliability of the analytical tools available for scanning microscopes.

The time has come to answer the question posed in my title : the end of an era ? Perhaps, but with no hesitation at all, I can assert that we are at the dawn of another.

REFERENCES[*]

Al-Ali LS 1975 EMAG pp 225-228
Al-Ali LS 1976 Optical and computer analysis of electron micrographs (Cambridge : Dissertation)

---

[*] References to the major conferences are given in the short form employed in Hawkes and Kasper (1989) ; see pp 1183-1187 for full details.

Andrews HC 1970 *Computer Techniques in Image Processing* (New York & London : Academic)

Born M and Wolf E 1959 *Principles of Optics* (Oxford : Pergamon)

Burfoot JC 1953 Proc. Phys. Soc. London B66 775

Cosslett VE 1940 J. Sci. Instrum. 17 259

Cosslett VE 1945 J. Sci. Instrum. 22 170

Cosslett VE 1946 *Introduction to Electron Optics* (Oxford : Clarendon)

Deltrap JMH 1964a Correction of spherical aberration of electron lenses (Cambridge : Dissertation)

Deltrap JMH 1964b Prague A 45

Deltrap JMH and Cosslett VE 1962 Philadelphia 1 KK-8

Drechsler M Cosslett VE and Nixon WC 1958 Berlin 1 13

Duffieux P-M (1946) *L'Intégrale de Fourier et ses Applications à l'Optique* (Besançon : privately published)

Frank J 1973 Optik 38 519

Frank J 1975 Optik 43 25

Frank J 1980 In *Computer Processing of Electron Microscope Images* ed PW Hawkes (Berlin & New York : Springer) pp. 187-222

Frank J and Al-Ali L 1975a Nature (London) 256 376

Frank J and Al-Ali 1975b EMAG 229

Gerchberg RW and Saxton WO 1972 Optik 35 237

Gerchberg RW and Saxton WO 1973 In *Image Processing and Computer-aided Design in Electron Optics* ed. PW Hawkes (London & New York : Academic) pp 66-81

Glaser W 1952 *Grundlagen der Elektronenoptik* (Vienna : Springer)

Glaser W 1956 Handbuch der Physik 33 123

Grinberg GA 1942 Dokl. Akad. Nauk SSSR 37 172 and 261

Grinberg GA 1943a Dokl. Akad. Nauk SSSR 38 78

Grinberg GA 1943b Zh. Tekh. Fiz. 13 361

Hanszen K-J 1966 Z. angew Phys. 20 427

Hanszen K-J and Morgenstern B 1965 Z. angew. Phys. 19 215

Hanszen K-J Morgenstern B and Rosenbruch K-J 1964 Z. angew. Phys. 16 477

Hardy DF 1967 Combined magnetic and electrostatic quadrupole electron lenses (Cambridge : Dissertation)

Hawkes PW 1964 Prague A 5

Hawkes PW 1965a Phil. Trans. Roy. Soc. London A257 479 and 523

Hawkes PW 1965b Optik 22 340

Hawkes PW 1965c Optik 22 543

Hawkes PW 1968 Optik 27 287

Hawkes PW 1970 Optik 31 213 and 592

Hawkes PW 1973 In *Image Processing and Computer-aided Design in Electron Optics* ed. PW Hawkes (London & New York : Academic) pp 230-248

Hawkes PW and Kasper E 1989 *Principles of Electron Optics* (London & San Diego : Academic)

Hibi T 1956 J. Electromicrosc. 4 10

Horner MF 1975 EMAG 209

Horner MF 1977 A computer system for processing electron images (Cambridge : Dissertation)

Howie A 1989 J. Microscopy 155 419

Kas'yankov PA 1956 *Theory of Electromagnetic Systems with Curved Axes* [in Russian] (Leningrad : University Press)

Kel'man VM and Yavor S Ya 1961 Zh. Tekh. Fiz. 31 1439 ; Sov. Phys. Tech. Phys. 6 1052

Lenc M 1992 Immersion objective lenses in electron optics (Delft : Proefschrift)

Lenz F and Scheffels W 1958 Z. Naturforsch. 13a 226

Linfoot EH 1955 *Recent Advances in Optics* (Oxford : Clarendon)

Linfoot EH 1960 *Qualitätsbewertung optischer Bilder* (Braunschweig : Vieweg)

Linfoot EH 1964 *Fourier Methods in Optical Image Evaluation* (London & New York : Focal Press)

Maclachlan MEC 1971 EMAG 98

Maclachlan MEC and Hawkes PW 1970 Grenoble 2 23

Melkich A 1947 Sitzungsber. Akad. Wiss. Wien, Math-Nat. Kl. Abt. IIa 155 393 and 439

Möllenstedt G 1956 Optik 13 209

Ohiwa H 1977 J. Phys. D : Appl. Phy. 10 1437

Oldfield LC 1971 EMAG 94

Oldfield LC 1973a Microwave cavities as electron lenses (Cambridge : Dissertation)

Oldfield LC 1973b In *Image Processing and Computer-aided Design in Electron Optics* ed PW Hawkes (London & New York : Academic) pp 370-399

Oldfield 1974 Canberra 1 152

Picht J 1939 *Einführung in die Theorie der Elektronenoptik* (Leipzig : Barth)

Preston HER 1968 Rome 1 179

Preston HER 1969 Proc. Roy. Soc. (London) A313 217

Preston HER 1970 Grenoble 2 41

Preston HER 1972 Manchester 82

Rosenfeld A 1969 *Picture Processing by Computer* (New York & London : Academic)

Saxton WO 1974 Computer Graphics & Image Proc. 3 266

Saxton WO 1978 *Computer Techniques for Image Processing in Electron Microscopy* (New York & London : Academic)

Scherzer O 1936 Z Physik 101 593

Scherzer O 1947 Optik 2 114

Seeliger R 1948/9 Optik 4 258

Seeliger R 1949 Optik 5 490

Seeliger R 1951 Optik 8 311

Seeliger R 1953 Optik 10 29

Septier A 1963 C.R. Acad. Sci. Paris 256 2325

Sturrock PA 1951a Proc. Roy. Soc. (London) A210 269

Sturrock PA 1951b Phil. Trans. Roy. Soc. London A243 387

Sturrock PA 1952 Phil. Trans. Roy. Soc. London A245 155

Sturrock PA 1955 *Static and Dynamic Electron Optics* (Cambridge : University Press)

Swift DW and Nixon WC 1962 Brit. J. Appl. Phys. 13 288

Thomson MGR 1966 Kyoto 1 207

Thomson MGR 1967 The aberrations of quadrupole electron lenses (Cambridge : Dissertation)

Thon F 1965 Z. Naturforsch. 20a 154

Thon F 1966 Z. Naturforsch. 21a 476

Vaidya NC 1972 Proc. IEEE 60 245

Vaidya NC 1975a Optik 42 129

Vaidya NC 1975b J. Phys. D: Appl. Phys. 8 368

Vaidya NC and Garg RK 1974 Canberra 1 150

Vaidya NC and Hawkes PW 1970 Grenoble 2 19

Vandakurov Yu V 1955 Zh. Tekh. Fiz 25 1412 and 2545

Vandakurov Yu V 1956 Zh. Tekh. Fiz 26 1599 and 2578 ; Sov. Phys. Tech. Phys. 1 1558 and 2491

Vandakurov Yu V 1957 Zh. Tekh. Fiz 27 1850 ; Sov. Phys. Tech. Phys. 2 1719

Zworykin VK, Morton GA, Ramberg EG, Hillier J and Vance AW 1945 *Electron Optics and the Electron Microscope* (New York : Wiley and London : Chapman & Hall)

Inst. Phys. Conf. Ser. No 130: Chapter 2
Paper presented at Int. Congr. X-ray Optics and Microanalysis, Manchester, 1992

Recent advances in electron microprobe analysis

S.J.B. Reed

Dept of Earth Sciences, University of Cambridge, Cambridge CB2 3EQ, UK

ABSTRACT: Recent developments in both instrumentation and analytical methods of EMPA are reviewed. Evaporated multilayers enhance the performance of WD spectrometers at long wavelengths, while doubly-curved crystals hold promise of higher intensities at 'normal' wavelengths. New robust thin windows for Si(Li) detectors are available and better resolution and throughput rates have become possible. Absorption correction methods based on improved 'phi-rho-z models' give better results, especially for light elements. The Monte Carlo method for electron trajectory simulation, which provides useful information on absorption and other corrections, has been refined with the benefit of more powerful computers, and has been applied to tilted specimens, particles, and layered samples.

1. INTRODUCTION

The bases of electron microprobe analysis (the analysis of solid samples by means of X-ray spectra generated by bombardment with a focussed electron beam) were established by Castaing (1951). The technique has since reached a fairly mature state, progress being mostly steady rather than dramatic. Recent advances described here include enhanced performance of both ED and WD spectrometers, especially for light elements, and other improvements flowing from developments in electronics and computers. Refinements of the correction procedures for quantitative analysis (especially for light elements) and application to samples of non-conventional geometry are also discussed.

2. WAVELENGTH DISPERSIVE SPECTROMETRY

Synthetic layered structures (e.g. lead stearate) with suitably large spacings have been used for many years for wavelengths beyond the range covered by normal crystals. More recently multilayers produced by vacuum evaporation of alternating light and heavy elements (e.g. C and W) have been introduced (Nicolosi et al 1986). These give considerably higher intensities (though the peaks are broad) and have the advantage that reflections of order greater than 2 are usually negligible. For optimal performance the device should be selected for the appropriate wavelength, hence several are needed to cover all the light elements.

In a conventional WD spectrometer with a cylindrically curved crystal the angle of incidence varies for rays lying out of the plane of the Rowland circle, hence Bragg reflection occurs over only a fraction of the area. This can be corrected in principle by using a doubly-curved crystal, the ideal geometry being obtained by rotation about the line connecting source and focus. The optimum second radius is then a function of wavelength, and for a given radius, high intensities are obtained only over a narrow wavelength range. However, an alternative form of double curvature, in which the second radius is equal to the diameter of the Rowland circle (Wittry and Sun 1990), gives substantially higher reflection efficiency than conventional geometry over the whole range of Bragg angles (fig. 1). Doubly-curved crystals are difficult to fabricate, but a possible solution is to use a stepped form (Wittry and Sun 1991).

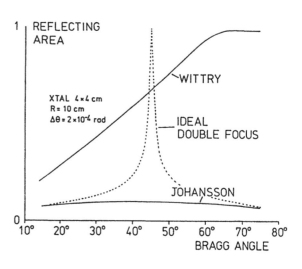

Fig. 1. Reflecting area of WD spectrometer crystal as function of Bragg angle, for different geometries (Wittry and Sun 1990).

3. ENERGY DISPERSIVE SPECTROMETRY

Advances in ED detector fabrication and electronics have yielded steadily improving energy resolution, <130 eV for Mn Kα now being attainable (at low count-rates). A 5-electrode FET preamplifier (Nashishibi and White 1990),

for example, gives less noise than
conventional FETs with pulsed optical feedback. The benefit of this is
felt mostly in the low energy region. In particular, the Kα peaks of
B (183 eV) and even Be (109 eV) are
well separated from the noise peak
centred at zero energy (fig. 2).

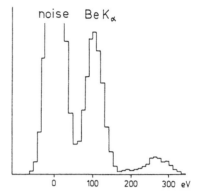

A long integrating time per pulse
is needed to minimise the effect of
noise originating in the detector
and preamplifier. This results in a
relatively long dead time, which

Fig. 2. Be Kα peak resolved by
low-noise ED system (L'Esperance
et al 1990).

limits the maximum rate at which pulses can be processed. A reduction in
integrating time allows faster throughput at the expense of broadening of
the peaks. Conventional systems commonly operate at up to about 10000
counts/s, but improvements in pulse processing electronics make 30000
counts/s possible while retaining reasonable energy resolution (Statham
1991). This is especially useful for mapping. Quantitative analysis is also
feasible at such count-rates, but sum peaks and pile-up losses need to be
taken into account (Reed 1990a).

The conventional Si(Li) detector has a Be window about 8 μm thick which
absorbs X-rays below approximately 1 keV (including K lines for Z<10).
Rotatable turrets allowing the window to be completely removed or replaced
by a thin plastic film have been available for many years. Relatively recently, robust thin windows which can withstand atmospheric pressure have
appeared. Different materials are used, containing elements such as B or C
(the absorption edges of which affect the transmission characteristics).
Thinner Be windows (e.g. 5 μm) have also been introduced.

4. THE IMPACT OF COMPUTER TECHNOLOGY

There is a continuing trend towards 'computerisation' of electron microprobes. At first, computer control of spectrometers, specimen stage, etc.,
was on a 'bolt on' basis, but increasing integration has been achieved.
As a result, knobs and dials have been supplanted by keyboards and monitor
screens. Computerisation offers many advantages including the ability to

store and recall operating conditions for different applications and the availability of automatic routines for setting up beam alignment, beam current, pulse height analysis conditions, etc. Fully automatic operation is also possible, allowing large numbers of points to be analysed without the presence of the operator. The computer is, of course, also used to process and store the analytical data.

The method of obtaining 'dot map' element distributions by setting the spectrometer to the appropriate peak and displaying the output (in pulse form) on a cathode-ray tube scanning synchronously with the beam has been superseded to a large extent by the use of computer-based image stores. One advantage of these is that the image can be monitored during accumulation, and may be manipulated afterwards to obtain the desired result, using the various facilities provided by standard image processing software (e.g. see Newbury et al 1991).

Electron microprobe instruments and ED spectrometers are produced by different companies, making integration difficult to achieve. This is not very important if the ED system is used merely for qualitative analysis. However, for quantitative ED analysis more interaction is involved, especially if combined ED and WD analysis is required, using ED for major and WD for trace elements (Ware 1991). It is therefore desirable that ED systems should be integrated with electron microprobe systems at a more fundamental level than has hitherto been the case.

5. ABSORPTION CORRECTIONS AND PHI-RHO-Z MODELS

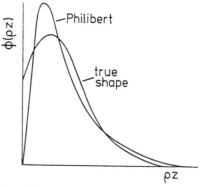

In order to calculate the absorption of X-rays emerging from the sample, knowledge of the depth distribution of X-ray production is required. This is represented by the function $\phi(\rho z)$, or 'phi-rho-z'. The absorption factor is given by $\int_0^\infty \phi(\rho z)\exp(-\mu\rho z\cos ec\psi)d(\rho z)$, where μ is the mass absorption coefficient and ψ is the X-ray take-off angle. The Philibert expression for $\phi(\rho z)$

Fig. 3.   Comparison between Philibert $\phi(\rho z)$ function and true shape.

used in the classical ZAF procedure is a considerable approximation, especially for $\rho z \to 0$ (fig. 3). This is unimportant when absorption is only moderate, since X-rays originating near the surface are not absorbed appreciably, but leads to significant errors when corrections are large (e.g. for
Z<10).

During the 1980s, more realistic phi-rho-z models were introduced, using
various mathematical functions including parabolic expressions (Pouchou and
Pichoir 1987), modified gaussian functions (Packwood and Brown 1981, Bastin
et al 1984) and exponentials (Pouchou and Pichoir 1988a). The absorption
factor obtained by integration is not very sensitive to the exact shape,
however, and even the quadrilateral function used by Love et al (1984)
gives satisfactory results. More critical is correct dependence of the
depth scale on the accelerating voltage and other variables.

The optimisation of these models is dependent upon input from basic
experiments on X-ray depth distributions (for example using tracers),
theoretical calculations (usually by the Monte Carlo method), and adjustment of parameters to minimise the mean error for large sets of
experimental data from known compounds. Useful papers on phi-rho-z models
can be found in the recent book edited by Heinrich and Newbury (1991).

Phi-rho-z models apply mostly to normal electron incidence, but modifications to allow for the effect of non-normal incidence (as in the case
of tilted specimens) have been developed (e.g. Sewell et al 1987, Pouchou
et al 1990). The $\phi(\rho z)$ function is also applicable to the analysis of layered samples: from measurements made at several different accelerating
voltages, information regarding the compositions and thicknesses of layers
in such samples can be deduced (e.g. Pouchou and Pichoir 1984).

An essential for calculating the absorption correction is the mass
absorption coefficient (m.a.c.) of the sample, which is derived from the
values for the constituent elements at the relevant wavelengths.
Experimental m.a.c. data are rather limited and have to be interpolated to
provide comprehensive coverage. Heinrich (1987) has produced revised tables
incorporating new data and using a somewhat different interpolation method
to that previously employed. In many cases absorption corrections are not
appreciably changed, but there are some significant differences.

6. MONTE CARLO CALCULATIONS

Direct experimental determination of $\phi(\rho z)$ is quite difficult and theoretical studies therefore have a useful role. The commonest approach is to compute electron trajectories by the 'Monte Carlo' method, using analytical expressions for electron scattering etc., with random numbers simulating the role of chance. Increased computing power enables more sophisticated models to be used, allowing physical reality to be approached more closely. In principle the Monte Carlo method can be applied directly to correction calculations, but this is not usual, owing to the computing time required. The method does, however, provide valuable data for incorporation into the equations used for on-line correction calculations. It is also useful for simulating electron trajectories in layered samples (Ammann and Karduck 1990) and for investigating absorption corrections for other non-standard geometries (e.g. particles), and can also be applied to backscattering corrections. The time required to compute large numbers of trajectories may be reduced drastically by using parallel processing (Romig et al 1990).

7. LIGHT ELEMENT ANALYSIS

There has been considerable progress in recent years in quantitative analysis of light elements (Z<10). Both ED and WD spectrometers give higher intensities (as described previously), enabling more precise measurements to be made. The peaks of light elements are subject to effects related to differences in chemical bonding etc., requiring special corrections, which have been investigated in some detail (e.g. Bastin and Heijligers 1986, 1989, 1991). Absorption corrections are particularly serious for light elements and the advent of improved phi-rho-z models has been helpful in achieving better accuracy.

A useful approach to absorption corrections is the 'variable voltage method'. If the dependence of the generated intensity on accelerating voltage V_0 is known, the variation in the absorption factor can be determined by measuring intensities at different values of V_0. Given a valid $\phi(\rho z)$ model, m.a.c. values for analysed samples can thus be obtained (Pouchou and Pichoir 1988b). Fig. 4 illustrates the application of this procedure. Using such an empirical value for the m.a.c. of a compound avoids uncertainties due to possible chemical effects on the elemental absorption coefficients (especially near absorption edges).

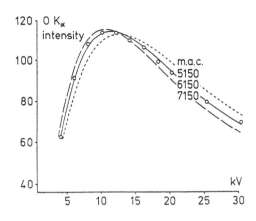

Fig. 4. Measured intensity of O Kα peak from $YBa_2Cu_3O_7$ versus accelerating voltage (circles), with curves calculated for different m.a.c. values (Mackenzie 1991).

8. ATOMIC NUMBER CORRECTIONS

To a first approximation the mass penetrated by the electron beam is independent of atomic number, but in reality penetration increases with increasing Z, because the 'stopping power' of heavy atoms is less. The correction required for this effect forms one part of the 'atomic number correction'. Stopping power corrections are calculated using the well known Bethe equation, which may be written: $S = const.(1/E)(Z/A)ln(1.166E/J)$, where E is the electron energy and J is the mean ionisation (or excitation) potential. This expression, however, becomes invalid at low electron energies (E<J). As an alternative to previously proposed remedies Joy and Luo (1989) have suggested a modification which behaves better at low energies: $S = const.(1/E)(Z/A)ln[(1.166E/J)+k]$, where $k \approx 1$.

9. FLUORESCENCE CORRECTIONS

In addition to characteristic X-rays generated by electron bombardment, there is also a contribution caused by fluorescence excitation involving other X-rays of energy greater than the critical excitation energy of the element concerned. The correction method of Castaing (1951) for fluorescence excited by other characteristic lines, modified and expressed in a more convenient form (Reed 1965), is still in general use. This underestimates the correction in some cases and appropriate modifications have been introduced (Reed, 1990b). Desirable further refinements include a more rigorous treatment of cases where L and M lines are involved and inclusion of the effect of continuum fluorescence, which is usually neglected.

REFERENCES

Ammann N and Karduck P 1990 Microbeam Analysis - 1990 ed J R Michael and
    P Ingrams (San Francisco: San Francisco Press) pp 150-4
Bastin G F and Heijligers H J M 1986 X-ray Spectrom. 15 135
Bastin G F and Heijligers H J M 1989 Microbeam Analysis - 1989 ed P E Russell
    (San Francisco: San Francisco Press) pp 207-10
Bastin G F and Heijligers H J M 1991 Scanning 13 325
Bastin G F, van Loo F J J and Heijligers H J M 1984 X-ray Spectrom. 13 91
Castaing R 1951 Ph D thesis, Univ. Paris
Heinrich K F J 1987 Proc. 11th ICXOM ed J D Brown and R H Packwood
    (London Ont.: Univ. W. Ont.) pp 67-119
Heinrich K F J and Newbury D E (eds) 1991 Electron Probe Quantitation
    (New York: Plenum)
Joy D C and Luo S 1989 Scanning 11 176
L'Esperance G, Botton G and Caron M 1990 Microbeam Analysis - 1990 ed J R
    Michael and P Ingrams (San Francisco: San Francisco Press) pp 284-6
Mackenzie A 1991 Physica C 178 365
Nashishibi T & White G 1990 IEEE Trans. Nucl. Sci. 37 452
Newbury D E, Marinenko R B, Myklebust R L and Bright D S 1991 Electron
    Probe Quantitation ed K F J Heinrich and D E Newbury (New York: Plenum)
    pp 335-69
Nicolosi J A, Groven J P, Merlo D and Jenkins R 1986 Opt. Eng. 25 964
Packwood R H and Brown J D 1981 X-ray Spectrom. 10 138
Pouchou J L and Pichoir F 1984 Rech. Aerospat. 5 48
Pouchou J L and Pichoir F 1987 Proc. 11th ICXOM ed J D Brown and R H
    Packwood (London Ont.: Univ. W. Ont.) pp 249-53
Pouchou J L and Pichoir F 1988a Microbeam Analysis - 1988 ed D E Newbury
    (San Francisco: San Francisco Press) pp 315-8
Pouchou J L and Pichoir F 1988b Microbeam Analysis - 1988 ed D E Newbury
    (San Francisco: San Francisco Press) pp 319-22
Pouchou J L, Pichoir F and Boivin D 1990 Proc. 12th ICXOM ed S Jasienska
    and L J Maksymowicz (Cracow: Acad. Mining and Metall.) pp 52-9
Reed S J B 1965 Brit. J. Appl. Phys. 16 913
Reed S J B 1990a Microbeam Analysis - 1990 ed J R Michael and P Ingrams
    (San Francisco: San Francisco Press) pp 181-4
Reed S J B 1990b Microbeam Analysis - 1990 ed J R Michael and P Ingrams
    (San Francisco: San Francisco Press) pp 109-14
Romig A D, Plimpton S J, Michael J R, Myklebust R L and Newbury D E 1990
    Microbeam Analysis - 1990 ed J R Michael and P Ingrams (San Francisco:
    San Francisco Press) pp 275-80
Statham P J 1991 Electron Microscopy and Analysis ed F J Humphreys (IoP
    conf. ser. no. 119) pp 425-32
Ware N G 1991 X-ray Spectrom. 20 73
Wittry D B and Sun S 1990 J Appl. Phys. 67 1633
Wittry D B and Sun S 1991 J Appl. Phys. 69 3886

Inst. Phys. Conf. Ser. No 130: Chapter 2
Paper presented at Int. Congr. X-ray Optics and Microanalysis, Manchester, 1992

75

A new mass absorption coefficient equation

Robert E. Ogilvie

Dept. of Materials Science and Eng. M.I.T. Cambridge, Ma. 02139

ABSTRACT: From the NBSIR 86-3461 publication, the "Bibliography of Photon Total Cross Section (Attenuation Coefficient) Measurements 10 keV to 13.5 GeV," published by J. Hubbell, H. Gerstenberg, and E. Soloman (1986), which covers the time period from 1907 to 1986, many of the elements were edited, corrected for coherent and incoherent scattering, then fit by the method of least squares to the following equation
$$\text{Log}(\tau/\rho) = A_K \, [\text{Log}(E / E_K)]^2 + B_K \, \text{Log}(E / E_K) + C_K$$
From this work and the results of McMaster et. al. (1969) show that each element must be evaluated for the coefficients A_K, B_K, and C_K. It should be noted that C_K is equal to $\text{Log}((\tau/\rho)_K)$.

1. INTRODUCTION

It was in 1914 when Siegbahn (1914) showed, that for a given element, the value of (μ/ρ) could be expressed as a function of the wavelength (λ) of the x-ray photon by the following equation

$$(\mu/\rho) = C \, \lambda^n \tag{1}$$

where C is a constant for a given material, which will have sudden jumps in value at critical absorption limits. Siegbahn found that n varied from 2.66 to 2.71 for various solids, and from 2.66 to 2.94 for various gases.

Bragg and Pierce (1914), during this same time period, showed that their results on materials ranging from Al(13) to Au(79) could be represented by the following

$$\mu_a = C \, Z^4 \, \lambda^{2.5} \tag{2}$$

where μ_a is the atomic attenuation coefficient, and Z the atomic number. Many investigators suggested that n should be closer to 3, and that the exponent of Z should be less than 4 for energies less than 100 keV.

In 1928 Jönsson (1928) carried out an extensive study of absorption coefficients. In his Thesis nearly 600 measurements are tabulated. However, the most interesting part of his Thesis is the "UNIVERSAL" absorption curve. Jönsson introduces a new true absorption coefficient, which may be called the electronic true absorption coefficient (τ_e), where

$$\tau_e = \tau_a / Z = (\tau/\rho) \, A / N \, Z \tag{3}$$

where A is the atomic weight, N is Avogadro's number, and (τ/ρ) is the normal true absorption coefficient. Jönsson proposed that τ_e is a function of $(Z \, \lambda)$. The $\text{Log}(\tau_e \, N)$ was plotted against the value of $\text{Log}(Z \, \lambda)$. This graph may be expressed in the following form

$$(\tau/\rho) = C \, Z^{n+1} \, \lambda^n / A \tag{4}$$

Unfortunately n decreases as $(Z \, \lambda)$ increases. Bearden(1966) made a "UNIVERSAL" plot for C, Al, Cu, Ag, and Au, which showed a smooth curve, to confirm within the limits of experimental error, Jönsson's conclusions that the electronic absorption coefficient is a simple function of $(Z \, \lambda)$. However, equation (4) has not proved to be in agreement with most experimental data.

The best analytical fit to the experimental data of (μ/ρ) as a function of Z and E was developed by Heinrich (1986), which is given by the following equation

$$(\mu/\rho) = C \, Z^4 \, (12398/E)^n \, \{ \, 1 - \exp[\, (-E+b)/a \,] \, \} / A \tag{5}$$

where E is the photon energy (eV), and C, n, a, and b are functions of Z. Equation (5) produces a curve that is concave downward for a given element and is limited to energies of less than 25 keV. It should be noted that (μ/ρ) for the lighter elements has a point of inflection, therefore the curve becomes concave upward. For Aluminum this occurs at about 12 keV.

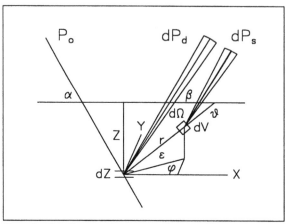

Figure 1. This illustrates how scattered x-rays will enter the detecting system from the volume dV.

It is well known that (τ/ρ) is concave downward for energies up to 100 keV. It can be illustrated that (τ/ρ) is the value that should be used in Electron Microanalysis. In Figure 1, we see that some of the photons produced in the layer dZ will absorbed, and that some will be scattered (coherent and incoherent) on the way to the detecting system. However, the volume dV will scatter some of the photons in a plane parallel to the x-y plane, which will enter the detector. If α is less than 90° and β is equal to 90° then the number scattered to the detector is greater then those that are scattered from the direct beam. Therefore, the ratio of the photons scattered to the detector to those that are scattered out of the direct beam is a function of α and β.

2. Procedure

From the Mass Attenuation Coefficient data in the NBSIR 86-3461 publication, which was prepared by Hubbell et.al. (1986), many of the elements were examined. The data was corrected for coherent and incoherent scattering, using the analytical equations given McMaster et al. (1969). Then the data was fit by the method of least squares to the following equation

$$\text{Log}(\tau/\rho) = A_K \, [\text{Log}(E/E_K)]^2 + B_K \, \text{Log}(E/E_K) + C_K \tag{6}$$

In equation (6) the value of E_K is the energy of the K-edge when E is greater then the energy of the K-edge, for energies less than the K-edge and greater than the L_I-edge, E_K is the Energy of the L_{III}-edge. For energies that fall between the L_{III} and the M_I edges, E_K is equal to the energy of the M_V-edge, etc. Because of the form of equation (6), the value of C_K is equal to $\text{Log}(\tau/\rho)$ at the low energy edge (E_K, E_{III}, M_V etc.). An example of the fit of equation (6) to the Aluminum and Gold data of (μ/ρ) is shown in Figures 2 and 3. The solid line gives the values of (τ/ρ), and for gold the we see the curve is extrapolated to the L_{III}, and the M_V edges. The values between the L_I and the L_{III} edges and between M_I and the M_V edges must be corrected by the appropiate jump ratios given by McMaster et al. (1986).

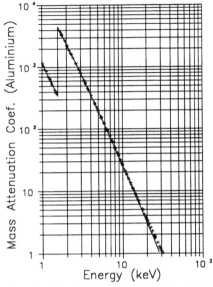

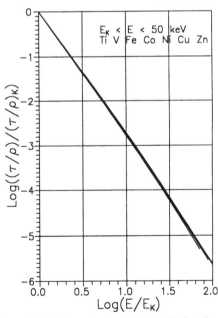

Figure 2. Aluminum (μ/ρ) data. Solid line is (τ/ρ)

Figure 3. Gold (μ/ρ) data. Solid line is (τ/ρ).

When the coefficients A_K, B_K, and C_K have been evaluated, it is an easy matter to compare the various elements. This is done with the following form of Equation (6)

$$\mathrm{Log}[(\tau/\rho)/(\tau/\rho)_K] = A_K \, [\mathrm{Log}(E/E_K)]^2 + B_K \, \mathrm{Log}(E/E_K) \qquad (7)$$

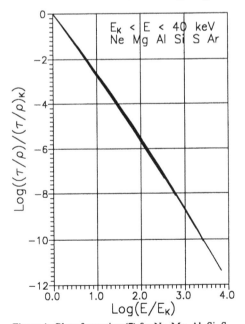

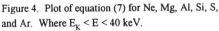

Figure 4. Plot of equation (7) for Ne, Mg, Al, Si, S, and Ar. Where $E_K < E < 40$ keV.

Figure 5. Plot of equation (7) for Ti, V, Fe, Co, Ni, Cu and Zn. Where $E_K < E < 50$ keV

3. Results

From a plot of Equation (7), the energy and atomic number dependence of (τ/ρ) is easily compared. It has been observed that no single equation that can predict (τ/ρ) or (μ/ρ).for a range of elements Unfortunately, this means that the coefficients A_K, B_K, and C_K must be determined for each element and the regions between absorption edges. An example of a plot of Equation (7) is illustrated in Figure 4 and 5. In Figure 4, the curve for Ne is the upper one and Ar is the lower one. The other elements fall in an orderly fashion. However in Figure 5, the curves do not fall in an orderly fashion. This is the rule for the 37 elements that have been examined. The same procedure has been applied to the data of McMaster et. al. (1969), and the results are the same. Table I illustrates the values of A_K, B_K, and C_K for a few of the elements examined. The standard error of estimate (S.E.E.) is the root-mean-square of the y-deviations about the fitted curve.

Table I

Z	Element	A_K	B_K	C_K	$(\tau/\rho)_K$	S.E.E
10	Neon	-0.09452	-2.60211	9.35163	11517.6	.0122
12	Magnesium	-0.11288	-2.55858	8.68899	5937.2	.0133
13	Aluminum	-0.10779	-2.58684	8.42057	4539.5	.0191
16	Sulfur	-0.10899	-2.60693	7.74386	2307.4	.0092
18	Argon	-0.10020	-2.65103	7.28490	1458.1	.0144
22	Titanium	-0.10199	-2.63882	6.60192	736.5	.0174
23	Vanadium	-0.05895	-2.69656	6.45003	632.7	.0138
26	Iron	-0.09478	-2.65172	6.05770	427.4	.0098
27	Cobalt	-0.12575	-2.63929	5.95025	383.8	.0235
28	Nickel	-0.10206	-2.63666	5.83095	340.7	.0132
29	Copper	-0.08143	-2.69049	5.69601	297.7	.0161
30	Zinc	-0.08243	-2.70806	5.62994	278.6	.0274

A correlation with atomic number for the coefficients A_K, B_K, and C_K has not been found. In fact because of the electronic nature of the elements, it should not be expected.

3. Conclusion

From the data published in the NBSIR 86-3461 bibliography and from the data published by McMaster et al., it has been found that a single equation for each region between edges must be evaluated, if the values of (τ/ρ) are to be better than 2%. It is also very important, that when absorption data is obtained for a particular element, that data be obtaind near the edges (±500 ev). This will insure a better value of C_K.

References

Bearden A J 1966 J. Appl. Phys. 37 1681
Bragg W H and Pierce S E 1914 Phil. Mag. 28 626
Heinrich K 1986 11th ICXOM 67
Hubbell J et al. 1986 NBSIR 86-3461
Jönsson E 1928 Thesis Uppsala
McMaster W et al. 1969 UCRL-50174
Siegbahn M 1914 Physikalische Zeitschrift 15 753

Inst. Phys. Conf. Ser. No 130: Chapter 2
Paper presented at Int. Congr. X-ray Optics and Microanalysis, Manchester, 1992

79

MAC measurements with a Borrmann crystal

Robert E. Ogilvie

Dept. of Materials Science and Eng. M.I.T. Cambridge Ma. 02139

ABSTRACT: The measurement of Mass Attenuation Coefficients is usually carried out using characteristic x-ray lines from x-ray tubes or an x-ray fluorescent sources. These techniques employ a single or double crystal spectrometer. However, some investigators use selected wavelengths from the Bremsstrahlung radiation. This latter technique has the problem of a lower peak-to-background ratio because of the coherent and incoherent scattered radiation by the crystal for the total Bremsstrahlung. In order to minimize the scattered radiation, a thick (5 mm) silicon Borrmann crystal was used. In order to obtain a suitable signal, a Rigaku rotating anode was used. Operating the x-ray tube at 20 keV and 200 ma, the intensity was measured in steps of 1° from 20° to 40° two-theta. This will cover the energy-range of 9.44 to 18.60 keV. This procedure was carried out with and without the foil.

1. Introduction

Anomalous transmission was first observed by Borrmann (1941) in single crystals of quartz, and latter in silicon and germanium by Batterman (1961). This technique has proved to be an important tool for the study the crystalline nature of solids. The phenomenon that is observed when a perfect crystal is oriented for the symmetrical Laue geometry, that is, when the diffracting planes are normal to the crystal face, one of the states of polarization of the incident beam is transmitted with very little absorption. That is, for a particular wavelength that satisfies the Bragg angle (θ), the energy of the incident beam flows parallel to the diffracting planes. When the beam reaches the back side of the crystal the two electromagnetic waves split into a forward diffracted beam and a diffracted beam. For a thick germanium or silicon crystal, when $\mu t > 20$, where μ is the linear absorption coefficient and t is the thickness, the forward diffracted and diffracted beam are surprisingly strong. For these conditions the incident monochromatic primary beam should be reduced to almost zero by the absorption factor $\exp[-20/\cos \theta]$. For a 5 mm crystal of silicon and a monochromatic beam of 10 keV x-rays, the absorption factor or (I / Io) will be 3.88E-19. However, for a 15 and 20 keV x-ray beam the absorption factor will be 4.78E-06 and 5.80E-03 respectively. Therefore, when selecting a particular energy from the Bremsstrahlung radiation it is important to select the proper thickness of crystal to reduce the scatter of the higher energy x-rays to the detector.

The Borrmann effect or anomalous transmission has been presented in an elegant manner by Batterman and Cole (1964), employing the dynamical theory, which is generally used whenever diffraction from large perfect crystals is being investigated. The concepts developed here have also been explained by Borie (1966), using the simpler Darwin theory. The method of Borie is treaded in detail by Warren (1968). The concepts developed in these publications is the explanation of the energy flow, standing waves, integrated intensity of the forward diffracted and diffracted beams, and absorption. This latter point, absorption, is of interest here. The effective linear absorption coefficient (μ_o(eff) can be expressed by

$$\mu_o(\text{eff}) = \mu_o \, t_o / \cos \theta \, [1 - \varepsilon \, P] \tag{1}$$

where t_o is the thickness of the crystal, P is the polarization factor and ε is F"(2θ) / F"(0), (F"(2θ) and F(0) are the imaginary part of the structure factor for F(2θ) and F(0) respectively). If we assume that ε = 0.95 and P=1, then the absorption term for the σ polarization-α branch in a 5 mm thick silicon crystal, and E = 10 keV, will be exp-[(80.2 0.5/cos 18.8)(1-0.95)], which is equal to 0.120.

2. Procedure

The experimental procedure is best explained using Figure 1. The x-ray source is a Rigaku rotating anode with a copper target. The single crystal of silicon was cut and polished with a (111) face. It as then oriented so that the [110] direction bisected the incident and diffracted beam. The divergent slit that was employed was 0.15 mm. Because of the width of the line source, ~0.1 mm, the divergence of the beam is about 0.15°. The crystal will therefore diffract an energy-band-width determined by the divergence of the beam and the 2θ value. That is, E_1, E_o, and E_2 will diffract simultaneously to the receiving slit. The forward diffracted beam will continue to diverge. The theta value of the crystal will determine the Energy of the

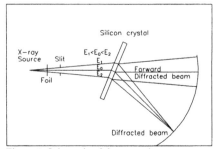

Figure 1. Schematic of the anomalous transmission experiment.

diffracted beam. If the theta value is fixed at a particular value and two-theta is scanned we obtain a peak that contains the energy-band-width ΔE. The Energy of the peak for the (220) planes in silicon is given by the following

$$E = 3.22896 / \sin \theta \qquad (2)$$

and the energy-band-width (ΔE) is given by the following

$$\Delta E = E \, ctn \, \theta \, \alpha \qquad (3)$$

where α is the angle of divergence of the incident beam on the crystal. For a 0.15 ° divergence and for a 10 15, and 20 keV beam, ΔE will be 76.7, 178.1, and 319.8 eV respectively. Figure 2 illustrates a scan at 20°, 25°, 30°, and 35° two-theta with theta fixed at 10°, 12.5°, 15°, and 17.5° respectively. The full widths at half maximum of the peaks are 0.297°, 0.251°, 0.228°, and 0.212° respectively. The technique used to obtain the data for the incident and transmitted beam for the nickel foil was to step scan two-theta in 1° steps with a counting time of 300 seconds. These result of these scans are illustrated in Figure 3

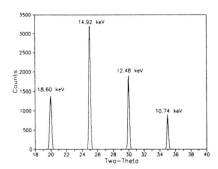

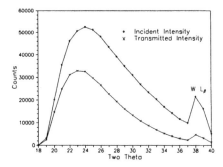

Figure 2. Two-theta scan at 20°, 25°, 30°, and 35° with theta fixed at 10°, 12,5°, 15°,17.5°.

Figure 3. Two-theta step scan of incident and transmitted intensity for a nickel foil.

The copper target will have some tungsten contamination from the filament and this spectra will be observed in a two-theta scan. The three strongest Tungsten Lβ characteristic lines, Lβ1, Lβ2, and the Lβ3 have two-theta values of 39.00°, 37.83°, and 38.40° respectively, which results in the peak at 38°. The maximum energy of the x-ray photons is 20 keV and the two-theta value should be 18.582°. The extrapolation of the data appears to be closer to 18.75°, which could mean that the voltage on the x-ray tube is 19.83 keV, or there is an displacement error in the two-theta scale. From data obtained with a LiF crystal in the reflecting mode it appears that the voltage on the x-ray tube is closer to 19.83 keV.

3. Results

From the data illustrated in Figure 3, which employed a nickel foil of 7.10E-3 grams per square centimeter, the mass absorption coefficient was determined as a function of energy. These results are plotted in Figure 4. The best fit to this data is a straight line. However, it was expected that the curve be slightly concave downward as found in the data of Hubbel et al. (1986). The values of the mass absorption coefficients were then corrected for coherent and incoherent scattering, using the data from McMaster et al., to obtain the mass attenuation coefficients (τ/ρ). The data was then fitted to Equation (4) by the method of least squares. These values were then compared with the results of the nickel data from Hubbel et al. (1986), which was corrected for coherent and incoherent scattering and then fitted by the method of least squares to Equation (4).

$$Log(\tau/\rho) = A_K [Log(E/E_K)]^2 + B_K Log(E/E_K) + C_K \qquad (4)$$

When A_K, B_K, and C_K were evaluated, the results were then plotted using Equation (5).

$$Log[(\tau/\rho)/(\tau/\rho)_K] = A_K [Log(E/E_K)]^2 + B_K Log(E/E_K) \qquad (5)$$

The experimental data and the NBSIR 86-3461 nickel data, which was put into the form of Equation (5), is shown in Figure 5. It should be noted that $C_K = Log(\tau/\rho)_K$.

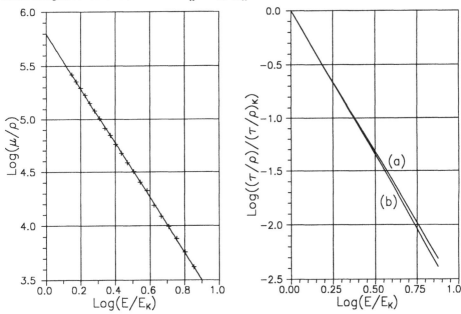

Figure 4, Mass Absorption Coefficient of Nickel as a function of Energy.

Figure 5. (a) NBSIR 86-3461 nickel data. (b) The experimental data using a silicon crystal

The experimental data illustrated in Figure 3 is very close to a linear function and when this data is corrected for coherent and incoherent scattering the values of A_K, B_K and C_K are -0.00817, -2.63098, and 5.81919 respectively, with a Standard Error of Estimate equal to 0.008. This was not expected, however, it was reproduced three times. The fact that it is reproducible and cannot be explained, does not mean that it is correct. Using x-ray fluorescence, there was no detectable impurities in the nickel foil. The surface and the uniformity in thickness of the foil appeared very good. The edited data presented in the NBSIR 86-3461 publication has a lower slope then the experimental data obtained in this work. From Equation 4 the values of (τ/ρ) for selected energies were calculated. The difference in (τ/ρ) for these calculations is given in Table 1

Table I

E (keV)	NBSIR	Ogilvie	% Difference
8.3328	340.68	336.70	-1.18
10	209.91	208.32	-0.76
12	128.48	128.84	+0.28
14	84.39	85.79	+1.63
16	58.40	60.30	+3.15
18	42.09	44.17	+4.71

The difference, or error, becomes to large at the higher energies but the value $(\exp(C_K))$ at the K-edge is in good agreement. A 1 mm germanium crystal was also examined, which gave good intensity at energies less than the germanium K-edge, but was not suitable for energies greater than the germanium K-edge.

4. Conclusion

The Borrmann crystal will produce a narrow band of energy with a low background, which is well suited for its use as a monochromator to measure mass absorption coefficients. However, the intensity is much lower then the conventional crystal systems. Therefore, it is necessary to use a high intensity source such as a rotating anode x-ray tube to obtain the intensity necessary for good counting statistics in a reasonable time. A tungsten rotating anode would give higher intensity, therefore , a better choice than the copper tube used in this work.

5 Acknowledgments

The author would like to thank Joe Adario for his help in obtaining the x-ray data on the Rigaku x-ray unit, Dr. Noel Thomas for the silicon crystal, and Prof. E. E. Haller for the germanium crystal.

References

Battermann B and Cole H 1964 Rev. of Modern Physics 36 681
Battermann B 1961 J. Appl. Phys. 32 998
Borie B 1966 Acta Cryst. 21 470
Borrmann G 1941 Zeit. f. Physik 42 157
Warren B 1969 X-ray Diffraction Addison-Wesley Publishing Co.

Inst. Phys. Conf. Ser. No 130: Chapter 2
Paper presented at Int. Congr. X-ray Optics and Microanalysis, Manchester, 1992

Determination of iron valence in oxides and sulphides by EPMA

H Kucha[1], S Jasienska[2], W Ciesla[2]

1 - Institute of Geology and Mineral Deposits, 30-059 Krakow, Ave Mickiewicza 30, Poland
2 - Institute of Metallurgy, 30-059 Krakow, Ave Mickiewicza 30, Poland

ABSTRACT: In simple iron oxides the FeLβ/Lα ratio clearly reflects the iron valence.In substituted Fe-oxides the ratio varies widely according to amount and type of substituted metal.In Fe monosulphides, carbides, phosphides and metallic iron the FeLβ/Lα ratio is consistently antipathetic to reflectivity.The relationship is represented by a stright line.It is suggested that the FeLβ/Lα ratio in Fe monosulphides, carbides, phosphides and metallic Fe measures a contribution of covalent (ionic) versus metallic bonding.

1.INTRODUCTION

Changes in X-ray emission spectra associated with oxidation of metals are usually studied using the L lines.The related wavelength changes are small, but the Lβ/Lα intensity ratio increases considerably in oxides as compared to the metals (Fischer, 1965).A systematic increase of the FeLβ/Lα ratio occurs as the ferrous/ferric ratio increases: magnetite-ulvite, hematite-ilmenite, hematite-magnetite-wustite and in anhydrous silicates (Albee and Chodos, 1970).However, in more complex oxides the FeLβ/Lα ratio provides only qualitative information on ferrous/ferric iron proportion (O'Nions and Smith, 1971). More recently it has been demonstrated that the Lβ/Lα ratio may provide quantitative values of the iron valence in Mg and Ca substituted iron oxides (Jasienska and Nowak, 1987).However, the FeLβ/Lα ratio is influenced not only by iron valence but also by coordination number, the bond length and angle as well as bond type.

2.EXPERIMENTAL CONDITIONS

The FeLβ/Lα values were measured at 15 kV on a Camebax microprobe using a KAP crystal.To prevent sample damage by heat released during electron bombardment a beam diameter of about 10 μm was used.

The FeLβ/Lα intensity ratios were measured using an automated procedure.In the first run an entire area of the FeLβ-FeLα spectrum was scanned using a step of 50 spectrometer units with a dwell time 1 sec.Based on this run, the background, the peak position and an initial Lβ/Lα ratio was calculated.Using peak positons obtained from the first run, the Lβ and Lα peaks were scanned with a step 50 and a dwell time of 20 sec.These measurements served as a basis for an accurate determination of peak centroids and intensity ratios.Peak centroids were calculated with the least square root procedure.To check the accuracy of the 20 sec measurements the Lβ and Lα intensities were remeasured with a dwell time of 30 sec at peak positions derived from least square root calculation.The measurements of the ratio were accepted as correct when the ratios derived from the 20sec and the 30 sec measurements were identical within 5% of the measured value.

Each measurement session started with determination of the Lβ/Lα ratio for the metallic iron.

The used samples consisted of: 1) synthetic wustite FeO, hematite , magnetite, cohenite Fe_3C, schrebersite Fe_3P, metallic Fe, cubic FeS_2, hexagonal FeS and 2) natural chalcopyrite $CuFeS_2$, bornite Cu_5FeS_4, marcasite orthorhombic FeS_2, monoclinic pyrrhotite Fe_7S_8, alabandite - ferroan alabandite series αMnS - α(Fe,Mn)S, pentlandite $(Fe,Ni)_9S_8$, goethite αFeOOH, lepidocrocite γFeOOH and siderite $FeCO_3$.The natural compounds were analysed by EPMA and XRD prior to the Lβ/Lα measurements. Reflectivity was measured in air at 546 nm using SiC and WC reflectivity standards.

3.SIMPLE IRON OXIDES

Simple iron oxides $FeO-Fe_3O_4-Fe_2O_3$ show a systematic decrease in the FeLβ/Lα ratio from 0.50 through 0.41 to 0.37 respectively.Similar or identical values were established in the studies of Albee and Chodos (1970), and Jasienska and Nowak (1987) proving a consistent change of the FeLβ/Lα ratio related directly to the change of iron valence.The diffrence of the FeLβ/Lα ratio between hematite (ferric Fe) and iron hydroxides (also ferric Fe) is small.In goethite it was measured to be 0.38 and in lepidocrocite it was found to be 0.39.

The FeLβ/Lα ratios are markedly higher for the Fe-O-Cr bond association than for Fe-O-Fe.This may be illustrated by the spinel $FeCr_2O_3$, where the FeLβ/Lα ratio changes from 0.45 to 0.55 depending on the substitution of Mg for Fe and Al for Cr. This relationship seems to be similar to the relationship observed in the Mg-doped wustite, where the FeLβ/Lα ratio increases with the increase of Mg concentration (Jasienska and Nowak, 1987).The change of the Fe bond type may cause a drammatic change in the FeLβ/Lα ratio despite the identical iron valence. The FeLβ/Lα ratio in siderite $FeCO_3$ equals 0.90-1.00 depending on the amount of the substituted Mg.

4.IRON SULPHIDES

Concerning iron sulphides the system is complicated by the presence of Fe^0 together with Fe^{2+} and sometimes also ferric Fe. The FeLβ/Lα ratio in sulphides is strongly affected by iron

valence.It changes from 0.09 and 0.08 in orthorhombic and cubic
FeS_2 to 0.60 in ferroan α-MnS.The plot of the FeLβ/Lα ratio versus reflectivity (Fig.1) shows the studied Fe-monosulphides,
carbides, phosphides, and metallic iron to project close to a
straight line.Hexagonal FeS and monoclinic Fe_7S_8 deviate from
the straight line probably because both are strongly optically
anizotropic.The $(Fe,Ni)_9S_8$ plots near the line depending on the

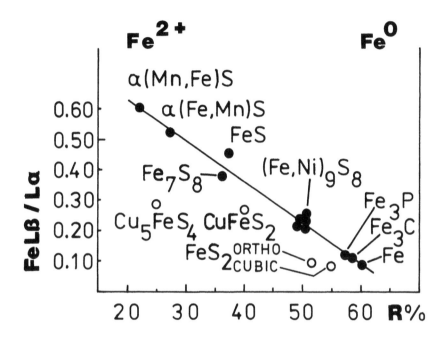

Fig. 1. A plot of the FeLB/Lo ratio versus reflectivity of
the studied minerals measured in air.

amount of Ni substituted for Fe.Cu-Fe sulphides plot away from
the straight line due to a big difference between pure Fe-S
and Cu-S bonding.Fe-disulphides (Fig.1) are also located away
from the Fe-monosulphide-carbide-phosphide-metallic iron line
due to a big difference in bonding between monosulphides and
disulphides (Burns, 1970).It appears that optically isotropic
minerals: αMnS-α(Fe,Mn)S, Fe_9S_8, Fe_3C, Fe_3P and Fe plot on the
straight line (Fig.1).This may suggest the existence of an important physical law explaining the observed relationship.The
FeLβ/Lα ratio seems to depend in the discussed case on the type of chemical bonding - covalent (ionic?) in the α(Fe,Mn)S,
metallic in iron, and mixed in other compounds (Fig.1).The reflectivity (R%) of the isotropic minerals depends on refractive index n and absorption coefficient K:

$$R\% = \frac{(n-1)^2 + n^2K^2}{(n+1)^2 + n^2K^2}$$

The absorption coefficient depends mainly on the number of con-

ductivity electrons per atom in the lattice (Weisskopf, 1968)
i.e. on the proportion of metallic bonding.The straight-line
relationship between the FeLβ/Lα ratio and reflectivity (Fig.1)
suggests that both values may have a similar physical interpretation, namely the proportion of covalent (low reflectivity,
high FeLβ/Lα ratio) versus metallic bonding (high reflectivity,
low FeLβ/Lα ratio).Measurement of reflectivity may thus allow
a prediction of the FeLβ/Lα ratios in iron sulphides and can
probably be used to study chemical Fe-bonding in microareas.

5.ACKNOWLEDGEMENTS

Authors wish to thank Prof.A.Wodzicki, Western Washington University, USA, for linguistic improvements to the manuscript.

6.REFERENCES

Albee, A.L., Chodos, A.A.(1970) Amer.Miner., 55, 491-501
Burns, R.G.(1970) Mineralogical Applications of Crystal Field
   Theory.Univ.Press., Cambridge, 224 pp
Fischer, D.W.(1965) J.Appl.Phys., 36, 2048-2053
Jasienska, S., Nowak, R.(1987)502-504.In: Brown, J.D. and Pack-
   wood, R.M.(eds) 11 ICXOM, 4-8 Aug., London-Canada 1987, 534 pp
O'Nions, R.K., Smith, D.G.W.(1971) Amer.Miner., 56, 1452-1463

Inst. Phys. Conf. Ser. No 130: Chapter 2
Paper presented at Int. Congr. X-ray Optics and Microanalysis, Manchester, 1992

87

Soft X-ray spectra of iron in silicates

G. Amthauer[*], M.K. Pavicevic[*], M. Bernroider[*], D. Timotijevic[**]

[*]   Institute of mineralogy, Hellbrunnerstr. 34, A-5020 Salzburg
[**]   Institute of Copper (Engineering-Belgrade, Pariske Commune 24/I, 11070 Belgrade

ABSTRACT: The $L_{III,II}$-X-ray emission spectra of iron in (i) some synthetic and natural pyroxenes within the solid solution series hedenbergite ($CaFe^{2+}Si_2O_6$) and acmite ($NaFe^{3+}Si_2O_6$) and (ii) some synthetic and natural garnets within the solid solution series almandine $Fe^{2+}_3Al_2Si_3O_{12}$, grossular $Ca_3Al_2Si_3O_{12}$, and andradite $Ca_3Fe^{3+}_2Si_3O_{12}$ have been measured using an electron microprobe. The L-spectra exhibit distinct shifts of the band maxima as well as changes of the L_{II}/L_{III}-intensity ratios with varying Fe^{2+}/Fe^{3+}-ratios, which has been determined independently by Mössbauer spectroscopy. The L_{II}/L_{III}-area ratio can best be used to determine the Fe^{2+}/Fe^{3+}-ratio in unknown samples of the same solid solution series.

1.   INTRODUCTION

There is still a need to determine the oxidation state of transition metal ions in minerals by a method with high spatial resolution, because of microtextures in rocks, zonation in minerals, small size of synthetic samples, etc. For instance, the Fe^{2+}/Fe^{3+}-ratio in minerals is dependent on the oxygen fugacity during the formation of ores or rocks and exhibits great variations in many minerals. This problem can principally be solved by taking the L-spectra of iron using an electron microprobe. In addition, information on the electronic configuration of the valence and the conduction band can be obtained from the $L_{III,II}$-X-ray spectra.

The aim of the present paper is (i) to develop methods to determine quantitative Fe^{2+}/Fe^{3+}-ratios from the L-X-ray spectra and (ii) to get information on the chemical bonding of iron in minerals. For that purpose, natural and synthetic pyroxenes within the solid solution series hedenbergite $CaFe^{2+}Si_2O_6$ and acmite $NaFe^{3+}Si_2O_6$, and natural and synthetic garnets within the solid solution series almandine $Fe_3^{2+}Al_2Si_3O_{12}$, grossular $Ca_3Al_2Si_3O_{12}$, and andradite $Ca_3Fe_2^{3+}Si_3O_{12}$ were studied. The Fe^{2+}/Fe^{3+}-ratios of these samples has been independendly determined by Mössbauer spectroscopy. In the pyroxenes Fe^{2+} as well as Fe^{3+} occupy the sixfold coordinated M1 cation positions. In the garnets Fe^{2+} occupies the eightfold coordinated X-positions and Fe^{3+} the sixfold coordinated Y-positions.

2.   EXPERIMENTAL METHODS

Mössbauer spectra were taken using a multichannel analyzer with 1024 channels, an electromechanical drive system with symmetrie triangular velocity shape, a 50mCi Co^{57}/Rh-source and the absorber at room temperature.

The $L_{III,II}$-X-ray emission spectra were taken between 707 and 745 eV with a computer controlled electron microprobe (JEOL JXA 8600, Link EXL) using electron beam excitation voltages of 4 KeV and 10 KeV, respectively, a TAP analysator crystal, and a step scanning system with a step size of 0.039 eV and a step time of 2 sec. The energy was calibrated with the energy E = 704.32 eV of the L_{III}-band of metallic iron.

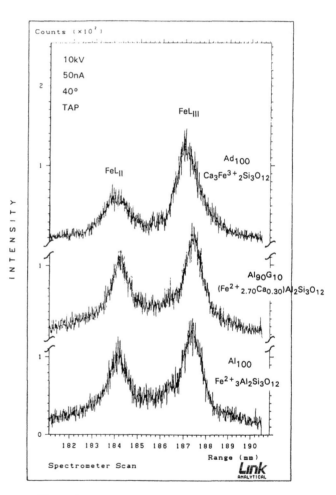

Figure 1:   L-X-ray emission spectra of iron in different garnets

The Mössbauer spectra were fitted by a least squares fitting procedure assuming Lorentzian line shape. The Fe^{2+}/Fe^{3+} ratio was determined from the area ratio of the corresponding Fe^{2+}- and Fe^{3+}-lines.

The $L_{III,II}$-X-ray emission spectra were evaluated using a computer program of Slavic et al. (1992) by fitting lines with Gaussian shape to the experimental spectrum. The program calculates position, peak heights, half-widths and area of each line as well as the total calculated spectrum and the residual.

3.   RESULTS AND
      DISCUSSION

In Figure 1 typical L-X-ray emission spectra of Fe in garnets are shown. There are distinct shifts of the L_{II} and L_{III}-peaks as well as distinct changes of the L_{II}/L_{III}-intensity ratio between andradite (Fe^{3+}) and almandine (Fe^{2+}). In Figure 2 the observed and fitted spectrum of almandine 100 is shown. The L_{II}-peak as well as the L_{III}-peak have been fitted by three Gaussian lines. The area of the L_{II} peak as well as the area of the L_{III}-peak is the sum of the areas of the corresponding lines. By this procedure the L_{II}/L_{III} area ratio (= intensity ratio) can be determined with high precision (σ = 6.1 %) much more better than from the peak heights of L_{II}- and L_{III}-peaks, respectively.

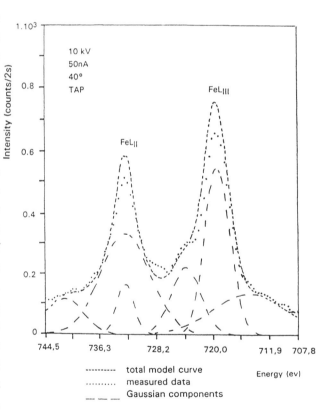

Figure 2:   L-X-ray emission spectrum of almandine 100 ($Fe^{2+}{}_3Al_2Si_3O_{12}$) after fitting.

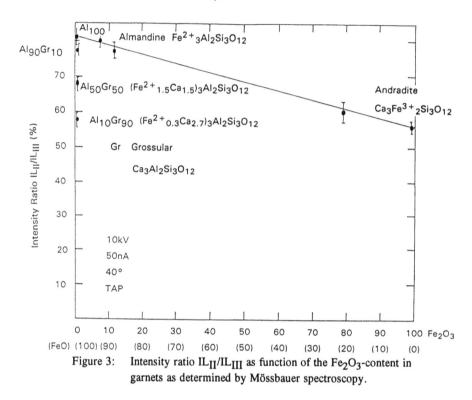

Figure 3:   Intensity ratio IL_{II}/IL_{III} as function of the Fe_2O_3-content in garnets as determined by Mössbauer spectroscopy.

In Figure 3 the L_{II}/L_{III} intensity ratio is plotted as function of the Fe_2O_3-content as determined by Mössbauer spectroscopy in different garnets. There is an almost linear decrease of L_{II}/L_{III} intensity ratio with increasing Fe^{3+}-content as well as with increasing Ca^{2+}-content. That means there is a strong influence on the spectra by the next nearest cation neighbours. In the hedenbergite - acmite solid solution series a similar trend is observed

4.   CONCLUSION

The Fe^{2+}/Fe^{3+} ratio can be determined with sufficient precision from the L_{II}/L_{III} intensity ratio in unknown samples within solid solution series, if (i) a calibration has been made of the same solid solution series similar to the procedure of this study (ii) an empirical or theoretical correction for other cations is performed.

5.   REFERENCES

Slavic I.A., Slavic J.I., Grzetic I.A., Pavicevic M.K. (1992) Mikrochim. Acta [Suppl.] **12**, pp 161-171.

Inst. Phys. Conf. Ser. No 130: Chapter 2
Paper presented at Int. Congr. X-ray Optics and Microanalysis, Manchester, 1992

91

Nicalon fibres: an EPMA and TEM study

G Love, SM Bleay, AR Chapman and VD Scott

School of Materials Science, University of Bath, Bath BA2 7AY, UK

ABSTRACT: EPMA studies showed that Nicalon fibres were composed of silicon carbide, a non-stoichiometric silicon oxycarbide and substantial amounts of free carbon. The ratio of carbide:oxycarbide was ~1.3:1 and the oxygen content of the oxycarbide was shown to increase towards the fibre surface. TEM work suggested that Nicalon was formed of microcrystallites of silicon carbide surrounded by amorphous oxycarbide and carbon.

INTRODUCTION

High-strength fibres are now available for reinforcing ceramics and metals and one such product is Nicalon, an inorganic fibre based upon silicon carbide. From a study of the literature it is evident that the exact nature of Nicalon remains unclear, although in addition to silicon carbide the presence of free carbon (Yajima *et al* 1976) and silica has been reported and a non-stoichiometric oxycarbide phase proposed (Laffon *et al* 1989). In the present paper electron probe microanalysis (EPMA) and transmission electron microscopy (TEM) are used to help characterise the fibre.

EXPERIMENTAL METHOD

Thin sections for TEM were produced from a metal-matrix composite plate, discs of 3mm diameter being cut from it using a hollow diamond drill. These were mechanically ground to a thickness of ~300μm and then dished on both sides using a VCR dimpling instrument. Finally, the central region of the specimen was thinned to perforation by bombardment of argon ions in a Gatan Duomill. To obtain transverse sections of fibre for EPMA, part of the plate was mounted in conducting resin and then ground and polished. Next, the surface was sputter coated with a thin layer (~10nm thick) of gold to render it electrically conducting. Individual fibres were employed for the EPMA study of the Nicalon surface. These were immersed first in dilute nitric acid to remove any polymeric size and then affixed to an aluminium planchette.

TEM analyses were undertaken in a JEOL 2000FX fitted with a LINK thin-window energy-dispersive x-ray spectrometer (EDS) and AN10000 spectrum analyser. A JEOL 8600M with wavelength dispersive spectrometers (WDS) was used to study bulk fibre sections and surfaces.

RESULTS AND DISCUSSION

A transmission electron micrograph taken from the centre of a fibre, fig.1a, shows microcrystallites a few nanometers in size. The corresponding electron diffraction pattern, fig.1b, exhibits broad rings and spacings consistent with microcrystalline β silicon carbide, the diffuse central spot suggesting a substantial amount of amorphous material is also present.

Fig.1(a) TEM image from fibre core and (b) the corresponding diffraction pattern.

Quantitative information on the fibres' composition was obtained using WDS by recording characteristic x-ray intensities from Nicalon and comparing them with those from standards containing known concentrations of the elements in question, thereby deriving 'k' ratios for silicon, oxygen and carbon. Here, silica was used as the standard for oxygen and silicon carbide for both silicon and carbon. For SiKα and O Kα radiation 'k' ratios were established from peak intensity measurements. However, the carbon 'k' ratio was 1.1 times higher if peak area measurements were substituted for peak intensities and when such discrepancies arise areas must be used. Chemical composition was determined from 'k' ratios by application of the Love-Scott ZAF correction (Scott and Love 1991) and data from Nicalon fibres within a given tow are shown in the table. When converted into atomic percentages, the results clearly show that there is more carbon than silicon .Consequently, there must be a minimum of 13 at % of free carbon in Nicalon, and probably significantly more than this.

Table 1. Quantitative EPMA Analysis of fibres in Composite Plate

Element	Av. wt.%	Std. Dev.	Normal. wt.%	Av. at. %
Silicon	54.87	±0.21	55.64	36.52
Oxygen	11.67	0.12	11.78	13.55
Carbon	33.12	0.4	32.57	49.93
Total	98.61	0.56	100.00	100.00

Peak shape changes in the carbon K x-ray spectrum associated with different carbon containing compounds have been mentioned above. Such effects arise when x-rays originate from transitions involving the valance band. For silicon the Kα x-ray, involving as it does an L to K transition, is independent of the nature of the chemical bond but the Kβ line (M to K transition) is not and the analysis of Nicalon can be taken a stage further by studying it closely. In fig.2, $K\beta$ peaks are shown from a range of silicon containing compounds and variations in its position and width are clearly evident. The position of peaks from silicon (Si-Si bonding) and silicon carbide (Si-C bonding) are similar although the latter shows a small peak at ~6.78Å. The spectrum from SiO_2 (Si-O bonding) however has its peak at a significantly longer wavelength and silicon monoxide (Si-Si and Si-O bonding) exhibits a doublet. Examination of the Nicalon $SiK\beta$ peak shows its position to be close to that of the carbide but it is somewhat broader and there is the hint of a satellite peak at ~6.8Å. These findings suggest that Nicalon has a high proportion of Si-C bonds but some Si-O bonding as well, and are consistent with our TEM investigations indicating β silicon carbide and a substantial amount of amorphous material - possibly an oxycarbide. To attempt to quantify these data let us assume an oxycarbide with the mid-range composition Si_2O_2C. We can synthesise a $SiK\beta$ peak for this compound by scaling the $K\beta$ peaks from the oxide and carbide.

$$Si_2O_2C = 1/2[0.501/0.7 \text{ x carbide peak} + 0.501/0.467 \text{ x silica peak}]$$

where 0.501, 0.7 and 0.467 are the weight concentrations of silicon in the oxycarbide, the carbide and silica respectively. Appropriate proportions of the oxycarbide and carbide peaks can then be added together to match the $SiK\beta$ peak from Nicalon using least squares analysis on some 180 data points.

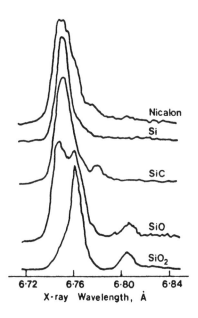
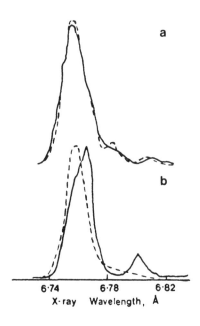

Fig.2 Normalised $SiK\beta$ peaks from different silicon containing compounds.

Fig3(a) $SiK\beta$ peak from Nicalon ——— and synthesised peak - - - - -

(b) $SiK\beta$ peak from fibre surface ——— and core – – – –

The best fit, fig. 3a, is obtained with 39.3% silicon carbide and 31.4% oxycarbide. This leaves 29.3% unaccounted for and this we must attribute to an element not bonded with silicon, namely free carbon. As a check on the validity of this approach the above figures may be converted into weight percentages to give 56.7% Si, 14.3% O and 28.9% C. Compared with the normalised microanalysis data in the table, the results agree reasonably well but a rather closer match can be achieved if, instead of the mid-range oxycarbide, an oxygen deficient oxycarbide is present and the oxycarbide/carbide ratio is somewhat higher. However, other confirmatory evidence would be required before pursuing this further.

To study surface composition of the fibres, oxygen Kα radiation was recorded as a function of electron beam energy and compared with data from the silica standard, fig. 4. The intensity ratio is seen to rise smoothly as the beam energy is reduced indicating enrichment of oxygen at the surface and confirmed by the shape of the SiKβ from Nicalon, fig. 3b, which now exhibits a much greater degree of Si-O bonding than previously. However, even at the lowest beam energy, corresponding to an excitation depth in the specimen of ~50nm, the intensity ratio is less than unity, indicating any silica surface film is very thin. A more plausible explanation of fig. 4 is that the stoichiometry of the oxycarbide changes towards the fibre surface with the oxygen : carbon ratio increasing.

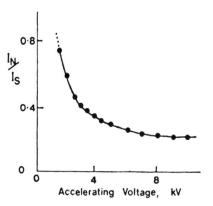

Fig.4. Ratio (I_N/I_S) of oxygen intensity from Nicalon fibre surface to that from silica plotted as a function of kV.

REFERENCES

Laffon C, Flank AM, Lagarde P, Laridjani M, Hagege R, Oiry P, Cotteret J, Diximier J, Miquel JL, Howard H and Legrand AP, 1989, *J . Mat. Sci* . **24**, 1503.
Scott VD and Love G, 1992, *X-ray Spectrom*. **21**, 27.
Yajima S, Kayano H, Okamura K, Omori M, Hayashi J and Akutsi K, 1976, *Am. Ceram. Soc. Bull.* **55**, 1065.

Inst. Phys. Conf. Ser. No 130: Chapter 2
Paper presented at Int. Congr. X-ray Optics and Microanalysis, Manchester, 1992

95

New possibilities in X-ray microanalysis of nitrogen and oxygen with the application of special types of multilayers

A Grudsky† and A Rudnev‡

† 'Seifert-Röntgen', St Petersburg, Russia

‡ Central Institute for Material Research, St Petersburg, Russia

Abstract. The results of the investigations of new types of multilayers for microanalysis are presented in this work. It is well known, that the region from oxygen Kα-line to carbon Kα-line is the most difficult region in the light elements analysis. Cr–Sc multilayer was made specially for nitrogen analysis. W–Sb multilayer has the best parameters near the oxygen Kα-line. We have made and studied a new dispersive element which consists of Ge (111) crystal and multilayer structure on the surface of this crystal. This double dispersive element gives the possibility to operate in the very wide wavelength region.

1. Introduction

At present, multilayer x-ray optics has become firmly established in (the) electron probe microanalysis (EPMA) and is widely used for analysis of light elements having ousted molecular crystals. The main advantage of multilayer optics when used in the EPMA is the high peak reflectivity and the possibility of selecting the period of the multilayer structure with due regard to peculiarities of the device and the spectral range. However, in our opinion, the possibilities of the multilayer optics are far from fully used in EPMA. Thus, for instance, in most produced microanalysers in the range of wavelengths of 2 to 12 nm, one structure was used. It is understood that with such an approach using, for instance, as in a device of Microspec WDX-3PC, a Ni–C pair with a period of 4 nm, best characteristics can be obtained in the Kα-line region of carbon.

In this paper, we try to demonstrate the maximum potential of multilayer artificial dispersing elements, paying attention to the requirements of light element EPMA. In particular, we study how to optimise reflectivity by choosing the best pairs of materials and the best periods of the multilayer structures. In so doing, we proceeded from the fact that the wide and practically important class of problems solved in the EPMA is the oxides and nitrides analysis, i.e. spectral lines of OKα and NKα. As it is known, precisely this spectral range in the soft x-ray field, even with the appearance of artificial multilayer structures, remains the most difficult to study. Bearing in mind that in practice it is difficult to fit the required number of dispersing elements in the holder of the microanalyser, we have also investigated possibilities of making multilayer coats on the surface of widely used crystal analysers of Ge and quartz.

2. Selection of parameters of multilayers

Calculations of multilayer reflectivities in the area of 20 to 35 Å show that carbon, traditionally used in this field as the best material for the light layer, does not allow reaching desired parameters. Therefore, the only possibility of raising the reflectance is the use of materials having the absorption edge just above the energy of oxygen and nitrogen $K\alpha$-lines respectively and, hence, anomalous decrease of the real and imaginary part of permittivity, which allows increasing considerably the reflectance of the multilayer. The optimal pair of materials appears W–Sb for the $OK\alpha$-line, as well as the pair Cr–Sc is the optimal for the $NK\alpha$-line.

When calculating the optimal multilayer period, we took into account the following factors: x-ray arrangement, coherent and incoherent scattering of the x-ray radiation as well as the considerations connected with the dependence of the range and the number of pairs of layers of the multilayer on the diffraction angle. As a result, we found out that the optimal range of diffraction angles, from the point of view of obtaining the maximum low limit of detection, is 25 and 40° from the multilayer surface.

3. Manufacture of multilayers

Cr–Sc and W–Sb multilayers were manufactured by 'Seifert-Röntgen & K° A/O' after the technology developed at the Institute of Applied Physics of the Russian Academy of Sciences. The results presented in this paper were obtained with multilayer structures with the following parameters:

Cr–Sc: – period of structured = 44.8 Å

– ratiod_{Cr}/d_{Cr+Sc} = 0.38

– peak reflectivity on the $NK\alpha$-line$R$ = 26.7%

– width of the reflection curve for the half of the height = 64°

W–Sb: – period of structured = 29.6 Å

– ratiod_W/d_{W+Sb} = 0.4

– peak reflectivity on the $OK\alpha$-line$R$ = 11.8%

– width of the reflection curve for the half of the height = 0.66°

Measurement of multilayers parameters were made by means of spectrometer-monochromator with the grating.

For checking the possibility of making the multilayer onto the surface of crystal analysers, we carried out superpolishing of Ge (111) and quartz (1011) surfaces. As a result, the root-mean-square surface roughness of Ge was from 5 to 7 Å, the surface of quartz, 3 Å. After this, on the surfaces obtained we deposited the most widely used Ni–C multilayer. Polishing of crystals surfaces and making of multilayers thereon were also carried out by the company 'Seifert-Röntgen & K° A/O'.

4. Results of measurements

The above described Cr–Sc and W–Sb multilayers were glued on the cylindrical holders and mounted in the microanalyser 'Camebax'. In figure 1, one can comprise the images of the set of corrosion cracks on the structural steel casting:

a - in secondary electrons

b - in $OK\alpha$ radiation with the use of LOD

c - in $OK\alpha$ radiation with the use of W–Sb multilayer

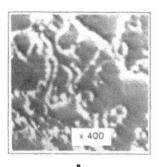
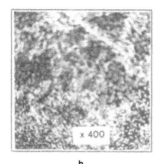
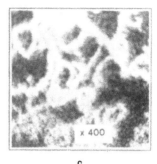

a   b   c

Figure 1. The set of corrosion cracks on the structural steel casting:
a - secondary electron image
b - oxygen distribution on the surface, LOD
c - oxygen distribution on the surface, ML 8 - 12.60

Table 1 presents characteristics obtained on sample BN with the use of the above said Cr - Sc
structure as compared with what can be obtained with the use of a now traditional Ni - C
multilayer. The data presented in Figure 1 and in Table 1 were obtained at accelerating voltage of
10 kV and probe current of 0.1 μA.

Spectral line/ standard	Sample	Structure	d spacing, [nm]	Peak intensity, Cps/μA	Peak to background, P/B
N$K\alpha$/BN	boron nitride	Cr - Sc	4.5	6.9 x 10^4	36
N$K\alpha$/BN	boron nitride	Ni - C	5.0	2.5 x 10^4	3

Table 1. Microanalyser "Camebax", the inclined spectrometer. Beam conditions: accelerating
voltage, 10 kV; specimen current, 0.1 μA.

Table 2 presents data obtained by Microspec's specialists for the Cr - Sc structure in comparison
with the structure of LSM-080 usually mounted in device WDX-3PC. In these measurements, the
Cr - Sc structure was used with a period d = 39 Å.

Spectral line/ standard	ML 7.80 Cr - Sc			Microspec reference		
	Peak intensity, Cps, μA	Peak/ background ratio	Halfwidth FWHM, eV	Peak intensity, Cps/μA	Peak/ background ratio	Halfwidth FWHM, eV
N$K\alpha$/BN	6.4 x 10^4	21	16	2.0 x 10^4	6.1	14
C$K\alpha$/VIT.C	2.1 x 10^5	102	12	7.6 x 10^5	82	13

Table 2. Beam conditions: accelerating voltage, 10 kV; specimen current 0.01 to 0.20 μA depending
on count rate.

Figure 2 presents theoretical (a) and experimental (b) curves of dependence of reflectivity on the incident angle on CuKα-line for the Ni - C multilayer on the surface of Ge (111). Experimental studies of this type of multilayers were carried out on a 4 axis diffractometer at the Danish Space Research Institute.

Figure 3 presents for comparison the images of the surface of X32H8AM2P alloy:

a - in secondary electrons
b - in CKα radiation with the use of molecular crystal LOD
c - in CKα radiation with the use of the said multilayer structure on the surface of Ge (111)

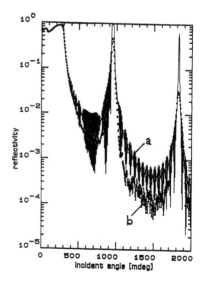

Figure 2. Theoretical (a) and experimental (b) curves of dependence of reflectivity on the incident angle on CuKα-line for the Ni - C multilayer on the surface of Ge (111).

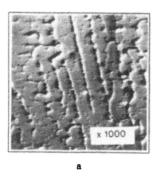

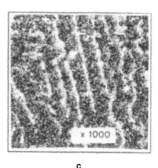

a   b   c

Figure 3. Chromium carbides in X32H8AM2P alloy surfacing:
a - secondary electron image of the alloy structure
b - carbon distribution, LOD
c - carbon distribution, ML 6.100 N on Ge (111)

5. CONCLUSIONS

The measurements have shown good prospects of new specialized multilayers for the analysis of oxygen and nitrogen and practical possibility of their manufacture on the surface of certain types of widely used crystal analysers.

We are grateful to Dr. F.E. Christensen for his assistance in the measurements on the CuKα-line. We are also grateful to Microspec's specialists for the measurements of new types of multilayers in WDX-3PC.

Inst. Phys. Conf. Ser. No 130: Chapter 2
Paper presented at Int. Congr. X-ray Optics and Microanalysis, Manchester, 1992

99

Quantitative and mapping application of chemical-shift spectra with EPMA

H Takahashi, T Okumura, Y Seo, A Kabaya, J W Critchell

JEOL Ltd, 1-2 Musashino 3-chome, Akishima, Tokyo 196, Japan
JEOL (UK) Ltd, JEOL House, Silver Court, Watchmead, Welwyn Garden City, Herts, AL7 1LT, United Kingdom

In conventional chemical-shift analysis with EPMA (electron probe microanalysis) peak shapes and peak positions of unknown spectra are compared with those of standard spectra. In the present study, the chemical-shift spectra were used for quantitative analysis by using a deconvolution (digital filter and least squares-fitting) method and observing the distribution of crystalline structures by using a ratio mapping method.

A JEOL JXA-8621 EPMA fitted with crystals of PET (pentaerythritol) and TAP (thallium acid phthalate) was used to analyse chemical shifts. A TAP crystal was used for OK analysis since it has a better resolution than a Pb-STE (lead stearate) crystal.

Fig. 1 and Fig. 2 show the results of deconvoluting the Si-Kβ spectrum of SiO_x specimen and the Fe-Lα,-Lβ spectrum of Fe_3O_4. For SiO_x, SiO_2 and Si were used as standards and for Fe_3O_4, Fe_2O_3 and FeO were used. Table 1 and table 2 show the comparisons of the quantitative analysis derived from the deconvoluted Si-Kβ and Fe-Lα-Lβ with those from the conventional ZAF method and it is seen that the % values are close. It can be estimated that the SiO_x spectrum is composed of SiO_2 and Si and the Fe_3O_4 spectrum of Fe_2O_3 and FeO.

The backscattered electron image of Fig.3 shows three superconductive phases and an incomplete reaction phase of a BiSrCaCuO specimen. The matrix 'a' corresponds to the 232 phases (Tc=80K), the dark grey phase to the 243 (Tc=110K), the white phase 'c' to the 221 (Tc=8~20K) and the black phase to an incomplete reaction phase. Fig. 4 shows the O-K$_\alpha$ peak shapes of these phases and the initial CuO material. We reported that O-K$_\alpha$ peak shapes reflect their crystalline structure (Takahashi et al 1989) Fig.6 shows the O-K$_\alpha$ from the perovskite-type structure. The superconductive O-K$_\alpha$ spectra of a, b, c phases in Fig.4 are close to those of perovskite-type structures in Fig.6. It is assumed that this incomplete reaction phase d is an intermediate phase between the initial CuO material and the last perovskite-type superconductor. At the peak position A and B shown in Fig.4, two colour maps were collected. Each map measured the X-ray intensities at 250 x 200 points and at each point, the ratio of X-ray intensities at the position A to that at B was calculated. The ratios of perovskite-type superconductors, the incomplete reaction phase and the CuO phases are close to 1.9, 1.5 and 1.0, respectively. Corresponding to these phases, therefore, the colour levels of white, grey and black were set. Fig.5 shows the result. From this ratio map, a two-dimensional distribution of chemical-shifts reflecting the crystalline structures could be obtained.

H Takahashi, Y Kondo, T Okumura, Y Seo: JEOL News, Vol. 27E, No.2, 2-7 (1989)

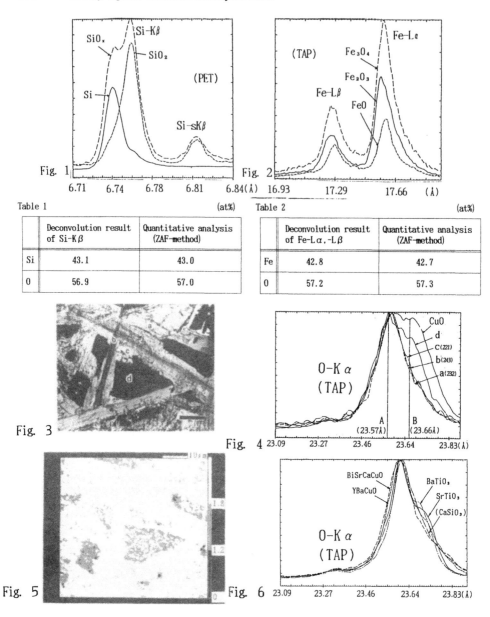

Table 1   (at%)

	Deconvolution result of Si-Kβ	Quantitative analysis (ZAF-method)
Si	43.1	43.0
0	56.9	57.0

Table 2   (at%)

	Deconvolution result of Fe-Lα,-Lβ	Quantitative analysis (ZAF-method)
Fe	42.8	42.7
0	57.2	57.3

Fig. 1 and Fig. 2. The results of deconvolution of the Si-Kβ and Fe-Lα,-Lβ spectra.
Table 1 and 2. Comparisons of quantitative analyses derived from deconvoluted Si-Kβ and Fe-Lα,-Lβ with the result from the conventional ZAF method.
Fig. 3. Composition image of BiSrCaCuO superconductor. Bar=10μm
Fig. 4. O-Kα spectra of four phases in BiSrCaCuO specimen.
Fig. 5. Chemical-shift distribution of O-Kα obtained from the ratio map made by dividing the x-ray intensity at A by that at B.
Fig. 6. O-Kα spectra of perovskite-type structures.

Inst. Phys. Conf. Ser. No 130: Chapter 2
Paper presented at Int. Congr. X-ray Optics and Microanalysis, Manchester, 1992

101

Electron configuration of the valence and the conduction band of magnetite (Fe_3O_4) and hematite (α-Fe_2O_3).

M.K. Pavicevic[*], D. Timotijevic[**], G. Amthauer[*]

[*]   Institute of Mineralogy, Hellbrunnerstr. 34, A-5020 Salzburg,
[**] Institute of Copper (Engineering-Belgrade), Pariske Commune 24/I, 11070 Belgrade

ABSTRACT: Soft $L_{III,II}$-X ray emission spectra of Fe in magnetite and hematite have been measured with different electron beam excitation voltages by EPMA between 690 eV and 730eV. On the basis of these data, the $L_{III,II}$-absorption spectra of Fe were calculated by the differential self-absorption method. By this procedure, information on the electron configuration of the valence band as well as on the conduction band can be obtained. The energy levels were determined using a computer program for fitting the observed emission and absorption spectra and compared with the result of MO-theory (SCF-X_α-method) in the literature.

1.   INTRODUCTION

The electron configuration of the valence and the conduction band determines a lot of different physical and chemical properties of 3d transition metal compounds. In iron oxides the completely occupied 4s-, the partially filled 3s-, and the empty 4p-orbitals of iron overlap with the completely filled 2s- and the partially filled 2p-orbitals of oxygen to the bonding, non-bonding, and antibonding molecular orbitals, which form the valence and the conduction band. The L-X-ray emission spectra are caused by electron transitions from the 4s-and 3d levels into the vacant L_{III}- or L_{II}-levels, respectively. The X-ray absorption spectra result from electron transitions from the $L_{III,II}$-levels into the vacant or partially filled energy levels of the conduction band. Therefore the L-X-ray emission spectra provide information on the valence band and the L-X-ray absorption spectra on the conduction band.

We studied the L-X-ray spectra of magnetite (Fe_3O_4) and hematite (Fe_2O_3). In the inverse spinel structure of magnetite one half of Fe^{3+} occupies tetrahedral sites and the other half

Table 1:   Observed energies of electron transitions (this work) and their assignments according to Tosell et al. (1974) in magnetite (Fe_3O_4).

electron transitions for different cluster units						
spectrum structure	$FeO_6{}^{10-}O_h$		$FeO_6{}^{9-}O_h$		$FeO_4{}^{5-}T_d$	
description	assignment	E (eV)	assignment	E (eV)	assignment	E (eV)
FeL_{III} peak	$2t_{2g}\uparrow \longrightarrow$ Fe $2p^{3/2}$	705,3	$2t_{2g}\uparrow \longrightarrow$ Fe $2p^{3/2}$; $2e_g\downarrow \longrightarrow$ Fe $2p^{3/2}$	705,8		
weak hump on the high energy side	$3e_g\uparrow \longrightarrow$ Fe $2p^{3/2}$	707,0	$3e_g\uparrow \longrightarrow$ Fe $2p^{3/2}$	707,8	$2e_g\downarrow \longrightarrow$ Fe $2p^{3/2}$; $7t_2\downarrow \longrightarrow$ Fe $2p^{3/2}$	707,3 ; 708,7
FeL_{III} absorption peak	Fe $2p^{3/2} \longrightarrow 2t_{2g}\downarrow$	709,0	Fe $2p^{3/2} \longrightarrow 2t_{2g}$	709,4	Fe $2p^{3/2} \longrightarrow 2e\uparrow$	710,3
weak hump on the high energy side	Fe $2p^{3/2} \longrightarrow 3e_g\downarrow$	710,7	Fe $2p^{3/2} \longrightarrow 3e_g\downarrow$	711,6	Fe $2p^{3/2} \longrightarrow 7t_2\uparrow$	711,3
weak peaks on the high energy side	Fe $2p^{3/2} \longrightarrow 7a_{1g}$; Fe $2p^{3/2} \longrightarrow 7a_{1g}$	713 ; 717	Fe $2p^{3/2} \longrightarrow 7t_{1u}$; Fe $2p^{3/2} \longrightarrow 7t_{1u}$	714 ; 718		
FeL_{II}-peak	$2t_{2g}\uparrow \longrightarrow$ Fe $2p^{1/2}$	717,5	$2t_{2g}\uparrow \longrightarrow$ Fe $2p^{1/2}$; $2e_g\downarrow \longrightarrow$ Fe $2p^{1/2}$	718,2	$2e_g\downarrow \longrightarrow$ Fe $2p^{1/2}$	718,8

Table 2: Oberserved energies of electron transitions (this work) and their assignment according to Tossell et al. (1974) in hematite.

spectrum structure description	assignment	E (eV)
weak hump on the low energy side	$2e_g \uparrow \longrightarrow$ Fe $2p^{3/2}$	704,0
FeL$_{III}$ peak	$2t_{2g} \downarrow \longrightarrow$ Fe $2p^{3/2}$ $2t_{2g} \uparrow \longrightarrow$ Fe $2p^{3/2}$	706,0
weak hump on the high energy side	$3e_g \uparrow \longrightarrow$ Fe $2p^{3/2}$	708,2
FeL$_{III}$ absorption edge		708,6
FeL$_{III}$ absorption peak	Fe $2p^{3/2} \longrightarrow 2t_{2g} \downarrow$	710,2
weak hump on the high energy side	Fe $2p^{3/2} \longrightarrow 3e_g \downarrow$	711,2
weak hump on the high energy side	Fe $2p^{3/2} \longrightarrow 7a_{1g}$ Fe $1t_{2g} \longrightarrow 7t_{1u}$	717,3 718,3
FeL$_{II}$-peak	$2e_g \downarrow \longrightarrow$ Fe $2p^{3/2}$ $2t_{2g} \uparrow \longrightarrow$ Fe $2p^{3/2}$	718,7

of Fe^{3+} together with all the Fe^{2+} octahedral sites. In the hematite structure all the iron is Fe^{3+} and occupies octahedral sites in a hexagonal closed packed structure of the oxygens. Both oxides have a relatively well defined composition and their structures and chemical bondings are well known (Tossel et al. 1974). Therefore they are very useful to test our methods.

2. EXPERIMENTAL

The L$_{III,II}$-X-ray emission spectra of Fe were taken with an ARL-SEMQ-electron microprobe using an electron beam excitation voltage of 4 KeV and 10 KeV, respectively, a TAP analyzing crystal, and a completely computer (PDP-11/04) controlled step-scanning method (0.15 eV/step) with 10s/step measuring time. The emission spectra were evaluated using a computer program of Timotijevic (1990). The absorption spectra were calculated from the normalized emission spectra taken at 4KeV and at 10KeV according to a method of Liefeld (1986). For more details the reader is refered to Timotijevic and Pavicevic (1992).

3.   RESULTS and DISCUSSION

The observed energies of the electron transitions as well as their assignment according to Tossell et al. (1974) are reported in Table 1 for magnetite and in Table 2 for hematite. For tetrahedral Fe^{3+} an FeO_4^{5-}-cluster is used, for octahedral Fe^{3+} a FeO_6^{9-}-cluster and for octahedral Fe^{2+} on FeO_6^{10-}-cluster.

Our data are in fairly good agreement with those calculated by Tossell et al (1974) and this encourages us to continue our investigations on more complicated compounds such as silicates and sulphides in order to obtain informations on their electronic structure. Our method is especially well suited for the study of small samples and samples with a microtexture both of natural and synthetic origin.

4.   REFERENCES

Liefeld R.J. (1968) Soft X-ray emission spectra at threshold excitation - Soft X-ray band spectra (London: Academic Press) pp 133-149.
Timotijevic D.M.; Pavicevic M.K. (1992) Microchim. Acta [Suppl] **12**, pp 255-259.
Timotijevic D.M. (1990) Magister Thesis, University of Belgrade.
Tossell J.A.; Vaughan D.J.; Johnson K.H. (1974) Amer. Mineral. **59**, pp 319-334.

Inst. Phys. Conf. Ser. No 130: Chapter 2
Paper presented at Int. Congr. X-ray Optics and Microanalysis, Manchester, 1992

105

X-ray absorption spectroscopy using an EPMA

H. KONUMA

National Research Institute of Police Science
6 Sanban–cho, Chiyoda–ku, Tokyo 102 Japan

ABSTRACT: The x–ray K–absorption spectra of Al and Ni are obtained in the EXAFS region using an EPMA with a sample holder which inserts the sample film between a target and a curved crystal. Data processing of the K–EXAFS are studied in these spectra, and the peak corresponding to the first–neighbor shell is clearly extracted by Fourier transform for the EXAFS oscillations.

1. INTRODUCTION

The phenomena of extended x–ray absorption fine structure (EXAFS) and x–ray absorption near edge structure (XANES) have had a long and valid history (Sandström 1957 and Azaroff 1974). Interest in EXAFS has been revived since Sayers, Stern and Lytle (1970) pointed out that EXAFS contains structural information which can be extracted using Fourier transform method, and that EXAFS can also be applied to structural analysis of materials in amorphous forms. Many applications of EXAFS measurements are found in reviews (Lee et al 1981, Teo et al 1981 and Teo 1986). However, most of those spectra are obtained with synchrotron radiation. Uses of synchrotron radiation put restrictions on experimental places and times.

It is useful and practical that x–ray absorption data can be measured in the laboratory and at the desired time (Konuma and Hayasi 1985). In this paper, we show measurements of x–ray K–absorption spectra of Al and Ni by an EPMA, and its EXAFS analysis.

2. EXPERIMENT

A whole system of EPMA (EXM–3500 ELIONIX INC.) were newly constructed for the purpose of standard EPMA analysis and for measurement of x–ray absorption spectra (Konuma et al 1989). The x–rays were emitted from a tungsten target by normal–incidence electron bombardment. The x–ray take–off angle is 35 degree. The focussing point of a fixed electron beam image, a curved crystal and a detector slit were set on a Rowland circle with a radius of curvature 140 or 160 mm. A sample holder, which inserts a sample film perpendicular to the x–ray path, is set between a target and a curved crystal. A thin Mylar film of 4 μm thick is set between the target and the sample holder to stop scattering electrons. A gas flow proportional counter with a polypropylene window of about 3000 Å thick was used, and P–10 gas (CH_4 10% and Ar 90%) flowed through the counter at an atmospheric pressure.

The spectrometer was set at a position corresponding to the starting photon energy $h\nu_1$, and intensity I_{01} without a sample film and I_1 with a sample film were measured one after another. So, we got an absorption coefficient $\mu_1 x = \ln(I_{01}/I_1)$ at the photon energy $h\nu_1$, where x is thickness of the sample film. Next, the spectrometer was moved by half the slit width, and intensity I_{02} without the sample film and I_2 with the sample film were measured similarly. An absorption coefficient $\mu_2 x = \ln(I_{02}/I_2)$ was got at the photon energy $h\nu_2$. These measurement processes were automatically repeated about 600 times to get a full K–absorption spectrum. It took about ten or twenty hours to finish a measurement. Error caused by fluctuations of an x–ray source and an x–ray measurement system can be neglected from the absorption coefficient μ or μx, because it takes at most 5 minutes to get both intensities I_0 and I at one photon energy point.   A probe condition was 20 kV and about 1.8 μA. To avoid x–ray intensity decrease by carbon contamination, the W target was automatically moved by 20 μm at intervals of ten to twenty minutes.

3. EXAFS ANALYSIS

The absorption spectrum μ in the K–EXAFS region is decomposed to three components of μ_b, μ_o and χ. A value μ_b is a background absorption caused by scattering, by excitation of core electrons without K–shell electron (mainly L–shell electron) of an absorbed atom and a smooth absorption by scattering atoms.   A value μ_o is a smooth K–shell absorption.   A value χ is the K–EXAFS oscillations. Then, χ is obtained by,

$$\chi = \{(\mu - \mu_b) - \mu_0\}/\mu_0. \tag{1}$$

In this equation, the value μ_b was determined by applying the least square fit to the Victoreen equation (Victoreen 1948) in the data of lower energy side from the absorption edge.   The value μ_0 was determined using a least square fit with a polynomial of forth order in the data above high energy side from the edge.

The single–scattering theory of the K–EXAFS oscillations is given by,

$$\chi(k) = -\sum_j \frac{N_j}{kr_j^2} |f_j(\pi)| \sin(2kr_j + 2\delta_1' + arg f_j(\pi)) \exp(-2\gamma r_j) \exp(-2\sigma_j^2 k^2) \tag{2}$$

where $f_j(\pi)$ is the backscattering amplitude from each of the N_j neighboring atoms located at a radial distance r_j with a Debye–Waller factor σ_j in the EXAFS. The derivation of this equation and the meaning of other notations are described in other papers (Lee and Pendry 1975, Lee and Beni 1977 and Lee et al 1981). The Fourier transform $F(r)$ of experimental data $\chi_{exp.}(k)$ is expressed according to Lee and Beni (1977), putting the inelastic term $\exp(2\gamma r)$ in this paper,

$$F(r) = \frac{1}{2\pi} \int \exp(2\gamma r) \exp(-i\phi(k)) k \chi_{exp.}(k) \cdot \frac{1}{|f(\pi)|} \exp(2\sigma^2 k^2) g(k) \exp(-i2kr) dk \tag{3}$$

where, Debye–Waller factors σ^2 were calculated by Beni and Platzman (1976) and Greegor and Lytle (1979). The inelastic factor γ, which is the inverse of the inelastic electron mean free path λ (Ashley et al 1976 and 1979) were used. The backscattering amplitude $f(\pi)$ and the scattering phase shift of photoelectrons, $\phi(k) = 2\delta_1' + arg\ f(\pi)$, were calculated using the Hermann–Skillman core wavefunction (Hermann and Skillmann 1963) following the program of Pendry (1974). The window function $g(k)$ was the same as that used by Lee and Beni.

4. RESULTS AND DISCUSSION

The Al K–absorption spectrum of 1.5 μm thick Al film, measured with the first order reflection on RAP (001) plane, is Shown in Fig. 1. The Ni K–absorption spectrum of 5 μm thick Ni film, measured with the first order reflection on LIF (200) plane, is shown in Fig. 2. The photon energy scale of the Al spectrum was determined in I_0 spectrum of a W target, with reference lines of $WM_\beta(6.757\text{Å})$ and $WM_3N_1(7.360\text{Å})$, and that of the Ni spectrum was also determined with reference lines of $WL_\eta(1.4211\text{Å})$ and $WL_{\alpha_2}(1.48743\text{Å})$ (Bearden 1967 and White and Johnson 1970).

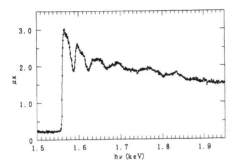

Fig. 1. X–ray K–absorption spectrum of Al.

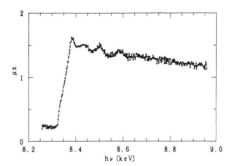

Fig. 2. X–ray K–absorption spectrum of Ni.

In the spectrum of Al, XANES and EXAFS structures are well appeared. The Al K–absorption edge of 7.944Å was measured. In the Ni spectrum, the edge structure is clear, and the K–absorption edge of 1.485Å was measured. But, the structure at high energy far from the edge is not distinguished, because of weak x–ray intensity for this Ni sample film.

EXAFS oscillations $\chi(k)$ are shown in Fig. 3 and Fig. 4. And the absolute Fourier transform for the EXAFS oscillations, $|F(r)|$, are in Fig. 5 and Fig 6.

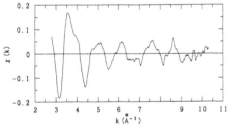

Fig. 3. Al K–EXAFS oscillations.

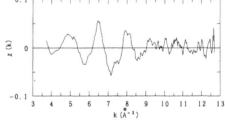

Fig. 4. Ni K–EXAFS oscillations.

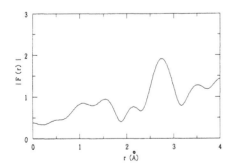

Fig. 5. Fourier transform for Al K–EXAFS.

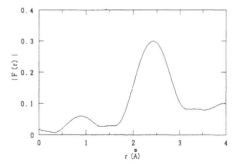

Fig. 6. Fourier transform for Ni K–EXAFS.

Physical constants used in the EXAFS analysis, such as σ_j^2, $\gamma=1/\lambda$, $f(\pi)$ and $\phi(k)$, were independently selected from theoretical calculations. Even so, absolute peaks of Fourier transform corresponding to the first–neighbor shell are clearly reconstructed for Al and Ni, respectively.

5. CONCLUSION

By using an EPMA with a point focus of the x–ray source, the x–ray K–absorption spectra of Al and Ni were obtained in the XANES and EXAFS regions. Wavelengths of the absorption spectra were determined with diagram lines along each emission spectrum of I_0. Physical constants used in EXAFS analysis, such as σ_j^2, $\gamma=1/\lambda$, $f(\pi)$ and $\phi(k)$, were independently selected from theoretical calculations. EXAFS analysis showed good results in the peaks corresponding to the first–neighbor shell.

REFERENCES

Ashley J C, Tung C J, Ritchie R H and Anderson V E 1976 IEEE Trans. Nucl. Sci. NS–23 1833.
Ashley J C, Tung C J and Ritchie R H 1979 Surface Sci. 81 409.
Azaroff L V and Douglas M P 1974 in X–Ray Spectroscopy ed. Azaroff L V (New York: McGraw–Hill) p284–337.
Bearden J A 1967 Rev. Mod. Phys. 39 18.
Beni G and Platzman P M 1976 Phys. Rev. B14 1514.
Greegor R B and Lytle F W 1979 Phys. Rev. B20 4902.
Hermann F and Skillman S 1963 Atomic structure Calculation (Englewood Cliffs, N. J.: Prentice–Hall).
Kiyono S, Chiba S, Hayasi Y, Kato S and Mochimaru S 1978 Proc. Int. Conf. X–Ray and XUV Spectroscopy (Jpn. J. Apl.Phys. 17 Suppl.17–2) p212–214.
Konuma H and Hayasi Y 1985 Nippon Kessho Gakkaishi 27 213 (in Japanese).
Konuma H, Ohyi H, Ono S and Miyazaki T 1989 Extended Abstracts (The 50th Autumn Meeting); The Japan Society of Applied Physics.
Lee P A and Pendry J B 1975 Phys. Rev. B11 2795.
Lee P A and Beni G 1977 Phys. Rev. B15 2862.
Lee P A, Citrin P H Eisenberger P and Kincaid B M 1981 Rev. Mod. Phys. 53 769.
Pendry J B 1974 Low Energy Electron Diffraction (London: Academic Press).
Sandström A E 1957 Handbuch der physik 30 (Berlin: Springer–Verlag) p78–245.
Sayers D E, Stern E A and Lytle F W 1970 Advan. in X–ray Anal. 13 248.
Teo B K and D. C. Joy ed. 1981 EXAFS Spectroscopy (New York:Plenum Press).
Teo B K 1986 EXAFS: Basic Principle and Data Analysis (Berlin:Springer–Verlag).
Victoreen J A 1948 J. Appl. Phys. 19 855.
White E W and Johnson G G 1970 X-ray Emission and Absorption Wavelength and Two–Theta Tables 2nd. ed. (Philadelphia: ASTM).

Inst. Phys. Conf. Ser. No 130: Chapter 2
Paper presented at Int. Congr. X-ray Optics and Microanalysis, Manchester, 1992

109

Simultaneous X-rays and Raman analysis of microsamples: state-of-the-art and preliminary results

Michel TRUCHET*, Michel DELHAYE** and Robert DEMOL**.

MUSEUM NATIONAL D'HISTOIRE NATURELLE, BIMM, URA CNRS 699 and UNIVERSITE PIERRE ET MARIE CURIE, LHFA, 12 rue Cuvier, 75005 Paris, FRANCE.

**UNIVERSITE DES SCIENCES ET TECHNIQUES DE LILLE-FLANDRES-ARTOIS, LASIR, UP CNRS 2631, 59665 Villeneuve d'Ascq Cedex, FRANCE.*

ABSTRACT. In numerous cases, the elemental analysis provided by the electron microprobe has to be complemented by the knowledge of the physical and/or molecular form of the sample at the probe point. This is now possible with our combined electron probe/laser-Raman probe, which has been successfully applied to various kinds of samples, for example biological, mineralogical, metallurgical, semi-conductors or synthetic materials.

1.INTRODUCTION.

There is a lot of evidence about the similarity of certain features of molecular analysis by vibrational laser-Raman microspectrometry and elemental microanalysis by electron probe stimulated X-Rays, such as : non destructive analysis, relatively poor detection limits requiring "great" concentrations in the explored microvolume, but the possibility of quantitative analysis and very good performance for qualitative analysis. This is why we undertook, a few years ago, originally for biological problems only, to built a combined microanalyzer, capable of simultaneous X-Ray and Raman analyses using a photon and/or electron microscope.

2.INSTRUMENTATION.

The electron microprobe, according to Castaing (1951) is originally equipped with a coaxial photon mirror objective (Castaing et Deschamps, 1958) which makes it adaptable to our purpose. As a first step, we choose to modify the light-box, outside the electron column, to both focus the laser beam and collect the diffused light.

The laser is the commercially available SPECTRA-PHYSICS 168 Argon Ion, emitting a green line at 514.5 nm, with the well known properties of coherence and good optical parameters in the TEM_{00} mode.

It is difficult to place the laser close to the electron column which is, moreover, mounted on rubber blocs but we took advantage of the optical fibers to connect the two.
The laser optical fiber is of the monomode type while the Raman spectrometer is

illuminated by a multimode type optical fiber.
Special devices have been utilized to : 1.-collect the laser beam at the output of the laser;
2.-give correct optical conjugation with the
Cassegrain-Schwarzchild mirror objective, in order to get the laser probe of approximately
$1\mu m$ in size. In fact, because of the mechanical fixing of the second miror in the objective
("spider"),a diffraction figure give a central spot of less than one micrometer in diameter
and 4 tails slightly longer than $1\mu m$. The main part of the laser power is concentrated in the
central spot, which can be considered as the "laser probe". A maximum of 1.5 W is
available with the laser; at the output of the illuminating system, approximately 30 mW
reaches the sample but, routinely, 1-2 mW are enough for good peak-to-noise spectra
without irradiation damage to the sample;
3.-focus the diffused light collected by the objective,
on the entrance face of the multimode transfer optical fiber;
4.-give correct optical conjugation with the Raman
spectrometer.
Because of the poor efficiency in light collection, due to the low numerical aperture of the
mirror objective (N.A. = 0.4), we choose as Raman spectrometer a special model of
DILOR X-Y, with its simultaneous multichannel detection (Barbillat, 1983) and Notch
filters (Carrabba *et al*,1990). In fact, unwanted laser light causes secondary effects with
optical fibers; on the other hand, these filters have too wide a cut-off limit, which makes,
in practice, Raman vibrations under 300 cm^{-1} quite inaccessible.

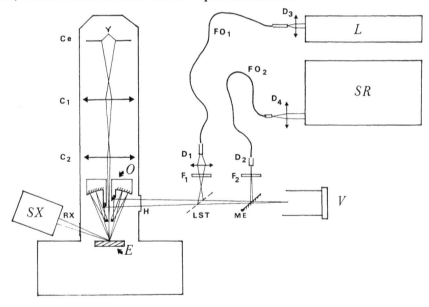

Figure 1 : *General schematic representation of the combined electron-probe/laser-Raman probe, using the coaxial Cassegrain-Schwarzchild mirror objective.*
Ce: electron gun; **C1/C2** : magnetic lenses (condensers); **E** : sample; **O** : Cassegrain-Schwarzchild mirror objective; **RX** : X-Rays; **SX** : Wavelength dispersive X-Ray
spectrometer (EDS if wanted); **H** : port-hole; **LST** : beamsplitter; **F1** : Notch filter; **D1** :
device for optical fiber (point 2 in the above text); **FO1** : monomode illuminating optical
fiber; **D3** : device for focusing laser light on the entrance face of the monomode
illuminating optical fiber (point 1 in the text); **L** : laser; **ME** : retractable mirror; **F2** :
Notch filter; **D2** : device for optical fiber (point 3 in the text); **FO2** : multimode transfer
optical fiber; **D4** : device for Raman spectrometer (point 4 in the text); **SR** : DILOR X-Y
multichannel Raman spectrometer; **V** : eye-piece.

3. APPLICATIONS.

The typical field of application is, whatever the sample, all cases where both elemental and structural analysis is needed on a micrometre or sub-micrometre scale.

For micrometre studies, the photon microscope is used to localize the area of interest and for Raman analysis; the electron column simply generates the electron probe. In sub-micrometre studies, the photon microscope is used only for Raman analysis; the localization of the point for analysis is obtained by electron imaging. Any sample gives surface scanning images (SEM, secondary and/or backscattered electrons), but transmission images (STEM) at low magnification are possible only when the sample is ultra-thin (100 nm or less). In this case, the peak-to-noise ratio of the Raman spectra is lowered by the reduction in the analyzed volume (Dhamelincourt *et al*,1991). In a "bulk" sample, the volume is approximately a few cubic micrometers (10^{-11}g) whereas in an ultrathin section, it is of the order of $10^{-2}\mu m^3$ (10^{-14}g).

3-1 : BIOLOGY.

Invertebrate animals often accumulate organic and/or mineral productions of their metabolism as intra- or extracellular concretions called *spherocrystals*. These structures size ranges typically from 0.1 to 10 μm and they are made of several concentric layers, often alternatively organic and mineral. When they are abundant in a given tissue, it is easy to extract and analyse them by biochemical means; but when they are scarce, or when a study concretion by concretion is required, only analytical microscopy can be performed. In a previous work (Ballan *et al.*, 1979), we have demonstrated that the use of electron microprobe and Raman-laser probe leads to a complete definition of the nature of insects spherocrystals. In these organic concretions, known to be made of purinic wastes, the electron microprobes detect the associated alkaline metals, Na and K, whereas the Raman microprobe detects a complex mixture of uric acid, monosodium- and monopotassium urates. These results, obtained on different concretions by the two methods, can now be achieved on the same single spherocrystal (fig. 2).

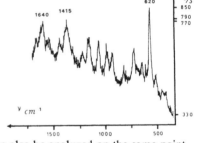

Figure 2. *Raman spectrum of a single spherocrystal in an insect fat body (Blatella). Laser, 514.5nm, 1 mW; sample : 10^{-11}g; integration time : 30 s; 16 accumulations; Na and K (K alpha) detected by electron probe at the same point. 620 cm^{-1}: ring-breathing mode; 1640 cm^{-1}: uric acid; 1415 cm^{-1}: Na and K urates.*

3-2. METALLURGY.

Surface coatings, oxidation or corrosion products can also be analyzed on the same point by the two methods; on figure 3, after surface treatment, the electron probe reveals S, Fe and Pb. Raman analysis demonstrates, in any point, a well crystallized lead sulphate.

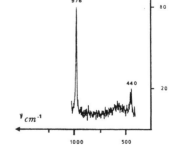

Figure 3. *Raman spectrum of a steel sheet surface. Laser, 514.5 nm, 1.5 mW; sample : 10^{-11}g; integr.time : 10s; 49 accumulations.*

3-3. MATERIALS.

Among new materials, diamond-like coatings are of increasing interest. In this example, we demonstrate that an heterogeneity in structure (diamond/graphite) is correlated to a variation in the intensity of the carbon K- alpha line: 45±5cp.s^{-1} (graphite)/ 65±10cp.s^{-1} (diamond).

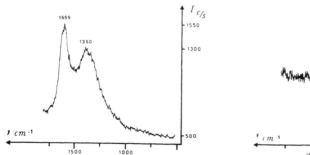
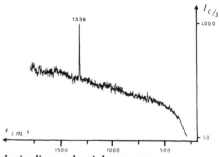

Figure 4. *Raman spectra of two regions of a synthetic diamond : right spectrum,*
characteristic line at 1336 cm⁻¹ of diamond; left spectrum, characteristic bands of
compacted graphite at 1350 and 1550 cm⁻¹. Laser, 514.5 nm, 2mW; sample : 10⁻¹¹ g;
integration time : 10 s; 25 accumulations.

3-4. MINERALOGY.

This sample of titanium dioxide (white pigment) as anatase was a powder embedded in resin
(as for biological samples in transmission electron
microscopy), sectioned at 0.5 μm and placed on
carbonated plastic support. Besides the very good
Raman spectrum, characteristic of this compound,
Ti and O (K-alpha) are easily detected.

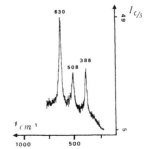

Figure 5. *Raman spectrum of TiO₂ as thin sample.*
Laser, 514.5 nm, 1 mW; sample : 10⁻¹³ g; integr.
time : 10s; 16 accumulations.

3-5. SEMI-CONDUCTORS.

In the monocrystal, the degree of crystallinity
can be controlled or dusts, defects can be
analyzed. In this example, an integrated circuit
is contaminated by oil; despite strong
fluorescence, the peak at 520 cm⁻¹ from the
Si monocrystal is clearly visible.

Figure 6. *Raman spectrum of an integrated circuit*
contaminated with oil. Laser, 514.5 nm, 1.5 mw;
sample : 10⁻¹¹ g; integration time : 20s;
16 accumulations.

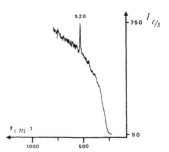

4.CONCLUSION.

Despite the low numerical aperture of the collecting objective, satisfactory Raman spectrum
are obtained with various kinds of samples. A better performance Raman probe is under
study,(Truchet and Delhaye, 1988) but will accept only small size samples (2mm).

5.BIBLIOGRAPHY.

BALLAN-DUFRANçAIS,C, TRUCHET, M. and DHAMELINCOURT, P., *Biol.*
  Cell., *36(1),* 51-58, 1979.
BARBILLAT, J. Thèse Doc. Etat, *Univ. LILLE (n°CNRS 605),* 1983.
CARRABBA, M., SPENCER, K., RICH, K. and RAUH, D., *Appl. Spectrosc.,*
  44(9), 1558-1561, 1990.
CASTAING, R. Thèse Doc. Etat, *ONERA Ed., PARIS,* 1951.
CASTAING, R. et DESCHAMPS, M. Revue Aero.,*ONERA Ed.,PARIS,* 1958.
DHAMELINCOURT, P., DELHAYE, M., TRUCHET, M. et DaSILVA, E.
  J. Raman Spectroscopy, *22,* 1-4, 1991.
TRUCHET, M. et DELHAYE, M. J. Microsc. Spectrosc. Electron.,*13*(2), 167, 1988.

Inst. Phys. Conf. Ser. No 130: Chapter 2
Paper presented at Int. Congr. X-ray Optics and Microanalysis, Manchester, 1992

Database and review of quantitative EPMA procedures

Kurt F. J. Heinrich, 804 Blossom Drive, Rockville MD USA

ABSTRACT: Currently used data reduction procedures are evaluated on the basis of a set of 1826 measurements on binary targets.

THE PROCEDURE

In the classic ZAF procedure the data evaluation ('correction') is done in three parts, which are repeated for specimen and standard, and recalculated in subsequent iterations: the atomic number correction (Z), the absorption correction (A) and the fluorescence correction (F).The combination of the Z, A and F corrections in an analytical procedure was formulated by Philibert and Tixier (1968) and is still widely used .

While the procedure of Scott and Love (1991) and one recently developed by Duncumb (1992) follow these lines, other recent procedures, such as the PAP and XPP procedures by Pouchou and Pichoir (1991) and the phi-rho-z method advocated by Brown (1991) and Packwood (1991) are developed along different paths. However, the three parts on which the classic ZAF procedure is based can still be identified and separated, although a differentiation of the stopping power effect and the backscatter effect is not explicit in the phi-rho-z method.
We use a test set of intensity measurements on binaries of presumably known composition and select subsets in which one or a few ZAF factors is predominant, to obtain the distribution of errors committed by the procedures.

THE DATA SET

The data set of 1826 measurements contains published as well as unpublished data obtained through the kindness of various authors. Values of doubtful usefulness were excluded, according to the following criteria:

Most data obtained on instruments having an x-ray emergence angle below 30°, which do not represent the present state of art, not only with respect to absorption losses, but also concerning other aspects of data collection.

Single values not corroborated by measurements of other compositions within the binary system under identical conditions, or of the same specimen at different operating voltages.

Sets of measurements that show lack of internal consistency.

Measurements of x-ray emissions of elements of atomic number below 12. The difficulties in measuring such bands and of obtaining reliable mass absorption coefficients as well as the electrostatic charging of poor conductors frequent in this group of data, reported by Bastin and Heijligers (1991), require considerations outside the scope of this work.

Measurements on specimens other than those of uniform composition within the range of x-ray excitation, limited by a flat surface normal to the electron beam (layered or coated specimens, particles, thin films and those measured at inclined beam incidence).

The sources of data used in the compilation are as follows:
Bastin's measurements of borides (Bastin GF and Heijligers HJM, Report ISBN 90-6819-006-7 CIP, University for Physical Chemistry Eindhoven Netherlands.),
Goldstein JI, Majeske FJ and Yakowitz H,1967 Adv. X-Ray Analysis10 431 ,
Colby J and Conley DK 1966 X-Ray Optics and Microanalysis eds Castaing et al (Paris: Hermann) 263,
Heinrich KFJ et al 1971 NBS Special Publication 260 28,
Pouchou 1991, compilation in Electron Probe Quantitation, eds Heinrich KFJ and Newbury DE (New York: Plenum Press),
Sewell DA, Love G and Scott VD 1985 J. Phys. D: Appl. Phys. 18 1245,
Bastin GF data base (personal communication, 1979),
Marinenko R (NBS) and Konuma H (Jap. Nat. Res. Inst. of Police Science, Tokyo) 1986 (unpublished),
Hlava PF, (Sandia) 1987 measurements (personal communication. See also:Romig AD Jr et al, Microbeam Analysis 1987 15.),
Schreiber TP and Waldo RA 1986, Proc. ICXOM11, ed:Brown JD, U London, Ontario, Canada 265.

The determinations of partner elements in oxides and nitrides reported by Bastin et al are not included in the set. These measurements appear to be affected by conductivity problems, and their error distributions are unusually wide.

THE TEST PROCEDURE

The test procedure SWITCH is written in TurboPascal. It offers menus for the selection of procedures and parameters, reads a file of data, and calculates the intensity ratios predicted according to the selections made. However, the analyst is interested in the error in the estimated weight fractions ('concentrations'). To perform the necessary transformation, the hyperbolic approximation is used. The reported errors include those in the assumed standard compositions and measurement errors, as well as those caused by the correction procedures.

SWITCH offers the following choices:
    for the calculation of the stopping power term: averaging the energy between Eo and Eq,
log. integral procedure of Philibert and Tixier (1968), numerical integration, Scott and Love (1991), Duncumb (1992), PAP (Pouchou and Pichoir,1991), Brown (1991) and Packwood (1991);
    for the backscatter correction term: Duncumb and Reed (1968), Scott and Love (1991), PAP;
    for the absorption model: Philibert-Duncumb-Heinrich (Heinrich 1970), Scott and Love (1991), PAP, PPS (Pouchou and Pichoir 1991), Heinrich's quadratic (1975) and duplex (1985) formulas, Duncumb (1992), Brown (1991) and Packwood (1991);
    for characteristic fluorescence: complete calculation (Heinrich 1987a), Reed and Long (1963), Reed (1990),or omit the correction;
    for continuum fluorescence: Heinrich (1987b), same with Small's (1987) continuum data , or omit;

for the choice of the J-factor: Wilson (1941), Bloch (1933), Duncumb et al (1969), Duncumb (1992), Berger and Seltzer (1964), Heinrich (unpublished: J=9.94 Z+19.52), Springer (1967), and Zeller (1975), as well as Joy and Luo's (1989) modification of the Bethe equation;

for the fluorescent yield: Burhop (1955), Heinrich (not published), Laberrigue, Wapstra (the last two from Bambynek (1972)), Reed and Long (1963) and Reed (1990);

for x-ray mass absorption coefficients: Heinrich (1966), Heinrich (1986), PAPMAC, by Pouchou and Pichoir (personal communication), and on-line numerical input..

THE ATOMIC NUMBER SUBSET (498 cases, Table I)

To discount the absorption and fluorescence effects, the following cases were purged from the data set in order to form the atomic-number subset:
Cases in which the absorption factor for the specimen was below 0.85.
Cases in which the difference between the two atomic numbers was below four (this excludes the strong K-K fluorescence).
Cases in which the overvoltage (Eo/Eq) was below 1.5 (see last entry in Table I).
This subset consists of 498 cases.

Discussion of table 1:

zaf1 - zaf4 (average energy): The stopping power is calculated at a single electron energy halfway between the operating beam energy and the critical excitation potential of the line in question.
zaf4b-7g: The series using the logarithmic-integral technique and those using numerical integration yield identical results. The choice of fluorescence yield is not significant in these cases, and the introduction of fluorescence due to the continuum worsens the statistics. (This correction would be important only in the determination of high-energy lines of small amounts of an element in a low-atomic weight matrix, such as 1% copper in beryllium). The choice of the absorption correction (Fa) and of the set of x-ray absorption coefficients (m) does not greatly affect the results, since absorption below 0.85 was excluded. Runs not included in the table showed that the choice of the backscatter model is not critical.

A significant reduction of the standard deviation is achieved with the method of Scott (zaf8) and the PAP model of Pouchou and Pichoir (zaf9). The best approaches to data reduction tend towards a standard deviation in the order of 1.1%. Since plots of the errors against parameters such as atomic numbers, operating voltages and absorption factors fail to reveal correlations, it may be assumed that the errors of measurement and of the estimate of specimen composition are in the order of 1%. A test of cases with overvoltage Eo/Eq below 1.5 shows that the standard deviation is somewhat higher in this group.

THE ABSORPTION SUBSET (812 cases, Table 2.)

To study the effects of the absorption of primary x-rays in the specimen a second subset was selected according the following criteria:
The absorption factor for the specimen had to be equal or lower than 0.85 (calculated by the duplex procedure).
Cases in which the difference between the atomic numbers was below four were eliminated (this excludes the strong K-K fluorescence).
Cases involving primary emission from elements of Z<10 were eliminated.

Table I: Atomic Number Subset (498 cases):

Name	S	R	Fa	flu	con	J	Joy	μ	w	ISI	mean	StD
zaf1	ave	dun	pdh	-	-	wil	-	fm			-. 0097	.0150
zaf2	ave	dun	pdh	s	-	wil	-	fm			- .0081	.0146
zaf4	ave	dun	pdh	s	-	dun	-	fm			+.0016	.0207
zaf4b	li	dun	pdh	s	-	dun	-	kh			+.0016	.0203
zaf4a	li	dun	pdh	s	-	bs	-	kh			+.0005	.0133
zaf7a	ni	dun	pdh	s	-	bs	-	kh		co	+.0005	.0133
zaf7b	ni	dun	phd	s	-	bs	-	kh		gr	+.0005	.0132
zaf7d	li	dun	pdh	c	-	bs	-	kh	bu		- .0001	.0139
zaf7e	li	dun	pdh	c	-	bs	-	kh	he		- .0001	.0139
zaf7g	li	dun	pdh	c	kh	bs	-	kh	bu		+.0001	.0152
zaf21	ni	dun	dun	s	-	bs	-	kh	bu	gr	+.0021	.0130
zaf22	ni	dun	dun	s	-	d2	-	kh	bu	gr	- .0065	.0152
zaf8 (Scott)	sc	sc	sc	s	-	wil	-	kh			-.0037	.0117
zaf8d	sc	dun	pdh	s	-	bs	-	kh			+.0009	.0114
zaf9 (PAP)	pap	pap	pap	s	-	zel	-	kh			- .0023	.0115
zaf9a	pap	pap	pap	s	-	bs	-	kh			+.0001	.0110
zaf10(PPS)	pap	pap	pps	s	-	zel	-	kh			- .0020	.0146
zaf15	pap	dun	sc	s	-	bs	-	kh			+.0005	.0112
zaf17	pap	my	dpl	s	-	bs	-	kh			+.0022	.0124
zaf10(PPS)	pap	pap	pps	s	-	zel	-	kh			- .0020	.0146
zaf19	br	-	br	s	-	-	-	kh			- .0128	.0161
zaf20	pw	-	pw	s	-	bs	-	kh			- .0082	.0155
Uo<1.5 (89 cases)	pap	dun	dpl	s	-	bs	-	kh			+.0008	.0145

Abbreviations used in the tables:

ave : average;   bl: Bloch;   br: Brown;
bs: Berger-Selzer (1964)   bu: Burhop;   con: contin. fluorescence;
co: Cosslett;
d2: Duncumb(1991);   dpl: duplex apprx.;   dun: Duncumb (1969);
Fa: absorption term;   fl: fluorescence;   fm: m from Frame program;
gr: Gryzinski;   he: Heinrich fluor. yield; ho:Howarth;
ISI: inn.shell ioniz.cross-section;
j-factor;   Jv: varying J-factor;   kh: Heinrich m.abs.coeff.;
li: log.integral;   mean: mean error;   ni: num.integration;
pap: Pouchou PAP;   pps: Pouchou PPS;   pw: Packwood;
quadratic approx;   R: backscatter;   S: stopping power;

RMS: root mean square (from zero error);
sct:: Scott;   StD: standard deviation (from mean);
spr: Springer;   w: fluorescence yield;   wil: Wilson;
zel: Zeller; μ: mass abs. coefficient;

The errors reported in Table II tend to exceed those in Table I. For the best methods, the root mean squares (RMS) converge towards a value of 0.013, which is probably close to the sum of experimental errors and errors in the estimate of composition of the binaries; this value compares well with that of 0.011 to which the data of Table I converge.

A probability curve of the error distribution of any of the above methods shows non-linearity in the extremes, suggesting outliers, particularly on the high side (fig 1). If they were removed, the RMS would be reduced almost to the level or the RMS of Table I. This result suggests that the residual error in the better absorption methods is small.

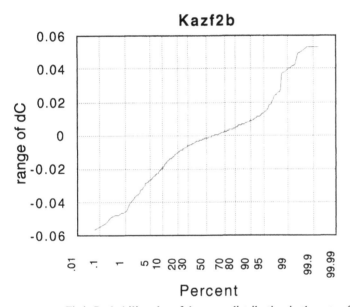

Fig1: Probability plot of the error distribution in the set azf2b.

   The lowest values of RMS were obtained with the method of Love and Scott, with PAP of Pouchou and Pichoir, and with the new method of Duncumb (1992), all of which give greater attention to the shape of the phi-rho-z curve and modify either the equation for stopping power or the factor J which figures in it. Joy and Luo (1989) have proposed a formula for the factor J in Bethe's equation, in which the effect of diminishing electron energy is taken into account. This approach is used by Duncumb. Love and Scott, and later Pouchou and Pichoir, chose instead to replace Bethe's equation by expressions which at higher energies give results close to Bethe's and tend to correct its failure at low energies. The choice of expression for J can strongly affect the accuracy of a given method. In the context of the making of a stopping power model for electron probe analysis, however, this fundamental parameter may be converted into a part of the empirical accommodation, since the energy-loss equation is adjusted after a model for J has been chosen. Each of the three most successful models uses a different expression for J, and in most cases changing the model's expression for J causes the error distribution to broaden.

The methods of Love and Scott and of Duncumb follow the traditional path of the ZAF procedure in which the different parts for atomic number, absorption and fluorescence calculation are clearly differentiated. The PAP method is formally derived in a somewhat different way, and its authors do not explicitly and separately calculate the absorption and atomic number expressions. However, PAP can be easily transcribed into the classical ZAF form, as was done in SWITCH. The explicit expression of the absorption correction factors is useful in practice: many measurements reported in the data set are of interest for the method evaluation, but they are not good examples for choosing adequate conditions of measurement. In particular, one should minimize the absorption correction if this were possible without producing other problems. This, however, may be difficult to judge if the absorption factors are not seen.

Table II: Results with the Absorption Correction Subset:

Name	S	R	Fa	fl	c	J	V	μ	w	II	mean	StD	RMS
f*<0.90, 1060 cases													
azf1-ex	pap	dun	pdh	s	-	zel	-	kh			-.0020	.0406	
f*<0.85, Z1>10 812 cases													
ZAF													
azf13	li	dun	pdh	s	-	bs	-	kh	he	co	+.0041	.0171	.0176
azf14	ni	dun	q	s	-	bs	-	kh	he	co	-.0056	.0184	.0193
azf15	ni	dun	dpl	s	-	bs	-	kh	he	co	-.0074	.0161	.0177
DUNCUMB													
azf18a	ni	dun	dun	s	-	d2	+	kh	he	co	+.0006	.0138	.0138
PAP:													
azf1a	pap	pap	pap	s	-	zel	-	kh	he	co	-.0016	.0134	.0135
azf10	pap	pap	pap	s	-	bs	-	kh	he	co	-.0002	.0186	.0186
PPS:													
azf24	pap	pap	pps	s	-	zel	-	kh	he	co	+.0005	.0150	.0150
SCOTT:													
azf2b	sct	sct	sct	s	-	wil	-	kh	he	co	-.0029	.0131	.0134
azf11	sct	sct	sct	s	-	bs	-	kh	he	co	-.0009	.0151	.0151
PACKWOOD													
azf3	pw	la	pw	s	-	wil	-	kh	he	co	-.0014	.0206	.0207
BROWN:													
azf21a	br	-	br	s	-	bs	-	kh	he	co	-.0130	.0190	.0230
azf12	br	-	br	s	-	wil	-	kh	he	co	-.0130	.0190	.0230

The statistic called root mean square (RMS) in tables II and IV is an indicator of the deviations from zero error. It thus combines information from the mean and the standard deviation which is calculated with respect to the mean. While these two statistics are of interest in the evaluation of the model, the RMS gives the indication useful to the user about the accuracy obtained, in one parameter.

THE FLUORESCENCE SUBSET

The full calculation of the fluorescence contribution requires the following steps:
  Determine which elements produce lines that cause appreciable fluorescence.
  Calculate the absolute intensities generated by these lines.
  Determine the emergent fluorescent contribution due to each line.
  Add all contributions and ratio them to the emergent primary intensity of the line which is
    measured (fluorescence correction factor).
The largest uncertainty is in the selection of the parameters which determine the emission of primary x-rays of both the emitting and the indirectly excited element (notably, the fluorescence yield.)

The excitation of K-lines by K-lines of other elements may cause very strong fluorescence (up to 30% of the primary emission.) However, the uncertainty in the primary emissions is not very significant since strong excitation is only obtained when the atomic numbers of the exciting and the excited elements are close. The same is true for the excitation of L lines by L lines, and of M lines by M lines. When the exciting lines are produced by a level other than that of the fluorescent emission, (e.g., K lines excited by L lines or viceversa), the relative errors in the calculation of primary intensities can be large, but the fluorescent intensities are lower.

Table III shows the cases of fluorescence which affect the result to more than 0.5% contained in our binary data collection, and which constitute the subset.

Table III: Cases of Significant Fluorescence (F>1.005, total:193)

Element Pair:	Emitter:	Excited by:	Cases:	Max. F:
- 13	12 Ka	13 K lines	14	1.08
- 26	24 Ka	26 K lines	56	1.36
- 29	28 Kb	29 Kb only	7	1.03
- 30	29 Ka	30 Kb only	1	1.006
- 42	14 Ka	42 L lines	7	1.009
- 78	28 Ka	78 L lines	4	1.03
- 79	29 Ka	79 L lines	63	1.07
- 13	29 La1	13 K lines	4	1.03
- 23	41 La1	23 K lines	14	1.015
- 26	50 La1	26 K lines	1	1.005
- 25	51 La1	25 K lines	5	1.012
- 47	79 La1	47 K lines	6	1.018
- 51	49 La1	51 L exc. La1	3	1.010
- 47	79 Ma1	47 L lines	8	1.016

A simplified procedure was first proposed by Castaing and modified by Reed and Long(1963), who avoids the full calculation of primary intensities by using simplifications to Cosslett's approach to calculating primary intensities. Recently, Reed (1990) proposed an improved procedure for the important K-K interaction, which is the the third alternative available. Reed's approaches do not consider cases in which only part of the K- or L- emission acts on the measured emission, and it neglects the effects from or on M-lines. These omissions are in most cases inconsequential.

Before entering in details, we compare the errors observed in the fluorescence subset with those observed in the absorption and atomic-number subsets (Table IV). The K-K and L-L fluorescence subset (fluor3) has a substantially smaller error distribution than the first two subsets. Most of the available data set for K - K interaction refers to the element pair 24-26, i.e., in the region of the periodic table for which the fluorescence corrections were designed and adapted. The set representing the element pair 12-13, which apparently was not used in establishing the correction procedures, is therefore of some interest.

Table IV: Comparison of Errors in the three Subsets

Subset	Entries	Mean	St. Dev.	RMS
ABSORPTION	1060	- 0.0005	0.0133	0.0133
ATOMIC NUMBER	98	+ 0.0056	0.0171	0.0176
FLUORESCENCE:				
entire fluor.subset	193	- 0.0016	0.0143	0.0143
Id.less At Nrs.51-25	186	- 0.0037	0.0083	0.0090
K-K and L-L only	81	+ 0.0002	0.0079	0.0079
one data pair excluded	80	- 0.0003	0.0066	0.0066
K-L, L-K, M-L	106	+ 0.0118	0.0146	0.0157

No benefit can be derived from the consideration of data in which the total errors are much larger than the expected fluorescent contribution. For this reason we have excluded the data from the element pair 51-25, as well as one data pair inTable IV. The fluorescence produced in pair 29-30, and in the cases of 14-42 and 50-26 could be ignored; several other cases are of marginal usefulness. Clearly, the experimental evidence on fluorescent excitation of L and M lines is weak and it will not be considered here in detail. In the interaction of different shells, the complete method is unreliable, and Reed's methods are recommended where applicable.

The weight of K-lines is not a significant source of error, particularly for low atomic numbers, since the relative intensity of the $K\beta$ lines falls rapidly below atomic number 20, being practically zero for atomic numbers below 12.

For the Mg-Al case, the choice of model for the fluorescent yield, ω, is important. Experimental values in this region of atomic numbers are scanty, and various models for ω produce a wide range of values. The fluorescent yield of Heinrich (unpublished) was an attempt to fit a curve to newer experimental values, and is probably too high for low atomic numbers.

Model	Mean error	S.D.	Root Mean Square
No fluor	- 0.0002	0.0065	0.0062
Wapstra	+ 0.0032	0.0057	0.0081
Heinrich	+ 0.0097	0.0054	0.0110

Surprisingly, the smallest errors are obtained when the fluorescence procedure is omitted. To the extent that the data (from a single source) can be trusted, they suggest that all proposed algorithms for fluorescent yield overestimate the yield for aluminum. Further measurements of the fluorescent yield for elements of atomic number below 26 would be useful.

Numerical Integration

Before modern on-line computation devices became available, the methods of data reduction were limited by the amount of calculation they required. The models for depth distribution had to be chosen among functions whose Laplace transform could be expressed formally. Such is the case in virtually every proposed method, from Philibert to Pouchou. However, the digital computer is built to perform at great speed simple operations in binary code. Hence, the formal integration is unnecessary; in fact, the numerical integration of some of the current models consumes less time than the calculation of the formal integration equivalents. Several proposed models lead to occasional failures in the computation, probably due to the use of numerical approximations of limited accuracy, such as that proposed for the error function, or to the inaccuracies of binary calculation. We therefore replace in most cases the formal integration formulae by numerical integration. A simple addition of 15-25 strips representing y.Δx, is quite satisfactory. The logarithmic integral over the energy interval of x-ray excitation can also be replaced advantageously by numerical integration.

Phi-Rho-Z Methods:

As can be deduced from Tables I and II, the performance of the phi-rho-z methods that were tested is inferior to the methods discussed above.The averaging of parameters of elements for the multi-element targets is often done haphazardly. Since tracer experiments for multi-element or binary matrices were never performed, no experimental proof of the averaging methods can be given. I believe that the atomic-number correction in the phi-rho-z methods could be improved to make them competitive with other methods.

Speed of Calculation:

Speed of calculation is nowadays a factor of interest only when a large number of measurements is converted in real time, as in quantitative area scans. In most instances, the time consumed in data collection exceeds that for the correction procedure. SWITCH has not been optimized for maximum speed, nor have the speeds of different procedures been compared systematically. Preliminary observations indicate substantial differences in the speed in which these calculations are processed, depending on the method used.

REFERENCES

Bambynek W et al 1972 Rev Mod Phys 44 4 716

Bastin GF and Heijligers HGM (1991) Electron Probe Quantitation eds Heinrich KFJ and Newbury DE (New York: Plenum Press) 145, 163

Berger MJ and Seltzer SM 1964 Nat. Research Council Publ.(Washington: National Acad. of Sciences) 205

Bloch J 1933 Z Phys 81 363

Brown 1991 Electron Probe Quantitation eds Heinrich KFJ and Newbury DE (New York:Plenum Press) 77

Burhop EHS 1955, J Phys Radium 16 625

Duncumb P and Reed SJB 1968, Quantitative Electron Probe Microanalysis, NBS Special Publ 298 (Washington DC: US National Bureau of Standards) 133

Duncumb P et al. 1969 Proc. 5th ICXOM (Berlin:Springer)146

Duncumb P 1992 (personal communication)

Duncumb P and Reed SJB 1981, reproduced in Heinrich KFJ, Electron Beam X-Ray Microanalysis (New York: Van Nostrand-Reinhold) 249

Green M 1962 Thesis (Cambridge U UK)

Gryzinski M 1965 Phys Rev 138 A336

Heinrich KFJ (1966), The Electron Microprobe, eds McKinley TD et al (New York: J. Wiley & Sons) 296

Heinrich KFJ 1970 NBS Tech Note 521

Heinrich KFJ and Yakowitz H 1975 Anal Chem 47 2408

Heinrich KFJ 1985 Microbeam Analysis (San Francisco: San Francisco Press) 79

Heinrich KFJ 1986 Proc ICXOM 11, Brown JD ed, (London Ont) 1986 67

Heinrich KFJ 1987a Microbeam Analysis (San Francisco: San Francisco Press) 23

Heinrich KFJ 1987b Microbeam Analysis (San Francisco: San Francisco Press) 24

Joy DC and Luo S 1989 Scanning 11 176

Packwood R 1991 Electron Probe Quantitation eds Heinrich KFJ and Newbury DE (New York: Plenum Press) 83

Philibert J and Tixier R 1968 Quantitative Electron Probe Microanalysis ed Heinrich KFJ (Washington: NBS) Spec Publ 298 13

Pouchou JL and Pichoir F 1991 Electron Probe Quantitation eds Heinrich KFJ and Newbury DE (New York: Plenum Press) 31

Pouchou JL 1979, personal communication

Reed SJB and Long JVPI 1963 ICXOM 3 eds H.H. Pattee et al (New York: Academic Press) 317.

Reed SJB 1990 Microbeam Analysis 109

Scott VD and Love G 1991 Electron Probe Quantitation eds Heinrich KFJ and Newbury D E (New York: Plenum Press) 19

Small J et al 1987 J Appl Phys 61 (2) 459

Springer G 1967 N Jahrb f Mineralogie, Monatshefte 9/10 304

Wilson MV 1941 Phys Rev 60 749

Zeller C, cited by Ruste J and Gantois M 1975 J Phys D, Appl Phys 8 872.

Inst. Phys. Conf. Ser. No 130: Chapter 2
Paper presented at Int. Congr. X-ray Optics and Microanalysis, Manchester, 1992

123

Evaluation and description of a new correction procedure for quantitative electron probe microanalysis

Claude MERLET

C.G.G, CNRS, Université de Montpellier II, Pl. E. Bataillon, 34095 Montpellier France.

ABSTRACT: The method of correction is based on a new description of the X-ray distribution depth ($\Phi(\rho z)$), which combines atomic number, and absorption correction [ZA]. Analytical expressions of required parameters for $\Phi(\rho z)$ are proposed: $\Phi(0)$, the surface ionisation; $\Phi(m)$ the maximum of the distribution; ρzm the position of the maximum, and the X-Ray range ρz_x. The accuracy of the procedure is demonstrated using a large number of measurements from the literature, particularly of light elements.

INTRODUCTION

As shown by Castaing (1951), the atomic number and absorption effects can be expressed by a single equation describing the intensity I_A^ϵ emitted from the element A in homogeneous sample:

$$I_A^\epsilon = cst.C_A \int_0^\infty \Phi_e(\rho z).\exp(-\chi_e \rho z)d\rho z \tag{1}$$

where C_A is the weight fraction of element A, ρz is the mass depth in the sample, $\chi = \mu/\rho \cosec\theta$, with μ/ρ being the mass absorption coefficient and θ the X-ray take-off angle, $\Phi_e(\rho z)$ is the depth distribution of X-ray. If there is no significant fluorescence effect, the general correction, which combines the atomic number and the absorption correction [ZA] is:

$$K_A = \frac{I_A^\epsilon}{I_A^s} = C_A[ZA] = C_A \frac{\int_0^\infty \Phi_e(\rho z).\exp(-\chi_e \rho z)d\rho z}{\int_0^\infty \Phi_s(\rho z).\exp(-\chi_s \rho z)d\rho z} \tag{2}$$

where $\Phi_s(\rho z)$ is the depth distribution of a standard. If the $\Phi(\rho z)$ curves are known, then the concentration C_A can be exactly calculated from the intensity ratio. The aim of this paper is to present a new correction procedure based on an accurate $\Phi(\rho z)$ distribution.

THE $\Phi(\rho z)$ DISTRIBUTION.

In this model (Merlet, 1992a), the X-ray distribution is approximated by a double partial Gaussian profile ($\Phi 1, \Phi 2$; Fig. 1) described by three coordinates: - at the surface, $[0, \Phi(0)]$, - at the maximum of the $\Phi(\rho z)$ curve, $[\rho z_m, \Phi(m)]$, - at the X-ray range, $[\rho z_x, \Phi(m)/100]$. Thus $\Phi(\rho z)$ may be expressed as:

for $\rho z \in [0, \rho z_m]$, $\quad \Phi 1 = \Phi(m)\exp\left\{-\left(\frac{\rho z - \rho z_m}{\beta}\right)^2\right\}$ $\quad$ with $\quad \beta = \rho z_m/[\ln(\Phi(m)/\Phi(0))]^{0.5}$ $\tag{3}$

and for $\rho z \in [\rho z_m, \rho z_x]$, $\Phi 2 = \Phi(m) \exp\left\{-\left(\frac{\rho z - \rho z_m}{\alpha}\right)^2\right\}$ with $\alpha = 0.46598(\rho z_x - \rho z_m)$

The double Gaussian profile was chosen because, it varies in a realistic way from light to heavy targets, as well as from high to low overvoltage (Fig.2).

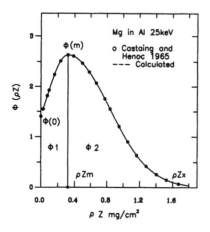

Fig. 1. $\Phi(\rho z)$ curve using the double Gaussian expression.

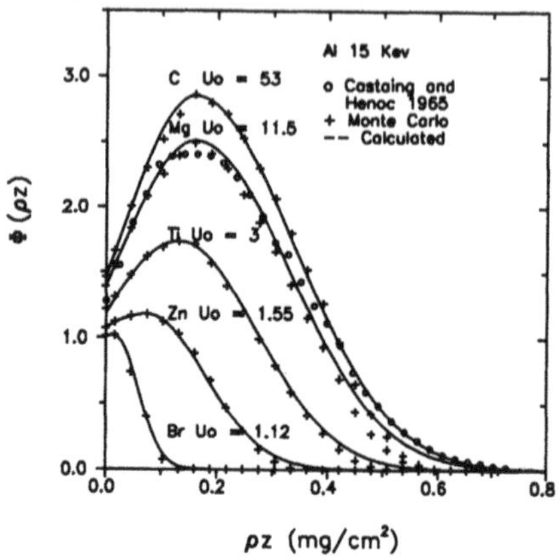

Fig. 2. $\Phi(\rho z)$ curves for various overvoltages using the double Gaussian expression.

Using expressions 3, equation 2 becomes:

$$[ZA] = \frac{\Phi(m)_e \left[\beta \exp(c)\left\{erf\left(\frac{\rho z_1}{\beta}\right) + erf\left(\frac{\chi_e \beta}{2}\right)\right\} + \alpha \exp(d)\left\{1 - erf\left(\frac{\chi_e \alpha}{2}\right)\right\}\right]_e}{\Phi(m)_s \left[\beta \exp(c)\left\{erf\left(\frac{\rho z_1}{\beta}\right) + erf\left(\frac{\chi_e \beta}{2}\right)\right\} + \alpha \exp(d)\left\{1 - erf\left(\frac{\chi_e \alpha}{2}\right)\right\}\right]_s} \quad (4)$$

with $c = (\chi \beta/2)^2 - \rho z_m \chi$, $\rho z_1 = \rho z_m - \chi \beta^2/2$ and $d = (\chi_e \alpha/2)^2 - \rho z_m \chi_e$

Similarly the atomic number correction factor [Z] is:

$$[Z] = \frac{\int_0^\infty \Phi_e(\rho z).d\rho z}{\int_0^\infty \Phi_s(\rho z).d\rho z} = \frac{\Phi(m)_e \left[\beta \cdot erf\left(\frac{\rho z_m}{\beta}\right) + \alpha\right]_e}{\Phi(m)_s \left[\beta \cdot erf\left(\frac{\rho z_m}{\beta}\right) + \alpha\right]_s} \quad (5)$$

THE $\Phi(\rho z)$ PARAMETERS.

In order to develop exact analytical expressions of the $\Phi(\rho z)$ parameters, experimental data (tracer measurements, transmitted and back-scattered electrons) and Monte Carlo simulations have been used.

The X-ray range ρz_x and the position of the maximum ρz_m have been obtained by fitting the Monte Carlo simulations. The proposed expressions are:

$$\rho z_x = \frac{A\left(E_o^{(2.2 - 0.0166E_o)} - E_c^{(2.2 - 0.0166E_o)}\right)\left(1 + 0.06[(0.05Z)^3 - 1]/U_0^{0.25E_o}\right)}{(1.078 - 0.015Z^{0.7})^{(1.2 - 0.04E_o)}} \quad (6)$$

$$\rho z_m = \rho z_x (0.11 + 0.41 \exp(-(Z/12.75)^{3/4}))/(1 + 80/(Z U_o^4)) \quad (7)$$

with $A = 0.001845(2.6 - 0.216E_o - 0.015E_o^2 - 0.000387E_o^3 + 0.0000501E_o^4)$

where Eo is the energy of incident electrons, Ec the energy of level c, and Z the atomic number. The parameter $\Phi(m)$ represents the amplitude of $\Phi(\rho z)$ at complete diffusion, and the analytical expression obtained (Merlet, in preparation) is:

$$\Phi(m) = \tau Q(U_d)/Q(U_o) + a.[0.28(1.2 - 1/Z^{0.5})b_1 + 0.258Z^{0.5}b_2] \tag{8}$$

with $b_i = (U_d/U_o)^{d_i}\ln(U_d)/\ln(U_o)(1/d_i)[1 - (1 - 1/U_o^{d_i})/(d_i\ln(U_o))]$, $\quad d_1 = 0.08$, $\quad d_2 = 6.28 - 20/Z$,

$a = 5(1 - 1/Z^{0.8})(E_o/30)^{p(Z)}$, $\quad p(Z) = -0.25 + 0.0787Z^{0.3}$ $\quad U_d = 1/[1 - (1 - 1/U_o^{1.61})/Z^{0.7}/(1 + 10/(ZU_o^4)]^{0.62}$

where τ is the transmission coefficient of Zeller and Ruste (1976) (with $t = \rho z_m/\rho z_x$), $Q(U) = \ln(U)/U^{0.92}$ the ionisation cross-section and $U = E/Ec$ the overvoltage.

The parameter $\Phi(0)$ represents the amplitude of $\Phi(\rho z)$ at the surface of the sample. The analytical expression of $\Phi(0)$ (Merlet, 1992b) has been obtained from a new expression of the energetic distribution of the backscattered electrons $d\eta/dU$, and may be written as:

$$\Phi(0) = 1 + a.[0.27b_1 + (1.1 + 5/Z)(b_2 - 1.1b_3)] \tag{9}$$

with $b_i = (1/d_i)[1 - (1 - 1/U_o^{d_i})/(d_i\ln(U_o))]$ $\quad d_1 = 0.02Z - m + 1$, $\quad d_2 = 0.1Z - m + 1$, $\quad d_3 = 0.4Z - m + 1$,

$a = 1.87Z(-0.00391\ln(Z + 1) + 0.00721[\ln(Z + 1)]^2 - 0.001067[\ln(Z + 1)]^3)(E_o/30)^{p(Z)}$

$p(Z) = -0.25 + 0.0787Z^{0.3}$ with m=1 for the K and L lines and m=.8 for the M line.

For multi-components specimens, $\Phi(m)$ and ρz_x are atomicaly averaged; $\Phi(0)$ and ρz_m are calculated using the average atomic number $(\overline{Z} = \Sigma C_i Z_i)$.

PERFORMANCE OF THE MODEL

In order to check the validity of the calculated $\Phi(\rho z)$ distribution, tests were carried out at various excitation energies on samples presenting high mass absorption coefficients (MAC) and wide range of atomic numbers.

For that purpose, the light elements measurements achieved by Bastin and Heijligers (1984,1986,1990) have been chosen. Table 1, Fig. 3 show that our procedure performs remarkably well up to accelerating voltages as high as 40 kV, and with high MAC's. The best results are obtained with the Oxygen measurements, in which, the discrepancy between the predicted and the experimentally K-ratio never exceeds 5% (relative). It is worth noting that equivalent RMS values are obtained at low and very high MAC, which is a proof of the accuracy of the $\Phi(\rho z)$ distribution. The MAC values selected for our study are close to those used by Bastin and Heijligers, and Pouchou and Pichoir.

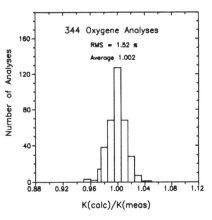

Fig. 3. Histogram obtained with our correction procedure on the data base of 344 Oxygen analyses.

Tests metal analyses of Carbides, Oxides and Borides, which show the ability of the model at very low overvoltages, are similar to those of Bastin (table 2).

TABLE 1. Relative RMS values and averages for calculated/measured K-ratio obtained with various correction programs for ultra-light elements analysis (Oxygen, Boron, Carbon) measured by Bastin and Heijligers (1984,1986,1990)

Procedure	Oxides 344 analyses		Borides 153 analyses		Carbides 117 analyses	
	Average	R.M.S.%	Average	R.M.S.%	Average	R.M.S.%
this model	1.002	1.52	1.004	3.55	0.997	3.36
Bastin et al. (1990)	0.999	2.48	1.008	5.17	0.998	3.86
Pouchou et al. (1986)			1.007	3.95		
Pouchou et al. (1988)			1.011	4.76		
Scott et al. (1992)			1.017	5.3	0.997	3.8
Rehbach et al. (1988)					0.964	5.76
Ruste et al. (1977)			1.039	9.59	0.946	11.94

TABLE 2. Relative RMS values and averages for calculated/measured K-ratio obtained with this model and Bastin's model for metal analysis in Carbides, Oxides, and Borides measured by Bastin and Heijligers (1984,1986,1990), and with Heinrich (1986) MAC's.

Procedure	Oxides 132 analyses		Borides 196 analyses		Carbides 145 analyses	
	Average	R.M.S.%	Average	R.M.S.%	Average	R.M.S.%
this model	1.006	1.67	1.006	2.50	1.000	1.50
Bastin et al. (1990)	0.998	1.93	1.024	2.43	1.011	1.48

REFERENCES

G.F. Bastin and H.J.M. Heijligers 1984, Report Eindhoven Univ. Of Techn. Netherlands, ISBN 90-6819-002-4
G.F. Bastin and H.J.M. Heijligers 1986, Report Eindhoven Univ. Of Techn. Netherlands, ISBN 90-6819-006-7
G.F. Bastin and H.J.M. Heijligers 1990, Report Eindhoven Univ. Of Techn. Netherlands, ISBN 90-6819-012-1
R. Castaing 1951, Thesis, University of Paris, publication ONERA n°55
R. Castaing and J. Henoc 1965, Proc. ICXOM 4, edited by R. Castaing, P. Deschamps, J. Philibert, Hermann, Paris 120
K.F.J. Heinrich 1986, Proc. ICXOM 11, edited by J.D. Brown and R.H Packwood, 67
C. Merlet 1992a, X-Ray Spectrom., (in press)
C. Merlet 1992b, Mikrochim. Acta, suppl. 12,107
J.L. Pouchou and F. Pichoir 1986, Proc. ICXOM 11, edited by J.D. Brown and R.H Packwood, 249
J.L. Pouchou and F. Pichoir 1988, Microbeam Analysis, edited by D.E. Newbury, San Francisco Press, 315
W. Rehbach and P. Karduck 1988, Microbeam Analysis, edited by D.E. Newbury, San Francisco Press, 285
J. Ruste and C. Zeller 1977, C.R. Acad. Sci. Paris B284: 507
V.D. Scott and G. Love 1992, X-Ray Spectrom., 21, 27
C. Zeller and J. Ruste 1976, Rev. Phys. App., 4, 441

Inst. Phys. Conf. Ser. No 130: Chapter 2
Paper presented at Int. Congr. X-ray Optics and Microanalysis, Manchester, 1992

127

The variable voltage method for calculating the absorption correction for soft X-rays

A.P. Mackenzie

IRC in Superconductivity, Madingley Road, Cambridge CB3 0HE

ABSTRACT: The use of a variable voltage method to improve the absorption correction for soft x-ray analysis has been studied. It is shown that the sample composition and the absorption correction can be deduced simultaneously, and that matrix-dependent changes to mass absorption coefficients are automatically taken into account. No more than three accelerating voltages need be used, and the method is insensitive to small errors in the shape of $\varphi(\rho z)$.

1. INTRODUCTION

One of the major areas of study over the past ten years in electron probe microanalysis (EPMA) has been analysis using soft x-rays. The use of soft x-rays is necessary for studying light elements, but it can also be useful when studying heavier elements, because lower accelerating voltages with better spatial and depth resolution may be employed, and fluorescence effects are small. The main difficulty is calculating the very large absorption corrections which often exist. For this purpose the distribution of x-rays generated in the sample as a function of mass depth ('$\varphi(\rho z)$') must be known. In the language of 'ZAF' correction procedures, the $\varphi(\rho z)$ function can be used to calculate the product of the atomic number, or 'Z' correction with the absorption, or 'A' correction:

$$ZA = \int_0^\infty \varphi(\rho z)e^{-\chi \rho z}d(\rho z) \tag{1}$$

where ρz is the mass depth and χ is the product of the mass absorption coefficient with the cosecant of the x-ray takeoff angle.

A great deal of progress has been made recently in producing analytical models which give realistic $\varphi(\rho z)$ distributions and thus improve the accuracy of the absorption correction for soft x-rays (Packwood and Brown, 1981; Pouchou and Pichoir, 1987; Bastin, Heijligers and van Loo, 1986). The ZA product as calculated via eq. 1 is combined with a fluorescence (F) correction (Reed 1965, 1990) to obtain the absolute concentrations via an iterative procedure in the standard manner:

$$c_{n+1,i} = ZAF(\{c_n\})k_i \tag{2}$$

where the subscript n denotes the cycle of the iteration, i denotes the ith element in the specimen and k is the experimental specimen/standard intensity ratio. The most serious remaining source of error is the absorption parameter, χ. Mass absorption coefficients for soft x-rays are not known accurately, and although the tabulated values have improved, quite large errors are still possible. A number of authors (Duncumb and Melford, 1966; Bastin, Heijligers and van Loo, 1986; Pouchou and Pichoir, 1988a) have pointed out that measurements taken at variable accelerating voltages can be used to check the accuracy of the absorption correction, because the average depth of x-ray production (and hence the amount of absorption) in the sample is a strong function of the primary electron energy. The purpose of this paper is to show how the variable voltage method can be used to yield an absorption-corrected concentration directly, and to point out that this approach has some key advantages over methods based on tabulated mass absorption coefficients. The sensitivity of the method to details of the $\varphi(\rho z)$ model used is also discussed.

2. VARIABLE VOLTAGE METHOD

In the variable voltage method of Pouchou and Pichoir (1988a), it is assumed that the $\varphi(\rho z)$ model used is accurate, and that any error that occurs when variable voltage data are modelled is due to an inaccurate mass absorption coefficient. The 'correct' mass absorption coefficient is obtained by minimising the sum of squared residuals:

$$S = \sum_{v=1}^{n} (I_v - \alpha J_v(\chi))^2 \tag{3}$$

in which the subscript v denotes each accelerating voltage used, I is the experimental intensity data, and J is the intensity predicted by the model within an unknown constant factor α. Pouchou and Pichoir referred to α as the machine efficiency, but it is really the product of machine efficiency with the concentration and several theoretical factors such as ionisation cross-sections and fluorescence yields which are not calculated by the analysis model. The correct value of α should also minimise S, so the sum is transformed by solving $\delta S/\delta\alpha = 0$ and substituting the resulting expression for α into eq. 3. The 'best fit' value of χ is obtained by solving $\delta S/\delta\chi = 0$ using Newton iteration.

Pouchou and Pichoir concentrated on the fitted χ value, but the value of α is also important. Consider the situation in which a sample of unknown composition is being analysed using a standard of known composition, and the mass absorption coefficients of the x-rays in both sample and standard need to be measured. Values of χ_s and α_s are obtained using the data from the standard, and then χ and α values are fitted for the specimen at each stage of the main iteration. For a given element, machine efficiencies, fluorescence yields and ionisation cross-sections should cancel between sample and standard, so at the nth iteration $\alpha_n/\alpha_s = 1$ if the concentration of the element concerned is correctly measured in the sample. If $\alpha_n/\alpha_s \neq 1$, the concentration for the next step in the iteration can be calculated using $c_{n+1} = \alpha_s/\alpha_n \cdot c_n$. Once the iteration has converged, the variable voltage method has produced a value for the concentration in the specimen which is insensitive to an erroneous mass absorption coefficient, since the magnitude of the absorption correction is obtained as well.

3. POTENTIAL ADVANTAGES OF THE METHOD

There is a strong case for adopting this kind of procedure routinely for soft x-ray analysis, because tabulated mass absorption coefficients are unlikely to be completely satisfactory. It is well known that the local chemical environment can have large effects on the peak position and line shape of soft x-rays, because they result from transitions involving outer atomic electrons.

Data Set	μ (cm²/g)	c (wt%)
a) Full	6232	16.39
(4, 12, 30 kV)	6115	16.07
(6, 14, 20 kV)	6280	16.56
b) (20, 25, 30 kV)	5758	14.97

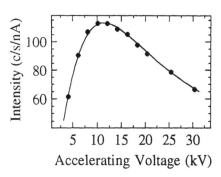

Accelerating Voltage (kV)

Table 1. Fitted mass absorption coefficients and O concentrations based on the data shown in figure 1, standardised to Fe_2O_3. The fitted parameters based on all eleven voltages are compared with the results of using limited subsets of the data are also shown.

Fig. 1. Variable voltage data for O K_α analyses of $YBa_2Cu_3O_{7-x}$. The solid line is the best fit obtained to the data using the PAP model (Pouchou and Pichoir 1987).

Likewise, low energy mass absorption coefficients can be expected to be matrix-dependent, so there is no reason to believe that calculating $\chi(\text{compound}) = \Sigma\, c_i\mu_i\text{cosec}\psi$ in the usual way will always be reliable. Absorption anomalies affecting the Ni L_α line have been reported by Pouchou and Pichoir (1985), and have also been observed by the author in analyses of CuO and Cu_2O using the Cu L_α line with pure Cu as the standard. One of the great advantages of the variable voltage method is that $\chi(\text{compound})$ is obtained directly, so this problem is overcome. In light element analysis, the fitted value of $\chi(\text{compound})$ is also much less sensitive to errors in the measured concentrations of the cations than a calculated value. A further advantage is that if the sample is tilted slightly (Mackenzie (1990) has pointed out that even tilts of 1° can be important in conventional analyses) the change in the path length is fitted out as a change in χ, and so does not affect the analysed composition.

If the variable voltage method is to be used routinely, it is desirable to take data at as few voltages as possible, in order to save time. The feasibility of using only three accelerating voltages is demonstrated using experimental O K_α data from one of the high temperature superconductors. The data are shown in figure 1, and the fitted parameters in table 1. The results of fits using the full data set and two separate subsets are shown in table 1a), and it is seen that the fitted parameters are insensitive to the use of only three accelerating voltages. Some caution is required, however, because if the measured intensity is nearly linear with voltage over the range used, large errors in the two fitted parameters can sometimes cancel to produce a good fit, as shown by the example in table 1b). This has been avoided in table 1a) by choosing the two sets of three voltages so that the fit through them has significant curvature.

3. SENSITIVITY TO THE SHAPE OF $\varphi(\rho z)$.

The main assumption made in the variable voltage method is that the $\varphi(\rho z)$ model is accurate. In practice, the most successful approach to $\varphi(\rho z)$ modelling has been to calculate the area of the curve (effectively the Z correction) and then to fit a curve with a chosen functional form using this area and a few other calculated parameters as constraints. Parabolae, exponentials and a modified Gaussian have all been used. Several of the most widely-used models

(Pouchou and Pichoir, 1987, 1988b; Bastin and Heijligers,1990) now employ the same calculation of the area (which appears to be satisfactory on the evidence of variable voltage measurements on systems for which absorption is low), so the only difference between them is a small shape change due to the different functions used in their construction. Since the tracer methods used to determine $\varphi(\rho z)$ curves experimentally are themselves sensitive to any error in χ, there seems to be little prospect of being able to determine which model predicts the shape of $\varphi(\rho z)$ for soft x-rays best. For this reason, a series of numerical tests was carried out to investigate the effects of shape changes in $\varphi(\rho z)$ on the variable voltage method. A series of deliberately distorted distributions was constructed from the PAP model by making changes of ±10% in the depth of maximum ionisation and maximum depth of ionisation, while retaining the original values for area and surface ionisation. Oxygen K_α data from several conducting oxides and cuprate superconductors (Mackenzie 1991) were analysed using the variable voltage method incorporating these distributions, and it was found that the results showed a spread only of approximately 2%, and no significant difference in the quality of the fit could be distinguished. These tests show that the compositions deduced by the variable voltage method are insensitive to small errors in the shape of $\varphi(\rho z)$.

4. CONCLUSIONS

In conclusion, the variable voltage method has several key advantages for the analysis of soft x-ray data. Since absorption in the specimen is calculated as part of the analysis, the chance of large errors from inaccurately calculated χ values is removed. Also, any specimen-dependent changes to the mass absorption coefficients are taken into account, and the concentration is fitted automatically. No more than three voltages need be used, as long as they are chosen appropriately.

REFERENCES

Bastin G F, Heijligers H J M and van Loo F J J 1986 Scanning **8**, 45
Bastin G F and Heijligers H J M 1990 Scanning **12**, 225
Duncumb P and Melford D A 1966 Proc IVth Conference on X-ray Optics and Microanalysis, ed. R. Castaing, (Paris: Hermann), pp 240-53
Heinrich K F J 1987 in X-ray Optics and Microanalysis, eds. J.D. Brown and R.H. Packwood, (Univ. of Western Ontario, Canada), pp 67-119
Mackenzie A P 1990 Proc. XIIth ICEM (San Francisco Press), pp 220-1
Mackenzie A P 1991 Physica C **178**, 365
Packwood R H and Brown J D 1981 X-ray Spect. **10**, 138
Pouchou J L and Pichoir F M A 1985 J. Microsc. Spectrosc. Electron. **10**, 291
Pouchou J L and Pichoir F M A 1987 in X-ray Optics and Microanalysis, eds. J D Brown and R H Packwood, (Univ. of Western Ontario, Canada), pp 249-53
Pouchou J L and Pichoir F M A 1988a in Microbeam Analysis, ed. D E Newbury, (San Francisco Press) , pp315-8
Pouchou J L and Pichoir F M A 1988b in Microbeam Analysis, ed. D E Newbury, (San Francisco Press), pp319-22
Reed S J B 1965 Brit. Journ. Appl. Phys. **16**, 913
Reed S J B 1990 in Microbeam Analysis, ed J R Michael and P Ingrams (San Francisco Press) pp109-14

Inst. Phys. Conf. Ser. No 130: Chapter 2
Paper presented at Int. Congr. X-ray Optics and Microanalysis, Manchester, 1992

Measurements and calculations of $\varnothing(0)$

J. D. Brown and A. Chan

Department of Materials Engineering, Faculty of Engineering Science,
The University of Western Ontario, London, Canada, N6A 5B9

ABSTRACT:A method is described for calculating $\varnothing(0)$ based on literature expressions for the ionization cross section, backscattered electron coefficients and backscattered electron energy distributions. Calculated values are compared with measured ratios based on thin films deposited on a variety of substrates.

1. INTRODUCTION

The value of $\varnothing(0)$ is a critical parameter for the equations of $\varnothing(\rho z)$ curves which accurately model both the shape and magnitude of this function. The definition of $\varnothing(0)$ is the ratio of intensity generated in a thin layer of an element deposited on a substrate divided by the intensity from the same layer isolated in space. The value of $\varnothing(0)$ is always greater than 1 since electrons backscattered from the substrate add to the generated intensity, a contribution which is absent from the isolated layer.

2. CALCULATION OF $\varnothing(0)$

Figure 1 illustrates electron trajectories which contribute to the $\varnothing(0)$ value. Duncumb and Melford(1966) first suggested that $\varnothing(0)$ could be calculated from the number, energy distribution and paths of backscattered electrons. They did generate some values based on the limited information then available. $\varnothing(0)$ can also be derived from Monte Carlo simulations of the trajectories of electrons entering solids (Karduck and Rehbach (1988)). The Monte Carlo method requires considerable computer time in order to follow a statistically significant number of electrons. For that reason a more rapid method of calculation is desirable so that individual values can be generated in a reasonable time (<1 second) for routine quantitative calculations. For that reason we have returned to the approach of Duncumb and Melford(1966) using the more extensive data and formulae now available.

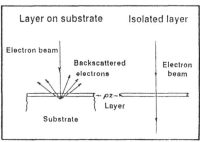

Figure 1 Electron paths through isolated and supported layers

$\varnothing(0)$ can be written as the number of ionizations resulting from the backscattered electrons, I_b, plus the ionizations from the direct beam, I_d, divided by the number of ionizations from the direct beam,

$$\varnothing(0) = (I_b + I_d) / I_d$$

I_d is proportional to the number of electrons in the beam, N_e, times the ionization cross section for the beam energy, $Q(E_0)$, times the layer thickness, ρz, since these electrons go straight through the thin layer i.e.,

$$I_d \propto N_e \, Q(E_0) \, \rho z.$$

I_b is the product of the number of backscattered electrons, N_e, the energy of those electrons, dN/dE, the ionization cross section, $Q(E)$, and the average length of path back through the thin layer, ρz. Using these parameters the contribution due to backscattered electrons is

$$I_b = \eta N_e \overline{\rho z} \int_{E_0}^{E_c} Q(E) \, \frac{dN_e}{dE} \cdot dE$$

A number of expressions for the total backscatter coefficient have been published in the literature (Reuter(1972), Heinrich(1981), Love and Scott (1978), Pichou and Pichoir(1986). Only the expression of Love and Scott has a dependence on electron energy which, although small, seems real based on experimental measurements. For that reason, this expression is used in the calculations below. Similarly, expressions for ionization cross section can also be found (Figure 2). Since a ratio is used in calculating Ø(0) only the energy dependence is important in these expressions which is fortunate since the absolute value for the cross section is somewhat uncertain. The back scattered energy distribution is the third parameter which is necessary in the calculation of Ø(0). Some early measurements were made by Kulenkampff and Spyra(1954) and more recent ones by Bishop (1966). These latter measurements are the ones most frequently quoted in discussions of the backscatter electron energy distributions. An expression to fit these data has been constructed by Cryzewski and Szymanski(1982). The predicted distributions are shown in Figure 3. It is their expression which has been

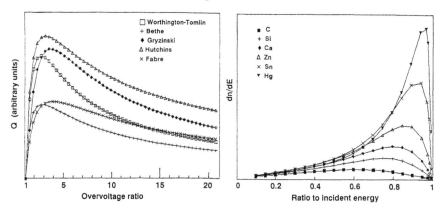

Figure 2 Plots of ionization cross sections

Figure 3 Backscattered electron energy distributions

used in the calculations. The final parameter which is necessary in the calculation is the average path length, ρz, of the backscattered electrons through the thin layer. Data in the papers of Niedrig(1984) and Fathers and Rez(1984) suggest that the backscattered electron angular distribution follows quite closely a cosine law for both low and high atomic number substrates which have very different energy distributions. Based on this fact the average path through the thin layer is π times ρz. Substituting the above expressions,

$$\emptyset(0) = \frac{\eta N_e \pi \rho z \int_{E_0}^{E_c} Q(E) \frac{dN_e}{dE} \cdot dE + N_e Q(E) \rho z}{N_e Q(E) \rho z}$$

A computer program was written to calculate $\emptyset(0)$ based on the above equation. The integration over the energy range for backscattered electrons was done by numeric integration in 100 steps from the absorption edge energy to the electron beam energy. Values calculated from this program are given in Table 1 using the Worthington-Tomlin ionization cross section.

Table 1. Calculated values of $\emptyset(0)$.

Line	Matrix	Electron energy, keV			
		20	23	26	30
Si Kα	Si	1.472	1.472	1.470	1.469
Ni Kα	Cu	1.576	1.630	1.667	1.682
In Lα	Ag	2.105	2.139	2.161	2.168
Bi Lα	Au	1.878	2.078	2.171	2.202

3. MEASUREMENTS OF $\emptyset(0)$

The basic definition of $\emptyset(0)$ is the ratio of intensties of a thin layer deposited on a substrate, Is, divided by the intensity from an isolated layer, Io i.e. $\emptyset(0) = I_s / I_0$.
To make and measure intensities from an isolated layer is difficult but relative measurements of $\emptyset(0)$ are rather simpler to make. If the same thin layer is deposited on a number of different substrates then measurement of the intensity from one substrate relative to another gives immediately the ratio of the $\emptyset(0)$ values. Thus if the substrates are A and B,
$$\emptyset(0)_A/\emptyset(0)_B = (I_{sA}/I_0)/(I_{sB}/I_0) = I_{sA} / I_{sB}.$$
Measurements were made for thin layers of silicon, nickel, indium and bismuth on carefully polished substrates of carbon, silicon, copper, silver and gold for electron energies from 20 to 30 keV. The measurements were performed on a JEOL JXA-8600 Superprobe under computer control. Values of ratios of $\emptyset(0)$ for one substrate relative to the value for a substrate of copper are listed in Table 2. For example, the first line of data represents the intensity from a thin layer of Si on carbon relative to the same layer on copper. Also listed in that table are ratios calculated from the above equation again using the Worthington-Tomlin ionization cross section.

4. DISCUSSION OF THE RESULTS

The values of $\emptyset(0)$ shown in table 1 are typical of the values calculated using other expressions for either the backscatter coefficient or the ionization cross section. Specific values can vary by about 10% depending on the specific equations chosen. In general, the $\emptyset(0)$ value increases with increasing electron energy and with increasing substrate atomic number. The exception is a tendency for $\emptyset(0)$ to become constant or even decrease for large overvoltage ratios, especially for low atomic number substrates. The measured $\emptyset(0)$ ratios in Table 2 have an accuracy of about 5%. Thus the

Table 2. Comparison of Measured and Calculated Ø(0) Values.

| Substrate | | Electron energy, keV | | | | | |
| | | 20 | | 26 | | 30 | |
Ratio	Tracer	Meas.	Calc.	Meas.	Calc.	Meas.	Calc.
C/Cu	Si Kα	0.66	0.638	0.63	0.639	0.63	0.639
Si/Cu	Ni Kα	0.69	0.787	0.75	0.784	0.73	0.781
Si/Cu	In Kα	0.81	0.782	0.81	0.781	0.80	0.780
Si/Cu	Bi Lα	0.89	0.829	0.83	0.788	0.86	0.785
Ag/Cu	Si Kα	1.08	1.168	1.17	1.171	1.18	1.173

agreement in Table 2 is within the limits of the accuracy of the equations and the experimental measurements. A comparison of the calculated values with the expression of Ø(0) of Karduck and Rehbach(1988) shows excellent agreement for low atomic number substrates but increasing difference between the values for high atomic number matrices with our values being approximately 10% higher.

ACKNOWLEDGEMENT

We would like to thank the Natural Sciences and Engineering Research Council of Canada(NSERC) for financial support for this project.

REFERENCES

Bethe H A 1930 Ann. Physik 5 325
Bishop H E 1966 4th Int. Cong. X-ray Optics and Microanalysis ed R Castaing et al (Paris: Hermann) pp153-8
Czyzewski Z and Szymanski H 1981 Proc. 10th Int. Cong. Electron Micros. Vol. 1 (Hamburg: Offizon Paul Hartung) p261
Duncumb P and Melford D A 1966 1st Nat. Conf. on Electron Probe Microanalysis College Park Md. paper 12
Fabre de la Ripelle 1949 J. Phys. (Paris) 10 319
Fathers D J and Rez P 1984 Electron Beam Interactions with Solids for Micros-copy, Microanalysis and Microlithography ed D F Kyser et al (AMF O'Hare: SEM Inc) pp193-208
Gryzinski M 1965 Phys. Rev. A 138 336
Heinrich K F J 1981 Electron Beam X-Ray Microanalysis (New York: Van Nostrand) p244
Karduck P and Rehbach W 1988 Microbeam Analysis-1988 ed D E Newbury (San Francisco: San Francisco Press) pp277-83
Hutchins G A 1974 Characterization of Solid Surfaces ed P F Kane and G B Larrabee (New York: Plenum) pp441-84
Kulenkampff H and Spyra W 1954 Z. Phys. 137 416
Love G and Scott V D 1978 J. Phys. D 11 1369
Niedrig H 1984 Electron Beam Interactions with Solids for Microscopy, Micro-analysis and Microlithography ed D F Kyser et al(AMF O'Hare: SEM Inc) pp51-68
Pouchou J L and Pichoir F 1986 Proc. 11th ICXOM, ed J D Brown and R Packwood (London Canada: UWO Graphic Services) pp249-56
Reuter W 1972 Proc. 6th ICXOM eds G Shinoda et al (Tokyo: U Tokyo Press) pp121
Worthington C R and Tomlin S G 1956 Proc. Phys. Soc. 69 401

Measurement of film thickness by EPMA

H.E. Bishop and D.M. Poole

AEA Technology, Building 393, Harwell Laboratory, Oxon. OX11 ORA.

ABSTRACT: The expressions for $\phi(\rho z)$ incorporated in many microprobe correction procedures may also be made the basis for a procedure to determine the thickness of surface overlayers. Although these expressions are not intended to take account of the effects of the difference in atomic number between an overlayer and the substrate, the error in the thickness determination is modest, reaching 50% in an extreme case such as a very thin carbon film on gold. An improved procedure to correct for the influence of the substrate should not be difficult to develop. For very thin overlayers crystalline effects must be considered.

1.INTRODUCTION

Although the EPMA has been used since the early days of the technique to measure the thickness of overlayers (Scott 1982), thickness measurements have not become viewed as a routine application of the instrument. A major reason for this omission is that no procedures to deal with thickness measurements are included in the standard correction software packages provided with an instrument by the manufacturer. Correction programmes now often contain explicit expressions for the depth distribution of characteristic X-rays generated in the specimen, $\phi(\rho z)$. Such expressions provide much of the basic information necessary to quantify film thickness measurements.

For many years we have used in our laboratory a simple graphical method of determining overlayer thickness based on $\phi(\rho z)$ distributions determined by Monte Carlo calculations (Bishop and Poole 1973). We have been able to achieve the same result more conveniently using a simple spreadsheet calculation, employing the expression for $\phi(\rho z)$ derived by Bastin (1984), the same expression as is used in our Tracor Northern Inc. correction programme. In this paper we discuss the limitations of applying an unmodified $\phi(\rho z)$ expression to film thickness measurements and what modifications are required to overcome them.

2. THICKNESS MEASUREMENTS

For convenience we will only consider the simple case of a surface film of one element on an elemental substrate. Extending the theory to include compound layers and standards is straightforward. The measured k ratio is related to the film thickness, t, by the expression:

$$k = \frac{\int_0^t \phi'(\rho z) \exp(-\chi \rho z)\, d\rho z}{\int_0^\infty \phi(\rho z) \exp(-\chi \rho z)\, d\rho z} \qquad (1)$$

where $\phi'(\rho z)$ is the ionization distribution in the film and $\phi(\rho z)$ that in a semi infinite slab of the same material, all other symbols have their usual meanings.

If the atomic number of the substrate is close to that of the substrate the ionization distribution within the film will be the same as that for the bulk and the relationship between k and t will be correctly predicted by equation 1, within the limitations of our expression for $\phi(\rho z)$. On the other hand if there is a large difference in atomic number between film and substrate, the ionization distribution in the film will be modified by the different electron scattering power of the substrate and equation 1 will only accurately reflect the relationship between k and t if we know the modified $\phi'(\rho z)$. In the absence of any model to predict the modified form of the ionization distribution, we have to make do with the unmodified form and estimate the error in the value of t obtained from equation 1.

We can get an estimate of the effects of a difference in atomic number by looking at how $\phi(\rho z)$ distributions and the integral distributions from equation 1 vary with atomic number. Figures 1 and 2 show sets of ionization and integral distributions for four elements covering the range of atomic numbers. These plots are all for a primary energy of 15 keV and a critical ionization potential of 1.5 keV so that any differences in the curves are due solely to atomic number. The different areas under the curves in figure 1 reflect the atomic number effect. The most remarkable feature of these two plots is how closely the integral plots overlay each other over the wide range of atomic numbers represented in figure 2. At first sight figure 2 suggests that we do not need to worry about differences in atomic numbers the integral curves for all atomic numbers are virtually the same, at least for k<4. Figure 2 does not however show how the similarity in the curves is achieved through a progressive change in the balance between ionizations caused by the ingoing flux of electrons and that by the backscattered flux.If this balance is upset by a change in atomic number we must expect a significant change in the nuber of ionizations produced in the overlayer.

Differences in atomic number between the film and the substrate will result in either an enhancement ($Z_f < Z_s$) or a reduction ($Z_f > Z_s$) of the k value relative to that for no difference in Z. The maximum effect will be for the limiting case of a thin surface layer where the change will be determined by the ratio of the $\phi(0)$ values for bulk samples of the substrate and the film materials. The effect will progressively be reduced as the thickness of the overlayer increases. From figure 1 we can see that in the extreme case of a thin carbon layer on a gold substrate the carbon X-ray signal will be enhanced by 45% by the extra backscattering from the gold substrate.

From the above discussion we see that the potential error in calculating film thicknesses using $\phi(\rho z)$ values for the material in the surface film and ignoring the effect of the substrate is limited to a maximum of the order of 50% in extreme cases (assuming our expression for $\phi(\rho z)$ is itself accurate). For a particular case the maximum error is determined by the ratio of the $\phi(0)$ values for film and substrate materials. Provided this limitation is clearly understood, a module for the determination of film thickness based on this approach would be a useful addition to EPMA software. Modification of this approach to incorporate the influence of the substrate should not be too great a task. The correction for the limiting case of a thin film is known, it is only necessary to predict how the required correction decreases with film thickness. Monte Carlo modelling should enable suitable procedures to be developed. Scott and Love (1991) report that they are currently working on this problem.

3. FLUORESCENCE EFFECTS

In our discussion we have ignored fluorescence effects. Scott and Love (1991) suggest that fluorescence by characteristic radiation from the substrate may be substantial and recommend a correction procedure (Cox, Love and Scott, 1979). Rather less is known about continuum fluorescence. Fluorescence of the surface film by the substrate continuum is, however, not likely to be a major problem as continuum fluorescence is normally only a few percent in bulk samples.

4. CRYSTALLINE EFFECTS

The crystal structure of the sample normally has no significant effect on the characteristic X-ray intensity observed from a bulk sample. This will no longer be the case for very thin films. Variations in X-ray intensity with incident beam angle may be expected to be similar to those observed in Auger Electron Spectroscopy (Bishop, 1990), as much as a factor of two in some cases. The main danger from crystalline effects may be expected from thin epitaxial films grown on semiconductor materials although strong texturing of a polycrystalline film could also provide an unexpected problem.

5.CONCLUSIONS

The electron microprobe provides a convenient and potentially accurate means of determining the mass thickness of surface overlayers. A reasonable estimate of film thickness may be obtained using a method based on expressions for the ionization distribution used in correction programmes. The prospects of modifying this approach to take account of the different electron scattering power of the substrate are good.

An incidental benefit of the increased use of the microprobe for film thickness measurements will be that it will provide an incentive to provide more accurate expressions for $\varphi(\rho z)$ which in turn will lead to improved absorption correction in bulk analysis.

This work is part of the Corporate Research Programme of AEA Technology.

REFERENCES

Bastin G F 1984 Scanning **6** 58.
Bishop H E 1990 Surface and Interface Analysis **16** 118.
Bishop H E and Poole D M 1973 J. Phys. D: Appl. Phys. **6** 1142.
Cox M G C, Love G and Scott V D 1979 J. Phys. D: Appl. Phys. **12** 1441.
Scott V D 1983 Quantitative Electron-Probe Microanalysis ed. V D Scott and G Love
        (Chichester, UK: Ellis Horwood) pp 283-292.
Scott V D and Love G 1992 X-ray Spectrometry **21** 27.

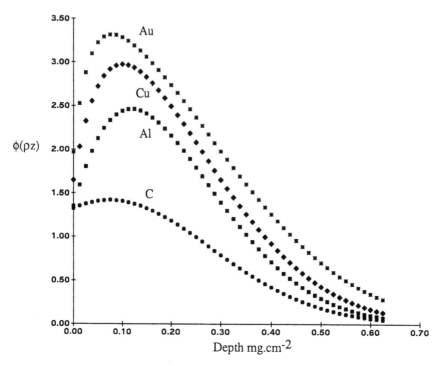

Figure 1. Bastin ionization distributions, E_0=15 keV, E_c=1.5 keV.

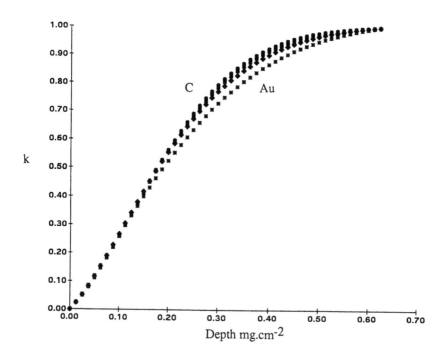

Figure 2. Normalized integral distributions.

Inst. Phys. Conf. Ser. No 130: Chapter 2
Paper presented at Int. Congr. X-ray Optics and Microanalysis, Manchester, 1992

139

Determination by EPMA and X-ray diffraction of the thickness, stoichiometry and crystallinity of thin films (Cr_xO_y) deposited on stainless steels

G. Auclair, C. Houpert and P. Choquet.

IRSID SNC Center of Unieux, B.P. 50, 42702 Firminy Cedex, France.

ABSTRACT : Characterization by EPMA and X-ray diffraction of ceramic coatings deposited on steels is described. The film thickness varied from 80 to 600 nm, which is thin with respect to these techniques. Such characterization is complicated by the presence of oxygen in the coating. These difficulties were solved by following the variations of k-ratios versus acceleration voltage for $K\alpha$ lines of Fe, Cr. In the case of O, a multi-layered crystal and full integral measurements of the peak were used. The crystallinity of these coatings were examined by X-ray diffraction.

1. INTRODUCTION

Ceramic coatings of chromium oxides deposited on stainless steels are very thin with respect to classical microprobe analysis and X-ray diffraction. In the examples described, the film thicknesses varied from 80 to 600 nm which is less than the expected excited volume in normal incidence conditions (1 mg/cm^2 i.e., 1 µm if the density of the material equals 10). Determination of the composition and the thickness of these coatings requires a realistic and reliable estimate of the so-called $\phi(\rho z)$ function representing the distribution in depth of the generated ionization. The extended form of the $\phi(\rho z)$ functions of the XPP model used in these investigations has been developed by Pouchou and Pichoir (1984, 1990). X-ray diffraction is commonly used to examine the degree of crystallization of materials. In such examples, grazing incidence is the only tool to both study the crystallinity and to evaluate the thickness of the coating. The model used for calculating the thickness has been developed by Brunel et al. (1988).

2. GENERALITIES OF THE MODELS

2.1 EPMA

When the target volume is uniform in composition or when the surface film is sufficently thick, it is commonly known that the composition analysis of the film can be performed using common procedures developed for bulk samples. In the case where the mass thickness is less than the ionization depth, the characteristic X-ray intensity coming from the film is expressed by a function of the physical characteristics of the coating material (Pouchou and Pichoir 1990). This intensity is compared to those of a bulk standard. In this way, the emerging intensity depends both on mass concentration C_A and on the film mass thickness μ_f. Taking into account that $\phi_A(\rho z)$ varies with the accelerating voltage, it is possible to separate the partial contribution due to the concentration and to the thickness with at least two measurements, one at high energy where R_x (maximum depth of ionization) $> \mu_f$ and one at lower energy where $R_x \leq \mu_f$.

2.2 X-ray diffraction

The classical parafocusing Bragg-Brentano diffractometry is not suitable for the analysis of thin films of thickness less than 0.5 µm due to the unfavourable peak-to-background ratio. Penetration depth can be considerably reduced by decreasing the incident angle α between the X-rays and the sample surface. The irradiated depth is a few nanometers at the critical value of reflection α_c, within some tenths of degree but can be increased up to one micrometer by adjusting the incident angle. The grazing incidence X-ray diffraction technique (GIXD) is applied to the analysis of crystallographic structure and gives information on the distribution profiles as a function of depth.

3. MEASUREMENTS

The arrangement of the coating is outlined in Figure 1. Cr is both present in the substrate and in the film. In this configuration, at least two measurements at two different accelerating voltages are necessary to approach both composition and thickness of the film by EPMA.

	Fe	Cr	O
i in nA	50	50	50
t in sec	20	20	**
standard	Fe pure	Cr pure	Cr_2O_3
crystal	LIF	PET	Ni/C

Table 1 : Analytical conditions for EPMA (for ** see Table 2). Composition in weight % of Cr_2O_3 are : Cr = 68.42 and O = 31.58.

line	i in nA	t in sec.	channels
O Kα	50	0.5	256

Table 2 : Analytical conditions for O Kα PHA setting : E = 1V, ΔE = 1.5 V.

---

thin film	:	Cr, O	(80-600 nm)

substrate

Cr, Fe, Mo, Ni

Figure 1 : Schematic description of the analyzed material

3.1 EPMA

The measurements of Fe Kα, Cr Kα and O Kα were performed on a microprobe analyzer SX 50 (CAMECA) in the conditions listed in Table 1. For Fe and Cr, at least 5 measurements were made on each sample and for each acceleration voltage. Calculations were performed with the mean of five measurements. The stability of the count rate was controlled by a Dixon test (Ancey et al. 1978). Although the peak shape alteration was less pronounced for oxygen than for carbon, boron and nitrogen compounds (Bastin and Heijligers 1989, 1991), the integral measurement of the peak intensity for O Kα was performed with a multi-layered monochromator Ni/C with a 2d spacing of 9.8 nm in the condition listed in Table 2. Under these conditions, the net count rates were 2.5 to 3 times higher than with a classical ODPB monochromator. The continuum was easily accessible on both side of the peak, and the interpolation of the background was easily performed by a linear cutoff (Fig. 2). The k-ratios for O Kα were calculated by comparing the O Kα area under the peak in the film to the O Kα area in the standard under the same conditions. An example of the O Kα spectra is shown in Figure 2. The reason for the choice of

	k-ratio O		k-ratio Fe		k-ratio Cr	
kV	mea.	calc.	mea.	calc.	mea.	calc.
5	.793	.785	-	-	-	-
10	.262	.280	.441	.440	.323	.315
20	.135	.130	.637	.650	.236	.230

Table 3 : Comparison of measured and calculated k-ratio for one sample relative to standards at 5, 10 and 20 kV and 40° take-off for Cr_xO_y ceramic coating film (5 density assumed). The calculated thickness is about 72 nm.

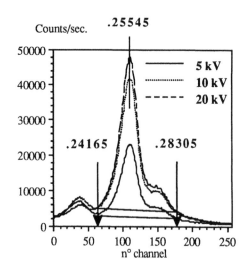

Figure 2 : O Kα spectra of Cr_2O_3 standard recorded at 5, 10 and 20 kV, with a Ni/C multi-layered monochromator. 50 nA, 0.5 sec/channel, 256 channels.

Cr_2O_3 for the standard was governed by the composition of the ceramic coating which was very close to this stoichiometry. In this way, deviation from the theoretical stoichiometry was easily observed.

Complete thin film characterization was performed using an iterative procedure which converges automatically to the mass thickness and the composition of the film. The software STRATA (SAM'x 1991) was used to perform the calculations given the following assumptions : 1-) the thin film density is very close to 5, 2-) O and Cr was the only elements present in the film, 3-) there was no other thin film between Cr_xO_y and the matrix and 4-) the take-off angle was θ = 40°. An example of the fitting between the measured and the calculated k-ratios is reported in Table 3 and illustrated in Figure 3.

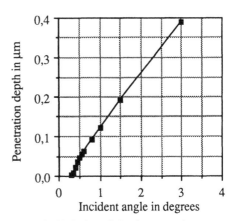

Figure 4 : Variation of depth penetration versus α calculated for Cr_xO_y with an assumed density of $\rho = 5$.

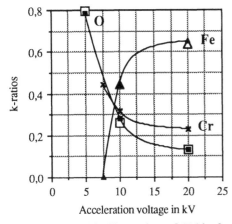

Figure 3 : Graphic illustration of Table 3. Evolution of k-ratios of Fe, Cr and O versus the acceleration voltage (open square : measured O k-ratio, full square : calculated O k-ratio, + : measured Cr k-ratio, x : calculated Cr k-ratio, open triangle : measured Fe k-ratio, full triangle : calculated Fe k-ratio).

3.2 X-ray diffraction

GIXD was carried out using a SIEMENS diffractometer (D5000) and its grazing incidence attachment which consists of soller slits and a plane monochromator. Detection was performed by a scintillation counter. Penetration depths τ have been calculated for the stoichiometric compound Cr_2O_3 (Fig. 4). The critical angle α_c is 0.37° with the Co Kα

radiation used for these experiments. For incident angles larger than α_c, the penetration depth varied linearly with α and was related to the absorption coefficient. Because of the small thicknesses investigated (80 to 600 nm) the incident angle was adjusted for each sample with respect to the calculated penetration depth.

4. APPLICATION AND DISCUSSION

The aim of this study was to perform the coating process while varying the process parameters on the stoichiometry, the thickness, the crystallinity and the corrosion resistance of the deposited films. In such studies, a complete experimental plan was formulated to examine the role of each parameter. In this article, the influence of one of these parameters (partial oxygen pressure : $P(O_2)$ in arbitrary unit) on the thickness, the stoichiometry and the crystallinity of the film is demonstrated. The correlation between $P(O_2)$ and 2 characteristics is shown in Figure 5. The higher the $P(O_2)$, the higher the weight % of O and the lower the thickness of the coating. X-ray spectra were collected on these samples and all of them with one exception, showed amorphous coatings. The most crystalized film was produced in a low partial pressure of oxygen. The validity of this data was established by comparison with data obtained by Glow Discharge Spectroscopy (GDS) for examining the elemental composition and with data collected by a surface profiler for an evaluation of the coating thickness The elemental composition did not drastically vary between the different methods. Argon was present in the coating but in less 1 % by weight. In this case, the microprobe analyzer results were not significantly affected by such a lower concentration. We also compared the intensity of the (100) peak of the substrate by X-ray diffraction and the calculated thicknesses (EPMA) to the measured thicknesses (Fig. 6) and concluded that the two methods were in good agreement. Indeed, the lower the (100) intensity, the thicker was the film. The discrepancy between the calculated thicknesses by EPMA, the relative intensity of the (100) peak by X-ray diffraction and the measured thicknesses by 2D rugosimetry can be attributed to 1-) the accuracy of such a mechanical measurement and 2-) to the roughness of the substrate surface. Determination of the thickness of the thicker coating was tried using several incident angles without success.

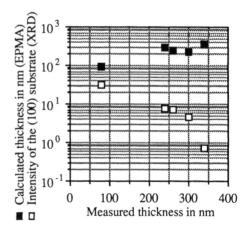

Figure 6 : Correlation between calculated thicknesses, intensity of the (100) peak of the substrate and measured thicknesses for the ceramic coatings deposited under different $P(O_2)$. Data series number 2.

Acknowlegments

We gratefully acknowledge G. Bardet and O. Valfort for helpful technical support during EPMA analysis and X-ray diffraction data collection.

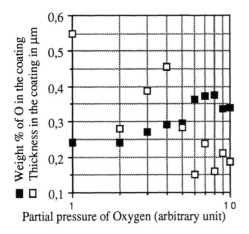

Figure 5 : Influence of the $P(O_2)$ on the thickness and on the composition of the coating ($P(O_2)$ units are arbitrary). Data series number 1.

5. CONCLUSIONS

The combination of physico-chemical investigations such as EPMA and X-ray diffraction are useful to characterize very thin ceramic coatings which is demonstrated in this article. The application of these techniques to control the deposition process parameters is proscribed and limited by the sample preparation.

References

Ancey M, Bastenaire F and Tixier R 1978
    Microanalyse et Microscopie à balayage (Les
    éditions de physique) pp 323-47
Bastin G F and Heijligers H J M 1989 Proc.
    12th. ICXOM, Cracow pp 104-7
Bastin G F and Heijligers H J M 1991 Scanning
    Vol. 13, 5 pp 325-43
Brunel M, Arnaud Y and Moncoffre N 1988
    Analysis, Vol 16, 5 pp 279-86
Pouchou J L and Pichoir F 1984 La recherche
    aérospatiale, n° 5 pp 349-67
Pouchou J L and Pichoir F 1990 Scanning Vol.
    12, pp 212-24
Sam'x 1991 Strata : Thickness and
    compositionnal thin film analysis software

Inst. Phys. Conf. Ser. No 130: Chapter 2
Paper presented at Int. Congr. X-ray Optics and Microanalysis, Manchester, 1992

A versatile computer program for improving the precision of quantitative electron probe microanalysis results

I Farthing[1,2], G Love[2], VD Scott[2] and CT Walker[1]

1. Commission of the European Communities, Joint Research Centre, Institute for Transuranium Elements, Postfach 2340, W-7500 Karlsruhe, Federal Republic of Germany.

2. University of Bath, School of Material Science, Claverton Down, Bath, BA2 7AY UK.

ABSTRACT: The correction program, written in QuickBASIC, is based on the quadrilateral model for X-ray absorption developed by Scott and Love. Tests of the program show that the RMS errors the for medium and heavy elements are 3.0-3.4% and 1.9%, respectively. For the light elements ($Z < 10$) the RMS error is less than 6% if Bastin`s absorption coefficients are used. Unlike most other methods a tilt factor is built in so that quantitative analysis can be performed on inclined specimens. A unique feature of the program is a self-test module consisting of EPMA data sets collated by Bastin, Sewell and Heinrich.

1. INTRODUCTION

If the precise measurements now obtainable with the modern electron probe microanalyser are to result in improved quantitative chemical data then correction procedures based on the ZAF method require updating. At the JRC Karlsruhe a new correction procedure for electron probe microanalysis called QUAD2 is being developed in co-operation with the University of Bath, UK. It is extremely versatile being capable of dealing with the analysis of elements ranging from atomic number 4 (Beryllium) to 96 (Curium) and with a wide range of experimental conditions including non-normal electron beam incidence. The correction program is designed to work off-line and is written in QuickBASIC for use on standard IBM compatible computers.

2. PHYSICAL BASIS

The X-ray absorption correction incorporated in the program is the Quadrilateral Model developed by Scott and Love (1992). This determines accurately the amount of X-ray absorption occurring in specimen and standard for all experimental conditions of practical interest, being appropriate for absorption levels of up to 95%. The mass absorption coefficients (MACs) employed are calculated from the latest algorithms of Heinrich (1987) and extended from element 92 to 96 using the line energies of Kleykamp (1981). Because uncertainties still exist in the MACs for the light elements ($Z < 10$), the program provides the option of manually inputting individual values.

The atomic number correction is as described in Scott and Love (1991). Here the backscatter factor (R) and the stopping power (S) are calculated separately. New empirical equations describe the variation of R with the electron backscattering coefficient and overvoltage. The S factor is determined from a modification (Bethe and Ashkin 1953) to the Bethe electron energy loss law (Bethe 1930), which renders it somewhat more physically realistic at low electron energies.

The correction dealing with characteristic fluorescence is essentially that of Reed (1965) but includes some minor revisions (Reed 1990) to improve its accuracy, and the continuum fluorescence correction (provided as an option in the program) is as described by Springer (1967).

3. PROGRAM DESCRIPTION

All the major X-ray lines can be used for analysis purposes and up to 99 analysis points may be input either as k-ratios or as X-ray intensities.   A special sub-routine for compound standards provides the k-ratio normally obtained with a pure standard.   In   addition,   element concentrations can be fixed or calculated by difference. A comprehensive database of physical constants (X-ray line energies, absorption edges, fluorescence yields) for each element is stored in a computer file system and these may be readily viewed and changed by the microanalyst.

The program is structured such that any part can be revised easily without extensive knowledge of computing or a detailed understanding of the operation of the program itself.   As seen from the flowchart in figure 1, the program is divided into a number of modules.   Each

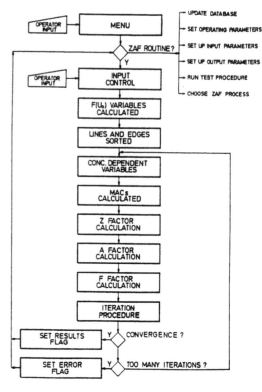

Fig.1 Flowchart showing the structure of the correction procedure.

module has a separate function and has a specific input and output. For example, the input for the module containing the algorithms to calculate the MACs is the X-ray energy and the atomic number of the absorber whilst the output is the MAC. All concentration independent variables are calculated outside the iteration loop to increase the processing speed A selection of output options is available so that the data can be presented in the appropriate form.

4. NEW AND UNUSUAL FEATURES

The correction program possesses several new and unusual features. Firstly, and most importantly, unlike most commercial programs, it can be modified and updated by the microanalyst himself. Secondly, a tilt factor is built in making it possible to perform quantitative analysis on specimens inclined by up to 45° from the horizontal. Although not widely employed, tilting the specimen can significantly improve the sensitivity of surface analyses. Finally the program contains a test module which enables any changes made to the correction procedure to be evaluated.   For elements of medium to heavy atomic weight the test module contains Sewell's database (1985), and a data set of 1825 binary systems collated by Heinrich (1991).   To assess the program's performance with light elements, Bastin's data sets for borides (1986a), carbides (1985), nitrides (1988) and oxides (1989) are included in the test module.

5. VALIDATION OF THE CORRECTION PROCEDURE

Figure 2 shows a histogram of the EPMA results for 1049 medium to heavy binary systems in the Heinrich database. This number of systems remained after the data had been filtered for repetition and suspect k-values. The term k' is the k-ratio predicted with the correction procedure and k is the measured k-value. It is seen that the distribution of the data is symetrical with a mean value of 1.007 and a root mean square (RMS) error of 3.0%. The actual error attributed to the correction procedure itself will be smaller than this, since there are undoubtly inaccuracies in the measured k-values. Experience with the Sewell database suggests that a conventional ZAF routine would give an RMS error of around 7%.

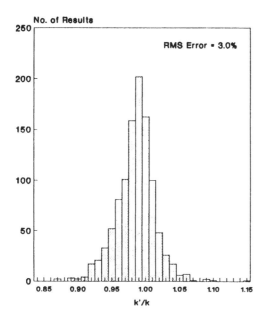

Fig.2 Histogram of EPMA results (1049 medium and heavy systems) in the Heinrich data set corrected with QUAD2.

When the elements in the 1049 binary systems were grouped according to the characteristic X-ray line used in their analysis it was found that the RMS error was 3.0% for the K_α lines, 3.4% for the L_α lines and 1.9% for the M_α lines.

An unexpected outcome from the test campaign is the finding that the introduction of the correction for continuum fluorescence results in a marked deterioration in the performance of the correction procedure. There was a shift in the mean k'/k value from 1.007 to 1.011 and the RMS error increased from 3.0 to 3.3%.

The test results for QUAD2 are collected in table 1. It can be seen that for the light elements the performance of the correction procedure is strongly dependent on the MACs used. For these elements the best results were obtained with Bastin's MACs.

Table 1 Test Results for QUAD2

System	MACs	%RMS Error	mean (k'/k)
	Medium and Heavy Elements		
554 systems[1]	Heinrich(1966)[3]	3.1	0.994
1049 systems[2]	Heinrich (1987)	3.0	1.007
	Light Elements		
Oxides	Bastin et al. (1989)	3.4	0.994
(321 systems)	Heinrich (1987)	6.8	0.999
Nitrides	Bastin et al. (1988)	5.1	1.004
(136 systems)	Heinrich (1987)	7.4	1.043
Carbides	Bastin et al.(1986b)	3.8	0.997
(117 systems)	Henke et al. (1982)	14.3	1.074
	Heinrich (1987)	12.3	0.999
Borides[4]	Bastin et al.(1986a)	5.3	1.017
(177 systems)	Henke et al (1982)	12.0	0.959
	Heinrich (1987)	18.7	0.919

1) Sewell's database (1985).
2) Heinrich's unpublished database (1991)
3) MACs of Henke et al. (1982) used for radiation <2 keV.
4) Ni and Co systems removed.

REFERENCES

Bethe H A 1930 Ann. Phys. Leipz. 5 325
Bethe H A and Ashkin J 1953 Exp. Nucl. Phys. (New York: Wiley) pp 252-253
Bastin G F and Heijligers H J M 1985 Quantitative Electron Probe Microanalysis of Carbon in Binary Carbides Rep. Univ. of Technology Eindhoven
Bastin G F and Heijligers H J M 1986a Quantitative Electron Probe Microanalysis of Boron in Binary Borides Rep. Univ. of Technology Eindhoven
Bastin G F and Heijligers H J M 1986b X-ray Spectrom. 15 143
Bastin G F and Heijligers H J M 1988 Quantitative Electron Probe Microanalysis of Nitrogen Rep. Univ. of Technology Eindhoven
Bastin G F and Heijligers H J M 1989 Quantitative Electron Probe Microanalysis of Oxygen Rep. Univ. of Technology Eindhoven
Heinrich K F J 1966 The Electron Microprobe eds T D McKinely, K F J Heinrich and D B Wittry (New York: Wiley) pp 296-377
Heinrich K F J 1987 Proc. 11th Int. Congress on X-ray Optics and Microanalysis eds J D Brown and A H Packwood (Univ. of Western Ontario) pp 67-119
Heinrich K.F.J 1991 National Institute of Standards and Technology, Gaithersberg, MD 20899 USA, unpublished work
Henke B L , Lee P, Tanaka T J, Shimabukura R L and Fujikawa B K 1982 Atom. Data Nucl.Data Tables 27 1
Kleykamp H 1981 Z. Naturforsch. 36A 1388
Reed S J B 1965 Brit. J. Appl. Phys. 16 913
Reed S J B 1990 Microbeam Analysis (San Francisco: San Francisco Press) pp 109-114
Scott V D and Love G 1991 Electron Probe Quantitation eds K F J Heinrich and D.E Newbury (New York: Plenum Press) pp 19-30
Scott V D and Love G 1992 X-ray Spectrom. 21 27
Sewell D A, Love G and Scott V D 1985 J Phys. D18 1245
Springer G 1967 N. Jahrb. Miner. Abh. 106 241

Weibull distribution as applied to EPMA

Andrzej Kuczumow

The Faculty of Chemistry, Maria Skłodowska-Curie University, 20-031 Lublin, Poland

Jef Helsen

Department MTM, Catholic University Leuven, B-3001 Belgium

ABSTRACT: The rules are presented which enable to create $\phi(\rho z)$ curves from very simple mathematical functions. Among them the Weibull function carries physical meaning for its connection with electron beam attenuation rules. In such approach $\phi(\rho z)$ curve can be described by the homogeneous linear differential equation of the second order, the same which describes the motion of pendulum in a viscous medium or the unloading of the condenser.

INTRODUCTION

There are three independent sets of data which inform us about the fundamental rules governing the transformation of electron energy loss into characteristic x-ray signals:

a) electron transmission curves;

b) thin film analyses;

c) $\phi(\rho z)$ curves.

Roughly, the electron transmission data, presented in the system: R - relative intensity as measured in relation to the intensity of primary beam versus parameter ρz, are declining, S-shaped curves. The results of thin film analyses seem at first sight to be the complementary curves for transmission measurements. At last, $\phi(\rho z)$ curves resemble the derivatives of thin film analyses. But the trouble begins when one compares the numerical values of data. Only the general kind of functions describing the curves is preserved but not the numerical values of coefficient sinvolved. Thus, years ago an idea evolved that $\phi(\rho z)$ curve was not expressed by a single mathematical expression but could be described assome combination of at least two simple functions.

HOW TO MODEL $\phi(\rho z)$ CURVE?

The simplest way to create the approximation of $\phi(\rho z)$ curve is by subtraction of two simple declining functions of the same kind with different coefficients. Sewell et al(1985b) exploited it with two straight lines of declining character:

$$\phi(\rho z) = [B-A(\rho z+\rho z_0)] - [B'-A'(\rho z+\rho z_0)] \qquad (1a)$$

where $A'>A$, $\rho z>0$ and $\phi(\rho z)>0$.

Pouchou and Pichoir (1987) proposed to apply the right wings of parabolae:

$$\phi(\rho z) = [B-A(\rho z+\rho z_0)^2] - [B'-A'(\rho z+\rho z_0)^2] \qquad (1b)$$

with the same conditions imposed on A, ρz and $\phi(\rho z)$. Now we have three versions of exponential functions. Philibert (1963) applied simple exponential functions:

$$\phi(\rho z) = B\exp[-A(\rho z+\rho z_0)] - B'\exp[-A'(\rho z+\rho z_0)] \qquad (1c)$$

here corrected for the term ρz_0.

From some obvious inconsistency of Buchner and Pitsch (1971, 1987) and Parobek and Brown (1978) approaches as to the electron transmission data it is clear that a more correct approach would be:

$$\phi(\rho z) = B\exp[-A(\rho z+\rho z_0)^n] - B'\exp[-A'(\rho z+\rho z_0)^n] \qquad (1d)$$

Although Rayleigh function (Kuczumow and Helsen 1989) was proposed in a simple version, nobody has tried to apply it in the described combination:

$$\phi(\rho z) = B\exp[-A(\rho z+\rho z_0)^2] - B'\exp[-A'(\rho z+\rho z_0)^2] \qquad (1e)$$

Of course, it is possible to have a great number of mixed combinations of that type and the approach by Packwood and Brown (1981, 1982) seems to be the best example of that kind of formalism.

PHYSICAL REALITY OF MENTIONED APPROACHES

All the mentioned approaches can give relatively good approximation of $\phi(\rho z)$ curve. It is a very difficult task to estimate the corectness of the above equations from the numerical point of view. Their physical reality is based on two tacit assumptions:

a) electrons are attenuated according to the simple rules   expressed by partial functions in equations (1);

b) processes involved in the emission of x-rays can be divided   into at least three independent stages: the attenuation of   electron beam ⇒ the spread of ionizations ⇒ the emission of   x-rays.

The ionizations are only the transition stage in the whole process. They are created in the process of the attenuation of electrons and disappear mainly in the process of radiant deexcitation. This allows to treat the transition stage as analogous to the process known in chemical kinetics, in which one replaces time parameter with ρz while concentrations with intensities. There are some indirect proofs indicating which partial functions mentioned in equations (1) should be most correct. The researches on transmission of electrons and positrons by Cosslett and Thomas (1964,5), Vyatskin and Makhov (1958), Valkealahti and Nieminen (1983,4), Mills and Wilson (1982) and on the distribution of implanted atoms in samples (Bodart and Deconninck 1982) strongly testify that the attenuation of a beam of such particles occurs according to the rule:

$$I = I_o \exp[-A(\rho z)^n] \qquad\qquad (2)$$

Thus, the first step in equation (1d) should result from equation (2) corrected for backscatter. The main problem is that the coefficients A and n are quite different for the electron attenuation function and for $\phi(\rho z)$ function (compare e.g. Buchner and Pitsch (1971) with Valkealahti and Nieminen (1983)). One can explain it by taking into account that the function resulting from the subtraction of two different functions of type (2) is approximately the Weibull distribution with other coefficients A and n. It is worth mentioning Philibert's (1963) paper in which the difference of two exponential functions gives the Weibull probability density function. Another explanation can be found in equations of chemical kinetics describing the consecutive reactions of the first order.

The distributions of electrons and positrons obey simple Weibull rules while $\phi(\rho z)$ curves are approximated by somewhat more complex Weibull rules. It is possible to describe $\phi(\rho z)$ curve by the use of differential equation of the second order:

$$d^2[\phi(R)]/dR^2 + (A+A')d[\phi(R)]/dR + AA'\phi(R) = 0 \qquad (3)$$

with $R = (\rho z + \rho z_o)^n$.

ACKNOWLEDGEMENTS

To the British Council and Stefan Batory Foundation for the support for one of the authors (A.K.) enabling him the participation in Conference.

REFERENCES

Bodart F and Deconninck G 1982 Nucl. Instr. and Meth. 197 59.
Brown J D and Packwood R H 1982 X-Ray Spectrom. 11 187.
Buchner A R and Pitsch W 1971 Zeit. Metall. 62 392.
Buchner A R 1987 Proc. 11th Int. Conf. on X-Ray Optics and Microanalysis ed Brown J D and Packwood R H (London - Canada: Graphic Services UWO) pp 262-4
Cosslett V E and Thomas R N 1964 Br. J. Appl. Phys. 15 235 + 883 + 1283.
Cosslett V E and Thomas R N 1965 Br. J. Appl. Phys. 16 779.
Kuczumow A and Helsen J A 1989 Proc. 12th Int. Congr. on X-Ray Optics and Microanalysis ed Jasieńska S and Maksymowicz J (Cracow - AMM Printing Office) pp 83-7.
Mills A P Jr. and Wilson R J 1982 Phys. Rev. A 26 490.
Packwood R H and Brown J D 1981 X-Ray Spectrom. 10 138.
Parobek L and Brown J D 1978 X-Ray Spectrom. 7 26.
Philibert J 1963 X-Ray Optics and X-Ray Microanalysis ed Pattee H H, Cosslett V E and Engstrom A (New York: Academic Press) pp 379-92.
Pouchou J L and Pichoir F 1987 Proc. 11th Int. Conf.on X-Ray Optics and Microanalysis ed Brown J D and Packwood R H (London -Canada: Graphic Services UWO) pp 249-53.
Sewell S A, Love G and Scott V D 1985 J. Phys. D: Appl. Phys. 18 1233 + 1245 + 1269.
Valkealahti S and Nieminen R M 1983 Appl. Phys. A 32 95.
Valkealahti S and Nieminen R M 1984 Appl. Phys. A 35 51.
Vyatskin Y and Makhov A F 1958 Zh. Tekh. Fiz. 28 740.

Inst. Phys. Conf. Ser. No 130: Chapter 2
Paper presented at Int. Congr. X-ray Optics and Microanalysis, Manchester, 1992

153

Characterisation of thin films using Monte Carlo methods

KG Laurie, G Love and VD Scott

School of Materials Science, University of Bath, Bath BA2 7AY, UK

ABSTRACT: A multiple scattering Monte Carlo method has been used to model x-ray emission from thin surface films of copper on substrates of carbon, titanium and tungsten. The x-ray intensity data (normalised with respect to bulk copper) has then been compared with experimental EPMA measurements. It is shown that whilst general trends (variation with film thicknes and substrate) are modelled reasonably well, the absolute intensity ratios can differ by up to 20% from the experimental values in certain cases.

INTRODUCTION

Electron probe microanalysis (EPMA) has unique possibilities for non-destructive investigations of surface films as it has excellent lateral resolution and is sufficiently sensitive to detect monolayers of atoms. Unfortunately, full characterisation (thickness and chemical composition) is complicated by the fact that for thin films (less than a micrometer thick) the substrate influences the x-ray emissions from surface regions and renders conventional EPMA correction procedures inappropriate. To account for such effects the x-ray distribution (or $\phi(\rho z)$ curve) in the thin film needs to be studied and the effects of different substrate/film atomic numbers, film thicknesses, electron accelerating voltages and measured x-ray line energies carefully assessed. If sufficient data is accumulated it should be possible to predict quantitatively how the x-ray distribution in the film is modified from that in the bulk material for any set of experimental conditions and hence, develop a correction procedure suitable for dealing with thin films.

The first step is to examine the effectiveness of Monte Carlo methods for generating $\phi(\rho z)$ curves for thin films. This is a stringent test of the accuracy of the x-ray distribution because, depending on the film thickness, different fractions are summed to establish the generated x-ray intensity. The present investigation is confined to x-ray emissions from different thickness copper films formed on substrates of carbon, titanium and tungsten. A Monte Carlo model is used to predict the x-ray intensity from the film and bulk copper and the results are compared with corresponding EPMA measurements.

EXPERIMENTAL METHOD

Essentially, Monte Carlo models are either of the single or multiple scattering type. This work has been almost entirely based upon the somewhat simplistic multiple scattering approach of Curgenven and Duncumb (1971) since it is considerably quicker than the single scattering method. The model has been extensively modified to deal with different combinations of film and substrate and the number of scattering events per electron trajectory has been increased to two hundred so that electron behaviour can be represented adequately in films with thicknesses less than a tenth of the electron range.

For the EPMA measurements high purity (>99.9%) substrates of carbon, titanium and tungsten were cut from 10mm diameter rods and mounted in bakelite. Surfaces were then ground and polished, the final surface finish being obtained with 1μm particle size diamond. After cleaning, the substrates were attached to a planetary holder inside an Edwards 12E6 vacuum coating unit and films of copper were deposited by thermal evaporation from a tungsten filament. Film thicknesses were measured with an Edwards FTM2 quartz crystal oscillator calibrated by multiple beam interferometry. X-ray intensity data were obtained using a JEOL 8600M electron probe microanalyser fitted with wavelength dispersive spectrometers, a lithium fluoride crystal being employed for the CuKα radiation and a crystal of thallium acid pthalate for the Lα x-rays.

RESULTS AND DISCUSSION

Before using the Monte Carlo model to predict thin film intensities its performance was first tested by generating the $\phi(\rho z)$ curve for bulk copper at 25keV (fig.1) and comparing it with x-ray distributions produced from tracer measurements of Brown and Parobek (1973) and the single scattering Monte Carlo model developed by Joy (1991). It is evident that the multiple scattering model agrees fairly well with the tracer measurements although the peak in the latter distribution is slightly closer to the surface. Certainly, the model fits the tracer data better than the single scattering method which generates a much lower peak than observed experimentally. It follows that the new programme should be adequate for predicting 'k' x-ray ratios (x-ray intensity from film compared with that from the bulk specimen) from copper films on similar atomic number substrates such as titanium. This is borne out in fig.2 where calculated values are compared with EPMA measurements obtained from five different thickness films.

In most cases the film and substrate will not be of similar atomic number and the k ratios will be influenced markedly by the substrate, fig.3. When films are thin, the plots tend to be linear with thickness but the gradient is a function of the substrate and, as may be seen, x-ray intensity ratios obtained from the carbon and tungsten substrates diverge as the copper films thicken. When the k ratio exceeds 0.5 and film thicknesses approach the electron range, the influence of the substrate lessens and the curves begin to merge again.

The Monte Carlo programme demonstrates the effect on the x-ray depth distribution as the film thickness increases.The 15keV copper on tungsten data, fig.4, shows that for thin films

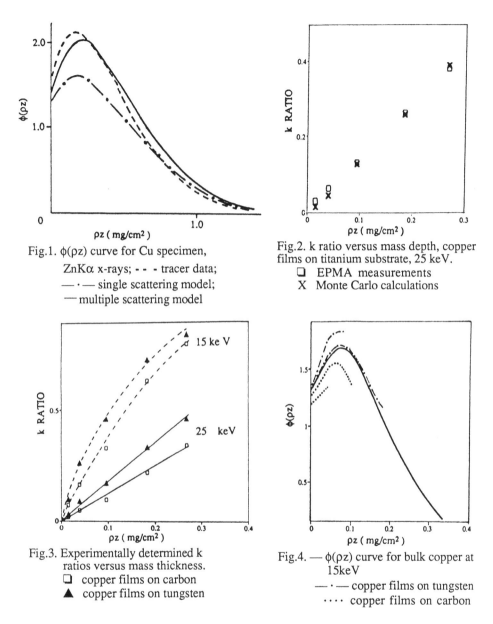

Fig.1. φ(ρz) curve for Cu specimen,
ZnKα x-rays; - - - tracer data;
— · — single scattering model;
— multiple scattering model

Fig.2. k ratio versus mass depth, copper
films on titanium substrate, 25 keV.
❑ EPMA measurements
X Monte Carlo calculations

Fig.3. Experimentally determined k
ratios versus mass thickness.
❑ copper films on carbon
▲ copper films on tungsten

Fig.4. — φ(ρz) curve for bulk copper at
15keV
— · — copper films on tungsten
· · · · copper films on carbon

the φ(o) value is substantially raised and the x-ray distribution is higher than that of copper.
As the film thickness increases φ(o) decreases, moving back towards the value for copper and
reaching it at a thickness ~ half the x-ray range. The peak maximum is also significantly
higher for thin copper films but drops steadily as thickness increases, so that for thicknesses
of about three quarters of the x-ray range the curve is virtually indistinguishable from bulk
copper. It is interesting to note that the maximum in the x-ray distribution does not appear to
be markedly affected by film thickness although there is a tendency for it to move to larger
depths when on a heavy substrate and *vice versa*, see fig.4.

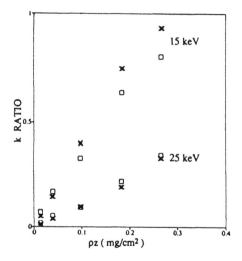

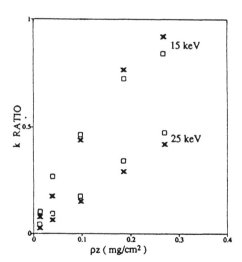

Fig.5. k ratios versus mass thickness for copper films on carbon.
      ❏ EPMA measurements
      X Monte Carlo calculations

Fig.6. As fig.5 but for copper films on tungsten

In fig.5, Monte Carlo predicted k ratios for thin copper films are compared with experimental values. For 25keV data on the carbon substrate agreement is excellent for all five film thicknesses. The corresponding calculations for the tungsten substrate, fig.6, give the correct trends but the Monte Carlo model consistently produces k ratios which are too low. The most likely explanation for the lack of agreement is that the model generates an x-ray depth distribution with the peak slightly too far from the surface, as suggested by the tracer data in fig.1. Results at 15keV on both carbon and tungsten substrates are similar in that the Monte Carlo results initially underestimate the k ratios but for thicker films overestimation occurs, again suggesting the peak in the predicted x-ray distribution may be too deep but also that the x-ray range is not sufficiently large.

Monte Carlo modelling can be very effective at predicting trends in the x-ray emission behaviour from thin films and showing how correction methods have to be modified to account for the influence of the substrate. However, thin film analysis is a very stringent test of the accuracy of the modelling technique and calculated x-ray intensity ratios must be treated with caution as errors of up to ~20% may be experienced.

REFERENCES

Brown JD and Parobek L 1973 *Adv. in X-ray Anal.* **16** 98.
Curgenven L and Duncumb P 1971 *Tube Investments Research Lab Rep 303*
Joy DC 1991 *Scanning Microscopy*, **5**, 329.

ACKNOWLEDGEMENTS
To SERC and AWE for support.

Inst. Phys. Conf. Ser. No 130: Chapter 2
Paper presented at Int. Congr. X-ray Optics and Microanalysis, Manchester, 1992

157

The spatial resolution of X-ray microanalysis experimental aspects

M. Żelechower

Silesian University of Technology, The Chair of Materials Science, Krasińskiego 8, 40-018 Katowice, Poland

ABSTRACT: If we try to study extremely small objects with commercial X-ray microprobe we will face the problem of the spatial resolution of the method. Some factors that determine the minimal volume of X-ray source have been well theoretically described but certain factors concern only a class of instruments or even an individual instrument (X-rays spatial aberration, take-off angle, quality of crystals). We propose the experimental routine in order to calibrate each individual instrument. The routine employing the convolution technique can be included to the instrument software package.

1. INTRODUCTION

The spatial resolution of X-ray microanalysis becomes essential when we try to study extremely small objects. It seems be common for the majority of cases associated with diffusion. That is why the determination of the volume of X-ray emission source for an excitation by the beam of charged particles (for instance electrons) was the subject of many papers. Fundamental papers by Bethe et al (1938), Castaing and Descamps (1955) should be mentioned first of all as well as the paper by Shinoda (1969). It is well known that the size of X-rays emission source is mainly affected by the energy of incident electrons and the mean atomic number of the target. However secondary effects (X-rays absorption, elastic scattering, fluorescence), characteristic for each experiment produce almost infinite number of experimental situations. Additionally we must take into account the experiment geometry: take-off angle for the EDS detection, the latter and an X-rays spatial aberration for the WDS detection. This considerable number of factors affecting every experiment makes the precise theoretical determination of the X-rays source volume extremely difficult. Moreover, with respect of secondary phenomena mentioned above, it seems to be better to discuss about the 'detection' area volume rather than about the emission area. In this way we would like to introduce the new term that determine the region from which we are able to measure X-rays signal. We should point out that its volume is different from that of being the emission source. In order to determine its lateral diameter in dependence on accelerating voltage and the target mean atomic number we propose the experimental routine followed by data processing.

2. THEORY

First of all we should say that each experimentally measured X-rays relative intensity profile from the mathematical point of view is the convolution of the true profile and so called 'window' function. This function reflects the instrumental broadening produced by objective and subjective factors mentioned above.

$$F(\alpha_1, ..., \alpha_N; x) = \int_{-\infty}^{+\infty} G(\alpha_1, ..., \alpha_N; x-t) A(t) \, dt \qquad (1)$$

where F(x) - measured profile
   G(x) - parametrised 'window' function
   A(x) - true profile
   $\{\alpha_j\}$ - set of 'window' function parameters
We should find the 'window' function solving the equation (1). One of possibilities is to apply the Fourier transform (FFT) however this method is not fully satisfied because it can produce some non-physical oscillations as shown by Rapperport (1969). It is supposed the better concept is to apply the convolution of parametrised 'window' function with the true profile followed by the least squares fitting (LSF) of the convolution to experimental data.

$$S^2(a_1, ..., a_N) = \min_{\alpha_1, ..., \alpha_N} \sum_{k=1}^{M} [F(\alpha_1, ..., \alpha_N; x_k) - y_k]^2 \qquad (2)$$

where $\{y_k\}$ - set of experimental data
   $\{a_j\}$ - found parameters of 'window' function
The next step is to establish the dependence of the 'window' function parameters on a target atomic number. Below we present the experimental routine in order to calibrate an individual X-ray microprobe.

3. EXPERIMENTAL

Several bimetallic junctions have been prepared in order to find a dependence between the target mean atomic number and the horizontal diameter of 'detection' area. Metals with neighboring atomic numbers Z have been mounted in each junction. Two perpendicular surfaces of each metal were finely polished then suppressed mechanically face to face. X-rays intensity profiles for each metal in junction have been measured. The experiment geometry and the corresponding profile is shown on Figure 1. It is clear that the true profile can be described by the staircase function while the 'window' function has been assumed to have the Gaussian shape. It is fair to assume that the size of the 'measurement' area is equal to four standard deviations of the latter. Results of measurements for several junctions have been shown on Figure 2. This is only the illustration for relatively narrow range of atomic numbers and only for K_α series but it is possible to find similar plots for another ranges of atomic numbers and X-ray series. The set of experimental points can be

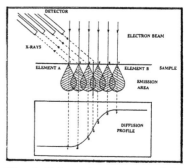

Fig. 1. Schematic geometry of the experiment.

approximated by the function C(Z):

$$C(Z) = c_0 \exp [c_1 Z] \tag{3}$$

Coefficients c_0 and c_1 have no physical sense so we should treat them as phenomenological. They are valid only for the individual instrument and for the atomic numbers range shown on Figure 2. Now we are able to determine the 'window' function for free bulk target that have the mean atomic number within this range. These results can be applied in order to restore the true X-ray profile (i.e. an element content profile) by fitting of the

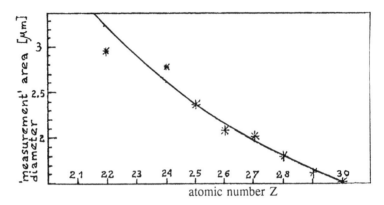

Fig. 2. 'Measurement' area diameter vs atomic number Z [*, experiment; ---, approximating function]

convolution of known 'window' function with parametrised true profile to experimental data. This routine can be described by equations (1), (2) however roles of integrated functions are reversed. The mathematical form of unknown profile should be assumed on the base of certain theoretical considerations. Its detailed shape will be determined by variable parameters. The example of such the routine is shown on Figure 3. This is the titanium line distribution in the MC-type (mixed) carbide in high-speed steel. In this case Gaussian shape has been assumed for restored true profile and two parameters determining width and height of the latter were varied. It is seen that the width of restored profile decreased considerably (as much as twice). Other examples can be found in the paper by Żelechower (1990). All these routines can be integrated into software package for any individual microprobe and successfully applied in cases we approach limits of the spatial resolution, especially in cases associated with diffusion in thin layers as well as grain boundaries studies in bulk materials.

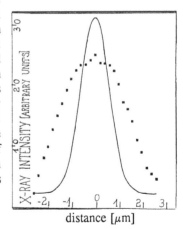

Fig. 3. Titanium line distribution in MC carbide [■, experiment; —, restored profile].

REFERENCES

Bethe H A, Rose M E and Smitt L P 1938 Proc. Am. Phil. Soc. 78 573
Castaing R and Descamps J 1955 J. Phys. Radium 16 304
Rapperport E J 1969 Electron Probe Microanalysis (New York and London: Academic Press) pp 117-36
Shinoda G 1969 ibidem pp 15-42
Żelechower M 1990 Proc. 12th. Int. Cong. on X-ray Opt. and Microanalysis pp 223-6

Inst. Phys. Conf. Ser. No 130: Chapter 2
Paper presented at Int. Congr. X-ray Optics and Microanalysis, Manchester, 1992

161

Quantitative electron probe microanalysis of thin III−V semiconductor layers

U. Zeimer*, J. F. Thiot**, C. Ständer***

*   Ferdinand−Braun−Institut für Höchstfrequenztechnik, Rudower Chaussee 5, Berlin 1199
**   SAM`x, rue Galilee 4, Guyancourt 78280
***   GETAG Instrumentebau, Am Obstmarkt 32, Mainz 6500

Quantitative Electron Probe Microanalysis is used to investigate thin single− and and multilayer structures by a deconvolution method after Pouchou and Pichoir. The results are compared to other analysis methods (X−ray Diffractometry and Low Temperature Photoluminescence).

1. INTRODUCTION

Modern III−V−devices often are based on thin layers within a heterostructure. Since thickness and composition of such thin layers determine the device performance, assessment of these properties is crucial for the optimisation of such a device. In our case a separated confinement QW laser (AlGaAs/GaAs/InGaAs) grown by LP−MOVPE is the device of interest.

To measure the composition of the various layers within this structure we have imployed Quantitative Electron Probe Microanalysis (EPMA). Since the layers of the device partly are thin (d< 0.1μm), a deconvolution of the measured signals has to be performed. For this convolution we have applied the method of Pouchou and Pichoir (1990).

2. ANALYSIS OF SINGLE LAYERS

To assess the capabilities of different analytical methods we have prepared by MOVPE $Al_xGa_{1-x}As$ layers of the same composition (x= 0.22) and various thicknesses (#1= 50nm, #2= 100nm, #3= 300nm).

2.1 X–ray Diffractometry

X–ray rocking curves have been measured by a Double Crystal Diffractometer (PW 1881 from Philips Analytical). Only in case of the thickest layer (#3, 300nm)a peak derived from the layer was resolved. Using HDR simulation software based on the Takagi–Taupin formalism, an excellent fit of the measured data for x= 0.228 was obtained.

2.2 Photoluminescence

The band gap and therefore the emission wavelength of $Al_xGa_{1-x}As$ is determind by the amount of Al within the layer.

Exiting the AlGaAs layer by an Ar^+–laser at a sample temperature of 10K the relative PL–intensity was measured. From the peak wavelength λ_{max} the Al atom fraction can be determined by the formula

$$x = \frac{E_g - 1.516}{1.4} \qquad E_g = \frac{1.239}{\lambda_{max}}$$

Due to the low PL intensity the evaluation of the thinnest sample #1 was not possible. For the thicker layers we obtained

Sample #	d[nm]	λ_{max}[nm]	E_g[eV]	x
2	100	684	1.811	0.210
3	300	676	1.833	0.226.

2.3 Quantitative EPMA

First conventional quantitative ZAF analysis at different beam energies was carried out to obtain the influence of very small beam energies and of the overvoltage ratio E_0/E_c on the results. The GaL_α, AsL_α and AlK_α lines were used in the measurement.

$$\text{Line energies:} \quad E_{AlK\alpha} = 1.487 \text{ keV}$$
$$E_{GaL\alpha} = 1.096 \text{ keV}$$
$$E_{AsL\alpha} = 1.282 \text{ keV}$$

If the overvoltage ratio is less than 4 the measurement error increases up to 20% at E_0= 3keV.

To overcome the limitation of the conventional approach we used the STRATA software packet of KEVEX based on the method of Pouchou and Pichour..

The measured k–values at different beam energies were analysed. If the elements in the thin layer are different from the substrate, the programme deduces layer thickness and composition.

In the case of AlGaAs/GaAs epitaxial layer and substrate have two elements in common. For a precise evaluation, measurement of the thickness by a different method is necessary. Using the independently obtained values for d the composition was the same (x= 0.22 + 0.005) for the three layers under study.

For sample #3 all three methods yield results which are in very good agreement.

Method	x–value
X–ray diffractometry	0.228
Low temperature PL	0.226
Quantitative EPMA	0.224

This demonstrates the capability of EPMA to give quantitative results down to a layer thickness of 50 nm.

Whether this method can be applied for even thinner layers currently is under study.

Besides the material system AlGaAs/GaAs we have also studied the quaternary system InGaAs/InP and obtained results which are in excellent agreement with the expected values. The presence of two elements (Ga and As) in the epitaxial layer which are not found in the substrate here allows for a determination of both composition and thickness by fitting of the experimental results.

3. MULTILAYER STRUCTURES

The device structure of special interest for us is the AlGaAs/GaAs/InGaAs SCH QW laser. For this complex structure X–ray diffraction yields rocking curves which are very difficult to interpret. PL only yields information on the InGaAs-QW. To separatly assess the layers of very different thickness (from 1000nm to 10nm) we used EPMA on a sample bevelled at a very small angle ($\alpha < 0.1°$) by ion beam etching.

Starting from the substrate the Al–content and the thickness of the single layers can be determind. That way, x and d of all layers underneath the one being currently measured are known and can be used for the deconvolution.

The accuracy of the determination of x and d is limited by

a)   the accuracy of the bevelling angle α and

b)   the precision of the spot position since the actual layer thickness depends on the spot position on the bevel.

An improvement of the accuracy is expected if in the deconvolution the nonperpendicular incidence of the electron beam is taken into account.

4. CONCLUSIONS

1.   The Al concentration, obtained using the deconvolution method of Pouchou and Pichoir is in a very good agreement to the results of other methods.

2.   The method is applicable even if common elements are present in the layer and the substrate.

3.   The method yields quantitative results for layers as thin as 50 nm.

4.   Using a bevelled sample complicated multilayer structures   with a wide range of layer thicknesses can be measured successive from substrate to top layer.

Literature:

Pouchou J. L., Pichoir F. 1990 Scanning, Vol.12, 212– 24

Inst. Phys. Conf. Ser. No 130: Chapter 2
Paper presented at Int. Congr. X-ray Optics and Microanalysis, Manchester, 1992

165

Nuclear microscopy—A novel technique for materials characterisation

G.W.Grime

University of Oxford Scanning Proton Microprobe Unit, Nuclear Physics Laboratory, Keble Road, Oxford OX1 3RH, UK.

ABSTRACT: The many interactions between high energy (MeV) light ions and matter form the basis of a number of different microanalytical techniques. Used simultaneously with a focused beam, they create a very powerful microanalytical technique, Nuclear Microscopy, which allows imaging, elemental analysis and mapping across the whole periodic table with sub-micron resolution, high sensitivity and quantitative accuracy and thick target capability. This paper outlines the principles of nuclear microscopy and describes the Oxford facility and some recent applications.

1. INTRODUCTION.

The use of MeV ions for materials characterisation is now common in certain specific applications. Two examples are the use of Rutherford backscattering analysis for the determination of impurity profiles in semiconductors and the use of particle induced x-ray emission (PIXE) for the analysis of airborne particulates. However, the limited availability of small MeV particle accelerators together with the fact that the majority of facilities produce beams with diameters of the order of millimetres means that ion beam analysis (IBA) is at present not routinely used outside these fields. However, recent developments in ion beam focusing technology and the increasing availability of small accelerators mean that it is now possible to consider establishing a nuclear microanalysis facility using resources comparable with those required for similar analytical instruments. Using a number of IBA techniques simultaneously allows a wide range of compositional and structural information to be obtained from a sample and this can be carried out with sub-micron resolution. This technique has been called **Nuclear Microscopy** and has potential applications in a wide range of scientific fields. In this article we describe some of the particle-solid interactions of importance to IBA, discuss the problems of focusing MeV ions to micron dimensions and present some recent applications of nuclear microscopy carried out using the Oxford Scanning Proton Microprobe (SPM).

2. ANALYTICAL TECHNIQUES USING MEV PROTONS OR LIGHT IONS

MeV ions interact with both the electron shells and the nuclei of sample atoms to produce observable effects which can be used for material characterisation. The interactions of most interest for analysis and imaging are described in the remainder of this section.

2.1 PIXE (Particle induced x-ray emission).

PIXE (Johanssen and Campbell 1988) is a process analogous to electron-induced x-ray emission (EPMA) in which inner shell vacancies created by the incident particle (usually protons) are filled by cascades from outer levels causing the emission of characteristic x-rays. The x-rays are detected using an energy dispersive Si(Li) detector so that all elements present in the sample above Na in the periodic table may be detected. The use of MeV protons ensures that the primary bremsstrahlung background (which limits the sensitivity of EPMA) is virtually absent and since the emission probability is high, PIXE is a fast, sensitive analytical technique with minimum detectable limits in the ppm region in favourable cases. Another feature of PIXE is that it is an inner shell process, so that the emission process is not affected by chemical binding and because of the long range of protons in matter, surface effects are small. This means that the x-ray yield in real samples is close to the calculated values and good quantitative accuracy can be achieved.

2.2 RBS (Rutherford backscattering).

Ions recoiling from direct elastic collisions with the nuclei of atoms in the sample lose energy according to the mass of the target ion; recoils from heavy nuclei having higher energy than recoils from light nuclei. If the target atom is not at the surface, energy is also lost during the passage through the material to the interaction site, so measuring the energy of recoiling ions gives information on the major element composition and depth distribution in the sample. This technique, known as Rutherford backscattering, is complementary to PIXE in that the greatest mass resolution occurs for light elements. The sensitivity is not as high as for PIXE, of the order of 0.1%, but RBS can be used for determining major element composition and depth profiles. In certain cases (heavy inclusions embedded in a light matrix) RBS can be used to carry out non-destructive three-dimensional mapping (Grime et al. 1991a). RBS is used in combination with PIXE to determine accurately the matrix composition used to calculate the thick-target yield corrections, permitting standardless PIXE analysis with good quantitative accuracy.

2.3 STIM (Scanning Transmission Ion Microscopy).

MeV ions in matter lose energy gradually by multiple collisions with atomic electrons so by measuring the energy of particles passing through thin (typically <30μm) samples a measure of the electron density along the ion path can be obtained. This can be used as a technique for mapping density fluctuations in samples thin enough transmit the beam. STIM can be carried out in two modes: with the detector mounted directly in the beam (bright field STIM), the contrast occurs solely due to proton energy loss in passing through the sample. With the detector mounted off-axis behind the sample (dark-field STIM) scattered particles are detected and contrast is due to a combination of small-angle scattering and energy loss in the sample.

With bright field STIM, each transmitted proton is a signal, so that the beam current must be reduced to the order of 2000 - 3000 particles/sec (0.1 fA) to avoid detector damage and saturation of the data acquisition electronics. This is normally carried out by reducing the apertures in the final lens, so an added benefit of using direct STIM is that the beam diameter is reduced and STIM imaging can be carried out at spatial resolutions of the order of 100nm or less (Bench and Legge 1989, Breese et al. 1992a). In addition, the beam dose to the sample is very low, so that sensitive samples can be studied extensively without damage. Dark field STIM can be carried out at normal beam currents used for analysis (100pA) and is a useful technique for mapping density variations in the sample during PIXE and RBS analysis.

2.4 IBIC (Ion Beam Induced Charge).

When an MeV charged particle passes through a semiconductor pn junction or gate region, electron/hole pairs are created which can be detected by charge-sensitive amplifiers connected to the device contacts. This can be used to map the active regions in semiconductor devices and investigate failure modes etc. This is analogous to the similar technique using electrons (EBIC), but because of the small penetration depth of electrons, junctions under investigation by EBIC must be close to the surface of the device. In contrast, the long range of MeV protons means that active areas buried under many um of

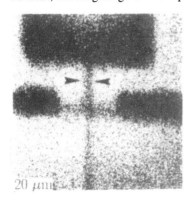

material (e.g. metallisations or passivation layers) can be investigated and this promises to be an important area of application for MeV ion microbeams. Like bright field STIM, each incident beam particle generates a signal and so IBIC must be carried out at very low beam currents with correspondingly low sample damage and high spatial resolution (Breese et al. 1992b). Figure 1 shows an IBIC image of a 0.8μm wide depletion region (arrowed) between two p-type implanted regions of a GaAs/GaAlAs High Electron Mobility Transistor (HEMT). This was imaged through the device passivation layers with an additional 10μm Al foil interposed to indicate the potential for imaging deep active regions.

Figure 1. An IBIC image of a 0.8μm depletion region (arrowed) of a HEMT. Scan size 20μm

2.5 Channeling

When the beam direction is aligned with an axis of a crystalline sample, many of the sample atoms are masked from the beam and the signal yield of any of the preceding interactions will be affected. This technique of channeling has been used with broad-beam RBS for some time to study crystal defects in semiconductors (Feldman et al. 1982), but combined with a focused beam, this allows crystal defects to be mapped with high spatial resolution. In particular, the use of channeling with bright-field STIM allows defects such as dislocations to be imaged in samples up to 50μm in thickness with high sensitivity at

Table 1. A summary of the capabilities of Nuclear Microscopy

Analysis
    PIXE: 1 - 10ppm MDL for Z>13.
        Multielemental.
        5 - 10% accuracy.
    RBS: 0.1 - 1% MDL
        Optimum for Z<10
        (light element stoichiometry)
        Depth profiling

Imaging
    STIM: Density mapping for thickness
        1 - 30µm.
    SEM: Surface topography mapping using
        secondary electrons
    IBIC: Semiconductor device mapping using
        beam-induced charge
    Tomography: 3-D imaging using STIM

Spatial Resolution
    0.4 - 1µm (PIXE, RBS)
    0.1 - 0.2µm (STIM, IBIC)
    Maintained in thick samples

Crystal studies (with channeling)
    Defect imaging in thick (<30µm) samples
    Lattice location of impurity atoms

---

resolutions of the order of 100nm (King et al. 1992).

2.6 Nuclear Microscopy (see table 1).

Used in combination and with a sub-micron resolution beam, these interactions form the basis of nuclear microscopy. Because of the penetration and low scattering of MeV ions the beam resolution is maintained even in thick samples, which simplifies sample preparation procedures.

3. FOCUSING MEV IONS - THE OXFORD NUCLEAR MICROSCOPY FACILITY

Focusing MeV ions to beam diameters which are of interest for microanalysis is difficult because of their high rigidity and the most successful technology in use at present is the magnetic quadrupole multiplet. Quadrupole lenses have a strong focusing action because the major components of the field are perpendicular to the beam axis. However a single quadrupole lens converges the beam only in one plane and diverges in a plane normal to this so two or more lenses of alternating polarity are required to form a point focus and combinations of 2, 3 and 4 lenses have been used. A triplet system (shown in figure 2) is in use at Oxford (Grime et al. 1991b).

Quadrupole lenses suffer from significant angle-dependant aberrations which increase the beam diameter. Apart from intrinsic spherical and chromatic aberration, numerical raytracing studies of quadrupole probe-forming systems (Grime and Watt 1984, Grime et al 1990) showed that the major aberrations in the Oxford triplet lens system were due to misalignment of the lenses and sextupole contamination of the quadrupole field. Lens misalignments are minimised by mounting the individual components of the system on mounts with micrometer adjustments of position and using careful alignment procedures. Sextupole field contamination is due primarily to departures from four-fold symmetry in the construction and assembly of the poles of the lenses and so the poles and yoke of the Oxford lenses are cut from a single piece of high purity magnet iron by a numerically controlled spark erosion technique with a precision of 2µm. Using a grid shadow

Figure 2. The quadrupole triplet probe-forming system used in the Oxford SPM . The beam enters from the right and the sample chamber is in the left foreground.

projection technique, the sextupole field contamination is undetectable (Jamieson et al. 1989) and a spot size of 0.3μm with sufficient beam current for ion beam analysis has been achieved (Grime et al. 1992).

Figure 3 shows the layout of the Oxford SPM facility (Grime et al. 1991b). 3MeV protons are generated by a 1.7MV Pelletron tandem accelator (National Electrostatics Corporation type 5SDH-2). After passing though a quadrupole triplet condenser lens, the beam is momentum analysed by a 90° double-focusing dipole magnet. A small (10 x 60μm) aperture placed immediately after the analysing slits acts as the object for the final probe-forming lens which forms a demagnified image of the aperture in the sample chamber. The beam can be rastered over the sample by means of magnetic deflection coils placed before the final lens. Detectors for x-rays, backscattered and transmitted ions and secondary electrons are located around the sample and signals from these are recorded together with the instantaneous beam position and this data is processed to construct elemental maps and composition profiles.

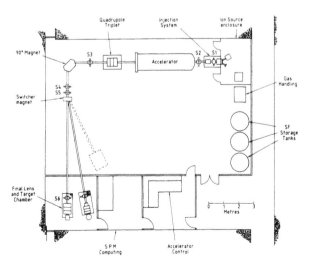

Figure 3. A plan of the layout of the Oxford SPM facility.

Table 2. A summary of recent applications of the Oxford SPM

Life sciences
   Metal uptake by micro-organisms
   Element gradients across membranes
   Defence mechanisms of plants
   Trace elements in neuropathology
   (Alzheimer's disease etc.)
   Metal toxicity (toxic elements in hair)

Archaeometry
   Studies of painting pigments
   Studies of pigments in glasses
   and ceramics
   Usewear analysis of flint tools
   Trace elements in buried bone

Geosciences
   Element zoning in rocks and minerals
   Composition of microcrystals
   and inclusions
   Studies of grain boundaries

Materials Science
   Channeling studies of crystals
   (dislocation imaging, strain
   mapping)
   Mapping complex devices with IBIC
   STIM mapping of multilayer devices
   Fabrication of high aspect ratio
   structures in photoresist

Environmental sciences
   Population studies of aerosol and
   fly ash particles
   Studies of lake sediment particles
   Trace element distribution in lake
   sediment porewater
   Bio-geochemical weathering of rocks

Industrial
   Metal diffusion in polymers
   Electrode surface studies
   Chemical effects in catalysts

---

4. APPLICATIONS OF NUCLEAR MICROSCOPY

Table 2 presents a list of recent and current fields of application of the Oxford SPM. Recent reviews have described the biological and medical applications (Watt and Landsberg 1992), microPIXE applications in materials science (Breese et al. 1992c) and archaeological and environmental applications (Grime et al 1992). We describe here some recent applications in medicine and environmental science in which the results depend primarily on PIXE.

4.1 Aluminium in Alzheimer's Disease.

Recent controversy over the role of aluminium in the aetiology of Alzheimer's disease has centred on the presence or absence of aluminium in senile plaques, a characteristic feature of the brains of sufferers from the disease. Several groups using a range of analytical techniques have reported significant levels of Al and other elements within senile plaques. A problem with this type of analysis is the necessity to carry out chemical staining on the tissue to

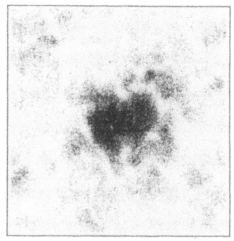

Figure 4. A bright-field STIM image of a senile plaque observed in an unstained frozen section of human brain tissue. Scan size 50μm.

visualise the plaques, which makes it difficult to eliminate the possibility of exogenous contamination as a source of the Al. This problem has been circumvented in recent work at Oxford by using STIM to localise the plaques. Because the amyloid protein which makes up the plaques is much denser than surrounding tissue, the plaques are visualised clearly in both bright field and dark field STIM, even in unstained frozen tissue sections (figure 4). Using STIM to localise plaques, we have demonstrated that the level of Al in plaques is less than the PIXE detection limit of about 15ppm and also measured levels of other elements including nitrogen (Watt and Landsberg 1992).

4.2 Single particle studies of ambient aerosols

Broad beam PIXE has been used for some time in the analysis of atmospheric aerosol filters. Using nuclear microscopy, single particles can be identified and analysed using PIXE to obtain trace element content and RBS to get light elements and particle size. This

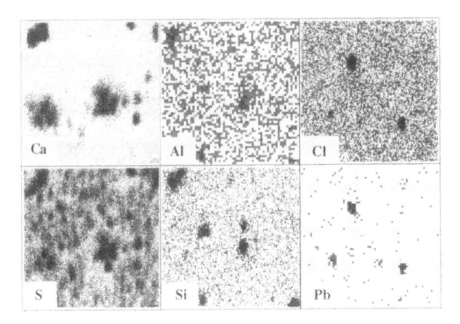

Figure 5. PIXE elemental maps of a filtered urban aerosol showing several different particle species. Scan size 50μm.

allows different particle populations to be identified much more precisely than with bulk analysis which in turn contributes to a much more accurate determination of the sources of ambient particles. An example of PIXE mapping of an urban aerosol filter is shown in figure 5. Even visually, different particle types can be identified (Ca/S, Al/Si, Pb/Cl) and using multivariate statistics on the composition of many particles, detailed population parameters can be obtained (see for example Artaxo et al. 1992). This type of analysis has been applied to other forms of particulates (power station and chemical plant fly ash, lake sediments) and promises to be a powerful technique in studying environmental particle chemistry.

With larger particles, nuclear microscopy can be used to investigate elemental distributions within single particles. This has been done for 30μm diameter fly ash particles and surface concentrations of volatile metallic elements have been observed (Jaksic et al. 1991).

5. CONCLUSION

Nuclear microscopy using a focused beam of MeV protons or light ions allows a wide range of sample properties to be determined on a micron or sub-micron scale. Although the technique is at present only available at a few central facilties, the increasing commercial availablity of accelerators and microbeam equipment means that nuclear microscopy will become more common and will take its place among the armoury of microanalytical techniques.

6. REFERENCES

Artaxo P, Rabelo M L, Watt F, Grime G W and Swietlicki E 1992 Proc 6th Int'l Conf. on PIXE, Tokyo 1992 (Nucl. Instr. and Meths: in press)

Bench G S and Legge G J F 1989 Nucl. Instr. and Meths. B40/41 655

Breese M B H, Landsberg J P, King P J C, Grime G W and Watt F 1992a Nucl. Instr. and Meths. B64 505

Breese M B H, Grime G W and Watt F 1992b Proc 3rd Int'l Conf. on Nuclear Microbeam Techn. and App., Uppsala 1992 (Nucl. Instr. and Meths: in press)

Breese M B H, Grime G W and Watt F 1992c Proc 6th Int'l Conf. on PIXE, Tokyo 1992 (Nucl. Instr. and Meths: in press)

Feldman L C, Mayer J W and Picraux S T 1982 Materials Analysis by Ion Channeling. (New York: Academic Press)

Grime G W and Watt F 1984 Beam Optics of Quadrupole Probe-Foriming Systems (Bristol: Hilger)

Grime G W, Watt F and Jamieson D N 1990 Nucl. Instr. and Meths. B45 508

Grime G W, Watt F, Duval A R, Menu M 1991a Nucl. Instr. and Meths. B54 353

Grime G W, Dawson M, Marsh M, McArthur I C and Watt F 1991b Nucl. Instr. and Meths. B54 52

Grime G W and Watt F 1992 Proc. 6th Int'l Conf. on PIXE, Tokyo 1992 (Nucl. Instr. and Meths: in press)

Jaksic M, Watt F, Grime G W, Cereda E, Braga-Marcazzan G M and Valkovic V 1991 Nucl. Instr. and Meths. B56/57 699

Jamieson D N, Grime G W and Watt F 1989 Nucl. Instr. and Meths. B40/41 669

Johanssen S A E and Campbell J L 1988 PIXE - A Novel Technique for Elemental Analysis. (New York: Wiley)

King P J C, Breese M B H, Booker G R, Wilshaw P W, Whithurst J, Grime G W and Watt F 1992 Proc 3rd Int'l Conf. on Nuclear Microbeam Techn. and App., Uppsala 1992 (Nucl. Instr. and Meths: in press)

Watt F and Grime G W eds 1987 Principles and Applications of High Energy Ion Microbeams (Bristol: Hilger)

Watt F and Landsberg J P 1992 Proc 3rd Int'l Conf. on Nuclear Microbeam Techn. and App., Uppsala 1992 (Nucl. Instr. and Meths: in press)

Inst. Phys. Conf. Ser. No 130: Chapter 2
Paper presented at Int. Congr. X-ray Optics and Microanalysis, Manchester, 1992

173

The use of high purity germanium detectors for X-ray microanalysis

D.A. Lock

Tracor Europa B.V., Milton Keynes.

ABSTRACT:High purity germanium detectors are now available as a choice for use in x-ray micro-analysis. They provide good light element detection and allow the analysing of heavy elements by K-line emission, provided the excitation energy is available (as is the case in a TEM). Superior inherent resolution capability, allows larger detectors to be manufactured, giving better solid angles and count-rates. Changes in identification and deconvolution routines allow for differences in escape peak positions.

The use of germanium as a material for manufacturing solid-state detectors is by no means new. Lithium-drifted, liquid nitrogen cooled germanium detectors have been in common use by physicists for gamma-ray detection since the early sixties. The manufacturing improvements which have led to their commercial availability as x-ray detectors have occurred over the last two to three years, and it is the intention here to review the current performance of these devices, compared to the well-known performance of lithium-drifted silicon (Si(Li)) detectors.

High-purity germanium (HPGe) detectors, without the necessity for lithium-drifting, were available in the very early seventies. Although still requiring liquid nitrogen cooling, they no longer suffered from the total failure which results if a lithium-drifted germanium detector is allowed to warm up (due to the inherent instability of lithium in germanium). However, resolution performance and other characteristics were such that they were poor contenders compared to the performance of Si(Li) detectors of that time, as described by Barbi & Lister (1981) and R.W. Fink (1981). Recent developments have meant that HPGe detectors specifically designed for micro-analysis are readily available and a direct comparison with Si(Li) detectors based on manufacturing experience, and usage is possible.

The requirements for a successful X-ray detector are; to respond to a broad range of x-ray emission energies to give analytical capability over a wide range of the periodic table, to offer good sensitivity and detection of low concentrations, to offer sufficiently good resolution such that element peaks in the x-ray spectrum do not obscure each other, and to be free of artifacts which may lead to mis-interpretation of analytical data.

1. Energy Range.

The ability of an x-ray detector to respond to a wide range of x-ray emission energies is governed by a) it's ability to absorb (and respond to) high energy x-ray emissions which otherwise might pass right through the detector crystal without interaction and b) it's ability to interact with low energy x-ray emissions which may only penetrate the front surface of the detector before being stopped.(The protective window in front of the detector will play a crucial role in determining whether these low energy x-rays arrive in the detector at all).

The HPGe detector offers a significant advantage over the Si(Li) detector, in that the germanium atoms, being larger and heavier than those of silicon, offer a greater chance for interaction with higher energy x-ray emissions. This greater stopping power leads to much increased detection efficiency for x-ray energies above 20kev. A direct comparison of Si(Li) and HPGe efficiency (absorbance) for 3mm thick detector crystals is shown in Fig.1. This performance opens up the possibility of detecting heavier elements by K-line emission, rather than by L- or M-line as is frequently necessary when using a Si(Li) detector. The use of this property of germanium is dependant on the ability to excite these higher energy x-ray emissions, as in the transmission electron microscope (TEM). The benefits often resulting from analysing by K-line are that potential peak overlaps between L- or M-lines from required elements and K-lines from other constituent elements can often be avoided, and peak-to-background may be better for the K-line emission. Fig.2 shows a logarithmic display of a spectrum taken

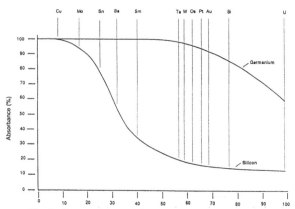

Fig.1. Comparison of Si(Li) and HPGe efficiency.

from a gold foil, mounted on a copper grid and analysed at 200kv. The gold K-lines are clearly visible in the 60-80 kev range. The tin k-lines in the region of 25kv are absorbed into the detector with 100%efficiency, compared to 75% efficiency available with a similar Si(Li) detector. In this situation quantification of tin impurity would be improved by analysis by K-line, rather than by the relatively poor peak to background of the L-lines.

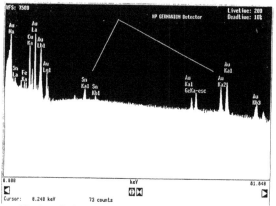

Fig.2.   Gold foil on Cu grid 200kv

At the other end of the energy-scale, the light-element detection performance of HPGe detectors is, at least, comparable to that for Si(Li) detectors. Direct measurement of dead-layers for a number of detectors indicates that the HPGe detector can be manufactured with a significantly thinner dead-layer. Consequently x-rays from light elements are able to successfully penetrate the detector.The inherently superior resolution (peak-width) of HPGe detectors also offers advantage in that similar or better resolution

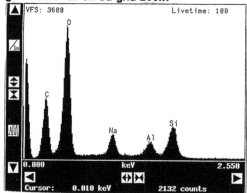

Fig.3. Specimen of albite at 5kv

specifications can be obtained on a germanium detector with three times the sensitive area, compared to a similar specification Si(Li) detector (30mm2 compared to 10mm2).
Both resolution and sensitive area are important for light element analysis. Good resolution aides separation of adjacent K-line emissions, and a larger sensitive area gives a greater solid angle subtended at the specimen, and higher count-rate of collection of the relatively poorly exited light element emissions. Fig. 3 is a spectrum taken at 5kv on a scanning electron microscope (SEM) from a specimen of albite, showing carbon and oxygen detection and the separation of carbon from the low energy noise peak. Additionally the

HPGe detector exhibits a reduced background at low energies. Compton scattering is reduced due to the higher stopping power of the germanium detector. Thus fewer x-ray photons are scattered within the detector, and fewer scattered photons are produced with x-ray energies which are still detected but no longer relate to the original emission from the specimen. This has been illustrated by McCarthy, Ales and McMillan (1990), by mounting the same Americium 241 source to both HPGe and Si(Li) detectors and comparing resultant spectra. Their result is shown in Fig. 4.

2. Resolution.

In attempting to specify detector performance, emphasis is always placed on the resolution specification which is always measured and specified as the full-width-half-maximum (FWHM), at manganese K alpha and usually at in-coming count-rates of from 1,000 to 3,000 counts per second. Insufficient emphasis is usually placed on resolution measurements elsewhere in the spectral range (particularly for x-ray

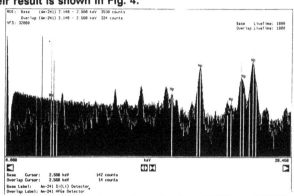

Fig.4. Comparison of spectra from HPGe and Si(Li) detectors, showing Compton scattering.

emissions from lighter elements), on peak shape and on peak-to-background.
The resolution of a peak (FWHM) is determined by the noise contribution from the associated electronics in the detection system and by the dispersion D of the detecting unit, defined as $D = \sqrt{5.52FeE}$, where F is the Fano factor, e is the mean number of charge carriers (electron-hole pairs) produced per unit energy and E is the energy of the x-ray emission being measured. Since F is similar for both HPGe and Si(Li) detectors, and e (the number of charge carriers produced) is lower for HPGe the result is a lower dispersion value and consequent superior inherent resolution performance for HPGe detectors. The current standard for a 30mm2 HPGe detector is 129ev FWHM at manganese K-alpha, compared to the best 10mm2 Si(Li) detectors specified at around 130-135ev FWHM.For a 30mm2 Si(Li) detector, specified resolutions in the range of 140-145 ev are typical. Fig. 5 shows the result of a survey of resolution measurements taken for 30mm2 HPGe detectors carried out over a twelve month period, when factory guaranteed resolution was 139ev. This shows the band of measured, on-microscope resolutions for manganese, oxygen and carbon, over a sample of 25 installations and indicates the light element resolution achieveable.

The adherence of a spectrum peak to a gaussian shape is important when closely adjacent peaks overlap and one is from a major, and one is from a minor constituent in the specimen. The measurement of peak width at full -width-tenth-maximum (FWTM) gives an indication of peak distortion, and for a true gaussian shape would be 1.82 times the FWHM value. A large deviation from this with a broad base to the peak indicates incomplete charge collection in the detector, and its dead layer. Historically this has been one of the significant problems with germanium detectors.

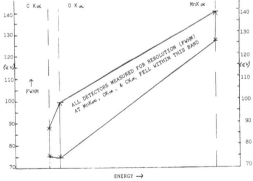

Fig.5. Survey of resolution (25 detectors)

The availability of germanium was such that impurities and dislocations in the crystal material acted as trapping centres for the charge carriers (electrons) with the result that a significant low energy shoulder was observable on most peaks in the spectrum, particularly in the region of 1 to 2 kev, just above the germanium L-line absorption edge.

Fig.6 shows the improvement which has occurred with different manufacturing techniques, comparing an aluminium K-alpha peak measured by detectors manufactured under the earlier and new processes.

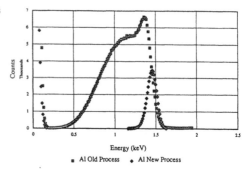

The use of germanium detectors requires attention to some practical considerations to avoid mis-interpretation of data.
The familiar escape peak observable from a Si(Li) detector at 1.74kev lower energy than the main peak, is substituted by the escape peak from the HPGe detector from the germanium

Energy (keV)

■ Al Old Process   ◆ Al New Process

Fig.6. Old Versus New Process

K edge at 11.1 kev. The fact that more true peaks from the K-lines from heavier elements can be detected also means that more escape peaks may be present in the accumulated spectrum. It is necessary to have access to tables and KLM markers that allow for the fact that an HPGe detector is in use, and these must include escape peak positions. This data set has been produced, together with an automatic least-squares-fitting routine to assist both in element identification and escape peak removal.
The improved sensitivity to higher energy X-rays extends (at reduced efficiency) above the most energetic K-line emissions and it is necessary to prevent stray radiation from being detected and processed by the electronics. An "upper-level" discriminator is incorporated in the electronics, similar to the familiar "low-level" noise discriminators for noise rejection. The "upper-level" discriminator simply rejects any detected events above the energy-range of interest.

3. Conclusion.

The conclusion is drawn that HPGe detectors offer superior resolution capability over Si(Li) detectors, and that advantage can be taken by using this property to build higher resolution detectors with larger sensitive area and consequent improved count-rate performance.
The useful energy range is much extended, with good light-element sensitivity and the ability to analyse virtually the entire periodic table by K-line emission.
Hardware and software developments have been completed to cater for the differing aspects of the spectrum produced from an HPGe detector. The extended analytical range and improved count-rate performance relate directly to the typical analytical requirements for TEM, and in specific instances can be of benefit for SEM.

REFERENCES:
Barbi N.C. and Lister D.B, "A comparison of silicon and germanium x-ray detectors" NBS Spec. Publ. 604, 1981, 35.
Cox C.E. "Small area high purity germanium detectors for use in the energy range 100 eV to 100 keV". IEEE Trans. NS-35: 28-32, 1988.
McCarthy J.J. et. al., Ales M.W. and McMillan D.J. "High purity germanium detectors for EDS", Microbeam Analysis, 1990.

Inst. Phys. Conf. Ser. No 130: Chapter 2
Paper presented at Int. Congr. X-ray Optics and Microanalysis, Manchester, 1992

177

Modern trends in energy dispersive X-ray emission analysis

M. Schiekel, P. Jugelt, J. Heckel[*]

Technical University of Dresden, Depts. Civil Engineering and Physics, Mommsenstr. 13, D-O-8027 Dresden
* Spectro Instruments GmbH, Ullsteinstr. 73, D-W-1000 Berlin

ABSTRACT: In the last few years a number of improvements have been introduced in energy dispersive x-ray emission analysis. Not only methodical investigations but also instrumentation are short reviewed.

1. INTRODUCTION

Nowadays analytical methods based on x-radiation are widely used in analytical applications both in research laboratories and industries, due to the advantages of these x-ray analytical procedures. The devices traditionally used were wavelength dispersive spectrometers (WDS), but during the past 20 years a number of methods have been developed exclusively based on energy dispersive spectrometers (EDS). EDS can be applied in all those investigations, in which elemental or structural analysis had to be performed: X-ray fluorescence analysis and microanalysis are well established EDS methods, polycrystal diffractometry using EDS is still a developing method.

2. INSTRUMENTATION

Three disadvantages in comparison to WDS limit comprehensive EDS application: the poorer energy resolution, the limited throughput rate and the fact, that high resolving EDS require permanent cooling. To achieve high resolution solid state planar detectors are usually used, based on Lithium doped Silicon [Si(Li)] or hyperpure Germanium crystals [HPGe]. Both detectors require cooling below 100 K to obtain best parameters. Usually cooling is performed using liquid nitrogen and only for special in-situ applications have different cooling methods such as gas cryostats or thermoelectric cooling been applied.
The most frequently employed detector in multielement analysis is still the Si(Li)-detector, but increasingly substituted by HPGe-detectors. Ultimate energy resolution of 120 eV for Mn-K_α-line will be obtained with HPGe-detector and 132 eV using Si(Li)-detectors. Improved technologies in HPGe-detector production have made these detectors also available for low energy photon detection below 1 keV (e.g. down to Boron). Simpler analysis,

where only a few elements have to be detected are an application domain of proportional counters mostly combined with filters.

Latest investigations in low energy photon detectors are concerned with superconductive detectors based on the determination of temperature pertubations occuring in liquid helium cooled absorbers made from compound semiconductors. Published results show excellent energy resolution comparable to WDS, as for instance 17 eV for Mn-Kα-line (Cammon 1987) using an ultracold HgCdTe-detector. Presently applications of these systems in the question are in space and heavy ion research. A wider application is prevented because of their small volume (below 1 mm^3) resulting in poor efficiency and the sophisticated cooling required for extremely low temperatures.

Pulse processing methods have not principially changed over the past decade. But development of special FET transistors and computer controlled time-variant pulse processors opened the door to high speed high resolution performances. Thus modern spectrometers realize at least a spectrometer resolution of 140 eV with throughput rates of more than 10,000 cps. Highest throughput rates are in the range of 100,000 cps (Link 1990) with an energy resolution of about 190eV.

3. ENERGY DISPERSIVE METHODS

3.1 Microanalysis

Every EDS nowadays includes a personal computer with extensive software for spectra evaluation, calculation of elemental concentrations, x-ray mapping and electron image handling. Using EDS more information is obtained than with WDS. Bremsstrahlung background has been used increasingly as an additional source of information for solving different problems of analysis, as for instance determining thickness and composition of biological samples (Hall 1972). Another example is the analysis of samples with unknown surface geometries by using peak-to-background (P/B) ratios compared to P/B ratios of suitable standard samples (Statham 1979).

When using empirical methods for background determination (e.g. linear approximation or digital filter technique), the principal disadvantage is that the absorption edge structure of bremsstrahlungs spectrum is neglected. To overcome this and to get true background intensity in EDS microanalysis of bulk samples, a method has been developed based on explicit calculation of the bremsstrahlung spectrum (Heckel 1984). This method also allows standardless analysis with a precision of a few percent, so that in some cases conventional ZAF corrections can be replaced by this method without any or with negligible loss of accuracy.

Electron microscopes require a vacuum. However recent advances in developing multilayer structures with high reflectivities for x-rays should enable analytical x-ray microscopes to be constructed which operate in air and have low x-ray backgrounds in the spectrum. Expected parameters achieved using Cu-tube primary x-rays are a local resolution of 10 μm and concentration limits at the ppm-level (Ryan 1988). There are a number of papers in this conference dealing in detail with this subject.

3.2 X-ray Fluorescence Analysis (EDXRF)

Most remarkable progress in x-ray fluorescence analysis has been made since the advent of EDS and excellent resolution at a reasonable price is available. Standard EDS allows rapid and simultaneous determination of major, minor and trace elements from sodium ($Z = 11$) upwards. Specially designed EDS enable analysis of low-Z elements down to B ($Z = 4$). The range of detectable concentration is routinely at the ppm-level or absolute mass of ng. Modern trace element analysis involves the problem to detect nano- or picogram amounts within bulk or matrix materials of various compositions. The main problem in EDXRF measurements is that the number of counts in the peak giving the signal must be clearly distinguished from statistical fluctuations of the background. Therefore any background reduction improves the detection limit.

In the last years efforts have been concentrated on reducing the radiation induced background and the methods have been developed:

- excitation in total reflection geometry,
- excitation with linear polarized x-rays,
- excitation with secondary target arrangement,

and combined techniques, as for instance application of polarized synchrotron radiation and excitation in total reflection geometry (Knoth 1978, Iida 1986).

Applying total reflection XRF a strong collimated x-ray beam imping-es on the smooth, plane and polished

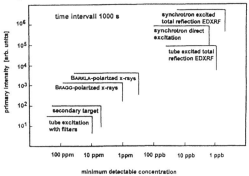

Fig. 1: Ultimate Energy Resolution of various EDXRF methods

surface of a suitable reflector material at angles below the critical angle of total reflection. The liquid sample material is evaporated onto this reflector material. When analyzing thin liquid samples absorption corrections may be neglected and the contribution from scattering to the background is drastically reduced giving best minimum detection limits (MDL) as seen in fig. 1. In the case of pulverized solid samples a previous decomposition is required giving MDL comparable to polarization methods. Background reduction by polarized x-rays is due to the fact that elastic photon scattering is strongly minimized in the case of an angle of 90° between primary beam and observation direction. A special problem is the choice of a suitable source of polarized x-rays. The ideal source, of course, is synchrotron radiation due to its features, e.g. perfect polarization, high brilliance and excellent collimation. A simpler but more easily available source is a x-ray tube in combination with a scatterer of amorphous or polycrystalline material with low atomic number (Barkla scatterer) (Heckel 1992), as for instance boron carbide. Another way generating polarized x-rays in the region below 10 keV is Bᴿᴬᴳᴳ-scattering (Wobrauschek 1988, Kanngiesser 1991) at an angle of 90°.

In the following years multilayer structures will also be applied in EDXRF. Using them as focussing broad band monochromators with high reflectivity in fluorescent radiation they allow an additional reduction of background by selective reflection in a small energy range. Thus concentration limits in the sub-ppm level can be expected in sufficiently short analyzing periods even by x-ray tube excitation.

3.3 X-ray Diffraction Analysis (EDXDA)

EDS can be applied in conventional diffractometers instead of proportional or scintillation counters and by doing so more information can be obtained due to the better energy resolution and the high efficiency. Thus complete separation of K_α- and K_β-lines can be realized (Pohlers 1992) with improved yield. On the other hand energy dispersive diffractometry can be performed using the same arrangement as for emission analysis. In this case the complete spectrum of scattered polychromatic x-radiation is simultaneously recorded and reflections due to crystal lattices occur as peaks overlapped by continuous background.

Applications of the technique have been the study of phase transitions under extreme environmental conditions, including radioactive materials. Also known are investigations of samples with large surfaces or volumes (Schiekel 1985) which require strong efforts to be made by conventional angular dispersive systems.

4. LITERATURE

Cammon D M et.al. 1987 Proc. 18th. Conf. on Low Temp. Phys.,
  Kyoto 1987 publ.in Jap. J. Appl. Phys., vol. 26 suppl. 26-3
Hall T A 1972 Micron 3 93
Heckel J, Jugelt P 1984 X-ray Spectr. 13 159
Heckel J, Haschke M, Brumme M, Schindler R 1992 J.Anal.Atom.Spectr. 7 281
Iida A, Voshinaga A, Sakurai K, Gohshi Y 1986 Anal.Chem.58 394
Kanngiesser B, Beckhoff B, Swoboda W 1991 X-ray Spectr. 20 231
Knoth J, Schwenke H 1978 Fres.Zt.Anal.Chem. 291 200
Link Analytical Ltd. 1990 Catalogue
Pohlers A 1992 Freiberger Forschungshefte B273 31
Ryon R W et.al. 1988 Adv. x-ray Anal. 31 35
Schiekel M, Jugelt P 1986 Proc. Third Work. Meet. Radioisotope
  Appl. and Radiat. Process. Leipzig 1985, pp 511-37
Statham P J 1979 Microchim. Acta, Suppl. 8 229
Wobrauschek P, Aiginger H, Owesny G, Streli C 1988 J.Trace and Microprobe
  Techn. 6 295

Microanalytical examination of segregation near the reduction front in the Mg, Mn and Ca doped wustites

J.KUSIŃSKI *, S.JASIEŃSKA *, C.MONTY **

* Academy of Mining and Metallurgy, Cracow, Poland.
** Laboratoire de Physique des Materiaux, CNRS
   Bellevue-Meudon, France.

ABSTRACT. Our paper reports the results of the microanalytical measurements of the elemental redistribution in Mg, Mn and Ca doped wustites after partial reduction. When reducing the doped wustite, iron is reduced first because the formation enthalpy of its oxide is much lower than that of CaO, MgO and MnO. Dynamic effects between Fe, Mg, Ca and cation vacancy fluxes and low solubility of Ca and Mg in iron cause segregation of these elements in initially homogeneous wustite near the Fe/doped wustite interface. Since the solubility of Mn in both phases (iron and FeO) is complete, the segregation effect at the reduction front in mangano-wustite is not so significant.

1. INTRODUCTION

Many studies have been devoted to non-stoichiometric iron monoxide (wustite) because of its interesting fundamental properties (high departure from stoichiometry, defect complex formation, problems of metastability, transport phenomena, etc.) and important applications. The complex scales produced in hot corrosion of iron have a large component of wustite in contact with metal. Wustite is also an important phase in the blast furnace (Gleitzer and Goodenrough 1985). Numerous studies have been also done on the influence of additions on the wustite because some of them change its defect structure, electronic and magnetic properties and play a role in the blast furnace during the ore (usually containing impurities) reduction process (Gleitzer 1980, Ishiguro and Nagakura 1982). Gleitzer (1980) suggested that modification of the wustite point defect structure is mainly the function of the substituant oxidation number and its ionic size. Our previous investigations (Kusiński et al 1989) showed that calcium, due to the relatively large ionic size, (as compared to iron,

manganese and magnesium) changes the structure of wustite
significantly. We proved that changes in the chemical
composition of wustite may significantly influence its
reduction process. Our present study is one among the series
of investigations devoted to the role of the iron substituted
elements (Ca, Mn and Mg) on the structure and reduction
behaviour of the doped wustite.

2.EXPERIMENTAL PROCEDURE

The Ca, Mg and Mn doped wustite samples have been prepared as
follows: oxidation of high purity polycrystalline iron plates
for 8 days at different temperatures ($1150^{\circ}C$, $1200^{\circ}C$ and
$1260^{\circ}C$) ; thermochemical treatment (in pure CaO powder at
$900^{\circ}C$ for 4 hrs; in pure MgO powder at $1000^{\circ}C$ for 6 hrs; in
pure MnO powder at $900^{\circ}C$ for 8 hrs), homogenization at $900^{\circ}C$
for 4 days. All these treatments have been carried out in the
$50\%CO/50\%CO_2$ atmosphere. Such treatment enabled us to obtain
concentration of 0.82-0.85% Ca; 2.5-2.8% Mn; 3.9-4.2% Mg,
uniformly distributed in the wustite (as measured by CAMEBAX
electron probe microanalyzer). The reduction of the
doped-wustites has been carried out at $900^{\circ}C$ and $1000^{\circ}C$ in CO
atmosphere. Before reduction the samples were annealed for
3 hrs in the atmosphere of $50\%CO$ - $50\%CO_2$; then the gas was
switched to pure CO and the samples were reduced for various
times (from 30s to 2h). After the reduction process the
samples were rapidly cooled down.

3.RESULTS AND DISCUSSION

Figures 1-3 represent the
structure of the partially
reduced, doped wustites. During
the reduction of the doped
wustites Fe is reduced first
because the formation enthalpy of
its oxide is much lower than that
of CaO, MgO and MnO. There was a
significant difference between
the microstructure of the iron
layers formed on the surfaces of

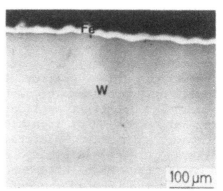

the mangano- magnesio- and calcio-
wustites. In the case of the
partially reduced mangano-wustite
a compact, thin, Fe-Mn layer was
observed at the surface (Figure 1).

Figure 1. Optical micro-
graph of the mangano-
wustite reduced at $900^{\circ}C$
for 30 min.

At $900^{\circ}C$ the solubility of Mn in the wustite and in the iron
is complete,so there is mainly one limitation of the reduction
process: the diffusion of the oxygen from the wustite through

the oxide/metal interface and the dense iron layer to the sample surface. The migration rate of the Fe, Mn and vacancies at the reduction front seems to play a less important role in the reduction process of the mangano-wustite. However, Mn diffusion coefficient in FeO is five orders of magnitude higher than in the iron. This causes Mn segregation at the

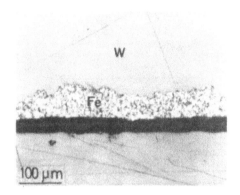

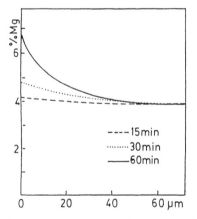

Figure 2. Optical micrograph of the magnesio-wustite re-duced at 900°C for 30 min.

Figure 3. Optical micrograph of the calcio-wustite reduced at 900°C for 1 hour.

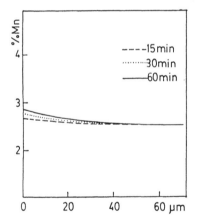

Figure 4. Mn profile near the reduction front in partially reduced (at 900°C) sample of the mangano-wustite.

Figure 5. Mg profile near the reduction front in partially reduced (at 900°C) sample of the magnesio-wustite.

reduction front (Figure 4). In the other two cases (magnesio- and calcio-wustites, Figures 2 and 3 respectively) the iron layers were more porous and thicker than that formed at the surface of the reduced mangano-wustite. Since oxygen diffuses readily from the reduction front to the surface through the porosity, the reduction rate of both calcio- and magnesio-wustite is much higher than that of mangano-wustite. Measurements of the Ca and Mg concentrations near the Fe/doped wustite interfaces has shown a large amount of segregation (Figures 5 and 6). Such an effect is due to the low solubility of Ca and Mg in the iron (the microprobe analysis indicated about 0,05% Ca and Mg in the as-reduced Fe layer) and a coupling between Fe, Ca or Mg and cation vacancy fluxes and the phenomenon is called "dynamic segregation" (Petot-Ervas et al 1990).

4.CONCLUDING REMARKS

During the reduction of the doped wustites Fe is reduced first. Increases in the Ca, Mg and Mn concentration were observed at the reduction front. In the case of magnesio- and calcio-wustites such a phenomenon is connected to the low solubility of Ca and Mg in iron and dynamic effects between Fe, Mg, Ca and cation vacancy fluxes. Segregation in mangano-wustite was not significant and we think, that this was the result of the differences between Mn diffusion coefficients in iron and wustite.

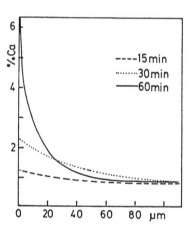

Figure 6. Ca profile near the reduction front in partially reduced (900°C) sample of the calcio-wustite

5.REFERENCES

1.Gleitzer C. and Goodenrough J.B. 1985, Mixed-Valence Iron Oxides in Structure and Bonding 61, Springer-Verlag, Berlin-Heildelberg, pp. 1-76.
2.Gleitzer C. 1980, Mat.Res.Bull., 15, p.507
3.Ishiguro T. and Nagakura S. 1982, Jap.Journ. of Appl.Phys., 138, 3, p.723
4.Kusiński J., Riviere A., Jasieńska S. and Monty C. 1989, Proceedings of VII Electron Microscopy Conference, Kraków - Krynica, pp. 405-408.
5.Petot-Ervas G., Mointy C., Dhalenne G., Leduigou J., Kusiński J., Klimczyk H., Kołodziejczyk M, Jasieńska S., Janowski J. 1990, IV-th Round Table Meeting on Transition Metal Oxides and Compounds, Mądralin.

Inst. Phys. Conf. Ser. No 130: Chapter 2
Paper presented at Int. Congr. X-ray Optics and Microanalysis, Manchester, 1992

Non-equilibrium states of eutectics in Mg-Zn alloys

R.Ciach, M.Socjusz-Podosek

The Institute for Metal Research, Polish Academy of Sciences,

30-059 Kraków, Poland

ABSTRACT: The process of non-equilibrium solidification of Mg-Zn alloys has been simulated theoretically based on Phann's equation and an evaluation done by Krupkowski. The maximum amounts of equilibrium and non-equilibrium eutectics appearing in as cast Mg-Zn alloys have been calculated and compared to microprobe results.

These eutectics have been observed in two forms due to various cooling rates during solidification. The (α + Mg_7Zn_3) - eutectics are obtained at higher rates and the (α + MgZn) - eutectoid at lower ones. Disadvantageous effects of the eutectics occurrence on the alloy mechanical properties were diminished with step by step annealing.

The microsegregation in Mg-Zn alloys has been described by an equation of non-equilibrium solidification by means of changes of zinc distribution across the grain as well as the amount of non-equilibrium eutectics. The calculations have been based on the assumption that there is such a theoretical rate of crystallization that it ensures full diffusion in the liquid and no diffusion in the solid state. The basic equation, after Pfann (1952) has been as follows:

$$k = \frac{N_{(Zn)}^x}{N_{Zn}^x} = \frac{dn_{(Zn)}(1-x)}{dx\, n_{Zn}^x}$$

After some rearrangements described in the papers of Krupkowski (1968) and Ciach et al (1980), the equation for the Mg-4wt% Zn alloy in its final form has been obtained:

$$\ln(1-x) = -0.2895\, N_{Zn}^x + 0.1248(\ln N_{Zn}^x)^2 + 0.6519\, \ln(-\ln N_{Zn}^x) - 0.4328$$

where

k - the distribution coefficient, x - the amount of the solidified

dendrite in 1 mole of the alloy at a given stage of solidification,
N^x_{Zn} - the mole fraction of zinc in the liquid at a given stage of
crystallization, $N^x_{(Zn)}$- the mole fraction of zinc in the solid formed   on
the dendrite surface at a given stage of solidification,
$n^x_{(Zn)}$- the amount of zinc in the dendrite at a given solidification
stage, and n^o_{Zn} - the amount of zinc in one mole of alloy.

The latter relationship may be used to calculate the changes of zinc
concentration across the solidified dendrite as well as the whole amount
of eutectics. Such a theoretical curve of maximum changes of zinc
concentration in the dendrite cross-section of Mg-4Zn alloy is presented
in Fig.1.

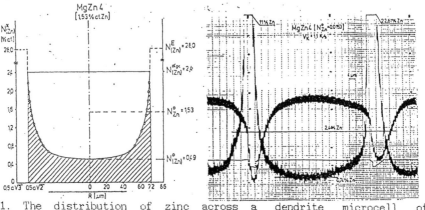

Fig.1. The distribution of zinc across a dendrite microcell of
Mg-4Zn alloy: a) theoretical changes across a spherical model,
b) microprobe experimental results.

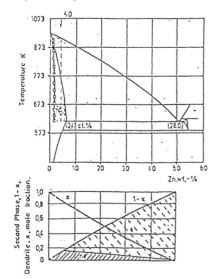

Fig.2 The general characte-
-ristics of the Mg-4wt%Zn
alloy.

a) the hypo-eutectic part
of Mg-Zn phase diagram with
the non-equilibrium solidus
of the Mg-4wt%Zn alloy.

b)   the   amounts   of:
eutectics (1-x), non-equilibrium eutectics (ne),
equilibrium one (e) and
solid solution (x) versus
Zn contents.

According to the equation, the theoretical "non-equilibrium solidus line" can also be calculated (Fig.2a). Its place suggests that the non-equilibrium eutectics should be expected in all alloys from the solid solution range as well as in hypoeutectic alloys. The general characteristics of the Mg-Zn alloys from this range showing amounts and relationships of the equilibrium (e) and non-equilibrium (ne) eutectics is presented in Fig.2.

Fig.3 shows the typical microstructures of the Mg-4Zn alloy cooled at various rates.

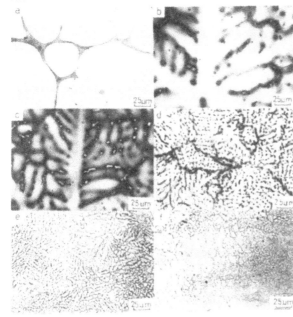

Fig.3. Microstructures of Mg-4Zn alloy solidified at: a) 0.03; b) 1; c) 10; d) 25, e) 150, f) 10^5 K/s. The changes of morphology from two phase, cell like structure to dendritic and eventually to equiaxial, single phase structure are visible at the increasing cooling rates.

In order to compare the theoretical and experimental values, the analysis of zinc contents across the dendritic grain has been done by means of the x-ray microanalizer and an example of it is shown in Fig.1b. The amounts of eutectics were determined metallographically and plotted versus a cooling rate (Fig.4).

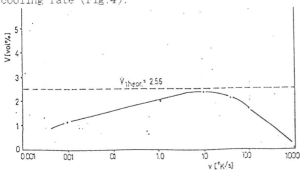

Fig.4. The dependence of the amount of eutectics on a cooling rate of Mg-4wt%Zn alloy

As it can be seen, there was a rate which provided the maximum contents of the (α + Mg_7Zn_3)-eutectics. Above and below it, the amounts of the precipitates decreased. However, at cooling rates, the grain size as well as the size of the particles enlarged (Fig.3), thus a decrease of mechanical properties was obvious.

The highest rate maintained with the rotating disc method gave as a result a completely one phase structure with no precipitates nor zinc segregation. The TEM micrograph shown in Fig.5 is the evidence for this.

Fig.5.TEM micrograph of the rapidly solidified Mg-4wt.%Zn alloy.

The microstructural investigations revealed two kinds of eutectics in Mg-Zn alloys. According to the phase diagram they are magnesium solid solution and either the intermetallic phase Mg_7Zn_3 or MgZn obtained after eutectoidal decomposition of Mg_7Zn_3 at 598K (Clark and Rhines 1957, Agarwal et al. 1992). The two "eutectics" differ from each other by structure and are shown in Fig.6. The respective diffraction paterns proved the different structure but were difficult to solve. At the same time the EDS analysis showed the same composition.

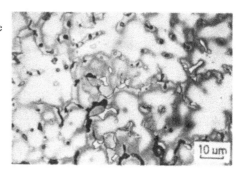

Fig.6. Two kinds of eutectics in the Mg-4wt%Zn alloy. White areas correspond to α+Mg_7Zn_3 while dark ones are the α+MgZn eutectoid.

The appearance of the two kinds of eutectics is very disadvantageous for alloy homogenization because it requires a step by step heat treatment to avoid a partial melting of Mg_7Zn_3 particles.

REFERENCES.

Ciach R, Zawadzka B and Podosek M 1980 Proc.Conf. on Solidification Technology (The Metals Society) pp 292-297
Agarwal R, Fries S, Lucas H, Petzow G and Sommer F 1992 Z.Metallknd 83 4
Clark J.B and Rhines F.N 1957 J.Metals 4 425
Krupkowski A 1968 Bull.acad.Pol.Sci.6 823
Pfann W.G 1952 Trans.AIME (J.Met.) 7 747

Inst. Phys. Conf. Ser. No 130: Chapter 2
Paper presented at Int. Congr. X-ray Optics and Microanalysis, Manchester, 1992

189

Composition of M_6C carbides formed in nickel-based hardfacing alloys

S. Hamar-Thibault*, N. Valignat* and S. Lebaili**.

*L.T.P.C.M., I.N.P. Grenoble, BP. 75 - F.38402 Saint Martin d'Hères, France

** Département de Métallurgie, I.G.M., U.S.T.H.B., BP 32 - El Alia, Alger, Algérie.

ABSTRACT: The solidification of some nickel-based hardfacing alloys containing large amounts of silicon, chromium and tungsten have been studied. These alloys show complex carbides (M_7C_3 and M_6C) and chromium boride. By means of quantitative EPMA the composition of the carbide and boride phases have been determined The M_6C carbides have a formula of type $(NiSi)_x(CrW)_yC$ with x and y varying respectively from 2.5 to 2.9 and from 2.5 to 3.6 while M_7C_3 carbides have a stoichiometric composition.

1. INTRODUCTION

Cobalt and nickel alloys are used as wear resistant materials for hardfacing. Their good resistance are due to the presence of hard phases (carbides, borides) in a fcc matrix. This study is a continuation of the investigation of carbides in cobalt (Hamar-Thibault et al 1982) and nickel wear resistant alloys. The Co-W-C system was the most thoroughly investigated owing to its importance in hard metals. The ternary system Ni-W-C was studied by Fiedler and Stadelmaier (1975) which show the formation of cubic ternary carbides of M_6C (Ni_3W_3C and Ni_2W_4C) and $M_{12}C$ (Ni_6W_6C) types. C was not analysed. Addition of chromium leads to the formation of other carbides (Cr_7C_3 and $Cr_{23}C_6$). Nasanori (1983) gave isothermal sections (1100°C) of ternary Ni-Cr-W system at constant carbon level (1 and 3%).

This paper presents the composition of carbides and borides formed in complex Ni-hardfacing alloys. The accurracy of quantitative electron probe microanalysis of the very light elements (C and B) was discussed.

2 EXPERIMENTAL PROCEDURE

A series of Ni-based alloys was prepared by different methods (casting in an air atmosphere, differential thermal analysis -DTA- with a cooling rate of 5°C per min and QIDS). The composition of the alloys have been chosen with a large range of carbon, boron and silicon concentration. The chromium content is maintained around 26at% while that of tungsten is around 3.4at%.

	C	B	Si	Primary phase
A	6.1	4.7	1.9	M_7C_3
B	4.7	4.9	5.8	M_6C
C	3.9	6.6	5.2	M_6C or M_7C_3

Table 1 : Non metallic element contents in the alloys (atomic percentage).

The specimens were carefully polished and analysed in an electron microprobe equipment CAMECA SX-50. The analyses were performed at 15keV using standarts of elemental Ni, Cr, Fe, W, Si and of compounds with known compositions for C (Cr_7C_3 with 8.71%wtC) and B (TiB_2 with 28.85wt%B). Carbon was analysed with a PC1 monochromator crystal (W-Si multilayers, 2d = 60Å) whereas boron was analysed with a PC3 monochromator crystal (Mo-B_4C multilayers, 2d ≈ 200Å). Neither significant shifts nor shape alterations of the carbon and boron peaks were observed, like those mentionned by Bastin and Heijligers (1990), between standart and analysed phases. Atomic number, absorption and fluorescence effects were corrected with the computer program CAMECA according to the PAP model (Pouchou and Pichoir, 1986). In spite of the number of elements, the total error on weight concentration was always below 3% for M_6C and 2% for M_7C_3 and Cr_xB.

3. EXPERIMENTAL RESULTS

3.1. Composition of the M_6C and M_7C_3 carbides.

These alloys were characterized by the formation as primary phase between 1300 and 1230°C of either the M_7C_3 carbide (alloy A) or a complex compound (alloys B and C). This latter phase appears in the micrograph as large whitish phase contrary to the carbide M_7C_3 (fig.1) which possesses a darkish contrast. These phases were surrounded by different eutectics between Ni(α), M_7C_3 and Cr_xB depending on the initial composition of the alloy.

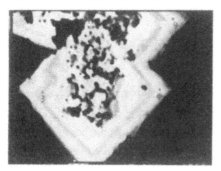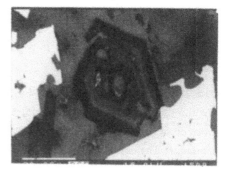

Fig.1: Morphology of M_6C (whitish) and M_7C_3 (black) carbides in alloy B.

The M_7C_3 carbides have an homogeneous composition with a W level slightly lower that the nominal composition of the alloy (1-3 at%). On the contrary, the whitish phase was characterized by high Cr, W and Si contents. Its average formula $(M_{4.9-6.5}C)$ approaches the carbide of M_6C type mentionned in the ternary Ni-W-C system.

The cubic M_6C carbides have a formula of type $(NiSi)_x(CrW)_y(CB)$ with x varying from 2.5 to 2.9 and y varying from 2.5 to 3.6. Si content was high around 12at%. W level decreased as Cr level increased so that the atomic percentage (Cr+W) remained approximatively constant between 42 to 45 at%.

	C	B	Si	Ni	Cr	W
quenched	13.4-15.9	0.8-1.1	11.8-11.9	27.5-28.1	25.5-27.3	16.0-16.2
DTA	11.9-15.3	0.0-1.0	10.6-13.3	28.8-29.9	26.8-30.7	13.2-18.0
annealed at 800°	12.3-15.1	0.1-1.2	8.2-12.2	28.4-30.3	27.3-31.9	11.3-15.2

Table 2: Composition of M_6C carbide (atomic percentage) after different treatments (alloy B).

The composition range of the M_6C carbides was different from that of the ternary carbides in the Ni-W-C system, where the W content of Ni_2W_4C (57 at%) lies inside the analysed composition range (51-62at%) whereas that of Ni_3W_3C (43 at%) is outside. On the contrary, the composition of M_6C carbides was very similar to the composition of these carbides formed in Co-based alloys, where, the atomic composition (W+Cr) varies from 43 to 46% and the formula of the carbides was of $(CoNiSi)_x(CrW)_yC$ type (2.6 <x< 3.8 and 3.2 <y< 3.6).

Thermal aging at high temperatures show that these M_6C carbides have a constant composition even after 10 days at 800°C with only some decrease in silicon level. We did not detect any noticeable loss in carbon while in M_6C carbides formed in Co-based alloys annealed at the same temperature, the composition of M_6C evolves to $M_{12}C$ (Hamar-Thibault et al 1982). During aging, the chemical composition of the M_7C_3 carbides does not change significantly.

3.2. Competing growth of the M_6C and M_7C_3 carbides.

For some compositions (alloy C), the primary phase formed after quenching in air was the carbide M_7C_3 indicating that the M_6C carbides were formed by a peritectic reaction from the M_7C_3 carbide as in ternary Ni-W-C system.

The formation of the thermodynamically stable phase with high tungsten content will therefore be favoured by a long stay in the hot zone or by a slow cooling. On the contrary, during quenching in air, the short stay in the hot zone is no longer sufficient and the formation of a phase type M_7C_3 carbide with tungsten content in the neighbourhood of the nominal composition of the alloy (4at%) outside its existence domain will be stabilized.

Moreover, the carbides (M_6C or M_7C_3) obtained from the solidification of alloy C show often some contrast variations in the form of parallel bands at the crystal faces In M_7C_3 carbides, the bands were large enough to allow significant analyses. The results show that the zones which present a bright contrast are rich in tungsten ($\approx 3.4\%$) and poor in chromium whereas zones with dark contrast are rich in chromium to the detriment of tunsgten ($\approx$ 2at%) so that (Cr+W) remained constant. These carbides have a stoichiometric composition and the contents of the other elements are constant whatever the contrast at the interior of the carbide.

On the contrary, the differences in contrast in M_6C carbides analysed in alloys B and C were not associated with the concentration variations as in the carbide M_7C_3. In fact the tungsten content is identical and it is only the carbon and boron contents that seem to be slightly affected. Giving the fixed composition of the M_6C carbide, its growth can only occur when the tungsten content in the neighbourhood of the crystal attains a high value by homogenisation of the neighbouring liquid.

3.3. Chromium borides formed in Ni-based alloys (Table 3).

The borides formed as eutectic during the solidification process have a fixed composition $(CrNiW)_5(BC)_3$ corresponding to the Cr_5B_3 binary phase. Since the carbon level was low, the influence of carbon on the boron analyses was low. As in carbides, (Cr+W) remained approximately constant.

	C	B	Si	Ni	Cr	W
alloy B	1.9	35.9	0.0	4.7	50.5	6.9
alloy C	1.6-2.6	35.2-35.6	<0.02	4.0-4.1	45.4-50.5	7.5-12.9
alloy C *	3.3	34.7	0.0	3.9	51.5	7.4

Table 3: Composition of borides (atomic percentage).

For very slow cooling rates and for some particular alloys (as in alloy C), the increase of boron level in the liquid results in a decomposition of the M_6C carbides giving a chromium rich carbide with the same characteristics (C*).

References

Bastin G.F. and Heijligers H.J.M. 1990 Scanning 12 225.

Fiedler M.L. and Stadelmaier H.H. 1975 Z. Metallkde 66 402.

Hamar-Thibault S., Durand-Charre M. and Andries B. 1982 Met. Trans. 13A 545.

Nasanori K. 1983 J. Iron Steel Inst. Japan 69 1455.

Pouchou J.L. and Pichoir F. 1986 J. Microsc. Spectrosc. Electron. 11 229.

Inst. Phys. Conf. Ser. No 130: Chapter 2
Paper presented at Int. Congr. X-ray Optics and Microanalysis, Manchester, 1992

193

Application of X-ray microanalysis to investigate adhesion of copper thick films

M Jakubowska, L Kaczyński, J Paduch

Institute of Electronic Materials Technology,

ul. Wólczyńska 133, 01-919 Warsaw, Poland.

ABSTRACT: The influence of diffused elements on adhesion of copper thick films contained different kinds of glazes has been examined using X-Ray microanalysis. The investigations   were performed on the traverse surface of samples and on the surface of the glaze near the substrate. The obtained results provide interpretation of the mechanism of adhesion of thick copper films to alumina substrates.

1. INTRODUCTION

Since the costs of noble metal pastes have been continuously increasing, it is necessary to replace them in mass production of microcircuits with other cheaper materials. Today, copper pastes fired in nitrogen atmosphere are becoming more popular in manufacturing hybrid microcircuits. The use of copper thick film conductors requires a detailed knowledge about their behavior on the substrate. Usually 96% Al_2O_3 is used as a substrate. Adhesion of the layers to the substrate is a very important parameter which determines the quality of the circuits. Copper films are fired in nitrogen with a very low oxygen content, typically 2 ppm, and this has a great influence on adhesion of the layers. The adhesion has been observed to be very sensitive to the conditions of firing and paste compositions [1-3].

In the present work the relationship between the changes of adhesion of copper layers during several firings and observed diffusion of elements contained in different glazes has been presented. This is related to the multilayer microcircuit technology, since the copper layer is to be fired several times. The obtained results provide interpretation of the mechanism of copper layer adhesion to the substrate.

2. SAMPLE PREPARATION

The three copper pastes contained copper powder and different lead-boron-alumina-silicate glaze were used for the investigation of adhesion strength. The glazes have different chemical composition which causes their different melting points and the wettability of the surface. These pastes were prepared in the Institute of Electronic Materials Technology in Warsaw. For comparison, the Du Pont QP-153 copper paste was used. All four pastes were screen printed and fired once, twice, three and four times in a nitrogen atmosphere with 2 ppm oxygen content. The adhesion of copper layers to alumina substrates was measured using a peel test. The results are shown in table 1. The measurements of adhesion are in $N/4mm^2$.

TABLE 1.   Adhesion of copper layer to alumina substrates
after one, two, three and four cycles of firing

No of cycles →	1	2	3	4
Paste 1	19.3	18.8	19.6	19.8
Paste 2	18.8	18.2	18.9	19.6
Paste 3	12.2	14.4	13.9	14.8
Paste 4	18.2	20.8	20.9	21.6

3. EXPERIMENTAL

The X-Ray microanalysis WDX and EDX and scanning electron microscopy were used for examining the surface of copper layer and the traverse surface of the samples. The surfaces of copper samples fired once, twice, three and four times were chemically etched in the mixture of the acids to examine the surface of glaze near the substrate.

In the pictures of SEI (see Fig. 1) the trace of the intermediate layer between the copper layer and the substrate can be seen as a non-continuous thin membrane which was not etched.

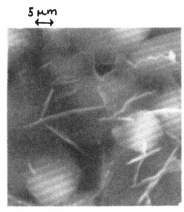

Fig.1 SEI of the intermediate layer of paste 4

Fig.2 Distribution of the
elements in the traverse
surface of paste 4. After
one(A) and four(B) firings

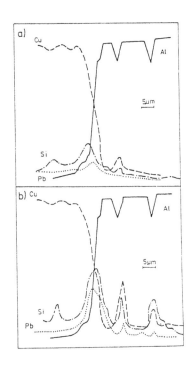

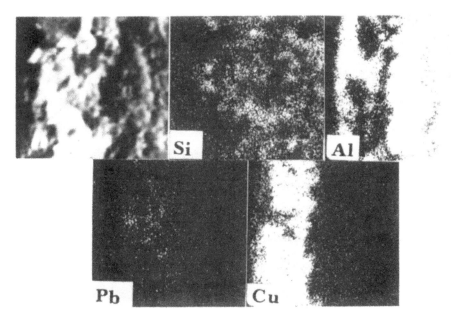

Fig.3 The structure of copper layer (Paste 1) after three firings   **x 1000**

The distribution and segregation curves of elements contained in the glaze were examined on the intermediate layer.

The following observations were obtained:

- The elements of the glaze stabilized into the structure of the substrate(Figs 2 and 3).
- Lead is concentrated near the intermediate layer.
- Silicon is also concentrated together with lead.
- The membrane between the copper layer and the substrate is not continuous and has a similar structure to the copper layer.
- The concentration of copper in the substrate is increased and the diffusion of copper is deeper with the number of firing cycles.
- The different adhesion of the samples is related to different kinds of glaze in the pastes. These glazes have different wettability of copper and the substrate. This is the reason of different tendency of stabilizing the glass into the substrate.

4. CONCLUSIONS

The binding of the copper layer with the alumina substrate is caused by stabilizing the glaze into the extended surface of the substrate and the copper diffusion into the substrate.

The stability and firmness of the glaze and copper in the substrate are directly dependent and increased with the number of firing cycles. This also would results in a better adhesion.

The membrane between the copper and the substrate differs in chemical contents from the copper layer but does not form a layer with different structure.

5. REFERENCES

[1]   Kuo C.Y., Adhesion of thick film copper conductors (1), Int. J. Hybrid-Microelectr., 4, 2, 70-78, 1981.
[2]   Kuo C.Y., Adhesion of thick film copper conductors (2), ibid., 161, 1, 17-22, 1988.
[3]   Kuo C.Y., Adhesion of thick film conductors (3), Proc. 8th European ISHM Conference, Rotterdam, 490-497, 1991.

Inst. Phys. Conf. Ser. No 130: Chapter 2
Paper presented at Int. Congr. X-ray Optics and Microanalysis, Manchester, 1992

197

Relative changes between the number and calcium content of atrial specific granules

Z YU, C.ZHONG and K. OGAWA*

Department of Biophysics, Shanghai Medical University, Shanghai 200032, China
*Department of Anatomy, Kyoto University, Kyoto 606, Japan

ABSTRACT: In this work, the release of atrial natriuretic peptide(ANP) was estimated by counting the number of atrial specific granule(ASG) through morphometry, and the alteration of Ca content in ASG was estimated by measuring the [Ca] through electron microscope X-ray microanalysis. The results showed that ASG number and its [Ca] increased or decreased parallelly during the alteration of water and salt load, and the lowest [Ca] value was reached at 30 min, while the lowest ASG number was reached at 60 min after adrenalin treatment. These findings suggest that the Ca releasing from ASG be a factor inducing the release of ANP.

1. INTRODUCTION

It is generally considered that in the stimulus-secretion coupling processes, an activator binding to the cellular surface receptor induces Ca mobilization, and then cellular secretion is followed. Zhong et al. (1990) proved that the ASG is an intracellular Ca store, which contains high concentration of Ca. The purpose of the present work is to study whether the Ca storing in the ASG plays a role in the secretion of ANP.

2. MATERIALS AND METHODS

Animal treatment: 59 adult Sprague Dawley rats weighing 200-250g were used. The rats were divided into seven groups. In these groups, the rats were treated in different ways to induce the change in ASG number. Group A, depriving of water for 5 days in 18 rats; Group B, drinking 2% NaCl solution for 4 days in 13 rats; Group C-F, injecting subcutaneously adrenalin 2mg/kg to each rat 15, 30, 60, 120 min before their being sacrificed, 2 rats in each time group; Group G, normal control in 18 rats. After the predetermined treatment, the rat was sacrificed by decapitation, and its right auricle was excised immediately and cut into blocks of 1 mm³ . The specimens were processed in the following two methods (see 2.1 and 2.2).

This work was supported by Grant 39070292 from the National Science Foundation of China

2.1 Preparation of ultrathin sections in routine method for morphometry

The ultrathin sections were observed under a JEM-1200 EX EM. Micrographs were taken at one pole of the nucleus at random in a magnification of 5,000. An array of short lines was super-imposed on each micrograph magnified 20,000 diameters simul-taneously. The numerical density(Nv) of ASG was calculated according to the formula of Nv=Na/D (Williams 1977). The numerical density on area(Na) and the mean diameter(D) of ASG could be calculated from the parameters obtained from the array of short lines superimposed on the micrographs. The datums were the average of 20 pieces of photos.

2.2 Ethanolic phosphotungstic acid(EPTA) method for micro-analysis

The EPTA method (Pimenta et al. 1980) was used to stain the ASG cytochemically. The auricle blocks of $1mm^3$ were fixed in 2.5% glutaraldehyde, and dehydrated in grade ethanol till 100%. Then, the blocks were immersed in 100% ethanol with 2% phosphotungstic acid for 5 hr at room temperature, rinsed with 100% ethanol 3 times, 15 min each, and embedded in Spurr's resin. The ultrathin sections were observed and [Ca] analyzed under the electron microscope.

2.3 Quantitative electron microscope x-ray microanalysis (EMMA)

A JEM-1200EX EM equipped with a Link AN 10000 energy dispersive x-ray spectroscope(EDS) was used for element analysis. The TEM mode of image was used with 80 kV accelerating voltage. The acquired spectra were processed with the Link Quantem/FLS quantitative analysis software. The element concentration was denoted in mmol/kg-dry-wt, and the datums were the average of about 20 measurements.

3. RESULTS

3.1 Alteration of Nv and [Ca] of ASG in rats with different water and salt load

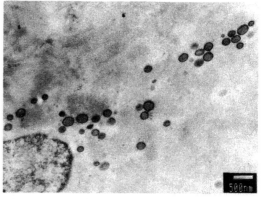

The morphometry results showed that the Nv of ASG was 6.02±2.30/um^3 in control Group G, and in-creased in Group A (de-priving of water) to 9.97±3.21/um^3 , decreased in Group B(drinking 2% NaCl) to 2.96±1.62/um^3 .

Fig 1 A micrograph of an atrial cardiocyte in EPTA staining section.

The differences between Group A and G, and between Group B and G were statistically significant(p<0.01). The [Ca] in ASG was analyzed on EPTA staining sections. In these sec-tions, the ultrastructure of the atrial cardiocyte could not be identified, but its nucleus and ASG could be seen clearly

(Fig.1). The element analysis results showed that the [Ca] of ASG was 92±18 in Group A, 38±22 in Group B and 64±16 mmol/kg-dry-wt in Group G. The differences of [Ca] between Group A and G, and between Group B and G were statistically significant (p<0.01).

3.2 Alteration of Nv and [Ca] of ASG in rats with adrenalin treatment

In the sections from Group C-F, in which the rats were sacrificed respectively at 15th, 30th, 60th, 120th min after the adrenalin treatment, the Nv of ASG were 4.32±2.31, 3.75±2.07, 3.14±1.39 and 5.19±1.84/um³, and the [Ca] of ASG were 40±16, 12±10, 31±14 and 54±20 mmol/kg-dry-wt respectively. The differences of both measurements between Group C, D, E and G were statistically significant

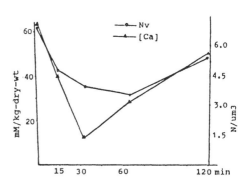

Fig 2 The alteration of the number and [Ca] of ASG after adrenalin injection.

(p<0.01), and the difference between Group F and G was not significant (p>0.05). The decrease in the Nv reached its lowest value at 60 min and the decrease in [Ca] reached its lowest value at 30 min after the adrenalin injection. The alteration of Nv and [Ca] in ASG was shown in Fig.2.

4. DISCUSSION

In our previous work (Zhong et al. 1990), it was observed that the ASG contained high Ca concentration and the Ca-ATPase existed on its membrane. On the basis of this observation, the ASG was named intracellular Ca store. The purpose of the present work was to explore the role of the Ca storing in the ASG in the secretion of ANP. Since De Bold (1979) and Cantin et al. (1982) proved that the number of ASG increased in water-deprived and in sodium-deficient rats, the number of ASG in a cardiocyte could be used as a criterior of ANP level in serum. It was observed that when the Nv of ASG increased in the water-deprived rats, an appearance of inhibition of ANP secretion, the [Ca] of ASG also increased, on the contrary, when the Nv of ASG decreased in the sodium-loaded rats, an appearance of increase of ANP secretion, the [Ca] of ASG also decreased. These results showed that a decrease in [Ca] of ASG was associated with a decrease in ASG number, which indicated an increase in ANP secretion.

Sonnenberg and Veress (1984) reported that adrenalin stimulated secretion of ANP. In the present experiment, the Nv of ASG decreased after adrenalin treatment and reached its lowest value at 60 min after the treatment. Meanwhile, the [Ca] of ASG decreased prominently at 15 min after the treatment and reached the lowest value at 30 min. Comparing the alteration course of

ASG number with that of [Ca] in ASG after the adrenalin injection, it was found that the alteration of [Ca] was earlier than that of ASG number. These findings suggested that the Ca released from ASG due to an activator become a messenger triggering the ANP secretion. The mechanism of the effect of ASG [Ca] on the release of ANP remains to be studied further.

The EPTA method was developed by Pimenta et al.(1980) to show eosinophil granules containing basic protein. Seidah et al. (1984) reported that ANP had been isolated from the rat atria, and sequenced and synthesized. It contains basic amino acids, such as arginine and lysine. Therefore, the ASG can be stained with EPTA, in which OsO_4 fixation and U and Pb staining is saved. The experimental data showed that the [Ca] of ASG in EPTA staining sections was slightly less than that in rapid freezing ultrathin cryosections (Zhong et al. 1990; Somlyo et al. 1988). It is believed that a large part of Ca in ASG is combined with the Ca-binding protein and can be measured by EMMA (Ornberg et al. 1988). Furthermore, the EPTA staining method is much simpler than the low temperature techniques in biological electron microscopy. Therefore, the EPTA staining method is suitable for measurement of Ca binding to protein.

REFERENCES:

Cantin M, Michelakis AM, Ong H, Ballak M, Beuzeron J and Benchimol S 1982 Advances in myocardiology eds E Chazov, V Smirnov and NS Dhalla (Plenum Medical Book Company, New York) pp 519-29
De Bold AJ 1979 Proc. Soc. Exp. Biol. Med. 161 508
Ornberg RL, Kuijpers GAJ and Leapman RD 1988 J Biol. Chem.263 88
Pimenta PFP, Loures MAL and Souza W de 1980 J Histochem. Cytochem. 28 238
Seidah NG, Lazure C, Chretien M, Thibault G, Garcia R, Cantin M, Genest J, Brady SF, Lyle TA, Paleveda WJ, Colton CD, Circerone TM and Veber DF 1984 Proc. Natl. Acad. Sci. USA 81 2640
Somlyo AV, Broderick R, Shuman H, Buhle EL,Jr and Somlyo AP 1988 Proc. Natl. Acad. Sci. USA 85 6222
Sonnenberg H and Veress AT 1984 Biochem. and Biophys. Res. Commun. 124 443
Williams MA 1977 Practical Methods in Electron Microscopy ed AM Glauert (Oxford, North-Holland Publishing Company)
Zhong C, YU Z, YU W and Ling Y 1990 Microbeam Analysis pp390-2

Inst. Phys. Conf. Ser. No 130: Chapter 2
Paper presented at Int. Congr. X-ray Optics and Microanalysis, Manchester, 1992

201

SEM and EDX studies of Na penetration in carbon and graphite materials

J Ogórek and J Wróblewska

R D Laboratory, of POLGRAPH SA, 33-300 Nowy Sacz, Poland

Abstract. One of the essential characteristics of cathode lining blocks used in AL reduction cells is their resistance towards electrolyte and especially Na penetration during electrolysis. Sodium penetrating a carbon body creates intercalation compounds the effect of which is an irreversible volume expansion. This can lead to material disintegration. SEM and EDX studies of two commercial carbon materials have been carried out to show their structure as well as sites of sodium concentration.

1. Introduction

The purpose of the cathode lining in AL reduction cells is twofold, it acts as a refractory container for molten metal and conducts electricity.

Carbon blocks are nowadays the basic constructional material for AL reduction cells. For their production carbon materials like anthracite, artificial graphite and coal tar pitch are used. During electrolysis, electrolyte and particularly Na penetrates into the carbon body. The initial reaction which leads to impregnation of Na is considered to be (Sørlie and Øye 1989):

$$Al(1) + 3NaF(in\ cryolite) = 3Na(in\ carbon) + AlF_3(in\ cryolite)$$

The interaction between Na and C is enhanced by bond forming through intercalation. The intercalation compounds $C_{64}Na$ and $C_{12}Na$ are known (Ascher 1959, Pflugmacher and Boehm 1970).

Sodium impregnation leads to dimensional expansion of the carbon materials and may cause dangerous cracks or distortions.

2. Experiment

The experiments have been carried out on two commercial carbon cathode materials: WAL62A and WAL7G (made in POLGRAPH SA). They have different composition and hence structure (see figure 1) WAL62A is composed of 70% anthracite (particle size 12 to 0.3 mm) and 30% graphite (particle size 12 to 0.3 mm). WAL7G is composed of 35% anthracite (particle size 3 to 0.3 mm) and 65% graphite (particle size 12 to

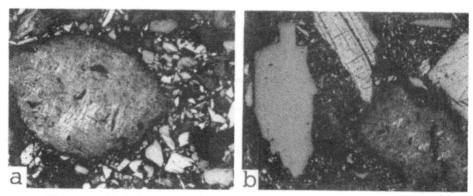

Figure 1. Typical structures of carbon cathode block materials: (*a*) WAL7G, (*b*) WAL62A.

0.3 mm). Coal tar pitch is used as a binder, after baking it forms a coke binder phase connecting particles of solid components.

Samples

Samples for studies were cut out from cylindrical carbon electrodes after experimental electrolysis in the so called Rapoport apparatus (Rapoport and Samoilenko 1957).

Samples were ground and polished with Cr_2O_3 to reveal particles of interest.

Equipment

The studies were performed using a polarized light microscope, Olympus PMG3 and a scanning electron microscope, Cambridge S120 equipped with a LINK EDX spectrometric system AN 10 000.

3. Results and discussion

The concentration of sodium in carbon samples after different time of electrolysis is listed in table 1.

SEM photos show characteristic material structure with big anthracite and graphite grains as well as binder coke phase. Corresponding x-ray maps of Na and Al (main element of electrolyte) show sodium and electrolyte penetration. On the basis of these results we found the heterogeneous character of sodium penetration. Artificial graphite was found to be very resistant. The Na concentration in this component was of a low and stable level (0.1 - 0.7%) even after long electrolysis. Anthracite due to its small porosity is quite resistant towards electrolyte penetration, but is easily impregnated with Na. Preferential electrolyte and sodium penetration take place through the binder coke phase. This phase formed from a mixture of coked pitch and small particles of graphite and anthracite has a lot of porosity which due to capillary attraction enables electrolyte penetration. The Na concentration in binder coke phase is significantly high even after a very short time of electrolysis and increases to above 2% after 8 hours.

4. Conclusion

The presented results show that techniques of SEM and EDX spectroscopy can be successfully used to support studies of Na and electrolyte penetration in carbon cathode materials.

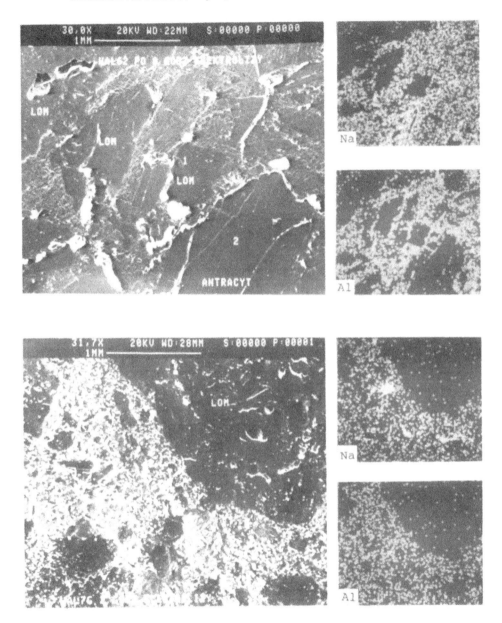

Figure 2. SEM micrographies and corresponding x-ray maps of Na and Al distribution: (*a*) WAL62A, 8 hours of electrolysis, (*b*) WAL7G, 2 hours of electrolysis.

Table 1.   Na concentration in carbon materials after experimental electrolysis.

Time of electro lysis (min)	WAL 62A					WAL 7G		
	Average sample	In carbon components			Average sample	In carbon comp.	
		graph.	anthr.	binder coke phase		graph. phase	binder coke
	wt%	wt%	wt%	wt%	wt%	wt%	wt%
5	0.75	0.12	0.29	0.77	0.36	0.05	0.34
15	1.39	0.63	0.37	1.42	0.60	0.12	0.75
30	1.04	0.16	1.09	1.67	0.65	0.10	0.68
45	1.23	0.19	1.02	0.99	0.78	0.16	1.07
60	1.25	0.13	1.33	1.30	1.13	0.20	1.02
120	0.93	0.17	0.77	1.15	1.38	0.33	1.65
240	0.70	0.14	0.92	1.00	1.50	0.25	2.03
480	1.85	0.18	1.32	2.82	2.25	0.26	2.31

References

[1]   Sørlie M and Øye H A 1989 *Cathodes in Aluminium Electrolysis* pp 86–9
[2]   Ascher R C 1959 *J. Inorg. Nucl. Chem.* **10** 238
[3]   Pflugmacher I P and Boehm H P 1970 *3rd Conf. Ind. Carbon Graph.* p 62
[4]   Rapoport M B and Samoilenko V N 1957 *Tsvet. Met.* **30** 44–51

Inst. Phys. Conf. Ser. No 130: Chapter 2
Paper presented at Int. Congr. X-ray Optics and Microanalysis, Manchester, 1992

205

Study of bioactivity on glass-ceramic for artificial bone in vivo and vitro

Zhang Xiao—Kai   *Chen Xiao—Feng   Huang Ping

Analysis and Testing Center , Shandong Normal University
* Department of Silicates Engineering , Shandong Institute of Light Industry

ABSTRACT: In the paper, bioactivity of glass—ceramic for artificial bone were investigated by soaking the material in SBF (in vitro) and implantation on animals (in vivo) , the surface of the glass—ceramic soaked in SBF for different times were observed and analysised by using the methods of SEM /EDX and thin—film XRD. The results indicate that the Ca, P—rich layer consists of carbonate —containing hydroxyapatite of small crystalltes . The hydroxapatite , carbonapatite and fluoroapatite can epitaxially grow from fluoroapatite presented on the material surface . The experiment on animals indicated that a strong bond can be formed between the surface apatite and the bone apatite . therefore , the glass—ceramics possessed good bioactivity , it can be used to make the parts of artificial bone .

In the paper , bioactivity of glass—ceramic for artificial bone were investigated by soaking the material in SBF (in vitro) and implantation on animals (in vivo) , the surface of the glass—ceramic soaked in SBF for different times were observed and analysised by using the methods of SEM/EDX and thin—film XRD .

1. SBF experiments

Specimens $(3\times6\times1mm^3)$ of the glass—ceramics polished with diamond paste were soaked in 40 ml of an acellular simulated body fluid (SBF) . The ion concentration of which were adjusted to be almost equal to those of the human blood plasma as shown in table 1. The fluid was buffered at PH 7.4 with triss—buffer , and its temperature was maintained at 37 ℃ .

Table 1:

Ion concentrations of simulated body fluid and human blood plasma

	Ion concent ration (mM)						
	Na^+	K^+	Mg^{2+}	Ca^{2+}	Cl^-	HCO_3^-	HPO_4^{2-}
SBF	142. 0	5. 0	1. 5	2. 5	126. 0	27. 0	1. 0
Blood plasma	142. 0	5. 0	1. 5	2. 5	103. 0	27. 0	1. 0

The surface of the specimens were subjected to scanning electron microscopic (SEM) observation , thin—film X—ray diffraction at various times after the immersion. Figures 1 and 2 shows SEM pictures and EDX spectrums of the glass—ceramics soaked in SBF at various times . Figure 3 show thin—film X—ray diffraction patterns of the surface of the glass—ceramic soaked in SBF for various periods .

It can be seen from fig. 3 that the surface of glass—ceramic is completely covered with a amorphous film rich in CaO , P_2O_5 and water in 20 days , the film was crystallized and transtated into a layer of hydroxyapatite microcrystal with the incorporation of F^- , CO_3^{-2} and OH^- in 45 days . This indicates that the Ca , P—rich layer consists of carbonate—containing hydroxyapatite of small crystalltes. The hydroxapatite , carbonapatite and fluoroapatite can epitaxially grow from fluoroapatite presented on the material surface .

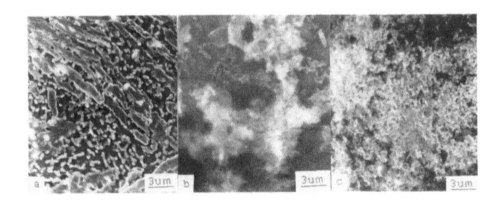

Fig. 1. SEM pictures of the surface of glass—ceramic
soaked in SBF for 0 ,   20 and 45 days

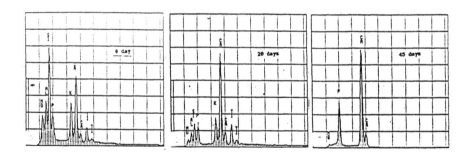

Fig. 2. EDX spectrums of the surface of glass—ceramic
soaked in SBF for 0 ,   20 and 45 days

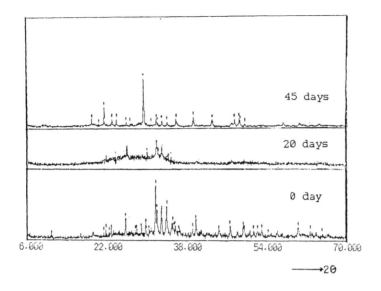

Fig. 3. Thin—film X—ray diffraction patterns of the
surface of glass—ceramic in SBF for various periods

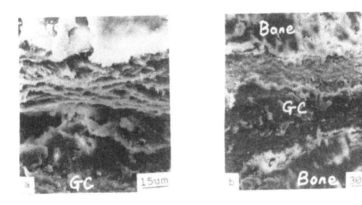

Fig. 4. Surface morphology of glass—ceramic implanted
into rabbit bone (a) twe weeks (b) ten weeks

2. Animals experiment

In order to investigate the role of the apatite phase present in glass—ceramics in the bonding of the glass—ceramic to the living bone . we have made experiment on animals .

Specimens $(2 \times 2 \times 5 mm^3)$ of the glass—ceramics polished with a diamond paste were implanted into defects of tibiae of rabbits , as shown in figures 4 and 5. At various times after the implantation , they were excised together with the surrounding bone and their bonding was examined by a detaching test in which an increasing tensile stress was applied to their interface and a fracture occurred. It was found that the glass—ceramic was so tightly bonded to the living bone in 10 weeks . This indicates that the apatite phase present in the glass—ceramic play the role of acceleration in forming the chemical bond of the glass—ceramic to the bone

It can be expected that osteoblast proliferates on the surface of the apatite layer in preference to the fibroblast , and conseguently the surrounding bone comes into direct contact with the glass—ceramic without the intervention of the fibrous tissue . When this occurs , a strong bond can be formed between the surface apatite and the bone apatite . The experiment on animals indicated that the glass—ceramics possessed good biactivity , therefore , it can be used to make the parts of artificial bone .

Fig. 5. SEM pictures of glass—ceramics after
separating implantation from bone forcibly
(a) plant—like tissue   (b) osteoblast

Inst. Phys. Conf. Ser. No 130: Chapter 2
Paper presented at Int. Congr. X-ray Optics and Microanalysis, Manchester, 1992

209

Effect of TiO_2 addition on crystallization of phlogopite and Apatite-containing glass ceramic

Chen Xiao—Feng, Zhang Xiao—Kai*, Teng Li—dong, Zhang Mei—Mei

Department of Silicates Engineering, Shandong Institute of Light Industry, Jinan—250100, PRC.
*Analysis and Testing Center, Shandong Normal University, Jinan—250014, PRC.

ABSTRACT: The role of TiO_2 in the crystallization of the glass in the system $K_2O-MgO-CaO-Al_2O_3-B_2O_3-SiO_2-P_2O_5-F$ has been investigated. It was found that TiO_2 addition was not absolutely necessary for the crystallization. However, it could change the feature of phase separation and influence the crystallization process to a certain extent, due to the characteristic and roles of Ti^{4+} ion itself. The phase separation was of primary importance to the crystallization of phlogopite and apatite crystals.

1. INTRODUCTION

Phlogopite and apatite—containing glass ceramic is a new type of biomaterials, which possesses good machinability and bioactivity due to its two main crystalline phases[1]. It is potentially important for surgical implantation. TiO_2 is one of the nucleation agents in the most common use for inducing crystallization. However, the recent research has indicated that TiO_2 can inhibit the chemical bond between the biomaterials and bone tissue[2]. In this work, the possibility to make the phlogopite and apatite—containing glass ceramic without TiO_2 addition was investigated.

2. EXPERIMENTAL PROCEDURE

2.1 Samples Preparation

The composition range of the base glass was (wt%): SiO_2 35—40, B_2O_3 5—8, Al_2O_3 10—15, CaO 10—15, P_2O_5 6—10, F 5—8. TiO_2 was added into the base glass in the following weight percent of the batches: 0, 2, 4 and 6 (wt%). The corresponding glass samples were named, respectively, T0, T2, T4 and T6. Well—mixed glass batches were melted at 1500℃ for 2h and quenched by pouring on to a steel plate. The glasses were annealed at 580℃ for 2h. The glass samples were in opaque state with pale yellow. They were crystallized by a three—stage heat—treatment method. Four white and untransparent glass ceramic samples were obtained, which were represented by T0′, T2′, T4′ and T6′, corresponding to T0, T2, T4 and T6.

2.2 Microanalysis of the Samples

SEM／EDAX method was applied to observe the microstructure and analyze the kinds and content of elements in different regions of phase separation in the parent glasses. To identify the kinds and study the feature of the main crystalline phases in the glass ceramics, XRD, SEM, TEM and microzone electron diffraction techniques were employed.

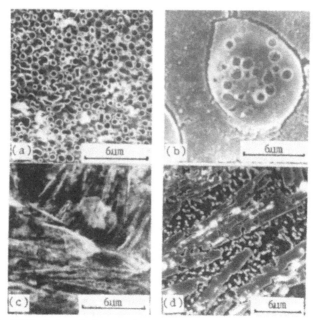

Fig. 1. SEM photos of the samples. (a), (b): the parent glasses T0 and T4; (c), (d): the glass ceramics T0' and T4'.

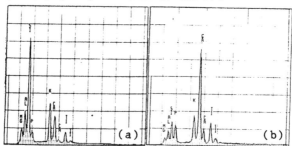

Fig. 2. EDAX spectra of the glasses T4. (a) the matrix phase; (b) the droplet phase.

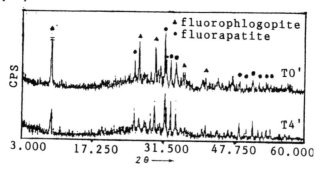

Fig. 3. XRD patterns of the glass ceramics T0' and T4' (the bulk samples with polished surfaces).

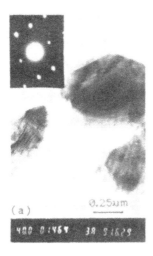

Fig. 4. TEM photos of the powder from the glass ceramic T0' and the patterns of micro — zone electron diffraction. (a) for the layer fluorophlogopite crystals, (b) for the granular fluorapatite crystals.

3. RESULTS AND DISCUSSION

SEM observation show that liquid (or glassy) phase separations occur in all the parent glasses. The microstructures of the glasses T0 and T4 are similar, respectivly, to T2 and T6. From Fig. 1. (a), (b) we can see that some spherical droplet phases are embeded in a continuous matrix phase. EDAX examination indicates that the droplet phases are comparatively rich in P, Ca, and Ti (for T2, T4 and T6), whereas the matrix phase contains relatively high Si, Al, Mg and K (see Fig. 2.). The phase separation is caused by contending for O^{2-} ions between Si^{4+} and P^{5+} ions. More Ca^{2+} and Ti^{4+} ions are drawn into the spherical phosphatic droplet phases to balance the negative charges of more O^{2-} ions attracted by P^{5+} ions. Appropriate TiO_2 addition don't change the mechanism of the phase separation and change obviously the elements distribution in each phase, but can inflence the feature of phase separation in the glasses, consequently, the process of crystallization.

In the heat treatment, the glasses T0, T2, T4 and T6 are all crystallized. Fig. 1. (c), (d) show that T0' and T4' contain two main crystalline phases: one possesses a layer structure, another are tiny crystal particles, which distribute densely about the layer crystals. XRD analysis indicates that the two main crystalline phases are fluorophlogopite and fluorapatite (Fig. 3.). By TEM method, we can see that two crystalline phases with differdnt feature exist in the powder of the glass ceramic T0' (see Fig. 4.). According to the result of XRD analysis, we can identify that the layer crystals are fluorophlogopite, and the granular ones are fluorapatite. The patterns of micro—zone electron diffraction of the two crystalline phases indicate that fluorophlogopite and fluorapatite belong, respectively, to monoclinic and hexagonal systems. The above analysis results imply that TiO_2 addition is not absolutely necessary to the crystalliztion of the glass ceramic.

For the glass T0, a great numbers of droplet phases rich in phosphate are in favour of the crystallization of dense and fine fluorapatite crystallites due to their similarity in composition and structure, which lowers the potential barrier for nucleation. The fluorophlogopite crystals are initially formed in the matrix phase, where CaF_2 (or MgF_2) plays an important role in inducing crystallization[3].

For the glass T4, TiO_2 doped can have the following effects: (1) Ti^{4+} as intermediate ion with higher field strength (Z/a^2) can form $[TiO_4]$ tetrahedron at a high temperature, which can be bonded to $[SiO_4]$ (the basic structural unit of the glass network). The bond degree of the glass network is increased and phase separation in cooling process can be inhibited to a certain extent. As a result, the quantity of the phosphate droplet phases is decreased and their sizes become much larger than T0 (Fig. 1. (b)), which lead to the reduction of the quantity of fluorapatite nuclei formed in the droplet phase, but the size of the crystals can be increased owing to the expansion of the growth space of the crystal particles; (2) As the melt are cooled, Ti^{4+} tends to take the coordination form of $[TiO_6]$. the structural difference between $[TiO_6]$ and $[SiO_4]$ makes Ti^{4+} to be excluded from the glass network with the form of $MgO \cdot TiO_2$, which forms even tinier liquid droplets separated from the original larger phosphate droplet phases (Fig. 1. (b)). This is a mechanism of secondary phase separation. The titanate droplets can be crystallized in heat treatment and foster the crystallization of fluorapatite. for the glass T4, the crystallization mechanism of fluorophlogopite is considerded to be similar to the glass T0.

4. CONCLUSSION

The fluorophlogopite and fluorapatite—containing glass ceramic in the system $K_2O—MgO—CaO—Al_2O_3—B_2O_3—SiO_2—P_2O_5—F$ can be prepared without adding TiO_2 nucleation agent. The droplet phases rich in phosphate in the glasses are of primary importance to the crystallization of the apatite crystals. the phologopite crystals are formed first in the matrix phase, where CaF_2 or MgF_2 plays an important role in inducing crystallization. TiO_2 doped can influence the process of the phase separation and the crystallization of the glasses in this system to a certain extent.

References

[1]W. Vogel, W. Holand and K. Naumann, J. Non—Cryst. Solids 80, 44—50(1986).
[2]Larry L. Hench, J. Am. Ceram. Soc. , 74 [7] 1494(1991).
[3]V. Saraswati, J. Non—Cryst. Solids 124,256(1990).

Microstructural studies of the production of BSCCO superconductors by the citrate gel route

A Gholinia and F R Sale

Manchester Materials Science, University of Manchester/UMIST, Grosvenor Street, Manchester M1 7HS

ABSTRACT: The production of Bi-Sr-Ca-Cu (BSCCO) high T_c superconductors by citrate gel route has been investigated. The compositions of the various phases formed in the sequence leading to the high T_c superconducting phases have been determined by EDX as a function of temperature and annealing time. Pb substitution for Bi has been shown to produce a high quantity of the high T_c (110K) superconducting 2223 phase. The volatilisation of both Pb and Bi during extended heat treatments has been studied. The superconducting properties of the samples have been measured by VSM.

1. INTRODUCTION

At least three superconductive phases which can be characterised by the number of CuO planes (n) in their perovskite crystal structure exist in the $Bi_2Sr_2Ca_{n-1}Cu_nO_{4+2n+\delta}$ system. The 2201 phase (n=1), which was first found by Michel et al (1987) has a $T_c \simeq$ 10K and a lattice parameter of c = 24.6 Å. The other two phases were discovered by Maeda et al (1988). The 2212 phase (n=2) with T_c = 90K and 2223 phase (n=3) with T_c = 110K have lattice parameters of c = 30.9 Å and c = 37 Å, respectively.

Schulze et al (1990) have shown that the single phase region of the 2223 is very small phase compared to those of the 2212 and 2201 phases in the phase diagram and hence explained the reason why the 2201 and 2212 phases are relatively easier to prepare compared to the 2223 phase. Takamo et al (1988) found that Pb-doping increased the volume fraction and stabilised the desired 2223 phase.

The complex composition of the 2223 phase has made it necessary to seek preparative techniques to overcome the problems of inhomogeneity and contamination associated with the conventional ceramic processing method using oxide or oxide/carbonate mixtures and repeated firing and subsequent grinding operations. In this study the citrate-gel route is used to obtain a precursor of high chemical homogeneity which is subsequently decomposed to give the final product. Effects of Pb-doping and the sequence of microstructures formed as a function of temperature, annealing time and as a result of volatilisation have been studied.

2. EXPERIMENTAL PROCEDURE

High purity nitrate salts of Bi:Sr:Ca:Cu with molar ratios 2:2:2:3 were dissolved in deionised water. The molar ratio of citric acid:metal nitrates was kept at 1:1, and the pH of the solution was kept at 6 by addition of NH_4OH solution. The solution was dehydrated to form the precursor. The precursor was calcined in two stages 300°C and 700°C. The decomposition behaviour of the calcined powder was studied by a SEIKO 320 simultaneous TG/DTA facility.

After ball milling, the powder was pressed into pellets of 3.5 mm diameter by 4 mm in a uniaxial die using pressures up to 370 MN m^{-2}. Annealing at selected temperatures was carried out in a horizontal cylindrical furnace.

The sintered products were characterised by X-ray diffraction using CuK_α radiation and a Philips diffractometer. The composition of the phases were identified by EDX analysis under Philips SEM and TEM. Onset critical temperatures were determined in an Oxford Instruments Vibrating Sample Magnetometer (VSM), where all the measurements were obtained on heating so that possible thermal drag would not create errors in comparing the results.

3. RESULTS AND DISCUSSION

3.1 Sintering

The STA curve of the powder calcined at 700°C, which initially contains mainly the 2201 phase, shows three major endothermic peaks in the region 800-880°C, as seen in figure 1. The onset of the first endothermic peak is at 814°C, the second is at 862°C and the third is at 880°C. By preparing samples at temperatures before and after the peaks the phases present were identified by EDX and XRD, as summarised in Table 1. The first peak is attributed to the phase transformation of 2201 to 2212 phase. The second peak is explained by partial melting, and the formation of Ca- and Cu-rich phases. The third peak is associated with the decomposition of liquid 2212 phase to yield the 2201 phase along with the fusion of the residual calcium cuprate.

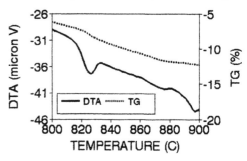

Fig. 1. Simultaneous TG/DTA curve of the 700°C calcined powder.

Transition temperature measurements show an increase in T_c with longer periods of sintering in the partial melting region, because of a high amount of the 2201 to 2212 transformation. Sintering in oxygen at 865°C showed a lower T_c than in air, suggesting that the 2201 to 2212 transformation is inhibited by higher oxygen partial pressure, as summarised in Table 2.

Table 1. Summary of the XRD and EDX studies

Sintering Temp.-Time-Atm	Main phase (XRD)	Main phase (EDX)	Trace phases (XRD)	Trace phases (EDX)
830°C-20h -air	2212		2 2 0 1, CuO, ?	
850°C-10h -air	2212	$Bi_{1.4}(Sr,Ca)_{1.9}CuO_x$	2 2 0 1, CuO, ?	
850°C-20h -air	2212	$Bi_{1.3}(Sr.Ca)_{1.7}Cu_{1.5}O_x$	2 2 0 1, CuO, ?	
865°C-0.5h -air	2212		Ca_2CuO_3 CuO	$(Sr,Ca)_{3.2}Cu_{1.7}O_x$ CuO
865°C-12h -air	2212		Ca_2CuO_3 CuO	
885°C-0.5h -air	2201	$Bi_{3.9}(Sr,Ca)_{3.9}Cu_{2.2}O_x$	Ca_2CuO_3 $CaCuO_2$	$(Sr,Ca)_{6.63}Cu_{3.28}O_x$ $(Sr,Ca)_{5.5}Cu_{4.4}O_x$
865°C-0.5h -O_2	2212		2 2 0 1, CuO, ?	

Table 2. Summary of VSM results

Sintering Temperature-Time-Atmosphere	$T_{c\ onset}$ (K)
830°C - 20 h - air	-
865°C - 0.5 h - air	74 K
865°C - 10 h - air	84 K
865°C - 12 h - air	110 and 90 K
865°C - 0.5 h - O_2	70 K

3.2. Pb-doping and volatilisation

The 2223 (T_c = 110K) phase is produced by 20% (Atomic percentage) substitution of the Bi by Pb and sintering at 850°C for 100h in air. SEM, EDX study of the sample, distinguishes the light grey matrix containing a lower amount of Ca- and Cu- (low T_c phase) relative to the dark grey needle shaped phase (high T_c phase) contained within the matrix, as shown in figure 2a. During the long period of sintering, the surface of the sample that was subjected to direct flow of air suffered enhanced volatilisation of Bi and Pb leaving phases rich in Sr, Ca and Cu behind, as seen in figure 2b. Thus the problem of volatilisation from Pb-substituted BSCCO superconductors is clearly demonstrated.

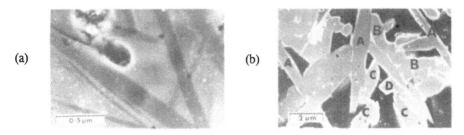

Fig. 2. Pb-doped sample 1.6:0.4:2:2:3, Bi:Pb:Sr:Ca:Cu, sintered at 850°C and annealed for 100 hours and air quenched (a) dense region of the sample (b) volatilized region (surface) A: $Sr_{0.17}Ca_2Cu_{1.1}O_x$; B: $Bi_{0.04}Sr_{0.65}Ca_2Cu_{0.27}O_y$; C: $Bi_{0.03}Sr_{0.25}Ca_2Cu_{1.14}O_z$; D: CuO.

4. CONCLUSION

1. The sequence of the phase transformation on heating a powder containing the 2201 phase, previously calcined at 700°C is:

814°C	2201 → 2212 phase
862°C	partial melting and occurrence of Ca- and Cu-rich phases to give the conditions needed for 2223 phase formation.
880°C	melting/decomposition (2212 and 2223 → 2201).

2. The 2223 (T_c = 110K) phase is only obtained with Pb-doping and long periods of sintering compared to the easily achievable 2212 (T_c = 90K) phase.

3. The higher stability of the 2212 phase makes possible the following decomposition reaction for the 2223 phase.

$$2Bi_2Sr_2Ca_2Cu_3O_{10+x} \rightarrow Ca_2CuO_3 + CuO + 2Bi_2Sr_2CaCu_2O_{8+y}.$$

4. The long periods of sintering have to be closely controlled because of the problem of volatilisation of Bi and Pb.

5. The samples sintered below the partial melting temperature show the high T_c phase by XRD but do not show superconducting properties. This may be because the densification of the sample by such solid state sintering has been so poor and the grains are either thinner than the London penetration depth (less than 100 μm) or contain less oxygen content than the stoichiometric composition. Conversely, the samples sintered at 865°C, because of liquid phase sintering, achieve a higher densification and larger grains such that a high $T_{c\ (onset)}$ is observed.

REFERENCES

Maeda H, Tamaka Y, Fukutomi M and Asomo T 1988 Jpn. J. Appl. Phys. 27 L209
Michel A, Hervieu M, Borel M M, Grandin A, Deslandes F, Provost J and Raveau B 1987 Z. Phys. B 68 421
Schulze K, Majewski P, Hettich B and Petzow G 1990 Z. Metallkde Bd. 81 836-842
Takano M, Takada J, Oda K, Kitaguchi H, Miura Y, Ikeda Y, Tomii Y and Mazaki H 1988 Jpn. J. Appl. Phys. 27 L1041

Inst. Phys. Conf. Ser. No 130: Chapter 2
Paper presented at Int. Congr. X-ray Optics and Microanalysis, Manchester, 1992

217

Atomic resolution imaging and analysis with the STEM

S. J. Pennycook, D. E. Jesson, M. F. Chisholm, and N. D. Browning

Solid State Division, Oak Ridge National Laboratory, Oak Ridge, Tennessee 37831-6030

ABSTRACT: Recent instrumental developments in the high-resolution scanning transmission electron microscope (STEM) offer the promise of combining incoherent imaging and spectroscopic analysis at atomic resolution, even in the presence of strong dynamical diffraction of low-order diffracted beams. This extends the realm of intuitive structure determination to substantially greater thicknesses and higher resolution than conventional methods. Recent applications to the study of interfacial structures and atomistic mechanisms of growth are described, along with indications of future directions.

1. INTRODUCTION

The attainment of an electron probe of atomic dimensions (2 Å) was demonstrated by Crewe and coworkers in the late 1960's and early 1970's, through the use of a high-brightness field emission gun and a high-resolution objective lens. Detecting scattered electrons with a wide-angle annular detector provided atomic-resolution images of small metal clusters and individual heavy atoms (Wall et al. 1974, Crewe et al. 1975). High sensitivity and strong atomic number- or Z-contrast was achieved, coupled with freedom from the contrast reversals characteristic of phase contrast imaging, as well as a significantly enhanced resolution. Despite this success in the biological field, early attempts to apply such techniques to materials science problems were rather disappointing (Donald and Craven 1980). Whereas biological samples tend to be thin and non-crystalline, materials science specimens are typically crystalline, so that dynamical diffraction is unavoidable in specimens of any practical thickness. Dynamical diffraction effects rapidly dominate the image contrast and destroy the intuitive nature of the image.

A solution to this problem was suggested by Howie in 1979. For samples which are not too beam sensitive, it is possible to sacrifice collection efficiency by significantly increasing the inner angle of the annular detector. The signal would then be formed primarily from diffuse scattering, which would be less sensitive to diffraction conditions while at the same time showing stronger compositional sensitivity as the scattering approaches the Z^2 dependence of unscreened Rutherford scattering. This resulted in greatly improved visibility for small catalyst clusters on amorphous or crystalline supports (Treacy and Rice 1989). Perhaps the greatest surprise has been the demonstration of atomic resolution incoherent images using such a detector even in the presence of strong dynamical diffraction of the low order diffracted beams (Pennycook and Jesson 1990, 1991). As discussed in the next section, provided we are concerned with signals generated very close to the atomic sites (i.e., integrated high-angle scattering or inner shell excitations detected spectroscopically), then we can consider the probe as effectively channelled along the atomic columns as a result of the strong dynamical diffraction itself. Therefore, for a probe focussed to atomic dimensions we illuminate the crystal column-by-column, and the image represents a directly interpretable map revealing the location and scattering power of each atomic column, as shown schematically in Fig.1. There are no contrast reversals with focus or specimen thickness, and we return to the intuitive interpretability suggested by the early biological results.

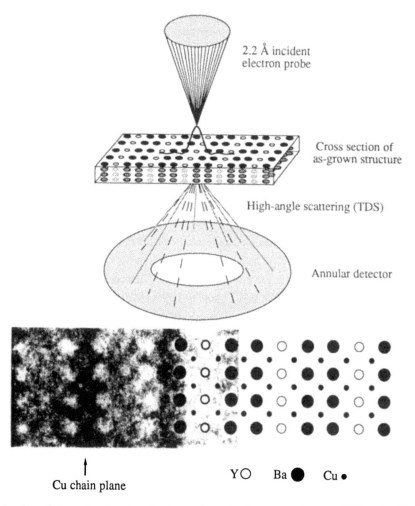

2.2 Å incident electron probe

Cross section of as-grown structure

High-angle scattering (TDS)

Annular detector

Cu chain plane

Y O   Ba ●   Cu •

Fig. 1.   Schematic showing the formation of a Z-contrast image of $YBa_2Cu_3O_{7-x}$ in the STEM.

2. SIGNAL GENERATION AND PROBE PROPAGATION

Due to the importance of the high-angle detector in restoring incoherent characteristics, we propose that it be referred to as the Howie detector. Its action is to separate diffraction effects of the incoming and outgoing electron wavefield. By choosing the inner detector angle sufficiently high (>50 mrad at 100 kV), the outgoing electrons are generated kinematically from the incident wave and play no part in the dynamical diffraction. Furthermore, as a result of thermal vibrations and the large angular range of the Howie detector, interference effects are averaged out, and each atom can be considered a weak point source as shown schematically in Fig. 2. This is similar to the situation for inner shell x-ray generation.

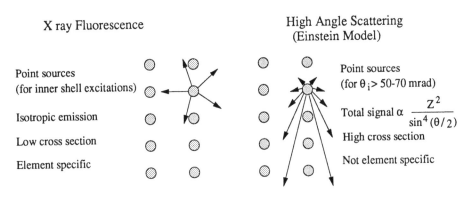

Fig. 2.   Incoherent signals for imaging and analysis.

The illumination of each atom is controlled by the dynamical diffraction of the incoming probe. Here again, the Howie detector offers a remarkable simplification. For point sources, we are concerned only with the intensity very close to the atomic sites, which naturally implicates tightly bound s-type Bloch states. Furthermore, these are precisely the states which add constructively as we integrate over a large incident angular range to form the focussed probe. Detailed calculations have shown that the image is due almost entirely to s-type Bloch states, which is the reason we see no interference fringes with increasing specimen thickness. Dynamical diffraction manifests simply as a channeling effect, an enhancement of the incident intensity along each atomic column, and serves to effectively confine the incident probe to the atomic columns as suggested by the schematic in Fig. 1. This quantum mechanical interference effect therefore results in a simple classical picture of current propagating along individual columns, and is the basis for column-by column imaging and analysis. It means that as far as *localized* signal generation is concerned, there is no beam broadening, and atomic resolution persists to thicknesses of several hundred Å before the flux at the sites is eventually depleted by the high-angle scattering.

The *s* states are themselves quite highly localized around each atomic column, with a diameter less than that of the incident probe. For most low-order zone axes, the *s* states are therefore insensitive to the nature and distribution of surrounding atoms. No supercell calculations are necessary, and images from interfaces can often be interpreted as simply as those from a bulk crystal. Excellent agreement with the incoherent imaging model has recently been demonstrated by Loane, Xu, and Silcox (1992) using multislice simulations. However, should a defect such as a dislocation be present, then this will alter the channeling effect of columns in its vicinity. Dislocations can appear dark or bright depending on their depth in the sample and the inner-detector angle (Cowley and Huang 1992 and Perovic 1992).

3. DIRECT IMAGING OF INTERFACE STRUCTURES

Conventional TEM provides intuitive interpretation only for column separations large compared to the resolution limit, and for thicknesses below a few tens of Å. Incoherent imaging provides an intuitive interpretation for at least an order of magnitude greater thickness, for column separations even below the resolution limit, together with strong compositional sensitivity over the whole thickness range and a 35% improvement in resolution. Its limitations are a rather poor signal to noise ratio, though this will improve with more efficient detectors.

Intuitive imaging of interfaces has provided some dramatic insights into not only unexpected interface structures but the atomic scale growth processes themselves. Figure 3 shows the amorphous/crystalline interface formed by room-temperature implantation of oxygen ions into a thin film of $YBa_2Cu_3O_{7-x}$ (YBCO). The macroscopic waviness seen in the low-magnification view is seen at higher magnification to resolve into discrete interface steps, the height of each step being the full 11.7 Å c-axis lattice parameter. Remarkably, the interface is atomically abrupt and shows an overwhelming tendency to reside at the Cu-chain planes in the crystal, which are seen as the dark vertical lines (see Fig. 1). This observation implies that the surface energy of the Cu-chain plane must be substantially lower than that of the many other possible crystal termination planes. It is the thermodynamically preferred crystal termination, which explains why the c⊥ *growth* mode is thermodynamically preferred, and also proceeds on a cell-by-cell basis. Images of single unit cell superlattices have provided significant insights into the growth mechanisms and transport characteristics of these materials (Pennycook et al. 1991 and 1992).

An interfacial transition zone just one or two unit cells in thickness can be seen directly in Fig. 4, which shows a YBCO film grown on a $KTaO_3$ substrate. The Ta columns of the substrate (left) show bright, while the K columns in between are not visible. In the interface region, however, the positions of the K columns are seen bright, which implies they have been replaced by a heavy species, presumably Ba from the superconductor. The structure projects as $BaTaO_3$, which does not preserve charge neutrality and is most likely stabilized by some appropriate vacancies. The ability to image unanticipated interfacial structures such as these is important in understanding interfacial properties such as tunnelling characteristics, and for many other systems, for example understanding the metal/ceramic interface.

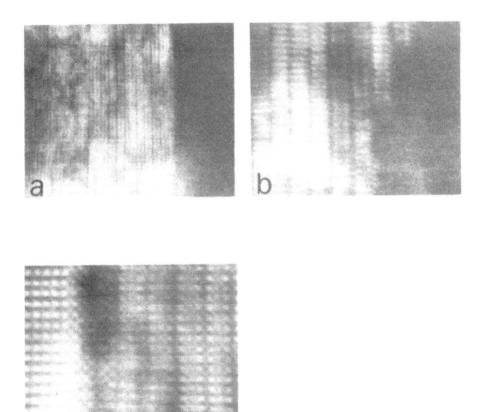

Fig. 4. A narrow transition zone seen at the interface of a YBCO film on $KTaO_3$.

Semiconductor materials have also shown unexpectedly complex interface structures arising from the highly non-equilibrium nature of molecular beam epitaxial growth. In fact, the atomic arrangements seen at the interfaces act as a fossil record of the actual atomic scale mechanisms that took place during growth. Figure 5 shows part of a $(Si_4Ge_8)_{24}$ superlattice that shows a different atomic arrangement at each interface (Jesson, Pennycook, and Baribeau 1991). Such a structure rules out interdiffusion and must reflect the growth behaviour. A model involving the exchange of Si and Ge atoms at particular active steps during the growth provides the basis for the schematic and simulated image also shown in the figure. It is important to realize that such structures have not been seen by scanning tunnelling microscopy or conventional TEM. Neither have they been seen in molecular dynamics or ab-initio studies. These structures are not the lowest energy interface structures, but they are what occur in real materials growth and will determine the actual electrical and optical properties.

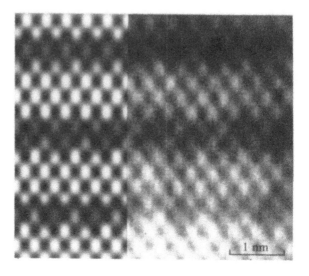

Fig. 5.   Complex atomic arrangements grown into a $(Si_4Ge_8)_{24}$ ultrathin superlattice.

4. FUTURE DIRECTIONS

The most obvious direction to pursue is to increase the microscope accelerating voltage to produce a finer probe. The FWHM probe intensity (the Rayleigh resolution limit appropriate for incoherent imaging) is $0.44 \, C_s^{1/4} \, \lambda^{3/4}$ where C_s is the objective lens spherical aberration coefficient and λ the electron wavelength. All images shown here were obtained with a VG Microscopes HB501UX STEM operating at 100kV with a probe size of 2.2 Å. Improving C_s has only a marginal effect on resolution, but increasing the accelerating voltage to 300kV reduces the probe size to 1.3 Å. Such an instrument is approaching final test at the time of writing. Since a large range of materials have spacings in this range, this will greatly increase the number of materials and zone axes which can be imaged.

A second attractive direction to follow is to utilize the advantages of incoherent imaging to enhance the image resolution by computer (Pennycook et al. submitted 1992). With incoherent imaging, as the spacing between two atomic columns is reduced, the image simply blurs into an elongated image feature. This can be seen directly in Fig. 5 where each dumbbell in the image comprises two columns 1.4 Å apart. There are no contrast reversals as occur in phase contrast imaging. Thus, simple incoherent deconvolution techniques can be used to restore the lost high-frequency information. Preliminary experiments with a maximum entropy method indicate that a factor of two resolution enhancement can be achieved in a remarkably simple and robust manner. Such an analysis also provides the relative intensity of each atomic column, thus quantifying the compositional information.

New approaches to three-dimensional reconstruction may even prove possible by combining information from two or more projections. A certain amount of three-dimensional information is already present in the image from the relative intensities of atomic columns, provided they have the same composition (Pennycook and Jesson 1991 and Jesson and Pennycook submitted 1992). It may be possible to determine interfacial roughness at a crystal/amorphous interface, for example, or to reconstruct the three-dimensional morphology of small catalyst particles, both of which have important technological implications.

Finally, it should not be forgotten that the channeling behavior, which extends intuitive imaging to the thick crystal regime, holds equally well for other localized excitations (see Fig. 2). Although spectroscopic signals are several orders of magnitude lower in intensity, and therefore not suitable for forming an image at atomic resolution, both x-ray fluorescence and electron energy loss signals are available simultaneously through additional detectors. The Z-contrast image can provide a reference image, enabling a probe to be held over an individual atomic column while a spectroscopic measurement is made. Due to its larger collection efficiency, the energy loss signal seems likely to prove more useful in this regard.

We are at the verge, therefore, of combining imaging and analysis at atomic resolution in the same instrument, of interpreting what we see intuitively to a first approximation, and performing simulations where necessary to refine the measurements. The fact that we no longer require model interface structures to interpret our images is revealing the remarkable complexity of real materials on the atomic scale.

5. ACKNOWLEDGMENTS

The authors are grateful to S. L. Carney, J. T. Luck, and T. C. Estes for technical assistance. This research was sponsored by the Division of Materials Sciences, U.S. Department of Energy, under contract DE-AC05-84OR21400 with Martin Marietta Energy Systems, Inc.

6. REFERENCES

Crewe, A. V., Langmore, J. P., and Isaacson, M. S. 1975 Physical Aspects of Electron Microscpoy and Microbeam Analysis ed B. M. Siegel and D. R. Beaman (New York: Wiley and Sons) pp. 47–62.

Cowley, J. M. and Huang, Y. 1992 Ultramicroscopy 40 171.

Donald, A. and Craven, A. J. 1980 Philos. Mag. A39 1.

Howie, A. 1979 J. Microsc. 117 11.

Jesson, D. E., Pennycook, S. J., and Baribeau, J.-M. 1991 Phys. Rev. Lett. 66 750.

Jesson, D. E., and Pennycook. S. J. Submitted 1992 P. Roy. Soc. Lond. A.

Loane, R. F., Xu, P., and Silcox, J. 1992 Ultramicroscopy 40 121.

Pennycook, S. J. and Jesson, D. E. 1990 Phys. Rev. Lett. 64 938.

Pennycook, S. J. and Jesson, D. E. 1991 Ultramicroscopy 37 14.

Pennycook, S. J., Chisholm, M. F., Jesson, D. E., Norton, D. P., Lowndes, D. H., Feenstra, R., Kerchner, H. R., and Thomson, J. O. 1991 Phys. Rev. Lett. 67 765.

Pennycook, S. J. 1992 Ann. Rev. Mater. Sci. 22 171.

Pennycook, S. J., Jesson, D. E., Chisholm, M. F., Ferridge, A. G., and Seddon, M. J. Submitted 1992 Proc. 10th Pfefferkorn Conference on Signal and Image Processing (Chicago: Scanning Microscopy International).

Perovic, D. D. 1992 Proc. Electron Microscopy Society of America (San Francisco Press) p. 1336.

Treacy, M. M. J. and Rice, S. B. 1989 J. Microsc. 156 211.

Wall, J., Langmore, J., Isaacson, M., and Crewe, A. V. 1974 Proc. Nat. Acad. Sci. 71 (USA) pp. 1–5.

Inst. Phys. Conf. Ser. No 130: Chapter 2
Paper presented at Int. Congr. X-ray Optics and Microanalysis, Manchester, 1992

225

Homogeneity of unsupported bimetallic Ni-M catalysts

S. Hamar-Thibault*, F. Janati-Idrissi**, J. Court**.
* Intitut National Polytechnique de Grenoble, ENSEEG, L.T.P.C.M.,
BP 75 - F.38402 Saint Martin d'Hères, France.
** Université Joseph Fourier, L.E.D.S.S, BP 53X- F.38041 Saint Martin d'Hères, France.

ABSTRACT: The aim of the present study was to characterize a series of bimetallic Ni-M catalysts (with M = Cr, Mo, Al) prepared by a new method of coreduction with a mixture of dry salts with naphthalene sodium. The repartition of the metals was investigated using STEM/EDX system and Auger spectroscopy. Quantitative microanalyses showed that the distribution of the dopant (Mo, Cr) at atomic level was homogeneous.

1. INTRODUCTION

The catalytic properties of a series of coprecipitated divided bimetallic Ni-M catalysts were studied in the liquid phase hydrogenation of citral (3,7-dimethyl-2,6, octadienal). The hydrogenation of citral gave citronellal, citronellol and 3,7-dimethyloctanol. Citronellal and citronellol are aromatic substances, widely used in the perfume industry. The presence of 3,7-dimethyloctanol detracts from the odor of citronellol which is its principal desired property.

It should be mentioned that only few works deal with this type of catalyst prepared with this reducing agent (Leprince et al 1976). Moreover, in these few cases, the physicochemical characteristics of the catalysts were not reported. In this paper, we try to link selectivity of the catalysts in the hydrogenation of citral to the catalyst structure analysed with physicochemical characterization methods at fine scale (nature and repartition of the dopant).

2. EXPERIMENTAL PROCEDURES

2.1. Catalyst preparation and reaction conditions.

We prepared $Ni_{100-x}M_x$ (M = Cr, Mo, Al) catalysts with x = 0, 3, 6 or 12 (at%) by coreduction of mixtures of dry salts with sodium naphthalene as reducing agent. The complete procedure was described in a previous paper (Bonnier et al 1989). Sodium naphthalene was prepared from sodium and naphthalene in THF. After the sodium had dissolved completely, anhydrous metal (promoter) salts (iodide or acetate for Al) of appropriate composition were added. A black precipitate was formed and at the end of the addition, the mixture was heated at its reflux temperature for 4 hours. The catalysts were washed with THF to remove naphthalene then either four times with water (WW) or four times with ethanol (EW).

Hydrogenations in the liquid phase (cyclohexane) were carried out at 353K in a static reactor under constant hydrogen pressure (1.MP).

2.2. Catalyst characterizations.

The bulk composition of each sample was determined by chemical analysis. The repartition of the metals was investigated using scanning transmission electron microscopy (JEOL 2000FX) equiped with energy dispersive X-ray analysis (STEM/EDX) and Auger spectroscopy (AES).

The total (S_{BET}) and the metallic surface area (S_{Ni}) were measured respectively by adsorption of N_2 at 77°K and the thiophene method (Sane et al. 1984).

3. RESULTS AND DISCUSSION

3.1. Physico-chemical characteristics of the catalysts (Table 1).

Compositions of the EW catalysts obtained by chemical analysis and STEM/EDX microanalysis were in good agreement with the desired compositions with the exception of Ni-Al catalysts. The distribution of Cr and Mo in Ni-rich particles was almost constant at any promoter level. Washings were efficient, because we are not able to detecte any iodide ions.

		Ni		$Ni_{88}Cr_{12}$		$Ni_{88}Mo_{12}$		$Ni_{88}Al_{12}$
Analyse M/Ni (at%)	nominal	-		13.6		13.6		13.6
	chemical	-		13.6	*12.5*	13.6	*11.1*	1
	STEM/EDX	-		15	*21*	11	*10*	<1
	AES	-		35	*30*	25	*10*	-
Surface m^2g^{-1}	BET	161	*62*	231	*130*	152	*115*	159
	metallic	24	*42*	40	*50*	44	*50*	37

Table 1 : Physico-chemical characteristics of the Ni-M catalysts

(ethanol (EW) and water (*WW in italic*) washing)

In the Ni-Al catalysts, Al level remained below 1at% in the Ni agglomerates while alumina particles could be detected in the liquid above the Ni agglomerates.

In the other catalysts, Ni, Cr and Mo peaks at the surface of the catalysts were well evidenced in Auger spectra. According to the large oxygen peak and to the shift observed for the Cr peak, the added metals seem to be in an oxidized state and segregated at the surface of the catalyst. Moreover, from electron and X-ray diffraction patterns, a large amount of oxidized Ni compared to the metallic Ni was detected.

In the WW catalysts, the Mo/Ni ratio was always below the nominal composition while Cr/Ni ratio remained constant. As shown in Auger spectra, oxygen level at the surface was low. In these Ni-Mo catalysts, Mo level was lower that the superficial composition in EW catalysts and of the same order as in the bulk. No significant accumulation of Mo was analysed at the surface. In Ni-Cr catalysts, the two forms of Cr (oxidized and metallic) were detected.

Washing with water eliminated a large amount of oxide (of Ni or other) and lead to catalysts with a large proportion of a metallic active promoted phase. WW catalysts were formed of metallic Ni doped particles with a homogeneous repartition of the dopant (Cr and Mo) explaining a high metallic surface area.

According to the nature of the promoter, the catalysts have different structures:
* Ni and Ni-Al catalysts are formed of the two phases Ni and NiO with very small amount of Al dispersed in the active phases A large amount of NiO was eliminated in WW catalysts.
* In Ni-Mo and Ni-Cr catalysts, Mo or Cr are dispersed in these two phases and in Mo-rich (or Cr-rich) agglomerates which are eliminated (or not) in WW catalysts.

3.2. Catalytic properties and catalyst structures

The hydrogenation of citral give citronellal, citronellol and 3,7-dimethyloctanol. Parallel reactions give by-products and result in a decrease of the maximum yield in citronellal (Y_{al}^m) and citronellol (Y_{ol}^m). The selectivities are characterized by Y_{al}^m and Y_{ol}^m. All steps are competitive hydrogenations of the C=C and C=O bonds.

$$\text{citral} \Rightarrow \textbf{citronellal} \Rightarrow \textbf{citronellol} \Rightarrow \text{3,7-dimethyloctanol.}$$

with branches to *nerol-geraniol* and *3,7-dimethyloctanal*

	Ni		$Ni_{88}Cr_{12}$		$Ni_{88}Mo_{12}$		$Ni_{88}Al_{12}$
Activity . 10^3 r0 mmol.min^{-1}m^2	9.4	1.2	16	23	4.5	11.5	22
Selectivity yield % citronellal Y_{al}^m	93	92	85	92	90	82	93
citronellol Y_{ol}^m	67	-	95	37	85	60	43

Table 2: Hydrogenation of citral with Ni-M catalysts (ethanol (EW) and water (WW) washing).

The initial hydrogenation rate of citral (r_0) expressed per m^2 of Ni (r_0) depends strongly on the washing process and on the nature of the promoter. The addition of Al and to a less extend Cr increases the activity while Mo reduced the activity without modifing Y_{al}^m.

The hydrogenation of the conjugated double bond is always very selective ($Y_{al}^m \geq 90\%$) and does not depend on the washing process or the nature of the promoter. The formation of nerol and geraniol remains always of minor importance.

On the contrary, Y_{ol}^m depends strongly on the washing process and on the nature of the promoter but the atomic composition of the promoter (in the bulk or on the surface) is of minor importance. WW catalysts lead to Y_{ol}^m significantly smaller than EW catalyts and for the latter, Y_{ol}^m is much lower for Ni-Al catalysts than for Ni, Ni-Cr or Ni-Mo catalysts.

In Ni-Al catalysts, the small amount of Al remaining in Ni agglomerates allows one to interpret the selectivities observed as arising from electronic charge transfers. For all steps, the C=C bonds are hydrogenated more selectively and more rapidly than the C=O bond. This result can be correlated with theoretical calculations performed on Ni-Al clusters where it was shown that electronic charge transfer between the cluster and formaldehyde or ethylene depend on Al level and has a great influence on the strength of chemisorption of these two compounds (Minglong et al 1989). This adsorption can strongly influence the selectivities of the steps of the reaction The presence of some amount of Al does not favor the chemisorption of the C=C, therefore the hydrogenation of the C=C bond may be easier (r_0 is also higher).

In the Mo-Ni catalysts, the hydrogenation of the C=C bond is always slow (r_0 very low) and therefore the hydrogenation of the C=O bond is possible. The hydrogenation is more selective in EW than in WW catalysts. The main difference between EW and WW catalysts is the elimination of a large amount of oxidized Mo. The same results are obtained in the Cr-Ni catalysts but no significant elimination of oxidized Cr is analysed.

Due to the presence of a M-rich phase, the behaviour of these catalysts is difficult to analyse:
* this phase can also favor or not the competitive chemisorption of reactants.
* electronic charge transferts can occur between M and Ni and also between Ni and the M-rich phase (oxide) with a variation of the Fermi level.
* the oxidized (oxide or hydroxide) phase can be different between EW and WW catalysts.

References

Bonnier J.M., Court J., Wierzchowski P. and Hamar-Thibault S. 1989 Appl. Catalyst 53, 217

Cerveny L. and Ruzicka V 1988 Kosmetica Aerosole Reichstoffe 114, 605.

Minglong Q., Barbier C and Subra R.1989 J. Molecular Catalysis 57, 165.

Leprince J.B., Collingnon N. and Normant H. 1976 Bull. Soc. Chim. Fr. 69 367.

Sane S., Bonnier J.M., Damon J.P. and Masson J. 1984 Appl. Catalyst 9, 69.

Inst. Phys. Conf. Ser. No 130: Chapter 2
Paper presented at Int. Congr. X-ray Optics and Microanalysis, Manchester, 1992

229

Quantification of grain boundary segregation in TiO$_2$ by STEM

Jeri Ann S. Ikeda, Yet-Ming Chiang and Anthony J. Garratt-Reed

Department of Materials Science and Engineering, Massachusetts Institute of Technology, Cambridge, Massachusetts, 02139 USA

ABSTRACT: A complete characterization of interfacial segregation involves knowing both the concentration of a solute or impurity and its spatial distribution, often at the subnamometer scale. However, in many materials problems the segregation per unit area of interface is a useful quantity. An example is the problem of space-charge segregation at interfaces where the aliovalent solute density per unit area of interface is equivalent to the charge-density per unit area. We have developed a new analytical method for measuring the solute segregation density at interfaces, here applied to the quantification of space-charge segregation at grain boundaries in TiO2. Application of space-charge theory, in essence the application of Gauss' law and Poisson's equation, allows the electrostatic potential to be deconvolved from segregation data.

1. INTRODUCTION

We have previously shown (Ikeda et al 1991) that donor- and acceptor-doped TiO2 behaves quite ideally from the viewpoint of space-charge segregation. Line-profiles across various grain boundaries in donor + acceptor co-doped TiO2 (Fig. 1) illustrate this point. When the net doping is varied at constant temperature (from acceptor to donor excess, left to right) or when temperature is varied at constant doping level (top to bottom), the species that segregate and deplete can be varied. The center profile in Figure 1 represents a grain boundary "titrated" in temperature and composition to the isoelectric (zero potential) point where no segregation is seen. However, it is well known that line profiles such as these are difficult to quantify. We have thus developed a new analytical method to accurately measure the solute segregation per unit area, and here apply it to the quantitative measurement and interpretation of space-charge segregation.

2. SAMPLE PREPARATION

Sintered pellets were made from co-precipitated powders of TiO2 doped with 1-2 mole% each of an acceptor (either Al^{3+} or Ga^{3+}) and a donor (Nb^{5+}). The solutes chosen are close in size to the host Ti^{4+} ion to minimize segregation due to strain relaxation. The relative amounts of the solutes were varied to give a wide range of net dopant compositions (0.5 mole% acceptor-rich to 1.4 mole% donor-rich). Sections of the pellets were then equilibrated at 1350°C in air for 4 hours. After that, the sections were ground and ion-milled (using a LN$_2$ cold stage) in preparation for analysis by STEM.

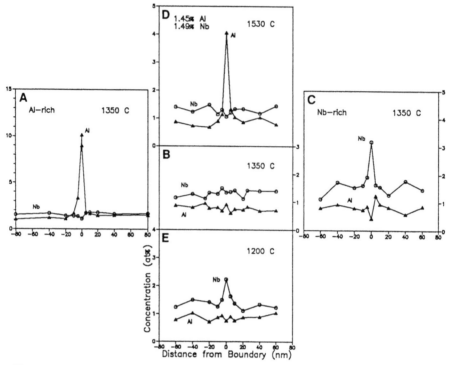

Figure 1: Segregation profiles across various grain boundaries illustrating that the effective charge of the segregated and depleted solutes (Al_{Ti}', Nb_{Ti}) vary with both composition and annealing temperature.

3. STEM ANALYSIS

The grain boundary analysis was conducted using a VG HB5 STEM operated at 100 kV with the current limit set at 10 μA. The STEM was equipped with a window-less x-ray detector (LINK Systems). Based on the results of a recent study by Furdanowicz et al. (1991) showing that beam broadening in nanoprobe STEM is much less than predicted by many conventional models, we used the VG HB5 in scan mode to analyze a volume containing the grain boundary as illustrated in Figure 1. The plane of the boundary was oriented closely parallel to the beam and centered in the rastered volume. The rastered region used is typically 25 nm x 35 nm, sufficiently large to contain the grain boundary and both adjacent space-charge layers (as determined by a line-scan across the boundary). Using the EDS analysis from like rasters in adjacent grains as internal composition standards, the excess solute per unit area of that boundary is obtained from the difference between the solute concentration in the volume containing the boundary and the volume

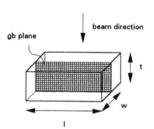

Figure 2: Schematic depicting the orientation of the grain boundary within the rastered volume.

in the bulk. The Cliff-Lorimer method was used to determine the solute concentrations with respect to Ti using the $Ti_{K\alpha}$, $Al_{K\alpha}$, $Nb_{L\alpha}$ and $Ga_{K\alpha}$ peaks. Acquisition times from 200-750s were used to obtain good counting statistics. The excess of solute i, C_i^{excess} is given by:

$$C_i^{excess} = \left\{ \frac{\left(\frac{C_i}{C_{Ti}}\right)^{gb} - \left(\frac{C_i}{C_{Ti}}\right)^{bulk}}{\left(\frac{C_i}{C_{Ti}}\right)^{bulk}} \right\} \cdot C_i^{bulk} \qquad (1)$$

where C_i^{bulk} is the concentration of i in the bulk, here taken to be the sample composition analyzed by Inductively Coupled Plasma spectroscopy. In quantification by the Cliff-Lorimer method, note that absorption corrections become unnecessary when this analysis is conducted in regions of constant thickness. Thus the excess solute per unit grain boundary area, σ, is given by

$$\sigma = \frac{\frac{1}{2} C_i^{excess} \cdot N \cdot V}{A} = \frac{1}{2} C_i^{excess} \cdot N \cdot w \qquad (2)$$

where N is the cation site density in the rutile lattice, V is the volume analyzed ($l \cdot w \cdot t$), and A is the area of the interface ($l \cdot t$). The excess solute density is equivalent to the excess charge density on the interface if it is assumed that all of the measured solute is contained in the space charge layer (i.e. no solute segregation to the boundary core) and that the valence of the solute is fixed.

4. RESULTS

The measured solute (charge) densities at one temperature are shown in Figure 3 as a function of doping at constant temperature. Each datum represents a single grain boundary. Analysis of the counting statistics error in solute densities shows that real differences exist between grain boundaries. In our sampling of approximately 100 boundaries, roughly one in ten showed no detectable solute segregation. In one case, such a boundary was identified (using high resolution TEM) as a {101} twin boundary.

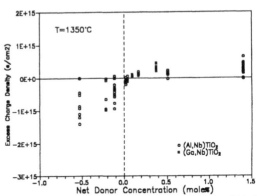

Figure 3: Measured excess charge densities at 1350°C

In the space charge layer, the local potential varies as a function of distance away from the interface (approximately exponentially), and accordingly the concentration of charged defects in the space charge layer varies with the local potential. Defining the potential as zero at the boundary and reaching some constant value ϕ_∞ in the bulk, and applying Gauss' law to the charged boundary plane and Poisson's equation to the spatial potential distribution, a general result can be obtained, relating the excess charge density of the interface, σ, to the potential difference between the interface and the bulk, ϕ_∞: (Ikeda 1992)

$$\sigma = -\left\{ \frac{\varepsilon kT}{2\pi} \sum_i n_{i,\infty} \left[\exp\frac{z_i e\phi_\infty}{kT} - 1 \right] \right\}^{1/2} \tag{3}$$

where ε is the static dielectric constant of the material, k is the Boltzmann constant, $n_{i,\infty}$ is the concentration of charged defect i in the bulk, and z_i is the effective charge of that defect. In our case, the solute concentrations make up the majority of the charged defects in the space charge layer and are the dominating species in Equation 3. Using the bulk solute concentrations from ICP analysis and the experimentally measured σ values, the potentials (ϕ_∞) resulting from the measurements are shown in Figure 4.

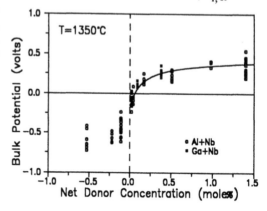

Figure 4: Potential values (ϕ_∞) obtained from measurements of the excess charge density.

On the donor-rich side, it has been shown (Ikeda 1992) that the bulk potential, ϕ_∞, is related to the lattice concentration of Ti vacancies by the following expression:

$$e\phi_\infty = \frac{g_{V_{Ti}}}{4} + \frac{kT}{4}\ln\left[V_{Ti}^{''''} \right]_\infty \tag{4}$$

where $g_{V_{Ti}}$ is the free energy of formation for moving a Ti ion from its position in the lattice to a particular surface, and $[V_{Ti}^{''''}]$ is the number of Ti vacancies per mole of TiO_2. From electrical conductivity data in Nb-doped TiO_2 by Baumard and Tani (1977), $[V_{Ti}^{''''}]_\infty$ can be calculated as a function of the net donor concentration. Thus, $g_{V_{Ti}}$ can be determined for each boundary measured. From the experimental data, the values of $g_{V_{Ti}}$ range between 2-3 eV. Using an average value of 2.4 eV in Equation 4, a theoretical curve superimposed on the experimental data in Figure 4 shows good agreement. The situation in acceptor-rich compositions (left-side of Fig. 4) is more complex and is discussed elsewhere (Ikeda 1992).

5. SUMMARY

A method of analyzing a volume of known dimensions containing an interface in order to determine the amount of segregant per unit interfacial area has been used to study the problem of space charge solute segregation in TiO_2. The results show this to be an accurate quantitative method which may be generally useful in the study of interfacial segregation.

6. REFERENCES

Baumard J F and Tani E 1977 J. Chem. Phys. **63** [3] 857
Furdanowicz W, Garratt-Reed A J and Vander Sande J B 1991 Microbeam Analysis 463
Ikeda J A S 1992 Ph.D. Thesis, Massachusetts Institute of Technology
Ikeda J A S, Chiang Y-M and Madras C G 1991 Ceramic Transactions, vol.24, ed T O Mason
  and J L Routbort, (Westerville, OH: The American Ceramics Society). pp. 341-8

Inst. Phys. Conf. Ser. No 130: Chapter 2
Paper presented at Int. Congr. X-ray Optics and Microanalysis, Manchester, 1992

233

The rate of mass loss in silicate minerals during X-ray analysis

R W Devenish * and P E Champness **

* Department of Materials Science and Engineering, University of Liverpool, P.O.Box 147, Liverpool L69 3BX.
** Department of Geology, The University, Manchester M13 9PL.

ABSTRACT: The rate of elemental loss from several silicate minerals during electron irradiation has been investigated as a function of dose, dose rate, accelerating voltage and specimen temperature. Semi-continuous monitoring of the X-ray peak intensities has shown that the mass loss can generally be represented by the sum of two exponentials, and critical doses for these processes have been determined. The main damage mechanism, which appears to have a temperature dependence, has been confirmed as radiolysis (ionisation damage).

1. INTRODUCTION

The current densities of the electron beams used for high-resolution analytical electron microscopy of materials in a TEM or STEM result in extremely high energy losses per unit volume and the dissipation of these losses often leads to significant structural and chemical changes. These changes ultimately limit the accuracy of an analysis. Silicates are known to suffer from radiolysis (i.e. electronic excitation leading to atomic displacement) during electron irradiation, although the mechanisms, apart from that in silica, are still unknown. The degree of sensitivity to damage depends on a number of factors (Veblen and Buseck 1983; Hobbs 1984; Champness and Devenish 1990) amongst which are the crystal structure, the amount of network-forming elements, the nature of the cations and the presence or absence of hydroxyl ions. However, as many of the parameters that influence the damage rate are anecdotal, our aim in the present study was to quantify the damage processes and to develop strategies to allow chemical analysis at high spatial-resolution, in spite of the limitations imposed by electron-beam damage. The silicates investigated were:- amphibole (tremolite) - a calcic double chain silicate, mica (margarite) - a hydrated, calcic sheet-silicate and plagioclase feldspar - a Na-Ca framework alumino-silicate. Chemical analyses of the specimens are given by Champness and Devenish (1992).

2. EXPERIMENTAL

Ion-thinned samples of the minerals were carbon coated on one side and were examined at a 100 kV in a VG HB501 field-emission STEM or at 50, 100, or 200 kV in a JEOL 2000 FX equipped with a LaB_6 filament. Cold stage runs on plagioclase used a liquid-N_2 cooled cryoholder on loan from Oxford Instruments. Other experimental details are described by Champness and Devenish (1992).

3. RESULTS

Decay curves in radiation chemistry are usually plotted as the logarithm of a normalised intensity versus dose. Data for elemental loss in mica and amphibole are presented in this way in Figure 1, where I is the characteristic X-ray intensity at dose D and I_0 is the intensity

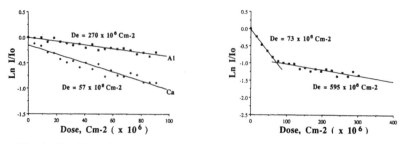

Fig. 1. Semi-log plot for mass loss from mica in the VG STEM. (a) (left) Al and Ca at a current density of 1.7×10^5 Am^{-2}. (b) (right) Al at a current density of 5.5×10^5 Am^{-2}.

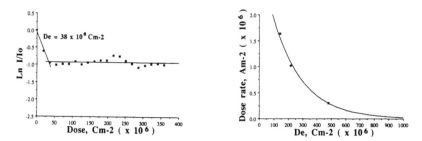

Fig. 2. (a) (left) Semi-log plot for loss of Al from plagioclase at a current density of 7.2×10^5 Am^{-2} in the VG STEM. (b) (right) Dose-rate dependence for loss of O_2 from mica in the VG STEM..

extrapolated to zero dose. As can be seen, the irradiation kinetics can be represented by either one or two exponentials depending on the value of the dose. For the loss of light elements in organic materials Egerton (1982) found that the data fitted an equation of the type:-

$$\frac{N}{N_o} = \exp\left(\frac{-D}{D_1}\right) + \exp\left(\frac{-D}{D_2}\right)$$

where N is the concentration after a dose D, and D_1 and D_2 are 'characteristic doses' equivalent to D_e, a critical dose representing the *initial* sensitivity to radiation damage for the process involving release of that element. For the silicates studied here a single, linear plot is observed at low doses (Figure 1a), whilst at higher doses two non-horizontal lines are seen (Figure 1b). Occasionally at intermediate doses the second line is horizontal, indicating a residual mass (Figure 2a). The value of the critical dose for the first loss process varies from about 1 to 70×10^6 Cm^{-2} at 100 kV. These values are four to five orders of magnitude higher than those measured for polymers (Egerton et al. 1987). Also, in polymers, there is no variation of the critical dose with dose rate, i.e. the loss of an element depends only on the accumulated dose, in contrast to sodium chloride (Egerton et al. 1987) in which the decomposition (for a given dose) decreases as the dose rate rises. In silicates however, the mass loss decreases as the dose rate falls. There is therefore a threshold value of the dose rate for a particular element in a given structure for which there is no mass loss (Champness and Devenish 1990). Figure 2b which shows the dose rate dependence for loss of oxygen from mica illustrates this point. As radiolytic processes are more efficient at lower energy, damage rates should be more pronounced at lower accelerating voltages. The results shown in Figure 3a confirm this expectation. The critical dose for loss of sodium in plagioclase at 50 kV is three times lower than that at 100 kV and 50 times lower than that at 200 kV under otherwise identical conditions. The temperature dependence for loss of sodium from plagioclase is illustrated in Figure 3b, which shows that D_c at the temperature of liquid N_2 is ten times that at ambient temperature. For Ca the ratio of the D_c's for the two temperatures was about two under the same conditions.

We also recorded micrographs of the damage structures produced in the minerals in order to gain more insight into the damage mechanisms. In amphibole several zones were observed (Figure 4a). The inner zone was intensely damaged, the middle region consisted of more electron-dense material and the outer zone was also damaged, but to a lesser extent than the central region. We believe the lighter damage here to be a consequence of the lower current density existing in the aberration tails of the electron probe. Figure 4b shows a series of damage columns produced in plagioclase at 100 kV by a focussed probe of current density 5 x 10^5 Am^{-2}. The specimen was subsequently tilted through 45° to reveal the damage profiles. Initially, shallow indentations are seen to form at the free surfaces, but these appear to stop growing at some small, limiting depth. A damaged region, which is presumably amorphous, then propagates through the specimen at equal rates from both the entrance and exit surfaces and eventually links to form a continuous column. A thin, dark band is observed around the periphery of each column.

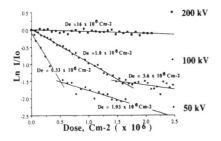 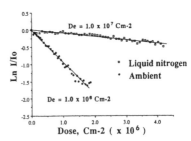

Fig. 3. Semi-log plot for loss of Na from plagioclase in the JEOL 2000 FX at a current density of 1.8 x 10^3 Am^{-2}. (a) (left) kV dependence. (b) (right) temperature dependence at 100 kV.

4. DISCUSSION

Radiation damage from the impact of a high-energy electron-beam occurs either by radiolysis or by direct displacement-damage. Displacement damage in the TEM is directional and the damage rate increases with increasing accelerating voltage. In contrast, the cross sections for radiolytic damage have an energy dependence proportional to the electron stopping-power and are favoured at lower beam energies. The kV dependence of our results confirms that the principle damage mechanism in silicate minerals is radiolysis. The compositional changes are initiated by atomic rearrangements and the resulting charge imbalance introduced into the specimen (Howitt et al. 1988), the high electric fields established being responsible for transport of ionic species away from the vicinity of the probe, thus causing depletion of the analytical signal. The overall kinetics of the damage will therefore be controlled by a balance between the radiolysis rate and these transport processes.

Inspection of the mass-loss curves suggests that elements are lost from two different structures and it is tempting to assume that the first loss process is concurrent with amorphisation, whilst the second process is equivalent to loss from a glassy phase akin to silica glass. The critical dose approaches saturation at high dose rates (Figure 2b) and may be a result of the induced fields repeatedly collapsing and building up as the specimen undergoes dielectric breakdown. There is also a threshold of the current density below which no elemental loss occurs and this may be a consequence either of charge neutralisation by recaptured secondary or backscattered electrons or of forward ionisation damage processes being offset by restorative back reactions such as the reformation of broken bonds, etc. The difference in rate of loss of sodium and of calcium from plagioclase at room and liquid-nitrogen temperatures confirms that the radiolysis mechanism has a temperature dependence, at least for this structure.

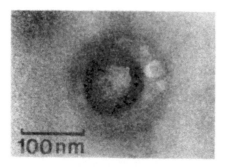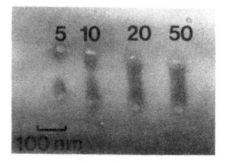

Fig. 4. (a) (left) Damage produced in amphibole at 200 kV after exposure to a dose of 3×10^8 Cm^{-2}; the current density was 1×10^5 Am^{-2}. (b) (right) A series of damage columns produced in plagioclase at 100 kV by a focussed probe after exposure for the times indicated in seconds; the current density was 5×10^5 Am^{-2}.

Examination of the micrographs suggest that, assuming that a critical current density is exceeded, material is initially removed atom plane by atom plane as occurs in Na-β alumina (Humphreys et al. 1990), with all elements being lost. This process then appears to slow down and stop, presumably as the remaining material becomes amorphous. In general, all elements other than silicon continue to be lost, albeit at a slower rate than before, leaving a residue of approximate composition SiO_2. The metallic elements appear to migrate to the sides of the damaged column (Figure 4), as occurs with Al in both crystalline Na-β alumina and in amorphous alumina (Humphreys et al. 1990); the oxygen appears to form bubbles which slowly coalesce and burst as they intersect the free surface.

5. CONCLUSION

X-ray analysis at high spatial resolution of necessity requires small probe sizes and high current densities and it is evident from this work that severe compositional changes can occur, even in a microscope with a standard thermionic source. There are several techniques that can be employed to ensure dependable quantitative analyses. Amongst these are: operating below the current-density threshold in combination with long analysis times and drift-correction software; the use of X-ray detectors with high-collection efficiency; maximising the signal / damage ratio by sensible choice of accelerating voltage and using low temperatures.

6. REFERENCES

Champness P E and Devenish R W 1990 Trans. Roy. Microsc. Soc. 1, 177
Champness P E and Devenish R W 1992 Proc. EUREM '92, Granada 2, 541.
Egerton R F 1982 J. Microscopy 126, 95
Egerton R F Crozier P A and Rice P 1987 Ultramicroscopy 23, 305
Hobbs L W 1984 Proc. 25th Scottish Univ. Summer School in Physics ed J N Chapman and
    A J Craven (SUSSP Publications, Edinburgh) pp 399 - 445
Howitt D G Medlin D L and Walker T M 1988 EMSA Bulletin 18, 69
Humphreys C J Bullough T J Devenish R W Maher D M and Turner P 1990 Scanning
    Microscopy Supplement 4,185
Veblen D R and Buseck P R 1983 Proc. 41st Annual Meeting Elect. Microscopy Soc.
    America ed G W Bailey (San Francisco Press, San Francisco) pp 350 - 353

Inst. Phys. Conf. Ser. No 130: Chapter 2
Paper presented at Int. Congr. X-ray Optics and Microanalysis, Manchester, 1992

237

Electro and thermomigration of metallic islands in the Si(100) surface

T. Ichinokawa, C, Haginoya, D. Inoue and H. Itoh

Department of Applied Physics, Waseda University, 3-4-1, Ohkubo, Shinjuku-ku, Tokyo 169 Japan

ABSTRACT: The electro and thermomigration of metallic islands of μm size formed by vapor deposition on the Si(100)2x1 surface have been investigated with an ultrahigh-vacuum scanning electron microscope (UHV-SEM) for various metallic islands. The migration speeds of metallic islands on a semiconductor surface are three-orders higher than those of vacancy or impurity in metal. The driving forces of the island migrations are discussed by use of the diffusion equations for electro and thermomigrations.

1. INTRODUCTION

The phenomena of current-induced mass transport, so-called electromigration and temperature induced mass transport, so-called thermomigration, have been investigated in the past 30 years for various metals. The most reliable explanations for electro and thermomigration were given by Huntington (1975) and Davies (1956), respectively. However, these phenomena are very complicated and have not been made clear quantitatively for the magnitude and sign of their driving forces. In the present study, it has been found that metallic islands of μm-order diameters formed on the Si(100) surface in UHV migrate by electric current or temperature gradient at temperatures higher than the melting points of islands. It should be emphasized that the electro and thermomigrations do not only occur on a atomic scale, but also on a μm-scale, such as islands. The characteristics of the island migration on the Si surface observed by a UHV-SEM are presented in this paper.

2. EXPERIMENTAL PROCEDURES

Experiments were carried out with the use of a UHV-SEM (JAMP-30). A p-type silicon wafer with a resistivity of about 10-25 Ω cm at room temperature and a size of 20x5x0.4 mm^3 was used as a substrate. The substrate crystal was held by tantalum electrodes and heated by passing a direct current through it for electromigration. On the other hand, thermomigration was observed by use of the temperature gradient on the substrate surface of the out side area of the electrode. The measurements of both effects were performed at temperatures higher than melting points of islands. The surface of the crystal was cleaned by flashing above $1100^{\circ}C$ for a few minutes at a vacuum pressure of less than 5×10^{-10} Torr. The local surface temperatures were measured by an infrared pyrometer with a spatial resolution of 0.1 mm. The various metal films of Au, Ni, Pd, Ag, In, Cu, Ga and Al were deposited from tungsten wire baskets onto the substrates at room temperature. All metal films were intermixed with the Si substrate at a temperature higher than $500^{\circ}C$, and islands were nucleated over the surface by further heating above $500^{\circ}C$ with a growth mode of the Stranski-Krastanov type. Au, Ag and Al islands consist of eutectic alloys and Ni, Pd, In, Cu and Ga islands are silicides. Concentration of the islands were determined through SEM observation by checking the melting points of islands and then referring to phase diagrams of Si-metal binary alloys. Dynamic motions of islands for several metals were monitored by TV-scan and stored in a video recorder.

3. EXPERIMENTAL RESULTS

3.1 Electromigration

Fig.1(a) is a secondary electron image showing electromigration of Au islands by passin a direct current downward and Fig.1(b) and (c) show electromigration in the opposite directions achieved by reversing the current.

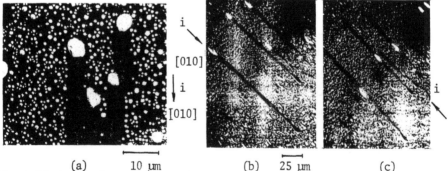

(a)   10 μm   (b)   25 μm   (c)

Fig.1 (a) A scanning electron micrograph taken after electromigration of Au islands by passing the electric current from top to bottom for 10 min at $800^{\circ}C$; (b) and (c) images taken after electromigration in the opposite directions achieved by reversing the current.

The images in Fig.1 were taken after cooling the surface down to room temperature and show the results of electromigration for 10 min at 800°C. The large islands migrated while coalescing with small islands and increasing their diameters. The larger the islands, the higher is the speed. It should be noticed that for Au islands the direction of the electromigration is opposite to that of the electric current. The velocity v increases exponentially with temperature T and is approximately proportional to the island radius. Ni and Pd islands migrate also in the same direction as that of the Au islands; Ag, In, Cu and Ga do not migrate; and Al migrates in the direction of the electric current. For heating with alternating current, no electromigration was observed and the stationary islands vaporized at above the evaporation temperatures. The effect of gravity is negligible because the island velocity is not influenced by the inclination of the substrate. The trace left behind migration is a clean Si surface and the bottom level of the trace is lower than that of the surrounding substrate surface. The depth of the bottom level is several hundred angstroms. The direction of the island migration is independent of the type of the Si substrates (p- or n-type) with different impurity concentrations.

3.2 Thermomigration

The islands on the substrate in the outer part of the electrode migrate due to the temperature gradient from low to high temperature independently of the type of metal. Fig.2(a) shows an example of Au-islands in a region of a 500 μm width and between 700°C and 800°C. The velocity of the Au islands is 0.1-1.2 μm/s depending on the island radius and substrate temperature. For thermomigration, it is also found that the larger the islands are, the faster is the speed. We observed furthermore the island migrations due to

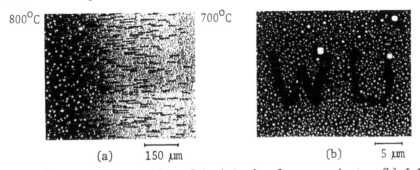

(a)   150 μm   (b)   5 μm

Fig.2 (a) The thermomigration of Au islands after one minute. (b) Island migration due to the electron beam effect. Letters of 'WU' are written by an electron beam scan.

the electron beam effect. If an electron beam of 0.1 μm diameter with a current density of $10^6 A/cm^2$ irradiates a fixed position on the surface at a substrate temperature above the melting point of islands by heating with an alternating current, the islands lying in a circle of several hundred μm diameter around the electron beam coalesce at its center and form a large liquid island of several tens of μm diameter. Moreover, if the electron beam scans slowly close to a given islands, the island moves following the electron beam, as shown in Fig.2(b). Thus, we can write letters of micron-size by using the electron beam. Such an effect is regarded as an island migration due to the temperature effect, because every island migrates to-wards the electron beam independently of the type of metal. Thus, we con-sider a main effect of the electron beam to be caused by the temperature gradient.

4. DISCUSSION

We regard electro and thermomigrations to be caused by diffusion phenomena of migrating atoms against the host lattice across the solid-liquid inter-face due to the electric and thermal effects. The average velocity v of the migrating atoms due to the concentration gradient, the temperature gradient and the electric current is given by

$$v = -D \frac{\partial \ln c}{\partial x} - \frac{D}{T} \frac{Q^*}{kT} \frac{\partial T}{\partial x} + B e Z^* \wp \, j, \qquad (1)$$

where D is the diffusion coefficient, c is a concentration of migrating atoms, Q^* is an "apparent heat of transport" for the migrating atom given by Davies (1956), k is Boltzmann's constant, B is a mobility, Z^* is the "ef-fective valence" of the migrating atom given by Huntigton (1975), $\wp$ is a resistivity of the metallic islands and j is the current density. From our experiment, it can be seen that Q^* is negative for every type of metals used in this experiment, because every island migrates from low to high temperature. The interpretation of thermomigration was described by Jaffe and Shewmon (1964) for several interstitial impurities in Cu, Au and Ag metals and they reported that almost all impurities migrate from the cold to the hot side. Here, it should be noticed that whether the electromigra-tion interacts or not with the thermomigration through a thermoelectric ef-fect. However, we experimentally confirmed that a local temperature differ-ence at the positions between the front and the rear of a large Au island during the migration at 800°C by a direct current is less than ± 5°. Thus, it can be seen that both phenomena are independent in this case. The third term in the right hand side of eq.(1) is given by

$$v = B \, e \, Z^* \, \wp \, j = (D_o/kT) \exp (-\Delta H/kT) eZ^* \wp j, \qquad (2)$$

where D_o is a constant and ΔH **is** the **activation** energy. Thus, the usual Arrhenius plots of $\ln(vT/j)$ versus $1/T$ are obtained as shown in Fig.3 for the three islands. From Fig.3, we can get the activation energy of 0.75 eV regardless of the island radius. On the other hand, the second term in eq. (1) is expressed by

$$v = - \frac{D_o \, Q^*}{k \, T^2} \frac{\partial T}{\partial x} \exp (-\Delta H/kT). \qquad (3)$$

Thus, Arrhenius plots of $\ln(v \, T^2)$ versus $1/T$ are obtained as shown in Fig. 4 for different values of $\partial T/\partial x$ with respect to the same island diameter. From Fig.4, we can get the activation energy of 0.6 eV independently of $\partial T/\partial x$. These values approximately agree with each other.

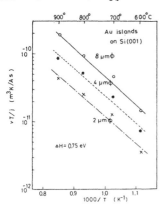 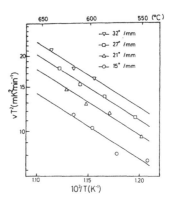

Fig.3 vT/j vs $10^3/T$ for electro-migration of Au islands as a parameter of island diameter.

Fig.4 vT^2 vs $10^3/T$ for thermo-migration of Au islands as a parameter of $\partial T/\partial x$.

According to the theory of electromigration, two origins of the driving force are considered: the first arises from the direct action of the external field on a charge of a migrating atom ("direct force") and the second from the scattering of conduction electrons by the impurities or vacancies under consideration ("wind force"). Therefore, the driving force $\overline{F}$ for the electromigration is given by the effective charge Z^* of the impurity or the defect under the existence of the macroscopic electric field $\overline{E}$.

$$\overline{F} = (Z + Z_{wind}) e \, \overline{E} = Z^* e \, \overline{E}, \qquad (4)$$

where Z and Z_{wind} are, respectively, a charge of the migrating atom and an apparent charge corresponding to the wind force. However, the estimation of magnitudes and signs of Z and Z_{wind} is very difficult for various metal alloys, because the calculation of the screening of the electric field at the site of an impurity by conduction electrons and the existence of imhomogeneities in the electric field and current flow near an impurity are impossible. For semiconductor alloys, since the role of the direct force is more significant than that of metal alloys, the difference of the electronegativities between silicon and metal becomes an indicator for characterization of the electromigration. On the other hand, for thermomigration, the velocity and direction of interstitial atoms are determined by the magnitude and the sign of the heat of transport of an interstitial atom, Q_I^*. In addition, in the case of metal alloys where diffusion occurs by a vacancy mechanism, the equilibrium vacancy concentration increases exponentially with temperature. Thus, if $Q_I^*=0$, the vacancy concentration gradient would make the interstitial atoms drift from low to high temperature as the vacancies drift from high to low temperature. Therefore, the total drift velocity is given by the sum of both effects. In the case of metal alloys, Q^* is almost negative such as reported by Jaffe and Shewmon (1964). Here, we should note that the total Q^* value of the semiconductor alloys is three-orders higher than that of metals, and velocities of electro and thermomigrations are much higher than the diffusion velocity due to the concentration gradient.

The size dependence of the island migration may be explained as follows: the driving force is proportional to r^2, because the effective volume to the island migration is probably the bottom sinking below the substrate surface. On the other hand, a horizontal component of the surface tension of a liquid island is proportional to the island radius and may act as a reaction force. Therefore, a large island migrates faster than a small island. Lastly, it is very interested to make clear a reason why the driving forces of the metallic islands on Si surface are quite larger ($\times 10^3$) than that of vacancies or impurities in metal.

References

Davies R O 1956 Rep. Progr. Phys. 19 327
Huntington H B 1976 Diffusion in Solids (New York: Academic Press) Ch.6
Jaffe D and Shewmon P G 1964 Act. Met. 12 515

Inst. Phys. Conf. Ser. No 130: Chapter 2
Paper presented at Int. Congr. X-ray Optics and Microanalysis, Manchester, 1992

243

An electroanalytical examination of the gas retention in a ramped (U,Pu)O$_2$ irradiated fuel

P.D.W. Bottomley, F. Daguzan-Lemoine*

Commission of the European Communities, Joint Research Centre, Institute for Transuranium Elements, PO Box 2340, 7500 Karlsruhe, Germany
*IPSN-DERS-SEMAR CEN Cadarache, 13108 St. Paul-Lez-Durance, France

ABSTRACT: An electron optical examination is used to measure or estimate the various forms (in solid solution, micro- and macro-porosity) of retained fission gas and is compared to that expected from the fuel's history. Reasonable agreement was obtained.

1. INTRODUCTION

As part of a program investigating fission gas release, (U,Pu)O$_2$ minipins irradiated to 1-2 at % burn-up underwent a rapid power ramp ($\sim$10 msec.) in a pressure cell under argon in the Silène reactor (Daguzan-Lemoine, 1987). The pins were open-ended to accomodate any fuel expansion. After the ramp, the volume of gas released was measured and analysed by mass spectroscopy. TP29 and TP36, were ramped at approx. 1400 J/g to reach the fuel's melting point. TP29 had a high fill gas pressure of 3.7 bar Ar (6.78 bar during ramp), whereas TP36 had a low pressure of 0.2 bar Ar (1.37 bar during ramp).

2. EXAMINATION METHOD AND RESULTS

Sections were taken of the minipin and prepared for electron optical examination including phase analysis (Quantimet) for porosity estimation. As a control, a non-ramped minipin (reference material No. 2) was also examined.

2.1 Optical and Scanning Electron Microscopy

The fuel has been dispersed and compressed into a ring of material inside the cladding during the ramp. The fuel initially occupied approx. 30 % of the available volume. This is the case in the interface sections of TP29 (Fig. 1a) and TP36 (Fig. 1b) but other sections show much less fuel (e.g. 5 % volume); because of dispersion and losses to the outer pressure cell. TP36 also shows radial, compared to TP29's more heterogeneous porosity. This probably results from the lower fill gas pressure and a higher melt fraction for TP36.

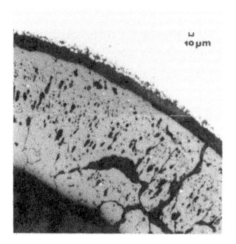
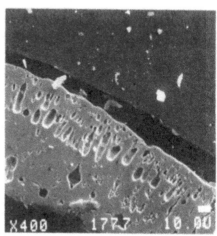

Fig. 1a   TP 29 Section (188x)   Fig. 1b   TP36 Section (400 x)

2.2   Microprobe Analysis

The microprobe (EPMA) analyses show that Xe is present in TP29 in solid solution or in porosity below 500 Å at the limit of detection (190/200 ppm Xe; Ancey et al., 1977) while in TP36 there was no detectible Xe (<100 ppm). Cs was barely detectible in TP36 (230 ppm Cs for 170 ppm Cs detection limit). This compares with average reference material values of 2700 ppm Xe and 2300 ppm Cs. The Xe values can be taken as the total fission gas (Xe + Kr) at these low concentrations without significantly affecting the calculation, since Kr is only 10 % of the Xe concentration (Mogensen, 1983).

2.3   Replica Examination

TEM micrographs of replicas indicate a dense structure in TP36 without intermediate (20 nm to 0.5 µm) gas bubbles but with a few grain boundary bubbles, whereas intragranular and grain boundary bubbles are noticeable in TP29.

2.4   Porosity Estimation

The gas contents are calculated from the bubble equilibrium equation:

$(2\gamma/r_b + P_{ext})V_b = C_g RT$

where γ (the bubble surface energy) is ~ 0.5 J/m^2 (Olander, 1976)

r_b = bubble radius $T = (U,Pu)O_2$ melting point

P_{ext} = external gas pressure $\lesssim 8$ bar   approx. 3000 K

C_g = gas concentration in the solid

V_b = volume fraction of bubble (cm^3/cm^3)

For TP29 taking a V_b value of 10 % for the intermediate porosity (20 nm- 0.5 µm) from the TEM micrograph and a mean diameter of 0.15 µm then $C_g \approx 120$ mm^3/g fuel (fuel density is approx. 10 g/cm^3). This will increase as bubble radius decreases or if pressure is higher, (Walker et al., 1988), therefore this will be a lower bound. For example there were pieces of limited melting with lenticular grain boundary pores after ramping. Taking 2 µm dia. and a maximum porosity of 10 % they would contain 10 mm^3/g under equilibrium conditions rising to ~ 1000 mm^3/g with 100 bar overpressure. TP36 has much lower intermediate porosity (not more than 2.5 %) with no grain boundary pores, and yields a retention of ~ 30 mm^3/g fuel.

For the larger bubbles (1 - 10 µm dia.) a mean porosity of 20 % was estimated for TP29 from the optical and SEM micrographs . This yields a residual retention of ~ 20 mm^3/g assuming equilibrium conditions (i.e. a lower bound). The retention in TP36 is lower (at ~ 17 mm^3/g) although the porosity estimation is similar (18 %), since the pores are bigger (r_b larger), moreover the external pressure (P_{ext}) is lower and is a more important factor for larger r_b values (0.5 - 5 µm).

The total gives a final retention of 150 mm^3/g rising to 300 mm^3/g in TP29 depending on bubble overpressure, and between 50 and 100 mm^3/g for TP36.

3.   COMPARISON OF REFERENCE MATERIAL ANALYSIS WITH PHÉNIX REACTOR DATA

The Xe content of the reference material No. 2 were used to revise the fuel's fission gas retentions after irradiation (Daguzan-Lemoine, 1987). These (with upper and lower bounds in brackets) are given in the table below, line 1, from which the values for gas release measured during the ramp (line 2i) are deducted to yield Silène final retention (line 2ii). It is seen that these retentions fall within the range of the post-irradiation examination (PIE) derived in the previous section (line 3).

			Minipin TP36 (mm^3/g)	Minipin TP29 (mm^3/g)
1)	Phénix irradiation	End-of-life retention	(385.2) 427.5 (471.9)	(267.75) 335.6 (408.2)
2)	Silène power ramp	i) gas release (measured)	325.5	47.4
		ii) final retention	(59.7) 102 (146.4)	(220.35) 288.2 (360.8)
3)	ITU Exam.	Retention estimates from PIE	50 - 100	150 - 300

4.   CONCLUSIONS

1.   Electron optical analysis and micrographs of (U,Pu)O$_2$ fuels TP29 and TP36 have yielded an estimate of fission gas distribution and retention values. These show a reasonable agreement with irradiation and ramp data.

2.   Total retention for TP29 is higher than for TP36. Gas release processes (e.g. bubble growth) appear dependent upon extent of melting and external pressure.

References

Ancey M, Bastenaire F, Tixier R, J. Phys. 1977 **D100**, p 817

Daguzan-Lemoine 1987 "Main Results of the Silène Programme" -Workshop on fission gas behaviour, 10/11 Dec. 1987 Cadarache France

Mogensen M 1983 J. Mass Spect. and Ion Phys. 48 p 389

Olander D 1976 "Fundamental Aspects of Nuclear Reactor Fuel Elements", Technical Information Center, ERDA (TID 26711-P1)

Walker C T, Knappik P and Mogensen M 1988 J. Nucl. Mater. **160** p 10

Inst. Phys. Conf. Ser. No 130: Chapter 2
Paper presented at Int. Congr. X-ray Optics and Microanalysis, Manchester, 1992

247

Micro-observation and microanalysis of ZrN films prepared by reactively RF sputtering

S. Maruno, Nagoya Institute of Technology, Showa-ku, Nagoya 466, Japan
P. Jin, Government Industrial Research Institute, Nagoya, Kita-ku, Nagoya 462, Japan
A. Nakamura, Asahi Chemical Industry Co. Ltd., 2-1 Samejima, Fuji 416, Japan

ABSTRACT: The composition, chemical bonding states and microstructure of ZrN films deposited on Si under with and without negative substrate bias (0 to 200V) have been precisely studied through the experiments of XPS and SIMS analyses and FE-SEM and STM observations. The bias effect results in the decrease in oxygen impurity as well as the formation of dense structure.

1. INTRODUCTION

The transition metal mononitrides TiN and ZrN have excellent properties such as golden color, high-grade hardness, metallic conductivity and high thermal stability [1]. Recently, it has been shown that thin films of TiN and ZrN can be used as diffusion barriers in several multilayer metallization schemes in LSI technology [2-4]. The barrier layers are usually deposited by reactive sputtering. It has been known that the sputtering conditions affect greatly the characteristics and physical properties of TiN films, leading to different diffusion barrier capabilities of TiN [5]. However, there is only limited information about the characterization of sputter-deposited ZrN films.

A systematic study of ZrN films deposited by reactively rf sputtering under various conditions (nitrogen flow ratio, total gas pressure and applied substrate bias) has been undertaken in order to examine the characterization and microstructure of films with advanced analytical techniques using thin film XRD, EPMA, AES, XPS, SIMS, FE-SEM and STM. In this paper, we report some experiment results of XPS and SIMS analyses, and FE-SEM and STM observations of the films deposited under with and without substrate bias.

2. EXPERIMENTAL

The ZrN films were deposited on silicon substrates in a rf diode reactive sputtering system. The gases were controlled by mass flow controllers to give the desired $N_2/(N_2 + Ar)$ flow ratios, and the total gas pressure was controlled by adjusting the opening of the main valve of the vacuum system. Table I shows the sputtering conditions. Films deposited under various nitrogen flow ratio, total gas pressure and applied substrate bias were subjected to XRD, AES, XPS, SIMS analyses, and FE-SEM and STM observations.

3. RESULTS AND DISCUSSION

Figure 1 shows FE-SEM micrographs of the ZrN films deposited under without substrate bias (a) and with bias of -90V (b). The cross section of film(a) shows a clearly defined columnar structure with much porosity. A maximum column width of 100-150 nm at near the film surface is estimated for the film of 1350 nm in thickness. The inset shows thtat the film surface is composed of tapered grains and open gaps of several tens of nanometers in width.

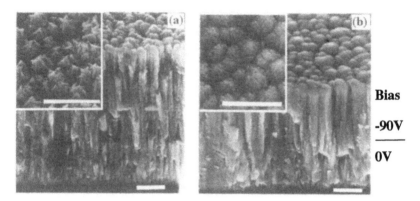

Fig.1 SEM micrographs of cross section of ZrN films deposited under without bias (a) and with bias -90 V (b). Insets show higher magnification surfaces of (a) and (b). Bar=300 nm.

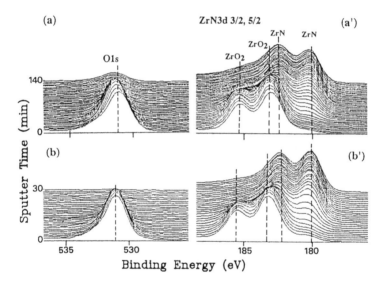

Fig. 2. The O1s and Zr3d XPS spectra of ZrN films deposited at without bias (a, a') and with bias -150 V (b, b') as a function of sputter time.

The grains appear to consist of tiny crystallites which have star-like habit. This relatively porous microstructure, which is similar to the zone 1 structure described in the Thornton's model [6], is considered to be caused by the self-shadowing effect of the deposition flux, since the energy of particles is decreased due to collisions which become pronounced at higher pressure. In contrast, the film (b) , which has a multilayer structure deposited under no-bias (the bottom layer) and subsequently under bias of -90V (the upper layer), exhibits a much denser microstructure with smoother surface consisting of domed grains. The ion-bombardment effect introduced by the applied negative

Table I. Sputtering conditions.

Target	Zr (100 mm ϕ)
Substrate	Si (100)
Sub.Temperature	Water-Cooled
T - S distance	50 mm
Base Pressure	4×10^{-4} Pa
RF Power	350 W
Ar-N$_2$ Flow	10 sccm
N$_2$ Flow Ratio	1 - 20 %
Total Pressure	0.67 - 2.67 Pa
Sub. Bias	0V to -200V

substrate bias increases the adatom mobility at the growing film surface, resulting in the transition of   film structure from a porous zone 1 to a relatively dense zone T.

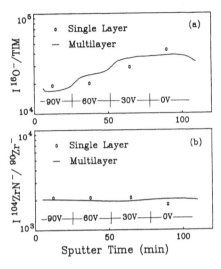

Fig. 3. SIMS depth profiles of the oxygen concentration, represented as I $^{16}O^-$/TIM (a) where TIM is the total ion monitor, and N/Zr ratio, represented as I $^{104}ZrN^-$/$^{90}Zr^-$ (b), for the ZrN film with a multilayer structure deposited stepwise at negative bias 0/30/60/90 V (the solid lines). The SIMS results for each of the bias-sputtered ZrN (single layer) are also shown as open circles. (c) is FE-SEM micrograph of cross section of multilayered ZrN.   Bar=100 nm.

XPS depth profiling was used to determine the in-depth changes in oxygen concentration and the bonding states of elements constituting the films. Figure 2 shows the O1s (a, b) and Zr3d (a', b') XPS depth profiles of the ZrN films deposited at without bias (a, a') and with bias -150V (b). The binding energy for O1s level appears at 531.1 eV. The peak intensity decreases with sputter time, and becomes almost undetectable at the inside of the biased film (b), while it still remains in the no-biased one (a). The shift of O1s from film surface in (a), toward high energy side by 0.28 eV, seems to be due to the Zr-O bonds resulting in the formation of ZrO_2. The Zr3d 5/2 and 3/2 levels for the surface of no-biased film (a) appear at 183.11 and 185.47 eV, respectively, indicating the film surface is in a heavily oxidized state. They are shifted toward lower energy side with sputter time and become 180.07 and 182.62 eV at the inside of film, respectively, indicating that the Zr3d has changed from an oxide bonding state to a nitride one, accompanied with a energy shift of about 3.0 eV. The Zr3d spectra of the biased film (b) show similar behavor as those of the no-biased one. The 3d 5/2 and 3/2 peaks from the surface appear at 183.26 and 185.63 eV, and those from the inside are 180.03 and 182.36 eV, with a relatively larger energy shift of about 3.3 eV. Considering the 3d core-electron peaks for Zr metal [7], it is indicated that electron charge is transffered from Zr atom to N and/or O when Zr-N and/or Zr-O bonds are formed, and the formation of Zr-O bonds with higher ionicity, possiblly to form only ZrO_2 in ZrN films, results in larger energy shift.

Figure 3 shows the secondary ion concentration profiles (using Cs+ as the primary ion beam) of the multilayered ZrN films (solid lines) deposited under varous bias, in comparison with the result measured from each of the single-layered films (open circles). As shown in (a), the oxygen concentration decreases significantly with increasing bias for both the single-layered and the multilayered films. The multilayered film exhibits step-like change in oxygen concentration at the 90/60V and 60/30V interfaces, as expected, while it shows no obvious change at the 30/0V. It is considered that the porous structure of films deposited at bias 30V, in which the etching ion flux can enter the no-biased layer, has contributed to give almost the same oxygen concentration. (b) shows the change in N/Zr retio with bias, where the N/Zr ratio is represented by I ZrN^-/Zr^- since nitrogen has a very low ionization effeciency. It is shown that the bias has little affect on the N/Zr rratio in the present experiment. (c) is a FE-

SEM micrograph of cross-section of the multilayered ZrN film. The bias effect on the microstructure can be clearly seen in the film deposited stepwise at various bias without interrupting the film growth.

Figure 4 shows the three-dimentional STM micrographs of the ZrN films deposited at bias 0V (a) and -90V (b). The no-biased film (a) exhibits a rough surface with fairly tapered cones, as expected from the FE-SEM micrograph in Fig. 1(a). The average surface roughness (Ra) is 3.9 nm and the mean of all the peak-to-valley values obtained within assesment length (Rtm) 16.7 nm. While the biased film (b) exhibits somewhat smooth surface with roundish-shaped cones (domed ones), in comparison with that of (a), and the values of Ra and Rtm are 2.0 nm and 7.1 nm, respectively. It was found that the great mumber of cones were formed even inside of the depression among each of the grains observed in Fig. 1. The results of STM observation demonstrate that the hyperfine structure on the film surface reflects the film growth process, and that there is a good accordance between the FE-SEM observation and the STM one with respect to the bias effect on the microstructure of ZrN.

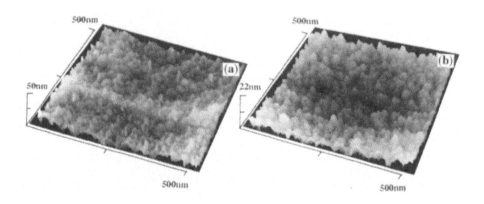

Fig. 4. STM micrographs of the surface of ZrN films (200 nm in thickness) deposited at without bias (a) and with bias -90 V(b). Note that vertical scale of (a) is 2.3 times as great as that of (b).

4. CONCLUSION

It has been found from the results obtained by the experiments of XPS, SIMS, FE-SEM and STM that the microstructure of reactively rf-sputtered ZrN films changed from a porous structure with much oxygen impurities to an extremely dense one almost free of oxygen as the applied negative substrate bias is increased.

REFERENCES:

1). L. E. Toth: Transition Metal Carbides and Nitrides (Academic Press, New York, 1971).
2). M. Ostling, et al.: Thin Solid Films 145 (1986) 81.
3). N. Kumar, et al.: Thin Solid Films 153 (1987) 287.
4). J. M. E. Harper, et al.: J. Vac. Sci. Technol. A7 (1989) 875.
5). P. Jin and S. Maruno: Jpn. J. Appl. Phys. 31 (1992) 1446.
6). J. A .Thornton: J. Vac. Sci. Technol. 11 (1974) 666.
7). H. Hochst, et al.: Phys. Rev. B., 25 (1982) 7183.

Inst. Phys. Conf. Ser. No 130: Chapter 2
Paper presented at Int. Congr. X-ray Optics and Microanalysis, Manchester, 1992

251

A microanalytical study of the gills of aluminium-exposed catfish

S Eeckhaoudt[1], J Landsberg[2], R Van Grieken[1], W Jacob[3], F Watt[2] and H Witters[4]

University of Antwerp (UIA), Dept. of Chemistry[1] and Medicine[3], Universiteitsplein 1, B-2610 Wilrijk-Antwerp, Belgium.
University of Oxford[2], Dept. Nuclear Physics, Keble Road, Oxford OX1 3RH, Great-Britain.
V.I.T.O., Dept. of Environmental Sciences[4], Boeretang 200, B-2400 Mol, Belgium.

ABSTRACT: Through a combined use of light- and electron microscopy with micro-analytical techniques, the aluminium distribution in gills of aluminium-exposed catfish was investigated. This study demonstrates that also with the "acid-resistant" catfish, exposed during seven days to acidified water (pH=4.3) with an elevated aluminium concentration (1 mg Al/l), aluminium can be found at, and even in, the gills.

1. INTRODUCTION

One of the consequences of the acidification of our environment is a higher mobility of metals in soils. They are washed out into the ground water and transported to rivers and lakes. Especially aluminium has been put forward as being toxic for fish gills. The gills are the primary organ for aluminium toxicity. By using bulk analysis one obtains an idea about the amount of aluminium accumulated in or on the gills, but not about the accumulation sites. Information about the subcellular localization of insoluble aluminium deposits on or in gill tissue is becoming accessible due to the availability of micro-analytical techniques. In this study, laser microprobe mass analysis (LAMMA), energy dispersive X-ray analysis in a scanning transmission electron microscope (STEM-EDX) and proton induced X-ray emission in a proton microprobe (PIXE, PMP), were used to study the (sub-)cellular distribution of aluminium at the gill level of aluminium-exposed catfish. The morphological information obtained by LAMMA and micro-PIXE is inferior to that from electron microscopy, but these techniques offer a greater sensitivity as compared to x-ray micro-analysis in an electron microscope.

2. MATERIALS AND METHOD

The fish used in this study are catfish (Ictalurus nebulosus). They were caught in the surface waters of the nature reserve "Zwart Water" (pH 4.3) located at Turnhout in the north-east of Belgium. The fish were stored in basins (1000 l) filled with synthetic water (pH 6.8, water temperature ± 16°C). To prepare this synthetic or control water, low concentrations of mineral salts (NaCl, KCl, CaCl$_2$, MgSO$_4$.7H$_2$O, NaHCO$_3$) were added to demineralised water. After an acclimation period of about five weeks, the fish were transferred and stored in a 150 l tank containing synthetic water (pH 6.8, water temperature ± 16°C), which was continuously recirculated and renewed. After one week, during which the fish were no longer fed and exposed to a constant day/night rhythm (12 hours/12 hours), the experiment started.

The fish were placed in individual containers with 0.5 l synthetic medium and divided in two groups. In a period of three hours the synthetic water of the control group was acidified to pH 4.3, by means of a mixture containing H$_2$SO$_4$ and HNO$_3$ (2 vol./1 vol.). The other fish were additionally exposed to 1 mg Al/l at pH 4.3. Aluminum was added to the acid water as AlCl$_3$.

At the end of the exposure period, the fish were killed and the second gill arches dissected. The wet weight of the gill tissues was determined immediately after dissection. They were dried at 105 °C, ashed in a furnace at 500 °C and the ashes were redissolved in a solution of nitric acid and bidistilled water. The total Al-content in the gill preparations was determined by plasma emission spectrometry (Jarell Ash, Atomcomp Model 750).

The gills, intended for micro-analytical investigations, were dissected from the fish and cut in small pieces (1 mm^3), fixed in a buffered 2.5 % glutaraldehyde solution, dehydrated in a graded ethanol series, passed through propylene oxide and embedded in epoxy resin (LX-112). A small part of the samples was post-fixated with osmiumtetroxide, for morphology purposes. Sections were cut to the required thickness with an LKB Ultrotoom III ultramicrotome, using glass knives (0.25-2µm) or a diamond knife (< 0.25 µm).

The samples were examined with a JEOL EX-1200 TEMSCAN Scanning Transmission Electron Microscope (STEM), equipped for EDX-analysis and coupled to a TRACOR TN 5500 and IBAS-KONTRON micro-analysis system. EDX-analysis was carried out on carbon-coated gill tissue sections, mounted on carbon coated nylon grids or Cu-grids with big grid openings (50-100 mesh) which were put in a graphite sample holder. Laser Microprobe Mass Spectrometry was performed using the LAMMA-500 instrument of Leybold-Hereaus (Köln, Germany). To enhance the visibility in the light microscope of the instrument, staining of sections with toluidine blue proved to be convenient. The resulting organic mass peaks did not interfere with the analysis. Proton induced X-ray emission was performed using the Oxford Scanning Proton Microprobe.

3. RESULTS

Through a combined use of light- and electron microscopy with micro-analytical techniques, the aluminium-distribution in gills of aluminium-exposed catfish was investigated. This study demonstrates that also by the "acid-resistant" catfish, exposed during seven days to acidified water (pH=4.3) with an elevated aluminium concentration (1 mg Al/l), aluminium can be found at, and even in, the gills.

Aluminium was identified in electron dense deposits. These deposits are situated at the borders of the secondary lamella, pointing to an association with the mucus layer covering the lamellae. Mucus cloggings with cell material and aluminium deposits can also be found in the embedding medium, in between the secondary lamellae.

Very locally, electron dense deposits were found inside cells, namely in epithelial cells, chloride cells and macrophages. They were mainly located at the basis of the secondary lamellae. The electron dense deposits consist of small spheres and seem to accumulate in delimited cell areas. The X-ray micro-analytical measurements also provide evidence for a relation between the sites of high aluminium levels and high levels of phosphorus in the gill epithelium.

The micro-analytical findings suggest that aluminium toxicity is associated with gill aluminium accumulation and gill surface interactions, but could also exert an effect through cellular uptake. Playle and Wood (1989) suggested that reduced aluminium solubility in the gill micro-environment (gills serve as an excretion organ for ammonia) can explain gill aluminium accumulation in the pH range 4.5-6.0. In more acidic water, below pH 5, adsorption and complexation of positively charged aluminium species onto branchial surfaces and mucus (negative charges because of carboxyl groups) would better explain aluminium accumulation on fish gills.

The exact mechanism and the way in which aluminium enters the gill cells are not known yet. A recent model for acute aluminium toxicity in fish (Exley et al. 1990), suggests that apically bound aluminium alters the permeability of membranes and allows the intracellular accumulation of aluminium. Aluminium accumulation taking place through bindings with ligands (e.g phosphate groups), would have severe biochemical repercussions. The distribution of aluminium in epithelial and chloride cells indicates the presence of such aluminium-binding groups. Aluminium present in chloride cells may interfere with the active transport of ions, resulting in a disturbed ion-regulation. Concentration of aluminium in macrophages on the other hand, points to a detoxification mechanism in which the metal is collected and stored.

5. ACKNOWLEDGEMENT

S. Eeckhaoudt wishes to acknowledge the financial support of the I.W.O.N.L.

6. REFERENCES

Exley C, Chapell J S and Birchall J D 1990 J. Theor. Biol. 151 417

Playle R C, Wood C M 1989 J. Comp. Physiol. B. 159 539

Inst. Phys. Conf. Ser. No 130: Chapter 2
Paper presented at Int. Congr. X-ray Optics and Microanalysis, Manchester, 1992

255

SEM/EDX study of the displacement of silver in $Cu_{2-x-\delta}Ag_\delta S$ grains equilibrated with copper

W. Barzyk

Institute of Catalysis and Surface Chemistry, Polish Academy of Sciences, 30-239 Cracow, Poland

F. Placido

Department of Physics, University of Paisley, Paisley PA1 2BE, Renfrewshire, Scotland

ABSTRACT: Scanning electron microscopy, energy dispersive x-ray analysis, and potentiometric measurements have been used to show that silver admixtures can be displaced from the bulk to the surface of grains of non-stoichiometric cuprous sulphide, by a simple electrochemical method. This process proceeds spontaneously as a result of equilibration of the sulphide grains, $Cu_{2-x-\delta}Ag_\delta S$, with metallic copper, Cu^0 in oxygen-free, acid solutions containing Cu^{2+} ions. The disproportionation reaction of cuprous ions plays the main role in the process.

1. INTRODUCTION

The present study shows that a silver admixture contained within the bulk of grains of non-stoichiometric cuprous sulphide, $Cu_{2-x}S$ (forming a solid solution or a phase of the Ag-Cu-S system) can be displaced from the sulphide bulk to its surface as a result of equilibration of the $Cu_{2-x-\delta}Ag_\delta S$ grains with metallic copper. The results obtained are important not only for a deeper understanding of the great ability of silver to migrate in sulphide deposits, but also for explaining the behaviour of silver admixtures during leaching of the sulphides using hydrometallurgical methods.

2. EXPERIMENTAL

Cuprous sulphide $(Cu_{1.86\pm0.02}S)$ grains of size 60-75μm, were synthesized and treated according to the procedure described in the earlier paper (Pomianowski and Barzyk, 1987). The experiments were performed in a special apparatus only slightly modified from the one presented by us earlier (Nowak et al., 1984; Barzyk et al. 1987, 1988). The $Cu_{2-x}S$ grains and Cu^0 foil, of comparable surface areas (approx. 0.1 m), were placed in different compartments of the apparatus and were in contact only through a circulating solution (0.1 M $CuSO_4$ + 0.1 N H_2SO_4). The process was monitored by measurement of the potential of platinum electrodes: one immersed in the grain sample [denoted as the sulphide fluidised-bed electrode, Pt $(Cu_{2-x}S)$], and the second in contact only with the circulating solution (denoted as Pt_{red-ox} electrode).

A scanning electron microscope with conventional secondary electron detector, solid-state backscattered electron detector and an energy dispersive X-ray analyser were used to examine changes in microstructure and surface compositions of the grains.

3. RESULTS

The equilibration stage was carried out for samples of grains of the pure sulphide $Cu_{1.86}S$, and of the silver-containing sulphide, $Cu_{2-x-\delta}Ag_\delta S$. The latter samples originated from the same stock of $Cu_{1.86}S$ grains into which various silver admixtures, ranging from 0.01 to about 0.1 mol Ag/mol $Cu_{2-x}S$, were introduced (at stage I of the experiments) by reaction of Ag^+ ions with the sulphide in solutions containing 0.05 M H_2SO_4. This fast reaction, taking about 15 minutes, was followed by a slower penetration of silver into the sulphide lattice, as was found in the earlier investigation (Barzyk, 1989). After stage I was finished, the solution in the apparatus was replaced with oxygen-free 0.1M $CuSO_4$ + 0.05 M H_2SO_4 solution. Then the equilibration process (stage II of the experiments) was started by inserting the copper foil in the circulating solution. The attainment of equilibrium in the system was indicated by equalizing of the potentials of the electrodes: $Pt(Cu_{2-x}S)$, $Pt_{red.-ox.}$ and Cu^0.

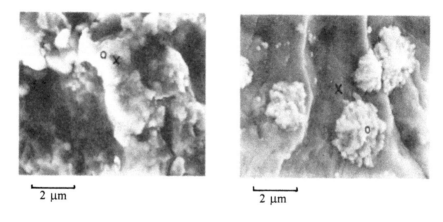

Fig. 1   **Surface structure of grains after doping with silver but before equilibration with copper.**

Fig. 2   **Surface structure after doping with silver and equilibration with copper.**

Typical SE images for the grains containing 0.04 mol Ag/mol $Cu_{2-x}S$ are shown in Figs. 1 and 2. The surface microstructure of such grains before and after equilibration with copper is shown in Fig. 1 and Fig. 2 respectively. Typical EDX spectra are shown in Figs. 3a,b and 4a,b, respectively. (Spectra "a" and "b" have been taken in the surface microregions denoted as "x" and "o" in Figs. 1 and 2, respectively).

It can be inferred from these EDX spectra that the product formed as a result of the reaction of the sulphide with Ag^+ ions was composed of Ag, Cu and S atoms. However, the agglomerates formed as a result of equilibration of the relevant sample with copper, consists of Ag atoms only. (This conclusion was also corroborated by x-ray diffraction results obtained for these samples).

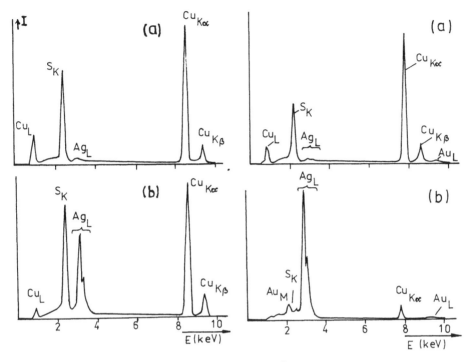

Fig. 3

a) EDX spectrum of grain after doping. Taken at point marked 'X' on Fig. 1.
b) EDX spectrum of grain after doping. Taken at point marked 'O' on Fig. 1.

Fig. 4

a) EDX spectrum of grain after doping and equilibration. Taken at point marked 'X' on Fig. 2
b) EDX spectrum of grain after doping and equilibration. Taken at point marked 'O' on Fig. 2

4. DISCUSSION

The mechanism by which the sulphide grains reach equilibrium with metallic copper, Cu^0, can be expressed by the following partial reactions:

A) For the system containing pure sulphide: $Cu_{2-x}S/Cu^{2+}$, Cu^+/Cu^0 :

a) at the surface of metallic copper, Cu^0:

$$Cu^0 + y\ Cu^{2+} \iff 2y\ Cu^+ \qquad /1/$$

b) at surface of pure cuprous sulphide, $Cu_{2-x}S$:

$$Cu_{2-x}S + 2y\ Cu^+ \iff Cu_{2-x}S/\ y_{ad}\ Cu^0 + y\ Cu^{2+} \qquad /2/$$

where $/\ Cu$ denotes an ad-atom of copper on its surface.

c) the solid state reaction following the reaction /2/:

$$Cu_{2-x}S/\ y_{ad}\ Cu^0 \iff Cu_{2-x+y}S \qquad /3/$$

The total reaction (summarising: /1/ + /2/ + /3/) can be written as follows:

$$y\ Cu^0 + Cu_{2-x}S \iff Cu_{2-x+y}S \qquad /4/$$

The balanced reactions, expressed by equation /4/ indicate that a finite amount of copper has been passed from the metallic phase, Cu^0, to the $Cu_{2-x}S$ grains, through the electrolyte solution, without changing its composition.

B) For the system containing sulphide doped with silver: $Cu_{2-x-\delta}Ag_\delta S/Cu^{2+}$, Cu^+/Cu^0:

   a) at the surface of the metallic copper, Cu^0:

$$Cu^0 \ + \ y \ Cu^{2+} \ \Longleftrightarrow \ 2y \ Cu^+ \qquad\qquad /1/$$

   b) at the surface of the cuprous sulphide doped with silver (which forms a solid solution or phases of the Ag-Cu-S system):

$$Cu_{2-x-\delta}Ag_\delta S \ + \ 2y \ Cu^+ \ \Longleftrightarrow \ Cu_{2-x-\delta}Ag_\delta S/ \ y_{ad} \ Cu^0 \ + \ y \ Cu^{2+} \qquad /5/$$

   where $/_{ad}$ Ag denotes an ad-atom of silver adsorbed at the sulphide surface.

   c) the solid state reaction following the reaction /5/:

$$Cu_{2-x-\delta}Ag_\delta S/ \ y_{ad} \ Cu^0 \ \Longleftrightarrow \ Cu_{2-x-\delta}Ag_\delta S/ \ y_{ad} \ Ag^0 \qquad\qquad /6/$$

Finally, the total reaction, summarizing the reactions: /1/ + /5/ + /6/, can be written as follows:

$$y \ Cu^0 \ + \ Cu_{2-x-\delta}Ag_\delta S \ \Longleftrightarrow \ Cu_{2-x-\delta+y}Ag_{\delta-y}S/ \ y_{ad} \ Ag^0 \qquad\qquad /7/$$

5. CONCLUSIONS

Silver admixtures can be displaced from the bulk of Cuprous sulphide grains, $Cu_{2-x-\delta}Ag_\delta S$, to the surface as a result of equilibration of the sulphide grains with metallic copper, Cu^0, in oxygen-free, acid solutions containing Cu^{2+} ions. The copper disproportionation reaction:

$$2 \ Cu^+ \ \Longleftrightarrow \ Cu^{2+} \ + Cu^0$$

which occurs at the surfaces of both copper and the cuprous sulphide phase, plays the main role in the equilibration process. According to the thermodynamic data for these systems, Cu^+ ions should be released at the surface of metallic copper and deposited at the surface of cuprous sulphide ($Cu_{2-x}S$ or $Cu_{2-x-\delta}Ag_\delta S$). The Cu^0 ad-atoms deposited at the sulphide surface diffuse easily into the lattice (due to the high mobility of copper in the rigid sulphur sublattice in the sulphides). Consequently, the silver atoms are displaced to the surface of the sulphide grains.

REFERENCES

Barzyk W., Wandzilak P. and Pomianowski A. 1987 Proc. 10th International Congress on Metallic Corrosion, Madras (India) November 1987, vol. 1 pp 299-317

Barzyk W., Nowak P. and Pomianowski A. 1988 Materials Science Forum 25-26 pp 565-568

Barzyk W. 1989 Physicochem. Probl. Miner. Process. 21 pp 141-156

Nowak P., Barzyk W. and Pomianowski A. 1984 J. Electroanal. Chem. 171 pp 355-358

Pomianowski A. and Barzyk W. 1987 Bull. Pol. Ac.:Chem. vol. 35 No.9-10 pp 461-470

Inst. Phys. Conf. Ser. No 130: Chapter 2
Paper presented at Int. Congr. X-ray Optics and Microanalysis, Manchester, 1992

Investigation of the space charge distribution and response characteristics of GaAs MSM Shottky barrier photodetectors with SEM-EBIC

O.V. Salata [*]

Department of Materials, University of Oxford, Oxford OX1 3PH, UK

ABSTRACT: SEM-EBIC technique was used for GaAs metal-semiconductor - metal (MSM) structures characterization. Dependence between space charge region (SCR) distribution and device response on the short electron irradiation pulse was observed. The origin of the slow component in the MSM structure time response is discussed.

1. INTRODUCTION

Planar MSM interdigitated photoreceivers are commonly used as photosensors in GaAs optoelectronic integrated circuits. The advantages of this type of photodetectors are planarity, simplicity and compatibility with GaAs FET processing technology.

One of the difficulties of the MSM photodetector practical usage is due to the presence of two different parts in the photoresponse. One of them is fast and determined by the drift times of charge carriers. Another is slow and controlled by the carrier diffusion and capture.

Our aim here was to reveal the possible reasons of such a behavior by means of the stationary SEM-EBIC and the time resolved SEM-EBIC as well.

2. EXPERIMENTAL

Original electrostatic beam deflection system has been designed and built for the application with the commercial SEM (Konnikov at al 1987). High gain low noise EBIC amplifier has been constructed for the steady state measurements.

* On leave from A.F. Ioffe Physical Technical Institute, St.-Petersburg, Russia

Several types of substrates were tested for MSM structures manufacture: 1) n-type Cr-doped GaAs substrate with donor concentration 10^{15} cm^{-3}; 2) n-type GaAs substrate with 10 μm epitaxial layer (LPE) with the same doping; 3) n-type GaAs substrate with LPE grown $Al_{0.4}Ga_{0.6}As$-GaAs double heterostructure, thickness of the top $Al_{0.4}Ga_{0.6}As$ layer was 0.1 mkm and thickness of the GaAs layer was 5 μm. After surface preparation an interdigitized metal structure was formed on the surface by liftoff lithography. Aluminium with Ti-sublayer was used for 0.1 μm thick metallisation. Metal fingers were 2 μm wide with the 3 μm spacing.

In order to reduce surface recombination and neutralise surface states two types of passivation layers were used: SiO_2 and Si_3N_4. Test structures were prepared for measurements of the space charge distribution under the electrodes .

3. RESULTS

It was found that the SCR uniformity is strongly depends on the GaAs surface preparation prior to the metal deposition. In fig.1 SEM-EBIC image of the MSM structure with badly treated surface is shown.

Fig.1. SEM-EBIC image of the MSM structure with poor SCR quality

One could see breakdowns and regions with a high gain. SCR position and width were measured as a function of applied voltage on a test structure. Under the zero bias the neighboring space charge regions were not in touch even in a best material. With the applied voltage SCR under the negative electrode expands towards the positive electrodes but only in the bulk GaAs. No SCR under the surface was observed.

The shortest response times with FWHM equal 58 ps were recorded for the GaAs-AlAs based structures and is shown in fig. 2.

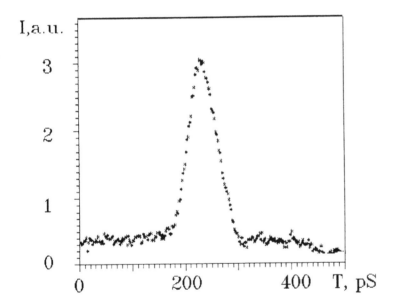

Fig.2. Fast pulse response of the MSM structure. Rise time 41 pS, fall time 46 pS.

Slow component in signal was observed during the irradiation by light pulses with duration time 10 μs. This component was due to the carrier capture on a deep levels. No slow component was revealed during the electron irradiation because it was no enough time to fill the traps. After the passivation response time of the structures was slightly reduced. The best result was obtained for Si_3N_4 layers.

4. CONCLUSIONS

The non uniformity of the electric field distribution between the MSM structure electrodes and the presence of the intermediate layers at the metal-semiconductor interface that prevented carriers to be swept out are the significant reasons of the slow components in the photoresponse. Space charge creation near the electrodes due to non ideality of the Schottky barriers leads to the carrier injection from the electrodes and because of that the nonuniformity of the electric field distribution is increased.

Thus the electric field structure and space charge distribution are of great importance for the slow component mechanism determination and MSM photoreceiver characteristics improvement. It was shown that with steady and temporary SEM EBIC modes it is possible to investigate electric field and space charge distribution and also to measure the spatial distribution of the photoresponse in GaAs MSM photoreceivers.

5. ACKNOWLEDGMENTS

I am very grateful to the XIII ICOM 92 Congress Organizers, Soros Foundation, Brasenose College and Department of Materials of the University of Oxford for financial support
    I would like to thank Dr. G.S. Simin for the MSM structures and useful discussions, Dr. V.E. Umasnsky for the initiation of this work , and Prof. S.G. Konnikov for the kindly provided research facilities.

6. REFERENCES

Konnikov S G Umansky V E and Lodyzinsky I I 1987 Pisma v ZTF 13 1183

Inst. Phys. Conf. Ser. No 130: Chapter 3
Paper presented at Int. Congr. X-ray Optics and Microanalysis, Manchester, 1992

263

The quantitative analysis of thin specimens

D B Williams

Lehigh University, Department of Materials Science and Engineering, Whitaker Laboratory, Bethlehem, PA 18015, USA

ABSTRACT: The accuracy of quantitative analysis of thin specimens using x-ray energy dispersive spectrometry has progressed slowly since its inception almost twenty years ago. Several of the existing barriers to improvement are now generally understood, but not implemented. Spatial resolution has improved considerably and nanometer level analysis is a reality along with single atom detectability. Electron energy-loss spectrometry has developed into a highly versatile analytical tool, but still requires sophisticated operational and computer skills. In combination, x-ray and electron spectroscopies in the electron microscope constitute a formidable materials characterization tool.

1. INTRODUCTION

Quantitative analysis of thin specimens in the analytical (transmission) electron microscope (AEM) is effected by x-ray energy dispersive spectrometry (EDS) for elemental analysis and electron energy-loss spectrometry (EELS) for elemental, chemical bonding, atomic environment and dielectric property analysis. EDS is relatively simple in concept and straightforward in practice, but limited in scope while EELS is complex, experimentally difficult but rewarding in the breadth of information it reveals about the specimen. Both techniques have been available commercially for ~15 years; EDS was pioneered by Cliff and Lorimer (1972, 1975) in Manchester and EELS was developed by Egerton (1974) in Oxford and Isaacson and Johnson (1975) in Chicago, inter alia. The current state of each technique is reviewed in this paper.

2. X-RAY ANALYSIS

The initial success of thin foil x-ray microanalysis was for two reasons. First the spatial resolution was 1-2 orders of magnitude below the $1\mu m$ level typical of bulk microanalysis. Second, quantification was achievable using the elegantly simple Cliff-Lorimer equation, for a binary system A-B:-

$$\frac{C_A}{C_B} = k_{AB} \frac{I_A}{I_B}$$

where C_A, C_B are the compositions of elements A and B and I_A, I_B are the characteristic x-ray intensities above background. The term k_{AB} is the Cliff-Lorimer sensitivity factor, determined experimentally using standards, or calculated empirically. Often k_{AB} ~1 ($\pm$ 0.2) so quantitative analysis with an acceptable error ($\pm 20\%$ relative) simply requires measurement of peak intensities. It is, however, considerably difficult to reduce the error in C_A/C_B to < $\pm$ 5% and $\pm$ 1% is currently unattainable because of several problems (Goldstein and Williams 1992):- (1) low x-ray counts (<<100,000) from thin specimens due to inefficient EDS detector geometry, (2) difficulties with obtaining suitable thin foil standards, (3) absorption corrections which require accurate, on-line foil thickness measurements, (4) uncertainties in ionization cross sections in the energy range 100-400 keV. Solution of any of the above four problems will bring a major advance in accurate quantification, which is essentially unchanged since the original work of Cliff and Lorimer.

Increased x-ray counts from a given thin specimen can be achieved by:- (i) Use of FEG sources, (ii) improved EDS-AEM interface giving detector collection angles up to ~0.3 sr (iii) multiple EDS detectors, (iv) controlled megahertz beam blanking to reduce detector dead times. Better standards can perhaps only be achieved through creation of stoichiometric thin film glasses that are stable under intermediate voltage electron beams. This has, to date, proven an unsatisfactory approach and the alternative route (calculation of k_{AB} from first principles) offers little help, given the poor state of knowledge of ionization cross sections and detector efficiencies. Serious attempts to try radically different methods such as pure element standards have not been made since this requires on-line specimen thickness determination and in-situ probe-current measurements. In-situ probe-current measurements are not routinely available in current AEMs. However, rapid determination of specimen thickness is available with EELS (Malis et al 1988). On-line thickness measurements can also help with the problem of the absorption correction. However, Horita et al's (1986, 1987) absorption-free k factor determination represents the best solution to this problem. This is well-illustrated by recent light element k-factor data (Westwood et al 1992) (see Figure 1), which would have been impractical to obtain several years ago.

Improved quantification is inextricably attached to improved AEM performance and this requires performance criteria. Initial attempts to establish criteria for peak to background ratio (P/B) in the x-ray spectra, on-column EDS resolution and peak shape and absolute efficiency of the EDS-AEM system have been reported recently

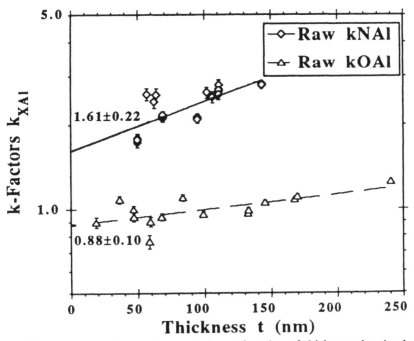

Fig. 1. Light element k-factor data plotted as a function of thickness showing how extrapolation to zero thickness gives an absorption-free k-factor. (Courtesy A D Westwood, reproduced by permission of Blackwells Scientific Publications)

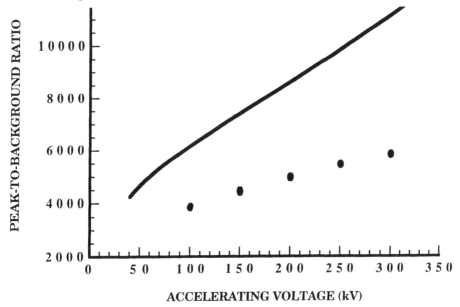

Fig. 2. Experimental (points) and theoretical (line) calculations of the change in P/B ratio for the CrKα peak as a function of accelerating voltage. The discrepancy indicates that AEM design can still be improved. (Courtesy S M Zemyan)

(Zemyan and Williams, 1992a,b). A factor of two discrepancy exists between experimental P/B and calculated P/B (see Figure 2) and P/B is sensitive to AEM operating conditions, indicating there is substantial room for improvement in instrument design.

Recently considerable progress has been made in defining the spatial resolution of x-ray microanalysis (Michael and Williams 1987), measuring it experimentally (Michael et al 1989) and simulating the experimental measurements (Williams et al 1992). FEG instruments permit measurement of composition changes over a few nanometers (see Figure 3) even for light elements (Westwood et al 1992). Improvements in spatial resolution usually bring a degradation in analytical sensitivity. However FEGs in combination with intermediate voltage AEMs produce sufficient increase in x-ray counts and improvement in P/B ratio that detection of single atoms in the analyzed volume is now feasible (Cliff and Lorimer 1992, Goldstein and Williams 1992).

However, expensive AEM designs and instrument improvements will be for naught unless sample preparation and in situ UHV cleaning techniques can be developed to give pristine foils with no surface chemical modification.

3. ELECTRON ENERGY-LOSS SPECTROMETRY

Improvement in the accuracy of quantification of EELS ionization-loss data is not a major issue since few users of the technique actually bother to quantify their spectra, despite the existence of Egerton's simple equation (Egerton 1974, 1986), analogous to the Cliff-Lorimer equation. Over the last 20 years, the technique of EELS has developed slowly. The microscope-spectrometer interface is not crucial as in EDS, but spectrometer design has improved and recent parallel-collection (PEELS) hardware has opened the door to many advances, particularly in the speed of spectral acquisition. Software developments have resulted in easier access to the detailed electronic structure information contained in the spectra. The elemental information in the ionization edges can now be more easily extracted using difference techniques (Shuman and Kruit 1985), especially in the case where small (<1%) amounts are present or where the high spectral background makes it difficult to discern the ionization edges (see Figure 4). Minimum detectability limits of <0.1 wt% are attainable which, given the inherent high spatial resolution of EELS, translates into single atom detectability (Krivanek et al 1991). It has been argued (Leapman and Hunt 1992) that EELS is superior to EDS over most of the periodic table, rather than just in the case of the light elements, for which EELS is traditionally thought best suited.

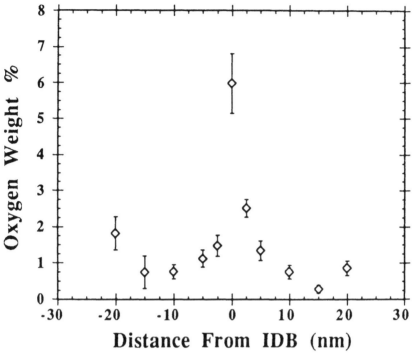

Fig. 3. High spatial resolution (<<5nm) composition profile showing oxygen enrichment at an inversion domain boundary in AlN. (Courtesy A D Westwood, reproduced by permission of Blackwells Scientific Publications)

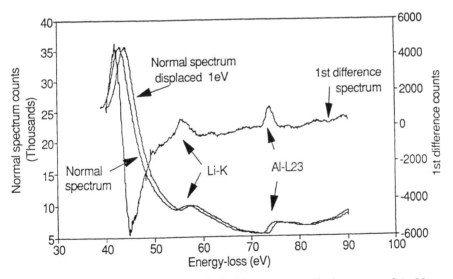

Fig. 4. Two energy loss spectra recorded with an energy displacement of 1 eV can be subtracted to give a difference spectrum in which small edges (e.g. the Li K edge) can be easily seen. (Courtesy J A Hunt)

Energy-loss near-edge fine structure (ELNES) analysis of the core edges has advanced to the stage where modeling software can not only predict the ELNES peak positions but make a good estimate of the expected intensity (Rez et al 1991). Thus, a detailed understanding of the density of states information contained in the ELNES can now be contemplated. Detailed ELNES changes on a scale of <1nm can be observed across interfaces, as shown in Figure 5a and used to infer the relative amounts of two phases contributing to the spectrum (Figure 5b).

The low-loss spectra (<50eV) are beginning to receive more attention. Kramers-Kronig analysis permits measurement of the dielectric constant on a local scale, (equal to the spatial resolution of a few nms), the band-gap in semiconductors can also be deduced on the same scale. Plasmon peak shifts can now be measured to <0.01 eV opening the possibility of effective electron density measurements in non-free electron type materials, and a whole new look at indirect plasmon loss microanalysis. Surface plasmons offer a sensitive method to probe the structure and chemistry of interfaces.

The combination of parallel-collection hardware and advanced spectral processing software has resulted in a most powerful thin foil analysis tool, that of spectrum imaging (Jeanguillaume and Colliex 1989, Hunt and Williams 1991). A complete (1024 or 2048 channel) spectrum can be stored at each pixel in a digital STEM image. Thus the spectral data are unbiased by the operator's choice of specific microstructural features. The presence of unsuspected elements can be verified after data acquisition. The same spectral data can be processed by different analysis routines and compared. Any of the electronic structure information in the spectrum can be selected and mapped - from simple elemental images to chemical bonding maps, dielectric constant maps, thickness maps, etc. Figure 6 is a series of spectrum images from an Al-Li alloy containing Al_3Li precipitates showing a) the zero loss peak position b) the first plasmon peak position c) the difference between b) and a) giving the plasmon loss image which is inversely proportional to the Li content. In d) the foil thickness is mapped, in e) the absolute Li content and f) the absolute Al content is imaged. A grey scale look up table permits direct quantification. The ability to store and access this large data base places considerable importance on the initial selection of the area of the microstructure to be mapped. Drift-correction routines are essential since maps can take considerable time (~1 hr for a 128 x 128 map with a 0.2 sec spectral acquisition time). Variations in the emission current of the FEG requires concurrent probe-current mapping to permit compensation for resulting signal intensity variations. Despite these problems, spectrum imaging is a reality and represents a major advance in the quantitative analysis of thin foils.

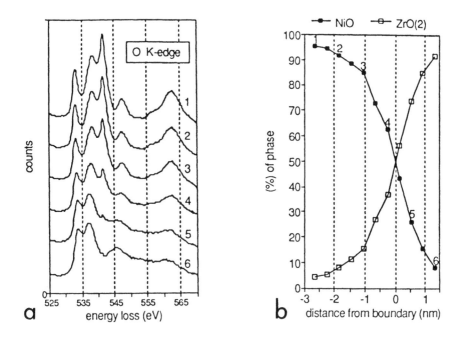

Fig. 5. a) ELNES variations in the O K edge as the electron beam is moved across an interface between NiO (1, 2, 3, 4) and ZrO_2 (5, 6). b) The relative contribution of each oxide to the O K edge signal. (Courtesy J A Hunt and V Dravid)

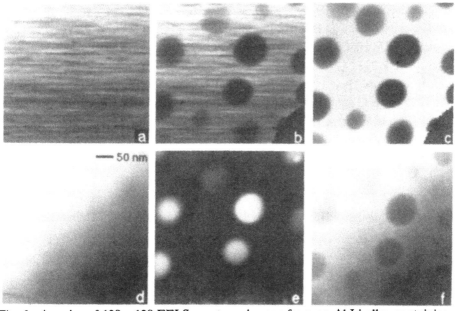

Fig. 6. A series of 128 x 128 EELS spectrum images from an Al-Li alloy containing Al_3Li precipitates: see explanation in text (Courtesy J A Hunt)

ACKNOWLEDGEMENTS

The author wishes to acknowledge extensive discussions with J I Goldstein and the financial support of NASA (NAG9-45) for the x-ray microanalysis studies. NSF (DMR89-05459) funded the electron energy-loss spectrometry aspects of this paper.

REFERENCES

Cliff G and Lorimer G W 1972 Proc. 5th EUREM (Bristol, The Institute of
  Physics) pp 140-1
Cliff G and Lorimer G W 1975 J. Microsc. 103 203
Cliff G and Lorimer G W 1992 Proc. 50th EMSA Meeting (San Francisco, San
  Francisco Press) pp 1464-5
Egerton R F 1974 Advances in Analysis of Microstructural Features by Electron-
  Beam Techniques (London, The Metals Society) pp 67-79
Egerton R F 1986 Electron Energy Loss Spectroscopy (New York, Plenum) pp
  265-6
Goldstein J I and Williams D B 1992 Microbeam Analysis 1 29
Horita A, Sano T and Nemoto M 1986 J. Microsc. 143 215
Horita A, Sano T and Nemoto M 1987 Ultramicroscopy 21 271
Isaacson M S and Johnson D E 1975 Ultramicroscopy 1 33
Jeanguillaume C and Colliex C 1988 Ultramicroscopy 28 252
Hunt J A and Williams D B 1991 Ultramicroscopy 38 47
Krivanek O L, Mory C, Tencé M and Colliex C 1991 Microsc. Microanal.
  Microstruct. 2 257
Leapman R D and Hunt J A 1991 Microsc. Microanal. Microstruct. 2 231
Malis T, Cheng S C and Egerton R F 1988 J. Elec. Microsc. Tech 8 193
Michael J R and Williams D B 1987 J. Microsc 147 289
Michael J R, Williams D B, Klein C F and Ayer R 1990 J. Microsc. 160 41
Rez P, Weng X and Ma H 1991 Microsc. Microanal. Microstruct. 2 143
Shuman H and Kruit P 1985 Rev. Sci. Instrum. 56 231
Westwood A D, Michael J R and Notis M R 1992 J. Microsc. In press.
Williams D B, Michael J R, Goldstein J I and Romig A D Jr 1992 Ultramicrosc.
  In press
Zemyan S M and Williams D B 1992a Proc. 50th EMSA Meeting (San Francisco,
  San Francisco Press) pp 1236-7
Zemyan S M and Williams D B 1992b J. Microsc. Submitted for publication

Inst. Phys. Conf. Ser. No 130: Chapter 3
Paper presented at Int. Congr. X-ray Optics and Microanalysis, Manchester, 1992

271

In-situ chemical analysis of dispersoid particles in two Al-Mg-Si alloys using analytical electron microscopy of thin foils

T. Sato[*] and G.W. Lorimer

Manchester Materials Science Centre, University of Manchester/UMIST, Grosvenor Street, Manchester, M1 7HS, U.K.

[*] now at Department of Metallurgical Engineering, Faculty of Engineering, Tokyo Institute of Technology, O-okayama, Meguro-ku, Tokyo 152, Japan

ABSTRACT: The technique of in-situ analysis of second phase particles within thin foils (Cliff et al, 1983) has been used to determine the composition of dispersoid particles in two Al-Mg-Si alloys which contained additions of Mn or Mn plus Zr.

1. INTRODUCTION

Small amounts of transition elements such as Mn, Cr and Zr are added to commercial aluminium alloys during melting. During homogenization of the cast billet these elements form fine dispersoid particles, which are retained during subsequent thermomechanical processing. The dispersoid particles prevent recrystallization during homogenization and exert significant control on the wrought microstructure produced during extrusion or rolling. The object of the present investigation was to study the nucleation and the chemistry of the dispersoids formed in Al-Mg-Si alloys containing additions of Mn or Mn plus Zr.

2. EXPERIMENTAL

Two 6082-type Al-Mg-Si alloys, compositions given in Table 1. were DC cast in 178 mm diameter moulds. The alloys were ramp-heated at a rate of $100Kh^{-1}$ (a practice essential in commercial alloy heat treatment to ensure that low melting compounds at grain

	Si	Fe	Cu	Mn	Mg	Cr	Zn	Ti	B	V	Zr	Al
Alloy A	1.09	0.19	0.005	0.64	0.80	0.001	0.016	0.006	0.001	-	-	Bal.
Alloy B	1.06	0.20	0.004	0.49	0.79	0.002	0.017	0.005	0.001	0.002	0.16	Bal.

Table 1 Composition of the alloys (wt %).

boundaries disperse before melting), isothermally aged at 623 to 843K for 3.6 to 180ks and finally water quenched. Thin-foil microprobe analyses of dispersoid particles were performed in Philips EM400T or EM430 analytical electron microscopes, or a Vacuum Generators HB501 scanning transmission electron microscope. Quantification of the thin foil microanalysis results was carried out using the ratio technique (Cliff and Lorimer, 1975) and calculated k-factors. Analyses of dispersoid particles embedded within thin foils were carried out using the extrapolation technique of Cliff et al (1983).

3. RESULTS AND DISCUSSION

Fig. 1 shows the microstructure of alloy A ramp-heated and aged at 773K for 7.2 and 108ks. Only one type of dispersoid particle was observed in alloy A. Extensive coarsening of the dispersoid particles occurred if the ageing time was extended at 773K or the ageing temperature was raised.

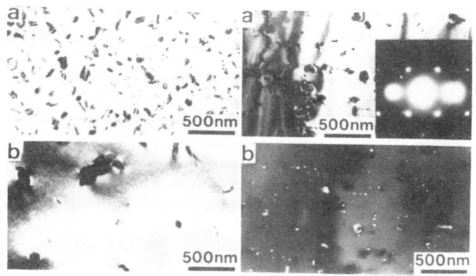

Fig.1: Electron micrographs of alloy A   **Fig.2:** Electron micrographs of alloy B
    ramp-heated and aged at 773K for   ramp-heated and aged at 773K for
    7.2 and 108ks showing coarsening.   7.2ks showing dispersoid particles.

Fig. 2 is a bright-field/dark-field pair of electron micrographs of alloy B after it had been ramp-heated and aged at 773K for 7.2ks. The large dispersoid particles in Fig. 2 are identical to the dispersoid particles detected in alloy A and are identified as type-1 dispersoids. The dark-field image in Fig. 2 was taken with a superlattice reflection from the small dispersoid particles. These small dispersoid particles are specified as either type-2 or type-3 dispersoids. In-situ analysis of the type-1 dispersoids in alloys A and B gave

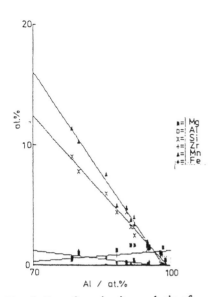

Fig. 3: Data from in-situ analysis of
of type-1 dispersoids in alloy B
ramp-heated and aged at 823K for
108ks. Extrapolated composition
$Al_{15}(Mn_{2.69}Fe_{0.22})Si_{2.09}$.

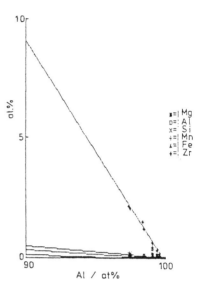

Fig. 4: Data from in-situ analysis of
type-2 particles in alloy B
ramp-heated and aged at 773K for
for 7.2ks. Extrapolated composition
$(Al_{2.95}Si_{0.05})Zr$.

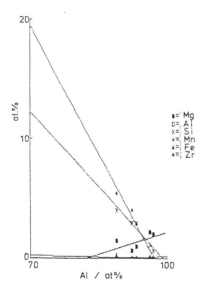

Fig. 5: Data from in-situ analysis of
type-3 dispersoids in alloy B
ramp-heated and aged at 843K for
7.2ks. Extrapolated composition
$(Al_{2.40}Si_{0.60})Zr$.

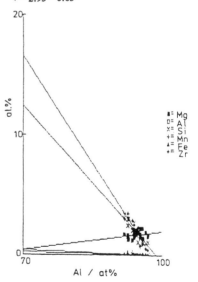

Fig. 6: Data from in-situ analysis of
type-3 dispersoids in alloy B
ramp-heated and aged at 823K for
108ks. Extrapolated composition
$(Al_{2.25}Si_{0.75})Zr$.

similar results irrespective of the ageing time and temperature. A typical set of data, for alloy B ramp-heated and aged at 823K for 108ks, is shown in Fig. 3. The analysis data can be extrapolated to a composition of $Al_{15}(Mn_{2.69}Fe_{0.22})Si_{2.09}$. Electron diffraction indicated that the type-1 dispersoids had a simple cubic structure with a lattice parameter of 1.26nm. These dispersoids were identified as the $Al_{15}(Mn,Fe)_3Si_2$ phase, a result consistent with the electron diffraction investigation of Westengen et al (1980) and the analytical electron microscopy of Strutt et al (1988). Electron diffraction studies of the small dispersoid particles formed in alloy B at 773K, Fig. 2, revealed that they had an Ll_2-type ordered structure (ordered fcc) and were coherent with the matrix. Fig. 4 is a set of analysis data from the small dispersoid particles observed after alloy B had been ramp-heated and aged at 773K for 7.2ks. These dispersoid particles contain mainly Zr and small amounts of Si. These coherent particles with the Ll_2 structure are essentially Al_3Zr and are specified as type-2 dispersoids. The amount of Si in the type-2 dispersoid particles increased as the ageing time at 773K was increased. When alloy B was ramp-heated and aged at 823K or 843K type-1 dispersoid particles were detected as well a small dispersoid particles, similar to those observed after ageing at 773K. In-situ analysis of the small particles indicated that they contained significant amounts of Si, Fig. 5, and that the amount of Si in the particles increased as the ageing time was increased, Fig. 6. Electron diffraction investigations of the small dispersoids showed that they had the DO_{23}-type structure with lattice parameters of a = 0.41nm and c = 1.73nm. These precipitates are specified to be type-3 dispersoids.

4. CONCLUSIONS

In alloy A (containing Mn) dispersoid particles of $Al_{15}(Mn, Fe)_3Si_2$ were formed. The crystal structure of this type dispersoids (type-1 dispersoids) was simple cubic with a lattice parameter a = 1.26nm. In alloy B (containing Mn and Zr) the type-1 dispersoids were also formed. Small particles of the Ll_2-type ordered structure were formed in alloy B aged at 773K (type-2 particles). Small amounts of Si were incorporated in these particles. The third type of dispersoids (type-3) found in alloy B aged at 823 or 843K contained a large amount of silicon and had the DO_{23}-type structure.

5. REFERENCES

Cliff, G., Powell, D.J., Pilkington, R., Champness, P.E. and Lorimer, G.W., 1983, Electron Microscopy and Analysis, Ed. P. Doig, IOP, London, pp. 63-66.
Cliff, G. and Lorimer, G.W., 1975, J. Microsc., Vol. 103, pp. 203-297.
Strutt, A.J., Lorimer, G.W. and Parson, N.C., 1988, EUREM 88, Inst. Phys. Conf. Ser., No. 93, Vol. 2, pp. 507-508.
Westengen H., Reiso, O. and Auran, L. 1980, Aluminium, Vol. 56, pp. 768-775.

Inst. Phys. Conf. Ser. No 130: Chapter 3
Paper presented at Int. Congr. X-ray Optics and Microanalysis, Manchester, 1992

275

Graphical manipulation of EDS data obtained by SEM

J. L. ROSSI[*], R. PILKINGTON[#], R. L. TRUMPER[‡]

[*] *Instituto de Pesquisas Energéticas e Nucleares IPEN - CNEN/SP , Caixa Postal 11049 (Pinheiros), São Paulo - Brazil*
[#] *Materials Science Centre, University of Manchester, Grosvenor St., Manchester M1 7HS - U.K.*
[‡] *A.R.E. Holton Heath, Poole, Dorset BH16 6JU - U.K.*

ABSTRACT: The microstructural characterization of a cast aluminium alloy reinforced with a boron fibre coated with silicon carbide was undertaken. The as received composite material showed a cast microstructure with the presence of intermetallic particles. EDS chemical analyses in the SEM were extensively used to identify the macro constituents. Many constituents were also chemically analysed in a TEM. Solute solvent plots were constructed from the analysed constituents and a graphical EDS data analysis manipulation technique enabled the chemistry of the particles to be inferred from the SEM data.

1. INTRODUCTION

A simple technique, based on the work by Feest (1980) for the identification of small particles, was described by Cliff et al (1983); this permits the in-situ identification by X-ray microanalysis of second phase particles. The technique, which was originally developed for bulk microprobe analysis was extended to thin foil specimens. The X-ray intensity, I^A, from a element A, is proportional to the weight fraction of that element, C^A, and the total X-ray intensity I^A_T from a thin sample which contains second phase particles is given by:

$$I^A_T \propto C^A_M L_M + C^A_P L_P, \qquad (1)$$

where C^A_M and C^A_P are the weight fractions of element A in the matrix and particle, respectively, and L_M and L_P the respective electron path lengths. The electron path length may be totally in the matrix, Fig.1 (a), totally in the particle, Fig.1 (c), or partially in both, Fig.1 (b). If element A, a solvent, is concentrated in the matrix, and B, a solute, is concentrated in the precipitate, then the analysis of A from the regions shown in Fig.1 can be expressed against B, graphically, as in Fig.2, where C^B_M and C^B_P are the weight fraction of element B in the matrix and particle, respectively. If both the particle and the matrix have fixed compositions, the data will plot on a straight line. Using similar solute-solvent plots from the analysed constituents and a graphical EDS data analysis manipulation, the technique enabled the chemistry of intermetallic particles present in an aluminium alloy to be inferred from the SEM data.

2. EXPERIMENTAL

Many second phase particles from an aluminium alloy were analyzed in polished samples using energy dispersive X-ray spectroscopy (EDS). The results were interpreted in terms of the solute-solvent plot method so that the matrix contribution to the chemical analyses results could be taken into account. Quantitative analyses (EDAX 91100 System) in the SEM were obtained using standard software. Thin foil chemical analyses were performed using energy dispersive X-ray spectroscopy, with the transmission electron microscope (TEM) operating in the nanoprobe mode. Quantitative analyses in the TEM were obtained using the ratio technique with fluorescence correction and K-factors previously obtained (Cliff and Lorimer 1975, Mehta et al 1979).

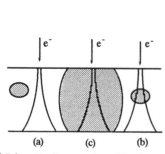

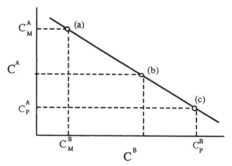

Fig.1 *Schematic diagram of possible electron paths in a thin sample containing second phase particles (Cliff et al 1983).*   **Fig.2** *Schematic diagram of the analytical data obtained from the sample shown in Fig.1 (Cliff et al 1983).*

3. RESULTS AND DISCUSSION

The liquid metal infiltration technique used in the fabrication of the composite material produced an as cast microstructure in the matrix material, i.e., primary aluminium rich dendrites, interdendritic areas of rich solute and coarse intermetallic compounds sometimes called constituents (Polmear 1989). Only two types of intermetallics were observed. The first type had a shape of thin curved platelets which in section appeared as long needles. The second type was of a globular shape. EDS analyses of the platelet type intermetallic with cross sections around 4 μm thick showed generally the presence of Al, Si, and Fe, and in some constituents small amounts of Ni were also detected. An average size of 10 μm was estimated for the globular constituents and EDS chemical analyses indicated the presence of Al, Si, Fe, and Cr, and in some constituents small amounts of Ni and V.

The constituents sizes were large enough to allow reliable in-situ chemical analyses by energy dispersive spectroscopy in the SEM. Usually the material activated by an electron beam (30 keV and a 100 nm diameter electron probe), which gives rise to characteristic X-rays, reaches a depth in an aluminium matrix of approximately 1.5 μm (Flewitt and Wild 1986). Thus, for constituents with dimensions larger than 1.5 μm, there is a large probability that some of them would yield chemical data from the constituent only with little information from the surrounding matrix. As for SEM it is not possible to visualize what is the constituent shape beneath the surface, some matrix interaction with the electron beam path is expected and somehow its effects must be taken into account. The matrix effect can be analysed when the EDS results are plotted in a solute-solvent plot as already explained for thin foils. In this work, EDS data were obtained in atom percent after ZAF (atomic number, absorption and fluorescence) corrections and normalized assuming that the detected elements total 100%. No attempts were made to take into account any ZAF correction errors, or raw data statistics acquisition errors.

Solute-solvent plots were constructed from 80 and 97 analysed constituents of platelets and globular types, respectively. Fig.3 shows a solute-solvent plot for the platelets type particles. The atom percent of Si, Fe and Ni were plotted against the Al atom percent and the fitted lines for Si and Fe extrapolated to zero at a common high Al composition. For this particular alloy as far as EDS chemical analyses is concerned, the matrix cannot be considered as consisting of only one type of element or phase. Rather it contains the aluminium primary phase and the silicon flakes. As both phases may be interacting with the electron beam during the constituent analyses and at indeterminate quantities, this implies that information regarding the second phase chemistry cannot be inferred from the solute-solvent plot. This fact also explains some of the data scatter. Alternatively, when the amount of Al and Si were plotted against the amount of Fe, Fig.4, the Al and Si variations seemed to attain a plateau, and then the amount of Si apparently started to decrease.

The occurrence of the plateau (or maximum/minimum) can be interpreted based on the following facts. First, the intermetallic can be larger than the volume of material which interacted with the electron beam. Second, there is a smaller chance to find another intermetallic containing Fe and interacting with the electron beam, than to find a silicon flake or the aluminium solid solution. Third, the intermetallics can exist at a range of homogeneity (Mondolfo 1976, Rivlin and Raynor 1981). It is not yet clear whether at the point where the Al and Si variation against Fe attained the plateau, the analysis was performed with a constant matrix interaction or only in the constituent without any matrix contribution. If the plateau analysis results were due to a constant matrix interaction, one would expect at some point a change in behaviour when the results would be yielded only from the intermetallics. Indeed, the results showed a tendency for the detected amount of Si to decrease after the plateau. The results, Fig.4, also indicated that the amount of Al and Si (after the plateau) in the platelets type intermetallics was around 67% and 18% atom percent respectively. Thus, it is deduced that these intermetallics contain Al, Si and Fe approximately in the ratio 67:18:15. For the globular constituents, similar solute-solvent plots were constructed. In this case it can be deduced that the globular constituents have Al, Si, Fe and Cr in the ratio 71:12:15:2.

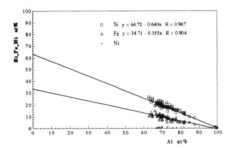
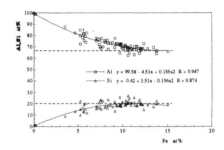

Fig.3 *Variation in atom percent of Si, Fe and Ni versus Al for in-situ SEM analyses of 80 similar second phase platelet type.*

Fig.4 *Variation in atom percent of Al and Si versus Fe for in-situ SEM analyses of second phase platelet type.*

Thin foils from different strips of the composite material (28% fibre volume fraction) were observed in the TEM. Around 15 constituents of each type, platelets or globular, were chemically analysed with the transmission electron microscope. For chemical analyses results see Table 1. It can be seen that the results obtained from the SEM analyses, following the procedure described above, agree very well with the results from the in-situ TEM analyses.

As far as is known, the above graphical EDS data analysis manipulation has not been previously tried on results obtained in the SEM and to some extent in the TEM. Two of the drawbacks of the energy dispersive spectroscopy results obtained from SEM are: first, the particle shape beneath the surface is not known and second, absorption correction factors are very large and prone to errors, at least for silicon. As the SEM results agreed with TEM results, which are obtained and calculated in a total different way and since the absorption effects are not significant due to the thinness of the the foil, there is a certain degree of confidence in the results yielded by this plotting technique. The EDS data can be collected in the SEM with relative easy and with little specimen preparation. It is possible to analyse in the SEM at least 15 particles per hour. However, in the TEM the achievement of such performance is improbable. The thin foil preparation is a very laborious process and due to the size of the constituents, very few particles can be found in condition that allow EDS analysis to be performed.

The energy dispersive results obtained in the SEM and the TEM indicates that the probable chemical composition of the platelet type particles is $Al_{67}Fe_{15}Si_{18}$ and with the aid of electron diffraction, the platelets were identified as having a tetragonal structure, a = 0.608 nm and c = 0.940 nm. The structure results agreed very well with published literature (Mondolfo

1976, Rivlin and Raynor 1981). Mondolfo (1976) reported the presence of a phase $FeSi_2Al_4$, in high-silicon alloys, with a tetragonal structure and lattice parameters: a = 0.612-0.616 nm and c = 0.948-0.949 nm. Rivlin and Raynor (1981) also reported a similar structure called τ_4. The published (Mondolfo 1976, Rivlin and Raynor 1981) chemical composition for the tetragonal phase however did not match the results obtained in this work. Although this phase is reported as having approximately twice the amount of Si compared to Fe, in this work it was established that the atomic concentration of Si is slightly higher than Fe, see Table 1. It should be pointed out that the chemical compositions mentioned in above literature were not obtained by in-situ chemical analyses and a large number of equilibrium and metastable Al-Si-Fe phases are commonly found in aluminium commercial alloys.

4. CONCLUSION

As an outcome, there is a certain degree of confidence in the results yielded by this technique. Since SEM analyses involve little specimen preparation, and are relatively easy to perform, this suggests that the cpllection of valuable information about phase chemistry could be simplified by use of this technique.

TABLE 1
EDS CHEMICAL COMPOSITION OF SECOND PHASES (at%).

	Al	Si	Fe	Cr	Ni	V
platelet type (SEM)	67	18	15	-	...	-
globular type (SEM)	71	12	15	2	...	...
platelet type (TEM)	67.1±1.4 [a]	17.2±0.4	15.4±0.8	-	0.3	-
globular type (TEM)	71.5±1.2	12.8±1.8	13.7±1.6	1.3±0.9	0.5	0.2

a) *indicated errors are standard deviation from the results.*
(-) *nil*
(···) *not determined.*

Acknowledgements - The authors acknowledge the support received from the Brazilian Government - CNPq/CNEN in the provision of a scholarship to J. L. R., and to the MOD U.K. for funding and material supply.

References

CLIFF G & LORIMER G W 1975 The quantitative analysis of thin specimens. J. Microscopy 103 203-7
CLIFF G, POWELL D J, PILKINGTON R, CHAMPNESS P E, LORIMER G W 1983 X-ray microanalysis of second phase particles in thin foils. Inst. Phys. Conf. Ser. No 68 (Guilford UK: EMAG) pp 63-6
FEEST E A 1980 Automated EPMA techniques applied to microstructures. Proc. conf. Solidification Technology in the Foundry and Cast House (Coventry UK: THE METALS SOCIETY) pp 188-194
FLEWITT P E J & WILD R K 1986 Microstructural Characterization of Metals and Alloys (London: THE INSTITUTE OF METALS) p 12.
MEHTA S, GOLDSTEIN J I, WILLIANS D B, ROMIG Jr. A D 1979 Determination of Cliff-Lorimer k calibration factors for thin-foil X-ray microanalysis of Na, Mg, and Al in the STEM. Microbeam Analysis. NEWBURY D E, ed. (San Francisco: San Francisco Press) pp 119-23
MONDOLFO L F 1976 Aluminum Alloys: Structure and Properties (Butterworths) pp 759-87.
POLMEAR I J 1989.Light Alloys, Metallurgy of the Light Metals. 2nd Ed.(Edward Arnold) p 39.
RIVLIN V G & RAYNOR G V 1981 Critical evaluation of constitution of aluminium-iron-silicon system. Int. Met. Rev. No 3 pp 133-52.

Inst. Phys. Conf. Ser. No 130: Chapter 3
Paper presented at Int. Congr. X-ray Optics and Microanalysis, Manchester, 1992

279

Study of carbon in a nickel-based superalloy by EELS

G. Blanche and G. Hug

LEM, ONERA–CNRS, BP72, 92322 Châtillon, France

ABSTRACT: A multiple-least squares method, fitting a set of appropriate reference spectra has been designed and used in order to resolve the C-K, Mo-$M_{4,5}$ overlap in the EELS spectrum. The quantification of carbon in different phases of an industrial superalloy is presented.

1. INTRODUCTION

Turbine engine manufacturers are constantly seeking ways of wringing extra power and efficiency out of gas turbine engines. One of the main means of improving efficiency is to increase the operating temperature and this requires the use of superalloys more heavily loaded with hardening elements. The N18[†] alloy, produced by pre-alloyed powder metallurgy, consists in an austenitic Nickel-Chromium-Cobalt-Molybdenum matrix, further hardened by precipitation of the γ' phase (Ni_3Al, Ti). Additions of soluble hardening elements such as Zirconium, Boron, Hafnium and Carbon provide extrastrength and creep resistance. However, after suitable thermo-mechanical processing, N18 exhibits a structural instability that gives rise to new complex phases which are poorly characterized. In the beginning of the aging treatment, carbide precipitates play an important role in the formation of the microstructure and in the properties of this alloy. After a long aging time, precipitates of carbides transform into other carbides with lower carbon concentration and they also provide sites for the germination of the undesirable σ and μ phases as well as for needle-shaped phases (Sims, Stoloff and Hagel 1987, Marty and Walder 1989). The σ and μ phases consist in a compact stacking of metallic atoms ; they are generally referred to as Topologically Close Packed phases (TCP). The needle-shaped phases, which are suspected to dissolve some quantity of carbon, grow to the detriment of carbides. They are called breakable because of their deleterious influence on the mechanical properties of the alloy.

The TCP phases are generated in Nickel–based superalloys through the following sequences of reactions :

$$MC \rightarrow M_7C_3 \rightarrow M_{23}C_6 \rightarrow \sigma$$

$$MC \rightarrow M_6C(M_{12}C) \rightarrow \mu$$

where M stands for the metallic elements : Mo, Cr, Co...

In these schemes, the relative Carbon concentration in the precipitates decreases as they transform into TCP phases. In addition it is not known where Carbon migrates during this process. The purpose of this study is to study the presence of Carbon in the needle-shaped phase by Electron Energy Loss Spectroscopy (EELS) (Shuman and Kruit 1985, Krivanek et al. 1987) in a Transmission Electron Microscope (TEM). EELS possesses an excellent energy resolution and a high sensitivity to light elements and in particular to Carbon. It has been found from EELS spectra that the presence of Carbon is likely in the needle-shaped phase (Marty et al. 1990). Conventional methods for quantification are however inappropriate to ascertain the presence of carbon because of the simultaneous presence of molybdenum and a strong overlap between the C-K and the Mo-$M_{4,5}$ edges. A multiple-least-squares (MLS) fitting has been introduced by Shuman and Somlyo (1987) and applied by Leapman and Swyt (1988) to quantify Calcium and Potassium in the presence of Carbon. A similar method including cal-

---

† : N18 is a patent of ARMINES/IMPHY/ONERA/SNECMA, n° 86.01604 (1986).

culated cross-sections and a simplex fitting has been applied by Manoubi and al (1990) and have proved to yield accurate results. In the present case, reference spectra generated from pure elements and compounds have been used since the Mo-$M_{4,5}$ edge is difficult to obtain by calculation.

2. EXPERIMENTAL PROCEDURES

The Time Temperature diagram (figure 1, Guédou et al. Honorat 1992) shows that the structural instability of N18 increases gradually. It consists in the growth of needle-shaped phases and in the reaction of carbides during aging. The aging treatments applied to the N18 alloy, were: 1165°C/4h + 650°C/24h + 760°C/4h. Thin foils suitable for TEM/EELS analysis were prepared by electropolishing under 25 V in a 45 vol. % 2-butoxye-thanol –45 vol. % acetic acid –10 vol. % perchloric acid at 0°C.

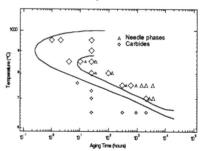

Figure 1 : Time Temperature Diagram. From Guédou et al. (1992)

The EELS analyses were carried out on a Jeol 4000FX TEM fitted with a Gatan 666 Parallel Detection Electron Spectrometer. A Beryllium holder cooled down to –165°C was used to avoid the contamination from the residual carbon originating from the microscope atmosphere and no carbon contamination was visible, even though small probes were held stationary for long periods. The EELS spectra were acquired in diffraction mode at 400 kV over the 0-1000 eV energy range. The collection angle (2β) was 10.34 mrad and the energy dispersion was 1 eV/channel. All spectra were corrected for dark current and channel to channel gain variation of the detector. In order to allow for subsequent mathematical treatments, special attention has been paid to collect the low-loss and core-loss spectra under the same experimental conditions and from the same sample region. The matrix contributions to the spectra has been eliminated by choosing a probe size smaller than the particle of interest. The typical time of acquisition was about 100 seconds.

3. DESCRIPTION OF FITTING METHOD

In case of an infinitely thin sample, the EELS characteristic inner-shell signal $S_x(\Delta E,\beta)$ (Egerton 1975) is given by :

$$S_x(\Delta E,\beta) = I_0 \cdot N_x \cdot \sigma_x(\Delta E,\beta)$$

where I_0 is the incident electron flux (electron·cm^{-2}·s^{-1}) passing through a volume of thin material containing N_x atoms of the x element. $\sigma_x(\Delta E,\beta)$ is the ionization partial cross-section for single scattering. For a specimen of finite thickness, inelastic multiple scattering alters the profile of the considered edges and diminishes the signal-to-background ratio (Egerton 1975,1986). In order to obtain a single-loss profile, a Fourier-logarithmic deconvolution method integrated into the EL/P application was applied (Leapman and Swyt 1981, Egerton and Whelan 1974). Prior to deconvolution, a complete loss spectrum which was obtained by combining (i) a spectrum of the zero-loss region, (ii) the energy region of carbon and of molybdenum and, (iii) a background extrapolated to the end of the multi-channel range with the classical power-law (AE^{-r}) (Egerton 1975,1986). The MLS procedure was then applied to the region of interest. The experimental data are fitted with the relation :

$$S = AE^{-r} + \alpha \cdot S_{Mo-M_{4,5}} + \beta \cdot S_{C-K}$$

where S_{C-K} and $S_{Mo-M_{4,5}}$ are the signals acquired on pure elements under the same experimental conditions as for the analysis of the needle shaped-particles (accelerating voltage, collection angle) and subsequently deconvoluted to remove plural scattering. The A, r, α, β coefficients are determined by the fitting procedure.

The limitation of the method may be found by considering the fine structure of the edges. Indeed, dissimilar atomic environments in the pure atomic samples and in the standards may result in different near-threshold fine structures. In addition, the extended fine structure (EXEELFS oscillation) may change in different compounds. These elements make the choice of the fitting window quite critical.

This procedure has been applied to a standard of known composition (Mo_2C in this case), in order to allow for the determination of a ratio k between the characteristic signals of the pure elements and of the reference compound. Considering the signals acquired on the pure elements :

$$S^P_{Mo} = I_{Mo} \cdot N_{Mo} \cdot \sigma_{Mo} = k_{Mo} \cdot \sigma_{Mo}$$

and
$$S^P_C = I_C \cdot N_C \cdot \sigma_C = k_C \cdot \sigma_C$$

The MLS fit is applied on the reference compound in order to find the coefficients α and β :

$$S_{Mo}(Mo_2C) = I_0 \cdot N_{Mo} \cdot \sigma_{Mo} = \alpha \cdot S^P_{Mo} = \alpha \cdot k_{Mo} \cdot \sigma_{Mo}$$

and
$$S_C(Mo_2C) = I_0 \cdot N_C \cdot \sigma_C = \beta \cdot S^P_C = \beta \cdot k_C \cdot \sigma_C$$

The ratio k is given by : $k = k_{Mo} / k_C = \beta / \alpha$.

The same procedure can be similarly applied to an unknown precipitate. The atomic Carbon to Molybdenum ratio is then deduced from :

$$S_{Mo}(unknown) = I_0' \cdot N_{Mo}(unknown) \cdot \sigma_{Mo} = \alpha' \cdot S^P_{Mo} = \alpha' \cdot k_{Mo} \cdot \sigma_{Mo}$$

$$S_C(unknown) = I_0' \cdot N_C(unknown) . \sigma_C = \beta' \cdot S^P_C = \beta' \cdot k_C \cdot \sigma_C$$

and the atomic ratio is : $N_{Mo}(unknown) / N_C(unknown) = \alpha' / \beta' \cdot k_{Mo} / k_C = k \cdot \alpha' / \beta'$. The ratio k has been determined to be : $k = 0.403$.

In order to determine the concentration of the other metallic elements in the precipitate, the fitted background and the fitted contribution of Molybdenum ($\alpha' \cdot S^P_{Mo}$) are removed from the initial spectrum. The atomic concentration of Cr, Co and Ni are then calculated by evaluating the partial cross–section through the classical routines SIGMAK and SIGMAL (Egerton 1979, 1981). A reduced chemical formula M_xC of precipitate hence can be deduced by addition of the atomic concentrations of all the metallic elements.

4. RESULTS

The MLS fitting procedure was applied to an intragranular precipitate (figure 2) which is known from its diffraction pattern to be a carbide. However, the diffraction analysis only informs on the fact that the precipitate have a fcc structure which is compatible with two types of carbides : M_6C or $M_{23}C_6$. The metallic elements Chromium, Molybdenum and Cobalt are detected by EELS and Energy Dispersion X-ray (EDX). They are present in higher concentration in the precipitates than in the matrix. Aluminium and Titanium are not found in the carbide. The results of the MLS fitting procedure are given in table 1. They are consistent with a M_6C carbide.

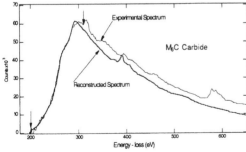

In a needle-shaped phase, EELS and EDS show that Nickel is the major metallic element. Chromium, Molybdenum and Cobalt are also present in lower concentrations. Aluminium and Titanium are not found. Diffraction analysis suggests that the needle-shaped phase is probably a μ phase with lattice parameters a~4.7-4.8 Å and c~25.7-25.8 Å. The results of the MLS proce-

Figure 2 : MLS fit on a M_6C carbide precipitate. The fit is restricted to the region defined by the arrows.

Mo/C	Cr/C	Co/C
2.60	2.66	1.33

Table 1

dure, in table 2, show that these precipitates contain Carbon. If confirmed, this result would highlight the present understanding of the nucleation μ phase since so far the TCP phases were not known to contain carbon.

Mo/C	Cr/C	Co/C	Ni/C
0.66	0.73	0.59	1.42

Table 2

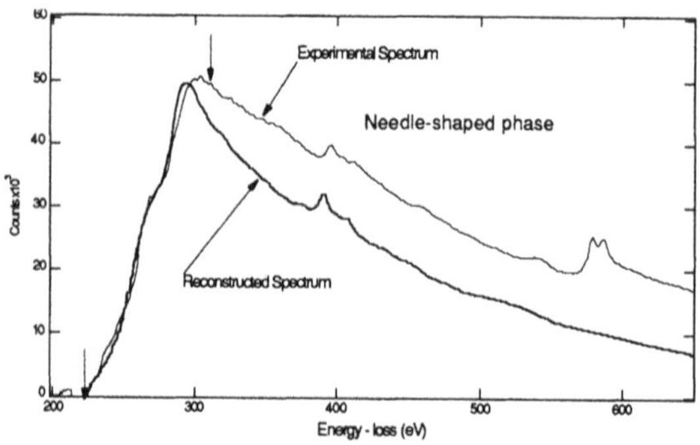

Figure 3 : MLS fit on a needle-shaped precipitate. The fit is restricted to the region defined by the arrows.

5. SUMMARY AND CONCLUSION.

The present MLS fitting method which makes use of reference standards, proves to be well suited to the quantification of Carbon in carbides even though it is used in difficult conditions such as overlapping edges and unknown cross-sections. This method has been successfully used to remove ambiguities leaved by standard diffraction analysis between the two carbides M_6C and $M_{23}C_6$ carbides. It has enabled us to show that μ phases, which appear as needle-shaped precipitates in the N18 alloy, contain a significant fraction of Carbon.

The following scheme can be proposed to clarify the structural instability of the N18 alloy. During the first period of aging, supersaturated Carbon forms carbides since it has the higher diffusion coefficient. As the metallic elements with higher atomic numbers segregate in the precipitates, the concentration of Carbon decreases in these either because the total volume fraction of precipitates increases or because Carbon tends to move back in the matrix where its solubility has increased due to the impoverishment in refractory elements.

ACKNOWLEDGEMENTS

The authors are indebted to Dr. A. Walder and M. Marty for their help and for helpful discussions. SNECMA is acknowledged for providing the heat treated N18 specimens. Dr. P. Veyssière is gratefully thanked for commenting the manuscript.

REFERENCES

Egerton R.F. 1975 Phil. Mag. 31 199.

Egerton R.F. 1979 Ultramic. 4 169.

Egerton R.F. 1981 in: Proc. 39th Annual EMSA Meeting, Atlanta, GA, Ed. GW. Bailey p. 198.

Egerton R.F. 1986 Electron Energy-Loss Spect. in the Elec. Mic. (Plenum, New York).

Egerton R.F. and Whelan M.J. 1974 Phil. Mag. 30 739.

Guédou J.-Y, Lautridou J.-C. and Honorat Y., 1992, Seven Springs Conference.

Krivanek O.L. Ahn C.C. and Keeney R.B. 1987 Ultramic. 22 103.

Leapman R.D. and Swyt C.R. 1988 Ultramic. 26 393.

Leapman R.D. and Swyt C.R. 1981 in: Analytical Electron Microscopy Ed. R.H. Geiss pp 164-172.

Marty M. Hug G. and Walder A. 1990, Spring Meeting of MRS, San Francisco.

Marty M. and Walder A. 1989 Technical report ONERA n° 32/1931M.

Manoubi T. Tencé M. Walls M. and Colliex C. 1990 MMM. 23-39.

Shuman H. and Somlyo A.P. 1987 Ultramic. 21 23.

Shuman H. and Kruit P. 1985 Rev. Sci. Instr. 56 231.

Sims C.T. Stoloff N.S. Hagel W.C. 1987 in : The superalloys, John Wiley and Sons.

Inst. Phys. Conf. Ser. No 130: Chapter 3
Paper presented at Int. Congr. X-ray Optics and Microanalysis, Manchester, 1992

283

Precipitate composition and morphology during reheating of low carbon, microalloyed steel for CCR

A K Ibraheem, C Zhou, and R Priestner

The University of Manchester/UMIST Materials Science Centre, Grosvenor Street, Manchester M1 7HS, UK.

ABSTRACT: The precipitates present in a controlled rolled, low carbon, Ti-Nb microalloyed steel have been studied on reheating to 850-1400°C. Below 1100°C, (Nb_{rich}, Ti) spheres dissolve and coarsen, and spherical caps and nodules of (Nb_{rich}, Ti) appear on the faces and corners of (Ti_{rich}, Nb) cubes and laths . AlN coprecipitates with many of the caps and cubes. At higher temperatures than 1100°C, Nb_{rich} spheres and caps dissolve completely, leaving only Ti_{rich} cubes and laths , which are stable to 1400°C. Characteristically, these are compositionally cored and laths are richer than the cubes.

1. INTRODUCTION

The composition, the morphology, and the kinetics of precipitation, dissolution and coarsening of microalloy precipitants play important roles in controlling the microstructure of heat treated and thermomechanically processed low carbon steel. Compositions of precipitates have been studied experimentally (Houghton et al 1982, Emenike and Billington 1989, Bowker et al 1989) and the likely sequence of precipitation has been predicted thermodynamically and compared with experiment (Houghton et al 1982, Keown and Wilson 1982, Subramanian et al 1985).

In preference to conventional controlled rolling (CCR), hot direct rolling (HDR) can save thermal energy, as-cast steel being rolled directly. Also, it has been shown that HDR of Ti- and Nb-containing steels improves strength relative to CCR (Kunishige and Nagao 1985, Matsumura and Sanagi 1988). However, only a few researchers have overcome the difficulties of simulating HDR on a laboratory scale. Instead, cooled material has been reheated to excessively high temperatures, in the hope that the disposition of microalloy precipitant would then be similar to that of solidified austenite.

The results reported here are part of a research programme aimed at investigating systematically the differences in the roles of precipitant in controlling microstructure during CCR and HDR processing. To provide a base line for these studies the morphologies and compositions of precipitates present on reheating a Ti-Nb microalloyed low carbon steel to 850-1400°C were investigated.

2. EXPERIMENTAL DETAILS

The composition of the steel (wt%) was:0.08C, 0.36Si, 1.4Mn, 0.003S, 0.51Ni, 0.037Al, 0.023Nb, 0.013Ti, 0.0048N
It was received as controlled rolled plate 18 mm thick. Specimens were subjected to heat treatment cycles described schematically in Fig.(1). Carbon and Aluminum replicas were prepared from polished specimens etched in 2% Nital; 10% Nital was used to release the replicas. Precipitates were analyzed using Phillips EM400T STEM/EDAX 9100 and EM430 STEM/EDAX 9900 fitted with a thin window detector.

3. RESULTS AND DISCUSSION

Microalloy precipitates of the shapes and morphologies shown in Fig.(2) were observed. The cubes and laths (a) and (b) were present at all reheating temperatures. Spheroidal precipitates (c) dissolved at temperatures above 1100°C. Below

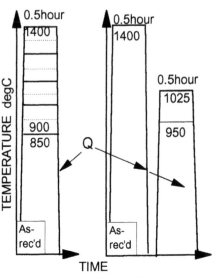

Fig.(1) Schematic representation of heat treatments.

this temperature, caps on lath and cube faces (d), and (e) and nodules on cube and lath corners (f) and (g) were also observed. Microdiffraction patterns of cubes and their attached caps showed that the caps were epitaxially oriented to the host particle, whereas the nodules were not.

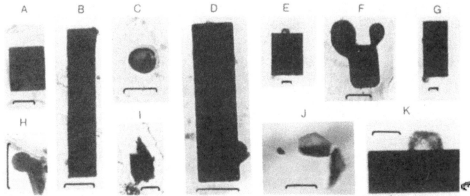

Fig.(2) Representative morphologies observed: A:Cube, B:Lath, C:Sphere, D:cap on lath, E:Cap on a cube, F:Nodules on a cube, G:Nodule on a lath, H:Al$_{rich}$ coprecipitated (copptd) with a cap which precipitated on sphere, I:Al$_{rich}$ (copptd) with a cap , J:Al$_{rich}$ (copptd) with a Nb$_{rich}$ precipitate, K:Faceted cap. Marker represents 100 nm.

A fraction of the spheres and caps were found to be two-phase , and analysis indicated the presence of Al as well as Nb and Ti (h), (i) and (j).

Caps were often faceted themselves (k) and were smaller than the host particles, whereas nodules were always rounded and often larger. These morphologies were also present in as received material, but coprecipitated Al-Nb-Ti particles were rare.

In specimens quenched from 1400°C only cubes and laths were present. Reheating caused the reappearance of all the morphologies described above. This experiment demonstrated that the appearance of the caps and nodules were associated with the coarsening of the spheroidal particles.

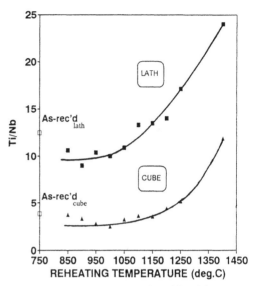

Fig.(3) Variation of Ti/Nb with reheating temperature.

Many MnS caps and nodules also appeared on cubes and lathes and fewer Nb/Ti caps on reheating quenched samples to 1025°C, but these are not considered in this paper.

Microanalysis of the cubes and laths showed that they were $(Ti_{rich}, Nb)(N_{rich})$ particles, their cores being richer in Nb than the edges, and the ratio Ti/Nb of the edges is about two times that of the cores. As shown in Fig.(3), the cubes and laths were distinguished as to their composition as well as their morphologies: this is a new result. The Ti/Nb ratio in both increased with increasing temperature. This is in agreement with Houghton's et al(1982) thermodynamic calculations of the equilibrium between (Ti,Nb)N and its austenitic matrix; however these authors did not recognize the presence of two types of such precipitates. The high Ti content of the laths suggest that they are also rich in nitrogen. This suggest that the two morphologies originated at distinct stages in the original production of the steel. Possibly, the laths

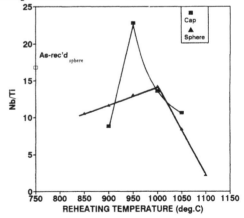

Fig.(4) Variation of Nb/Ti with reheating temperature.

form before or during solidification, and cubes in the solid state after solidification. The thermodynamic stability of both ensured their persistence throughout processing, although as shown here, their composition adjusts to temperature.

The spheroidal particles were the result of particle coarsening at temperatures below 1100°C, and their composition varied with temperature as shown in Fig.(4). This is in contrast to calculations by Houghton et al(1982) for (Nb,Ti) carbides, which showed that the Nb content should increase continuously with temperature. The presence of N_2, and probably C, was confirmed in the Ti_{rich} cubes and laths, but not in Nb_{rich} spheres, caps and nodules. This questions whether these particles are carbonitrides, as usually

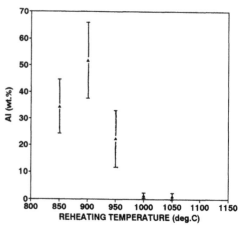

Fig.(5) Al wt.% coprecipitated Vs. reheating temperature.

assumed, or metallic in nature. Further work is needed to confirm such a proposal.

Microanalysis showed that Al coprecipitated significantly with Nb and Ti in spheres, caps and cubes at temperatures below 1100°C, significant coprecipitation were observed between 850 - 950 °C, Fig.(5). The results here confirmed the sequence of precipitation suggested by Houghton et al(1982), Emenike and Billington(1989). Fig.(2,i) shows a Nb_{rich} cap precipitated on a Ti_{rich} cube, and Al coprecipitated on the cap.

4. CONCLUSIONS

1. Two types of (Ti_{rich},Nb) precipitates were observed, cubes and laths ; the laths were richer in Ti than the cubes.
2. Nb_{rich} spheres, caps, and nodules dissolved above 1100°C. C and N_2 were not detected in these particles.
3. Al coprecipitated with many of the spheres, caps, nodules, and cubes.

REFERENCES:
Bowker J T, Ng-Yelim J and Malis T F 1989 Mat. Sci. Tech. 5 pp1034-1036
Emenike C O I and Billington J C 1989 Mat. Sci. Tech. 5 pp450-456
Houghton D C, Weatherly G C and Embury J D 1982 Proc. Int. Conf.on Thermomechanical Processing of Microalloyed Steel ed A J De Ardo et al (The Met. Soc. of AIME) pp267-292
Keown S R and Wilson W G 1982 Proc. Int. Conf.on Thermomechanical Processing of Microalloyed Steel ed A J DeArdo et al (The Met. Soc. of AIME) pp343-359
Kunishige K and Nagao 1985 Trans. Iron Steel Inst. Jpn. 25 pp315
Matsumura Y and Sanagi S 1988 Tetsu-to-Hagane 74 No.7 pp150-157
Subramanian S V, Shima S, Ocampo G, Castillo T, Embury J D and Purdy G R 1985 Proc. Conf.on HSLA Steels 85 Beijing (ASM Int.) pp151-156

ACKNOWLEDGMENT: The authors would like to thank the SERC for their generous financial support for this research programme.

Inst. Phys. Conf. Ser. No 130: Chapter 3
Paper presented at Int. Congr. X-ray Optics and Microanalysis, Manchester, 1992

287

Precipitation and recrystallization in aluminium alloys

D. Hamana, S. Nebti, A. Boutefnouchet and S. Chekroud

Research Unit Of Materials Physics

University of Constantine, Algeria.

ABSTRACT : With the ultimate goal of being able to predict and control microstructure after thermo-mechanical processing, and to develop a superplastic aluminium alloy it is vital to have a detailed understanding of the microstructure changes which take place during aging of a deformed supersaturated solid solution.The development of recrystallization and precipitation microstructures has been investigated in Al-8 wt % Mg and Al-3,5 wt % Cu alloys, using TEM, SEM and X-ray diffraction. It has been shown that the aging of a deformed supersaturated solid solution can lead to a very fine polycrystalline structure.

1. INTRODUCTION

Common superplastic alloys are usually of eutectic or eutectoid composition because both recrystallization and grain growth tend to be restricted by the presence of the small separate grains of the two phases (Davies et al 1970). However, it is important for general applications to develop a superplastic aluminium alloy by addition of a third element or by thermomechanical treatment if the recrystallized grains can be remarkably refined (Hamana, Nebti, Hamamda 1990, Paton and Hamilton 1985). Precipitation and recrystallization are the two main solid-state transformations of aluminium and copper based alloys which condition the quality of the semi-finished product. It is well established that dispersed, hard incoherent particles can either decelerate or accelerate recrystallization of a metallic matrix. On the other hand the precipitation reaction in aluminium alloys is strongly dependent on the deformation rate (Gernov, Sirenko, Hamana 1984, Hamana and Choutri, 1991). The aim of this paper is to study the interaction between recrystallization and precipitation in Al-8 wt % Mg and Al-3,5 wt % Cu alloys and to try to obtain a very fine recrystallized structure without any addition.

2. EXPERIMENTAL PROCEDURE

The Al-Mg alloy contains 8 wt % Mg and the total impurity percentage measured by electron microprobe does not exceed 0,95 wt %. Samples are sectioned from as-received cast material , homogenized at 708 K for 17 hours after cold rollling (35 % reduction), and rapidly quenched in iced water. Then, some quenched samples are deformed (5 - 35 % reduction) and aged at 523 K for various periods of time to develop precipitation and recrystallization processes. Al-Cu alloy is prepared from 99,999 % Al and Cu by melting and contains 3,5 wt % Cu. After homogenization and quenching, samples are deformed (5 - 35 % reduction) and aged at 473 K. X-ray diffraction, optical microscopy, transmission electron microscopy (TEM) and microprobe analysis are used to investigate the alloys.

3. RESULTS AND DISCUSSION

Macroscopic deformation processes which traverse many grains and contain several local lattice rotations lead to the formation of shear bands. In this study, the degree of cold deformation required to get extensive macroscopic shear band development in Al-8 wt% Mg alloy was in the order of 35 %. The decoration of shear bands with Al_3Mg_2 precipitates was obtained by aging the deformed samples 2 hours at 523 K. After 4 hours of aging a high grain size is observed (Figure 1a), and seems to be equal to that of undeformed sample (Figure 1b). However, X-ray diffraction indicates that both structures are very different (Figure 2). The deformed sample has a very fine structure confirmed by TEM micrograph (Figure 1c) on which we can see clearly the presence of very fine grains. The latter give continuous diffraction rings, from the first to the last aging stages, due certainly to a high misorientation angle of their boundaries.

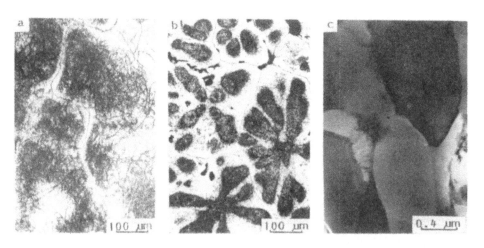

Fig. 1. Optical micrographs of Al-8 % Mg alloy quenched and aged 4 h at 523 K with (a) and without (b) precedent cold working (35%).TEM micrograph of the deformed sample (c).

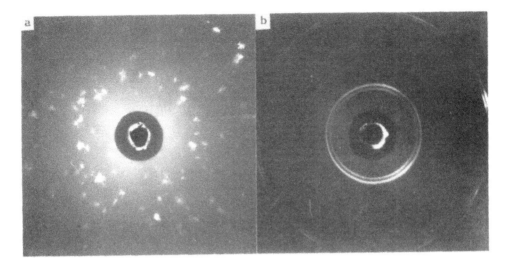

Fig. 2. Debye - Scherrer back plane films of quenched and aged (4 h at 523 K) Al-8 wt % Mg alloy without (a) and with precedent cold working (35 %) (b).

On the other hand, after a small amount of plastic deformation (5-10 %) and aging at 523 K during 2 minutes, nucleation and growth of new grains were observed at structure defects (Figure 3). The grain size is smaller than that of undeformed sample. However, the final structure is less fine than that obtained after a high amount of plastic deformation, as confirmed by X-ray diffraction (Figure 4). Both structures were very stable (stabilized by precipitate particles) and exhibited aligned diffraction spots (Figure 4a) and a continuous diffraction ring (Figure 4b), even after 300 hours aging at 523 K .The same observations were made in Al-3,5 % wt Cu alloy. The strong dependance of the superplastic behavior of metal on grain size has been adequatly demonstrated numerous times (Paton and Hamilton 1979), and it is now well known that a fine grain size is a primary requirement for superplasticity. Homogenization followed by rapid quenching, cold working (35 %) and aging at 523 K is a new way for imparting a fine grain structure to aluminium alloys which have precipitating constituents, finer than that obtained after Zr addition.

4. CONCLUSION

During conventional processing of aluminium alloys, recrystallization occurs in a largely uncontrolled fashion, and resulting grain sizes are still large. However, aging of highly deformed samples yields to a very fine polycrystalline structure, revealed by X-ray and electron diffractions.

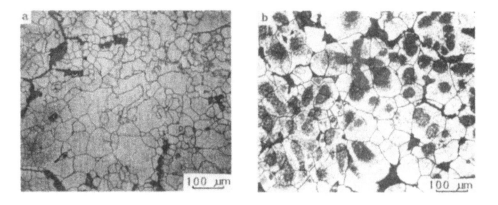

Fig. 3. Microstructure of quenched, deformed (10 %) and aged at 523 K for 2 minutes (a) and 4 hours (b) Al- 8 wt % Mg alloy.

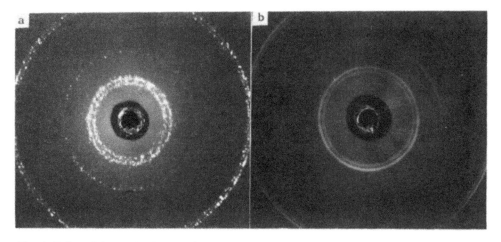

Fig. 4. Debye-Scherrer back plane films of quenched and aged (300 hours at 523 K) Al-8 wt % Mg alloy after 10 % (a) and 35 % (b) precedent cold working.

REFERENCES

Davies G J, Edington J W, Cutler C P and Padmanabham K A 1970 J. Mater. Sci. 5 1091

Hamana D, Nebti S, Hamamda S 1990 Scripta Met. Mater. 24 2059

Paton N E and Hamilton C H 1985 United States Patent I19I B1 4, 092, 181

Gernov S A, Sirenko A F, Hamana D 1984 Ukrainian Phys. Jour. 29 3 436

Hamana D and Choutri H 1991 Scripta Met. Mater. 25 4 859

Paton N E and Hamilton C H 1979 Met. Trans. A 10A 241

Inst. Phys. Conf. Ser. No 130: Chapter 3
Paper presented at Int. Congr. X-ray Optics and Microanalysis, Manchester, 1992

291

Stability and microstructure of ferrites containing aluminium

U. Wege, W.M. Dawson[*] and F.R. Sale

Manchester Materials Science Centre, University of Manchester/UMIST, Grosvenor Street, MANCHESTER, M1 7HS

[*] Philips Components, Crossens, SOUTHPORT

ABSTRACT: The effects caused by additions of aluminium on the sintering of conventionally prepared MgO-based ferrites have been studied using optical microscopy, transmission electron microscopy and microanalysis along with XRD. Dilatometry was used to study the kinetics of sintering of the ferrites. Magnetic properties have been determined as a function of aluminium additions and microstructure. The results have been related to the level of aluminium addition which gradually changed the overall stoichiometry of the ferrite from one of iron deficiency to one of iron excess.

1. INTRODUCTION

The magnetic properties of ferrites depend strongly on lattice defects and second phases as well as porosity and grain size. Hence, the improvement of magnetic and electrical properties are dependent upon the engineering of the correct microstructure. [1] During production the microstructural control is achieved by control of the processing conditions and the levels of impurities and small amounts of additives, which affect the sintering behaviour. [2] In addition, the effects of impurities and additives upon anisotropy compensation and are important factors. Finally, an appropriate sintering atmosphere has to be chosen to prevent a coarse grain structure or heämatite precipitation. MgO based ferrites are characterised by a very high electrical resistivity compared to MnZn or NiZn ferrites. Mg-Zn ferries are used almost exclusively as deflection yokes for televisions. In the present study additions of aluminium were made to a commercial iron-deficient material which is used as a magnetic yoke ring. The various levels of Al addition were selected to change the overall composition of the ferrite from one of iron deficiency to one of iron excess, as the Al^{3+} ions decrease the Me^{2+} excess and hence reduce the amount of Wüstite phase (Mg0.Zn0) produced as a result of the Fe-deficient nature of the ferrite.

2. EXPERIMENTAL PROCEDURE

Powders with the composition of $Mg_{0.63} Zn_{0.37} Mn_{0.1} Al_x Fe_{1.78} O_y$ were prepared using the standard technique of mixing oxides. The values of X used were X=0; 0.05; 0.1; 0.2; 0.5 atomic additions to the basic formula. Toroids were pressed and sintered in air at

1250°C. The sintering behaviour was studied using dilatometry. Analytical transmission electron microscopy using a Philips EM430T, fitted with an EDAX energy dispersive X-ray spectrometer was used to investigate the structures and composition of the sintered ferrites.

3.   RESULTS AND DISCUSSION

The relationship between shrinkage and temperature is shown in Figure 1. It is evident that with increasing Al additions the shrinkage reduces. These data may be interpreted in terms of sintering theories related to iron-deficient and iron excess ferrites. [3]

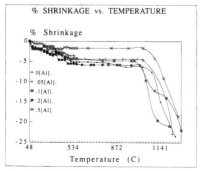

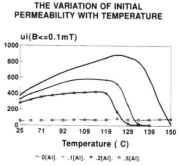

Figure 1:   **Figure 2:**

Two completely different microstructures are developed during sintering (Figure 3). The iron deficient standard sample contains Wüstite 2nd-phase which is rich in MgO-ZnO. [4] Also Si and Ca are present which are almost all situated in the triple points and along the grain boundaries. With increasing Al up to the stoichiometric composition ($X = 0.12$), the Wüstite phase disappears. With further additions above $X = 0.12$ excess Fe_2O_3 is present which is accompanied by the formation of cation vacancies and divalent ions ($2Fe^{3+}$ ¾V_c + O^{-2}). As shown in Figure 3 the porosity changes continuously from intergranular to intragranular and the average grain diameter decreases from $15\mu m$ (standard sample) to $3\mu m$ with the hightest amount of Al addition. Grain size effects the magnetic properties and the permeability is seen to increase linearly with increase in grain size. [5]

The variation of initial permeability with increasing temperature is shown in Figure 2. The maximum in each curve shifts slightly towards lower temperature when Al is added in the range of $0.12 < X < 0.22$. It is assumed that in this composition range the Mn^{3+} of the ferrite is reduced to Mn^{2+}. The sample with .5 atom % content shows within this temperature range only a straight line showing no maximum. This behaviour is concomitant with a significant increase in the Curie Temperature from 141°C (standard sample) to 215°C for the sample with .5 at% Al. This behaviour is because of the increase of the Fe^{2+}/Fe^{3+} ratio necessary to maintain the stoichiometry of the spinel phase as the higher amounts of Al are added. The excess iron may dissolve into the spinel as γ-Fe_2O_3 with the formation of cation vacancies which explain the change in the sintering behaviour since the sintering of ionic components is rate limited by the diffusion of oxygen via anion vacancies.

XRD showed extra lines for the sample with the highest addition of Al extra lines were found which were identified as belonging to MgO Al_2O_3. TEM analysis reveals, at grain boundaries, an increase in Mg from 6.7 to 12.7 wt% compared with an increase in Al from 5.4 to 9 wt% and a decrease in Fe and Zn content. Concentration of Al and Mg and wüstite precipitation of MgO Al_2O_3 affects the grain boundary mobility so that a smaller grain size is obtained. Figure 4 shows an area of a triple point where two different second phases were found. These phases contained Al which appeared in various combinations with SiO_2 or CaO. Figure 4 also shows a [001] zone diffraction pattern from the sample with the highest amount of Aluminium. The lattice constant increases with increasing $[Fe^{2+}]$ content and it is assumed that the Mg^{2+} ions occupy octahedral sites and magnetically disturb the lattice.

CONCLUSION

(1) Al additions to the MgO-based ferrites generally decreased the total % shrinkage and the sintered density.

(2) The microstructure changed systematically with increasing Al additions from one of iron-deficient to one having iron excess appearance. The grain sizes of all samples containing Al additions were smaller than that of the standard composition.

(3) The sample with the highest amount of Al produces a second phase MgO Al_2O_3 at grain boundaries and grain junction. Al was always found at higher concentrations at the grain boundaries.

REFERENCES

(1) Development in soft magnetic ferrites for power applications, Electronic Ceramics Edited by B.C.H. Steele Elsevier Applied Science, pp. 147-168, 1992.

(2) F.J.C.M. Toolenaar, Journal of Materials Science, 23 (1988), 3144-3150.

(3) P.J.L. Reijnen, Science of Ceramics, 4 (1968).

(4) Mahloojchi, F. and Sale, F.R., Proc. VI CIMTEC World Conference on High Tech. Ceramics, Milan, 1986. In High Tech. Ceramics, ed. V. Pincenzini, Elsevier Amsterdam, 1987, pp. 1038-48.

(5) Perdurjn, D. et al, Proc. Brit. Ceram. Soc., 10, 263 (1968).

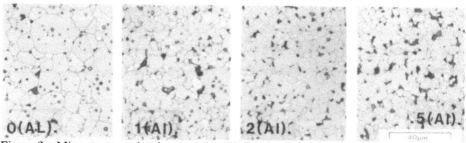

Figure 3 : Microstructure development in a MgZn ferrite containing Al.

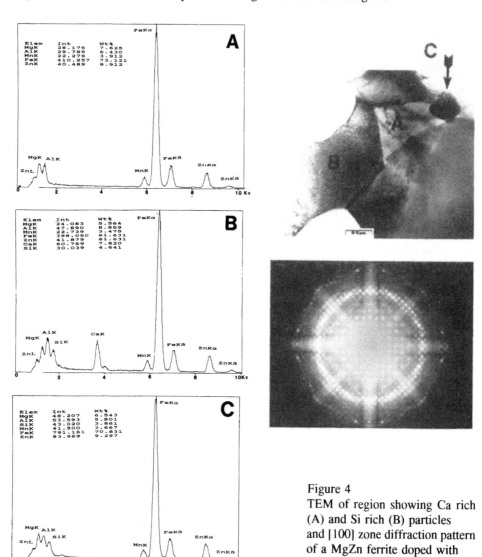

Figure 4
TEM of region showing Ca rich (A) and Si rich (B) particles and [100] zone diffraction pattern of a MgZn ferrite doped with .5 atom% Al showing Kikuchi lines.

Inst. Phys. Conf. Ser. No 130: Chapter 3
Paper presented at Int. Congr. X-ray Optics and Microanalysis, Manchester, 1992

295

New energy filtering TEM—principles and applications

W.Probst, G. Benner, J. Mayer*

Carl Zeiss, Electron Optics Division, D-7082 Oberkochen, Germany
* Max-Planck-Institut für Metallforschung, D-7000 Stuttgart 1, Germany

ABSTRACT: An EFTEM featuring an Omega-type imaging spectrometer which is integrated in the optical ray path is described. The instrument provides optimum simultaneous use of the signals which are contained within the spectrum of inelastically scattered electrons. For ease-of-use the optical ray path is designed for using fixed conjugate imaging planes and planes of symmetry. Fast recording of large area elemental distribution images, high quality EELS, and low background filtered diffraction patterns become feasible and examples will be shown.

1. INTRODUCTION

The penetration of an electron beam through a thin specimen causes numerous interactions between beam electrons and specimen atoms. The variety of resulting signals which can be recorded via the appropriate detectors provides different amounts of information concerning specimen structure and/or chemistry. The inelastic scattering process and the inelastically scattered electrons in particular potentially reveal a maximum of simultaneous information, concerning e.g. microstructure, mass, specimen thickness, elements, and even molecules. In a conventional TEM (CTEM), however, inelastic scattering events in thicker specimen areas reduce contrast and resolution whereas the spectroscopic information in the inelastic scattering background is lost in the conventional imaging and diffraction modes. These limitations can be overcome by the integration of an imaging electron energy loss spectrometer into the TEM. An energy filtering transmission electron microscope (EFTEM) enables the utilisation of those electrons offering decisive benefits, and indeed give new impetus to the field of TEM (Mayer 1992).

2. METHODOLOGY

An electromagnetic prism in combination with an electrostatic mirror integrated in the optical ray path was described by Castaing and Henry (1964) for spatially separating the inelastically scattered electrons from the unscattered and elastically scattered ones. The

EM 902 launched by Carl Zeiss seven years ago, and now also the EM 902 A, are the only instruments to feature such a system, which has been integrated with great success in the imaging beam path.

Carl Zeiss now presents the EM 912 OMEGA which has been developed on the basis of calculations performed by Rose (1978) and Lanio (1986) in co-operation with Krahl (1990). In this instrument, the electrons are separated via four sector magnets instead of a prism and directed on an Ω-shaped beam path onto the optical axis of the microscope

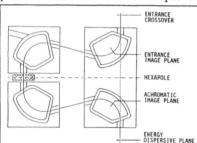

(Bihr.1991;Fig.1). As this makes the electrostatic mirror superfluous, the limitation to acceleration voltages smaller than 100 kV also no longer exists. This is a major benefit for the microscope's use in the material sciences. The design and construction of the filter microscopes are such that the advantages offered by the filtering process can be used as elegantly and reproducibly as possible. Both spectrometers are thus designed in accordance with the principles of symmetry using conjugate imaging planes in the optical ray

Fig.1. Schematic drawing of an Omega type spectrometer using four sector magnets (>90°) and a hexapole corrector in the plane of symmetry.

path (Fig.2). Most of the aberrations which are caused by the use of non-circular lenses in particular and which require a complex correction procedure are therefore eliminated

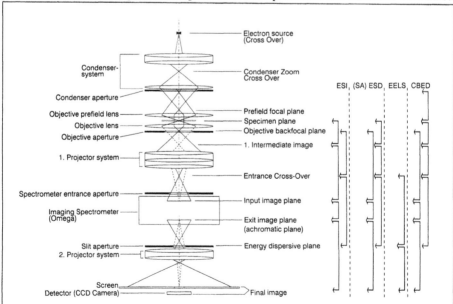

Fig.2. Schematic drawing of the EM 912 Omega's optical ray path. The conjugated imaging planes are indicated for different modes of operation.

without any need for further corrections. The user does not notice in any way that the electron beam follows such a crooked path and can use the instrument in the same way as a conventional instrument. In the case of the EM 912 OMEGA, whose Omega spectrometer is integrated in the future-oriented electron optics of the EM 910, symmetrical conditions and conjugate planes are also utilised in the illumination system. Koehler illumination and the many advantages which it offers can therefore be used without complicating the adjustments required of the user.

If electron-spectroscopic imaging (ESI) is required, i.e. if electrons of a selected energy are to be used for image formation, the characteristic energy loss values can be easily selected manually or automatically by computer-controlled digital adjustment of the acceleration voltage in precise 0.2 V steps. The spectrometer magnetic fields or the position and setting of the slit aperture remain unchanged and therefore stable.

3. APPLICATIONS

Simultaneously with the two-dimensional image information an EFTEM provides extensive electron spectroscopic information which can be utilised for the rapid production of elemental distribution images. To do this, two images with a defined window of energy loss electrons are recorded either both in front of, or one in front of and one behind a characteristic element absorption edge. With the aid of the appropriate algorithms (power law $A*E^{-R}$ in Fig.3), these images are used to compute the background image pixel by pixel for the element image also recorded and are then subtracted from the latter. The entire process required to obtain the net element distribution image takes about 60 seconds where image acquisition time is only 10 to 20 seconds for 512 x 512 pixels (Fig.3). This is faster by a factor of about 1000 than comparable methods using a Field Emission STEM. The microscope is controlled by the image processing system during this procedure. Pixel resolution or imaged area, respectively, can easily be increased to 1024 x

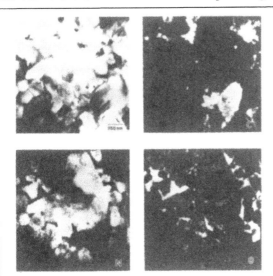

Fig.3. Qualitative characterisation of sintered poly-mer-derived Si-C-N powder ceramics: Elastic bright-field image (upper left) and net element distribution images of carbon (C; for SiC), nitrogen (N; for Si_3N_4), and oxygen (O; for O containing inter-face).Bar represents 250 nm.

1024 pixels or larger without increasing acquisition time if appropriate slow-scan CCD cameras are used.

Such element distribution images produced with ESI are easily confirmed via an EEL spectrum with serial or parallel acquisition (Fig.4). Instrument and detector provide optimum signal to background ratio as is indicated by a jump ratio of >15 at the C_K-edge.

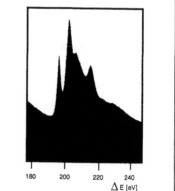

Fig.4. EELS spectrum showing boron K-edge in BN. Parallel recording via a slow-scan CCD camera. Jump ratio of >3.

An EFTEM allows not only all possible combinations of different analytical data such as ESI, EELS, and EDX, but also diffraction patterns to be recorded in an ideal way. Due to the removal of the disturbing inelastic component, diffraction patterns which are elastically imaged are markedly sharper and richer in fine details, especially in the immediate vicinity of the zero beam. High-voltage microscopes would be necessary to obtain a noise background of a similar low intensity, but the signal is reduced at the same time. The filter allows the use of thicker specimens for diffraction in the convergent beam mode (CBED), for example. Direct quantitative evaluation is also possible for such filtered diffraction patterns (Fig.5).

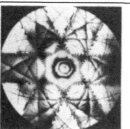

Fig.5. Conventional (left) and Zero-Loss filtered (right) Si (111) Large Angle CBED pattern. In case of filtering contributions from thermal diffuse and inelastic scattering are both removed. HT: 120 kV.

An EFTEM like the EM 912 OMEGA offers many more modes of operation, e.g. element- or structure contrast, plasmon loss imaging, image EELS and others more. In principal the quality of all results obtained in conventional modes of operation of a TEM can substantially be improved by filtering.

4. REFERENCES

Bihr J, Benner G, Krahl D, Rilk A, Weimer E 1991 Proc 49th An. Meet. EMSA pp 354-5
Castaing R, Henry L 1962 Compt. rend. Acad. Sci. (Paris) 255 pp 76-78
Krahl D, Paetzold H, Swoboda M 1990 Proc. XIIth Int. Congr. Electr. Microsc. pp 60-1
Lanio S 1986 Optik 73 pp 99-107
Mayer, J 1992 Proc 50th Ann. Meet. EMSA pp 1198-9
Rose H 1978 Optik 51 pp 15-38

Inst. Phys. Conf. Ser. No 130: Chapter 3
Paper presented at Int. Congr. X-ray Optics and Microanalysis, Manchester, 1992

299

The effect of arsenic and grain size on carbide composition in a low alloy steel

Z. Larouk and R.Pilkington.

Univ. of Constantine, Algeria, and Univ. of Manchester.

ABSTRACT The paper reports changes in carbide compositions after changing trace elements and grain size for a 1% Cr 1% Mo 3/4% V steel. For fine grained material, the addition of As appears to result in less Mo in carbides after ageing at 550°C, but this trend is reversed under creep conditions. However coarse grained material, containing only the addition of Ti, shows little change under ageing or creep.

1. INTRODUCTION

Durehete 1055 is a Cr/Mo/V low alloy steel used as a bolting steel in electrical power stations operating at 565°C under low stresses, where creep is the major deformation mode. This steel contains 1% Cr 1% Mo 3/4% V + 0.2% C, together with small additions of Ti (0.08%) and B (0.005%). The effect of trace elements in changing carbide composition and degrading microstructure is still under debate (Yu and Grabke 1983, Larouk and Pilkington 1990), since the precipitation of new carbides during service exposure was recently observed in Cr/Mo/V steels (Carruthers and Collins 1983, Collins 1989). The present work is a further attempt to study this problem by examining the effect of microstructure, ageing time and stress on the carbide compositions for two vacuum melted casts of Durehete 1055, one cast produced with the sole addition of Ti, the second with Ti + 0.045% As.

2. EXPERIMENTAL

The material was examined in two different microstructural conditions: fine grained material (2h at 980°C W.Q. + 4h at 680°C A.C.), and coarse grained material (1/2h at 1250°C A.C. + 4h at 680°C), giving prior austenite grain sizes of 8μm and 150μm respectively. Creep tests were carried out at 565°C for various times on the heat treated material. Carbon extraction replicas were prepared for examination and analysis in a

Philips 400T electron microscope equipped with EDAX analysis facilities. These replicas were obtained from heat-treated material prior to testing, from the heads and from the gauge lengths of creep specimens after testing.

3. RESULTS AND DISCUSSION

The microstructure of the fine grained material for both casts consists of a fine tempered bainite (Fig. 1) with a hardness of 300 H_{v30}. For the coarse grained material (Ti cast only), the tempered bainite is also on a coarser scale, with the hardness significantly higher at 370 H_{v30}. A low volume fraction of relatively large particles (100-200 nm) are present at grain boundaries in the coarse grained material (Fig. 2); these particles are not observed in the fine grained samples.

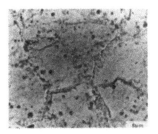

Fig. 1. Transmission electron micrograph of carbon extraction replica of fine grained Ti cast.

Fig. 2. Transmission electron micrograph of carbon extraction replica of coarse grained Ti cast.

Microanalysis/diffraction of these carbides shows that they are MC cubic, based on TiC/VC but containing Mo, with V being the major element. No differences are detected in compositions between carbides in the matrix or on the grain boundaries of fine grained material. However, although most of the particles in the coarse grained material contain mainly Ti, some grain boundary carbides are Fe rich (M_3C type) containing Cr, Mn, Mo and V. Analysis of the carbides can be represented on pseudo-ternaries (Cliff et al 1984), in this case based on Ti-Mo-V, and it is found that there is little variation in carbide compositions between the casts before test (Larouk 1990).

For the fine grained material, the variation in carbide composition as a function of ageing time (up to 9528h) has already been reported for the Ti cast in comparison with casts containing Ti+P (0.02%) and Ti+Sn (0.02%) respectively [2]. It was found that the mean composition lines on pseudo-ternary diagrams became displaced further away from the Mo corner after ageing at 565°C. In the present work, the presence of As appears to reverse this trend (Figs.3,4), and this behaviour is enhanced under conditions of combined ageing and stress (creep) - the mean composition line after creep for 110h is located in a similar position to that after ageing for 3641h. Hence, for fine grained material, it seems that As has the opposite effect to that of P and Sn under ageing conditions, but is similar to that for P under creep conditions.

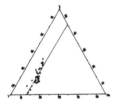 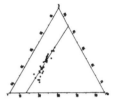

Fig. 3. Carbide compositions for (Ti+As) cast; fine grained material aged 110h.

Fig. 4. Carbide compositions for (Ti+As) cast; fine grained material aged 3641h.

For the coarse grained material, the position of the mean carbide composition line for the Ti cast before test (Fig.6)is significantly different from that for the fine grained material (Larouk and Pilkington 1990) in that the carbides contain much less Mo.

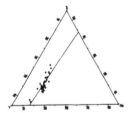 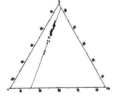

Fig. 5. Carbide compositions for (Ti+As) cast; fine grained material crept at 300MPa for 110h.

Fig. 6. Carbide compositions for Ti cast; coarse grained material before test.

Ageing or creep produce little change in the mean composition lines (Figs.7,8) - it is found that creep in fine grained material increases the Mo content of the carbides (Larouk and Pilkington 1990). It should also be noted that a few particles , rich in Mo, are found to occur in the Ti cast crept for 3674h (Fig.9).

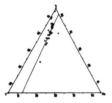

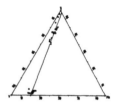

Fig. 7. Carbide compositions for Ti cast; coarse grained material aged 1620h.

Fig. 8. Carbide compositions for Ti cast; coarse grained material crept at 380MPa for 1620h.

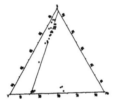

Fig. 9. Carbide compositions for Ti cast;
coarse grained material crept at 360MPa for 3674h.

It can therefore be concluded that carbide compositions change significantly under ageing and creep conditions at 565°C - a temperature that is well below the tempering temperature of 680°C. The presence of 0.045% As increases this effect in fine grained material. Grain size may also influence this behaviour in that little compositional change was noted in the carbides for coarse grained material, contrary to that for fine grained material. It is also possible that new carbide precipitation (M_2C) may occur in coarse grained material under creep conditions.

4. REFERENCES

Carruthers R B and Collins M J 1983 Met.Sci. 17 107
Cliff G et al., 1984 Electron Microscopy and Analysis (Inst of Phys Conf. No. 68 P Doig Ed.) 63
Collins M J Mater. Sci. & Tech. 1989 5 323
Larouk Z 1990 Ph.D. thesis, University of Manchester
Larouk Z and Pilkington R 1990 Proc. XIIth Int. Congress of Electron Microscopy (San Francisco Press) 2 420
Yu J and Grabke H J 1983 Met.Sci. 17 389

Inst. Phys. Conf. Ser. No 130: Chapter 3
Paper presented at Int. Congr. X-ray Optics and Microanalysis, Manchester, 1992

303

Study of formation of Al-Fe alloys by mechanical alloying

J. Paduch, H. Krztoń, J. Wojtas, E. Barszcz

Institute for Ferrous Metallurgy,
44-100 Gliwice, 12 K. Miarki Str., Poland

ABSTRACT: Four elemental powder mixtures: 50 wt % Ni-Fe, 70 wt % Al-Fe, 33 wt % Al-Fe and 14 wt % Al-Fe were mechanically alloyed in a planar type ball mill. The microstructure and chemical composition of powders as a function of milling time were studied by means of X-ray diffraction and electron proble microanalysis.

1. INTRODUCTION

Mechanical alloying (MA) is a technique for synthesis of alloys that are difficult or impossible to produce by other means (Benjamin, 1970). The MA process consists of repeated welding, fracturing and rewelding of powder particles in high - energy ball mill. Starting from the elemental, crystalline powders, ball milling produces the homogeneous alloys and in several system forms the amorphous state (Schultz, 1988). The MA process creates wide possibilities of designing new alloys including FeAl intermetallics for high temperature applications.

The aim of the present research was to investigate changes of the chemical composition of the powder particles during the processing of alloy from elemental metals by MA. For this purpose the microstructure and chemical composition of Ni-Fe and Al-Fe powders as a function of milling time were studied by means of X-ray diffraction, electron probe microanalysis and light microscopy.

2. EXPERIMENTAL METHODS

The mechanical alloying was carried out in a planetary ball

mill (Retsch). Four metallic compositions were examined: 50wt% Ni - Fe, 70wt% Al - Fe, 33wt% Al - Fe and 14wt% Al - Fe. The spectrally pure powders of Ni, Al and Fe were mixed to give the desired average composition and placed in cylindrical steel containers. The steel balls had a diameter of 10 mm and the ball to powder weight ratio was 6:1. All handling was done in a purified argon atmosphere. The milling process was performed in intervals of 30 minutes milling and 30 minutes rest in order to keep the milling temperature low without artificial cooling. After different times the milling process was interrupted and a small quantity of the milled powder was removed for study by means of optical microscopy, X-ray diffraction and electron probe microanalysis.

X-ray diffraction was used to study the structural changes. The measurements were carried out on Philips diffractometer PW 1140 using Co K_α radiation. The diffraction patterns were scanned in steps of 0.05° (2θ) and fixed-time counting (4 s) was employed. Crystal parameters of the examined alloys were fitted by the Rietveld method (Rietveld, 1967). The DBWS-9006 PC programme of A. Sakthivel and R. A. Young was used for calculations. The pseudo-Voigt function was used as the profile function.

Specimens for optical microscopy and microanalysis were prepared by pressing the powders with copper. Qualitative and quantitative microanalysis were carried out on an ARL SEMQ microanalyser. Quantitative microanalysis was carried out using spectral standards of Fe and Al. The correction of microanalysis results was done by a computer program based on Monte Carlo simulations (Paduch and Barszcz, 1986).

3. RESULTS AND DISCUSSION

Behaviour of the powders and the morphological changes in the course of experiment were typical for the MA process, which was described in details, e. g. by Schultz (1988). The diffraction patterns of the powders after various times of milling are shown in Fig. 1. After several hours of milling the new non-ordered two-constituent phases started to form in

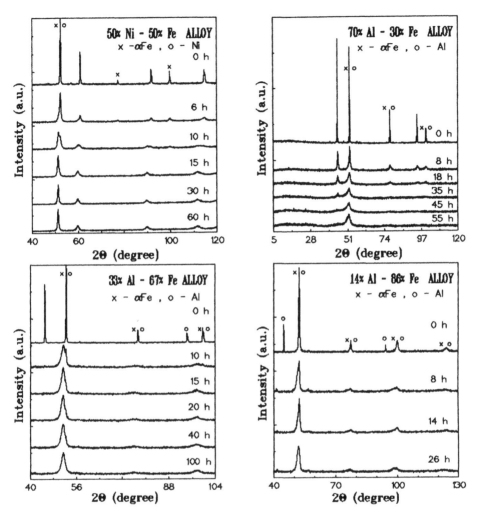

Fig. 1. X-ray diffraction patterns of Ni-Fe and Al-Fe powders
after various times of milling.

all of the mixtures. The NiFe phase with a fcc structure for
Ni-Fe mixture and AlFe phase of α - Fe type for Al-Fe mixture
have been observed along with pure metals. The concentration
of pure Fe in Ni-Fe mixture and pure Al in Al-Fe mixture
decreased with increasing time of milling. Simultaneously the
increase in lattice constant value of two-constituent phases
and broadening of diffraction lines were observed. Broadening
effect gives evidence of decreasing of crystallite size.

For 70 wt% Al - Fe system the amorphization was evident after 45 hours of milling.

Formation of two-consituent phases has also been observed by X-ray microanalysis. In Ni-Fe system during 5-15 h of milling homogenous phase was formed. The chemical composition of this phase was estimated as 49,9 wt% Ni and 50,1 wt% Fe. Standard deviation for 23 results was 0,58.

The homogenization of Al-Fe powders took place at a slower rate and the pure metals were present even after some tens of hours. The mean chemical composition of Al Fe phases was found to be almost constant.

In 70 wt% Al - Fe system the mean chemical composition of AlFe phase was estimated after 8 h of milling as 70,6 wt% Al and 29,4 wt% Fe, after 45 h as 69,8 wt% Al and 30,2 wt% Fe. Standard deviation in the former case was 7,29 and in the latter one 0,27. The relevant values for the remaining phases are as follows:

- for 33 wt% Al-Fe system:
  after 15 h: 33,4 wt% Al; 66,6 wt% Fe, sta. dev. 1,98
- for 14 wt% Al-Fe system:
  after 8 h: 16,7 wt% Al; 83,3 wt% Fe; sta. dev. 3,38
  after 14 h: 15,7 wt% Al; 84,3 wt% Fe; sta. dev. 2,61

4. CONCLUSION

The investigation has shown that from the point of view of the classical X-ray microanalysis, mean chemical composition of new phases produced by mechanical alloying, was stable after formation and refinement of lamellar structure.

REFERENCES

Benjamin J S 1970 Metall. Trans. 1 2943
Paduch J, Barszcz E 1986 J. Microsc. Epectrosc. Electron 11 81
Rietveld H M 1967 Acta Cryst. 22 151
Schultz L 1988 Mat. Sci. Eng. 97 15

Inst. Phys. Conf. Ser. No 130: Chapter 3
Paper presented at Int. Congr. X-ray Optics and Microanalysis, Manchester, 1992

307

Characterization of SiAlON ceramics by EPMA and TEM

J. Vleugels, T. Laoui, R. Wouters and O. Van Der Biest

Department of Metallurgy and Materials Engineering (MTM), Katholieke Universiteit
Leuven, De Croylaan 2, B-3001 Leuven, Belgium.

ABSTRACT: Five Si_3N_4-Al_2O_3 ceramic grades with widely varying Al_2O_3 content were
prepared by hot pressing at 1650°C. The general features describing the microstructure
and the distribution of the various phases formed in the above ceramics have been
investigated by EPMA and TEM. SEM and TEM revealed four different phases that
could be identified as β'SiAlON, O'SiAlON, X SiAlON and α-Al_2O_3. Quantitative
chemical analysis of these phases was done by EPMA.

1. INTRODUCTION

During the last 20 years, Si_3N_4 based ceramics have attracted much attention as structural
ceramics, because of their excellent thermal and mechanical properties. In the late 1970s,
Si_3N_4 ceramics entered the field of cutting tools. Their applicability nowadays is situated in
turning cast iron and Ni based alloys at high speeds. For machining of steels however, they
cannot be used. In order to solve this problem, the chemical interactions between Si_3N_4
based materials and iron alloys are studied in our laboratory. Within this framework, a
number of SiAlON ceramics with widely varying alumina content have been prepared. The
development of these ceramics, their microstructures and the chemical composition of the
different phases formed in them are reported in this paper.

2. PREPARATION OF SIALON CERAMICS

Commercial Si_3N_4 powder, grade HCST LC 12-SX, (97% α, submicron particle size, 1.8-
2.1 wt% Oxygen), and Al_2O_3 powder, grade Baikowski SM8, (95% α, submicron particle
size), were used to prepare powder mixtures with a composition along the Si_3N_4-Al_2O_3 line
in the phase diagram, shown in Figure 1. The five powder mixtures were chosen so that
there is at least one composition in each two or three phase region. The oxygen content of
the Si_3N_4 powder was not taken into account so that in reality the starting compositions are
above the Si_3N_4-Al_2O_3 line.

Powders were mixed in propanol on a multidirectional mixer for 72 hours in a polyethylene
bottle containing alumina milling balls. The weight loss of the alumina balls was
negligible. After evaporation of the propanol, samples were hot pressed under vacuum for
one hour at 1650°C and a pressure of 30 MPa in a graphite die, coated with boron nitride.

3. MICROSTRUCTURAL CHARACTERIZATION

3.1. X-Ray diffraction

No α Si₃N₄ could be detected by X-Ray diffraction, indicating that the $\alpha \rightarrow \beta$ Si₃N₄ conversion is complete in all SiAlONs. Excellent agreement between the identified phases and those that might be expected from the subsolidus diagram (Figure 1) is found for all SiAlONs, except the appearance of Al₂O₃ in SiAlON 4. This implies that the maximum solubility of Al₂O₃ into Si₃N₄ at 1650°C is less than might be expected from the phase diagram at 1650°C given by Jack (1978) and the subsolidus diagram at 1450°C given by Naik et al (1978).

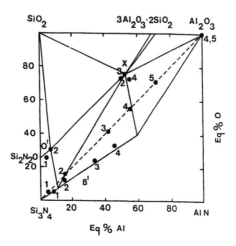

Figure 1: Chemical composition of starting powder mixtures and phases formed, given in the subsolidus diagram reported by Naik et al (1978). The numbers on the phase diagram refer to the five different SiAlON grades (SiAlON 1 through 5) and each indicate the starting composition of a grade or the chemical composition of the phases found in it.

3.2. Scanning electron microscopy (SEM)

Backscattered electron micrographs, taken on a Jeol 733 microprobe, of the five different SiAlON materials are given in Figure 2. The micrographs of SiAlON 1 and 2 (Figure 2a and 2b) show small dark elongated O' (Si₂N₂O) grains embedded in a bright β'SiAlON matrix. The aspect ratio of the O' grains in SiAlON 2 is larger than in SiAlON 1. In addition to the O' and β' phases, darker irregularly shaped grains can be distinguished in SiAlON 2. These grains were identified as X phase SiAlON. SiAlON 3 (Figure 2c) consists of β'SiAlON phase and X phase. In SiAlON 4 (Figure 2d), isolated bright α Al₂O₃ grains can be seen. In between these grains, X phase can still be distinguished from β'SiAlON phase. SiAlON 5 (Figure 2e) is dominated by large isolated α Al₂O₃ grains. In between these grains there is β'SiAlON, the presence of X phase is not clear in this micrograph.

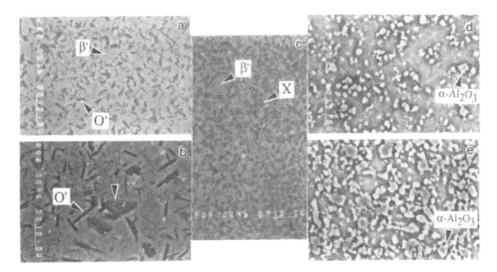

Figure 2: Backscattered electron micrographs of the different SiAlON grades: SiAlON 1 (a), SiAlON 2 (b), SiAlON 3 (c), SiAlON 4 (d) and SiAlON 5 (e).

3.3. Electron probe microanalysis (EPMA)

Quantitative chemical analysis was done by EPMA, using a pure SiO_2 and Al_2O_3 standard. Samples and standards were coated simultaneously with a very thin gold layer, to ensure a homogeneous coating thickness. Quantitative analysis was performed for Si, Al and O. N was calculated by difference. The measured compositions are indicated in Figure 1.

Chemical analysis of β'SiAlON phase in SiAlON 5 could not be obtained because of the impossibility to exclude alumina grains out of the analyzed volume.
Maximum solubility of Al_2O_3 in Si_2N_2O was reached in SiAlON 2. This point of maximum solubility is in perfect agreement with the phase diagram of Naik et al (1978).

The measured chemical compositions of X phase in SiAlON 2 and 3 are very close to the composition, $Si_{12}Al_{18}O_{39}N_8$, reported by Naik et al (1978). This being one of the eight different compositions found in literature. The composition found in SiAlON 4 is somewhat different. This might be caused by the size of the interaction volume of the electron beam. We might conclude that X phase is situated on the Si_3N_4 - $3Al_2O_3.2SiO_2$ (mullite) line in the phase diagram.

3.4. Transmission electron microscopy (TEM)

TEM characterization of thin foils was performed on a JEOL 200CX microscope operating at 200 kV. Within the resolution limits of the microscope, no intergranular amorphous phase could be observed at the grain boundaries or the triple junctions in the five different SiAlON materials. The existence of a very small grain boundary phase however, a few atomic layers thick and formed by impurities in the starting powders, is not excluded.

The O' grains in SiAlON 1 and 2 can be easily distinguished from the hexagonal β'SiAlON grains, not only by their elongated morphology but also by their characteristic striations (Figure 3a), caused by a high stacking fault density parallel to the (100) planes of the orthorhombic structure according to Lewis et al (1985).

The X phase grains in SiAlON 3, 4 and 5 show a characteristic twin pattern (Figure 3b). The crystal structure of X phase is triclinic as determined by electron diffraction analysis and with lattice parameters similar to those proposed by Zangvil (1978).

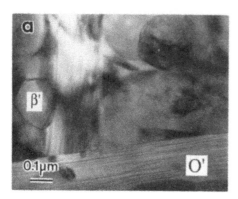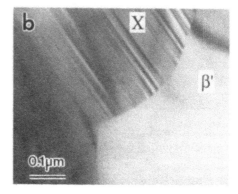

Figure 3: TEM micrographs of argon ion thinned SiAlON samples. (a) shows the presence of elongated O' grains in SiAlON 1, containing a high stacking fault density. (b) shows a grain boundary between a β'SiAlON grain and an X phase grain in SiAlON 3.

4. CONCLUSIONS

The chemical composition of X phase SiAlON was found to be very close to the formula, $Si_{12}Al_{18}O_{39}N_8$, reported by Naik et al (1978), one of the eight compositions found in literature, and is situated on the line between silicon nitride and mullite in the phase diagram.

At 1650°C, the maximum solubility of Al_2O_3 in Si_3N_4 is lower than might be expected from the reported phase diagrams.

5. ACKNOWLEDGEMENTS

J.V. thanks the Belgian I.W.O.N.L. fund for a research fellowship. This work was also supported by the Commission of the European Communities under project BREU-0096-C.

6. REFERENCES

Jack K H 1978 Phase Diagrams, Materials Science and Technology London: Academic Press) 5 pp 241-285
Lewis M H, Reed C J and Butler N D 1985 Mater. Sci. Eng. 71 87
Naik I K, Gauckler L J and Tien T Y 1978 J. Am. Ceram. Soc. 61 332
Zangvil A 1978 J. Mater. Sci. Lett. 13 1370

Inst. Phys. Conf. Ser. No 130: Chapter 3
Paper presented at Int. Congr. X-ray Optics and Microanalysis, Manchester, 1992

311

Investigations of titanium dioxide particle size changes during thermal treatment by TEM method

M. Wiśniewski, H.Ratajska and A. Goryńska

Applied Inorganic Chemistry Centre of Polish Academy of Sciences,
Kuźnicka 1, 72–010 Police, POLAND

ABSTRACT: In this paper the results of investigations of hydrous titanium dioxide particle size changes during calcination are presented studied by TEM method.

1. INTRODUCTION

Titanium white is the most important white inorganic pigment. Its properties are in high degree depended on the particles size (Latty 1958, Sullivan and Cole 1959, Sheynkman and Kasperovich 1974). The aim of this work was to investigate the particles size changes of hydrous titanium dioxide during calcination and the effect of presence of some additions controlling the anatase-rutile phase transformation such as rutile nucleus, KOH or H_3PO_4.

2. EXPERIMENTAL

The raw material for investigations was hydrous titanium dioxide which can be described by the empirical formula $TiO_2 \cdot xH_2O \cdot ySO_3$, where x = 6.94 to 8.95 and y = 0.064 to 0.067. It was prepared by the hydrolysis of the industrial solutions of titanium sulphates and washed from the metallic impurities.

3. INSTRUMENTATION

The investigations of the particles sizes were carried out with the transmission electron microscope TESLA BS-500. The samples for investigations were prepared by a "wet method" in a form of suspensions dispersed ultrasonically in ethyl alcohol and deposited on Formvar films.
The pictures were taken with a JEOL JSM 6100 scanning electron microscope.

4. RESULTS

Hydrous titanium dioxide before thermal treatment after being dispersed ultrasonically in ethyl alcohol has an average particles diameter 0.04 to 0.06 μm. After drying to the constant mass in the room temperature the particles size is 0.04 to 0.10 μm. Thermal treatment in the temperature range from 100 to 500 ºC results in increasing the particle size to 0.11 to 0.18 μm. During calcination in the temperature 600 ºC, which is characteristic for the begining of anatase crystallization, a decrease of particles size was observed in all samples to 0.05 to 0.07 μm. The similar observation was described by Goldshteyn and Sheynkman (1973). Further temperature increase during calcination without any additions results in the following particle sizes:

800 ºC	0.13 to 0.19 μm
900 ºC	0.16 to 0.22 μm
1000 ºC	0.23 to 0.44 μm

The changes of the average particles size of hydrated titanium dioxide during thermal treatment is shown on the Figure.

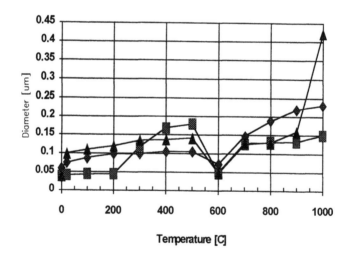

Sample 1 - $TiO_2 . 6.935 H_2O . 0.064 SO_3$
Sample 2 - $TiO_2 . 7.694 H_2O . 0.066 SO_3$
Sample 3 - $TiO_2 . 8.952 H_2O . 0.067 SO_3$

Effect of thermal treatment on particles size

Addition of 5 % of rutile with the average particles size of 0.023 μm to the hydrous titanium dioxide before calcination results in less increase in particle size during calcination at higher temperatures:

$$800 \ oC \quad 0.13 \ \mu m$$
$$900 \ oC \quad 0.13 \ \mu m$$
$$1000 \ oC \quad 0.15 \ \mu m$$

In the Table the results of the particles size measurements of calcinates obtained at the temperature range 20 to 1000 oC without any promotors and at the temperatures 850, 900 and 950 oC in the presence of promotors K_2O, P_2O_5 and rutile nucleus are presented. The higher the temperature the larger the particles size is. At temperature 950oC the average particles size is almost 0.2 μm and is similar to the earlier obtained results for the industrial calcinates of 0.205 ± 0.005 μm. This is the optimal value for the white pigments to have the best optical properties.

Temperature	Average	particle	diameter	[μm]	
[C]	Sample 1	Sample 2	Sample 3	Sample4	Sample5
wet	0.061	0.040			
20	0.074	0.041			
100	0.088	0.044			
200	0.098	0.045			
300	0.099	0.118			
400	0.105	0.169			
500	0.106	0.178			
600	0.072	0.048			
700	0.148	0.125			
800	0.188	0.131	0.118	0.066	0.158
900	0.220	0.133	0.167	0.187	0.176
1000	0.231	0.151	0.186	0.213	0.216

Sample 1 - TiO_2 . 6.935 H_2O . 0.064 SO_3 without promotors
Sample 2 - TiO_2 . 7.694 H_2O . 0.066 SO_3 with 5 % rutile nuclei
Sample 3 - TiO_2 . 7.361 H_2O . 0.067 SO_3 with 0.2 % K_2O
Sample 4 - TiO_2 . 7.361 H_2O . 0.067 SO_3 with 0.1 % P_2O_5
Sample 5 - TiO_2 . 7.361 H_2O . 0.067 SO_3
with 5 % rutile nuclei, 0.2 % K_2O and 0.1 % P_2O_5

Results of particle size determination af calcinates after thermal treatment

5. REFERENCES

Latty J.E., 1958, J.Appl.Chem., 8, 96
Sullivan W.F. and Cole S.S., 1959, J.Am.Ceram.Soc., 42, 127
Sheynkman A.J., and Kasperovich V.M., 1974, Zh.Prikl.Kchim., 47, 1715
Goldshteyn L.M. and Sheynkman A.J., 1973, Zh.Prikl.Kchim., 46, 299

Inst. Phys. Conf. Ser. No 130: Chapter 3
Paper presented at Int. Congr. X-ray Optics and Microanalysis, Manchester, 1992

315

A TEM-EDX study of sol-gel derived PLZT thin films

G Dražič, T Beltram, M Kosec

"J. Stefan" Institute, University of Ljubljana, 61000 Ljubljana, Slovenia

ABSTRACT: Using analytical electron microscopy the microstructure, degree of crystallisation, particle size and chemical composition of the phases present were investigated in sol-gel derived PLZT thin films deposited on different substrates (Si, Si/Pt, Si/Ti/Pt, MgO). At firing temperatures above 750°C, several μm sized, monocrystalline, preferentially oriented perovskite PLZT rosettes surrounded by fine grained Pb deficient pyrochlore phase were observed. In the case of Si substrates, an amorphous Pb-Si-O reaction layer at the substrate/film interface was detected.

1. INTRODUCTION

Sol-gel derived ferroelectric films of submicron sizes, deposited on various substrates, are recently becoming very important materials used for the fabrication of high quality electronic and electrooptic components. Recently (Myers and Chapin (1990), Huffman (1990), Tuttle et al (1991)) it was reported that in sol-gel derived P(L)ZT thin films sphellurite-like rosettes of perovskite phase and small grains of pyrochlore phase are formed during the firing process. The presence of pyrochlore phase was explained by the loss of PbO at elevated temperatures.

The aim of this work was microstructural investigation of spin-coated, sol-gel derived PLZT thin films deposited on various substrates. The structural relationships, chemical composition of the phases present and possible reasons for pyrochlore phase formation are reported and discussed.

2. EXPERIMENTAL DETAILS

PLZT (9.5/65/35) sol was prepared from Ti and Zr propoxides and Pb and La acetates in methoxyethanol solution. Details of the preparation were reported elsewhere (Beltram et al (1991)). Samples were prepared by solution spin-coating on Si(100), Si/Pt, Si/Ti/Pt (Pt and Ti were cold sputtered on Si(100)) and MgO substrates. The number of coatings varied between 1 and 3. After each coating the film was dried at 400°C for 30 min. The samples were fired for 5 min at temperatures in the range between 500 and 850°C. Samples were examined by scanning (Jeol 840A) and transmission electron microscopy (Jeol 2000 FX, operated at 200 kV). The chemical composition of phases was determined using a Link AN-10000 EDX system with UTW detector connected to the TEM. The Cliff-Lorimer method and absorption corrections were used for quantitative analysis. Plan view and cross-section TEM specimens were prepared by mechanical thinning, dimpling and ion milling with 4 keV argon ions.

3. RESULTS AND DISCUSSION

After firing samples deposited on Si(100) monocrystal substrates at temperatures up to 500°C, the PLZT film remained amorphous. The chemical composition of the film was not changed and no reaction layer was detected. On firing these samples at temperatures above 650°C, a fine-grained pyrochlore (Py) phase in the PLZT film and an amorphous Pb-Si-O layer at the film/substrate interface were detected. In samples with Pt covered Si(100) substrates in addition to Py phase in the film perovskite (P) rosettes and a Pb-Si-O amorphous layer between the film and the substrate were found at firing temperatures above 700°C. The number and size of the perovskite rosettes increased with the firing temperature, while the size of the Py phase remained practically unchanged (10 nm). Firing the samples on Si/Ti/Pt substrates at 850°C caused the formation of a well crystallised PbTiO$_3$ layer between the Pt layer and PLZT film. Samples of PLZT film deposited on MgO(100) monocrystal, fired at 800°C, consisted only of up to 5 μm sized perovskite rosettes, surrounded by fine grained Py phase. All these results and the values of the diameter of the perovskite rosettes and pyrochlore grains are summarised in Table 1.

Tab.1. Phases present in samples using different substrates, fired at various temperatures. D(P), D(Py) - max. diameter of perovskite rosettes and pyrochlore grains. In all cases the firing time was 5 min.

substrate	firing temp.	phases	D(P)	D(Py)
Si(100)	≤ 500°C	PLZT(am.)		
	650 - 850°C	Pb-Si-O, Py		10nm
Si(100)/Pt	≤ 500°C	PLZT(am.)		
	700, 750°C	Pb-Si-O, Py, P	0.1μm	10nm
	800, 850°C	Pb-Si-O, Py, P	0.3, 2μm	10nm
Si(100)/Ti/Pt	850°C	Pb-Si-O, PT, Py, P	2μm	10nm
MgO(100)	800°C	Py, P	5μm	10nm

PLZT(am.) - amorphous film with starting PLZT composition, P - perovskite,
Pb-Si-O - amorphous layer at substrate/film interface, Py - pyrochlore,
PT - PbTiO$_3$ layer between Pt layer and PLZT film.

Rosettes were monocrystalline, preferentially oriented (textured) in the <100> direction and consisted of slightly misoriented (from 2° to 6°) subgrains. The size of the subgrains was of the order of 10 nm in the case of samples on Si/Pt substrate, fired at 800^oC, and 100 nm in the case of samples on MgO substrate, fired at the same temperature. In Fig.1 a SEM (BSE) micrograph of perovskite rosettes (a) and an SAED pattern of a rosette (b) in PLZT film on MgO substrate, fired at 800^oC for 5 min are shown. In Fig.2 an HRTEM micrograph of a perovskite rosette in a sample on Si/Pt substrate, fired at 850^oC for 5 min is shown.

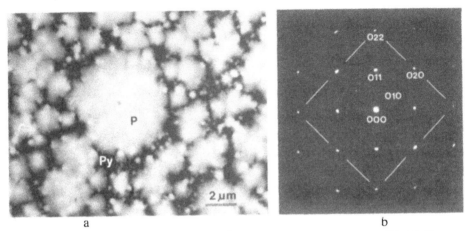

Fig.1. SEM (BSE) micrograph (a) and SAED pattern (b) of a thin PLZT film on MgO(100) substrate, fired at 800^oC for 5 min.

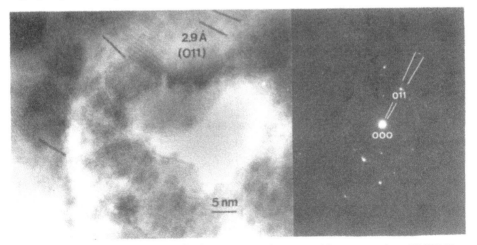

Fig.2. HRTEM micrograph and SAED pattern of a perovskite rosette in a PLZT film on Si(100)/Pt substrate, fired at 850^oC for 5 min.

The appearance of an amorphous Pb-Si-O layer and a large amount of Py phase after firing PLZT films on Si-based substrates could be explained by the reaction between PbO from the PLZT film and the SiO_2 layer at the substrate. We did not succeed in making a defect-free Pt layer on Si substrates and consequently also in samples with coated Si substrates, a reaction layer was detected. Due to defects (cracks) in the PLZT film and pores in the Pt coating, oxidation of the Si surface took place at higher firing temperature. By forming an amorphous Pb-Si-O layer the PLZT film became PbO depleted and a large fraction of Pb-deficient Py phase was formed. This pyrochlore phase also had a higher Zr /Ti ratio (70/30) compared to the starting composition (65/35). As a consequence, the Zr/Ti ratio in the perovskite phase decreased to 55/45 at a firing temperature of 850°C. In Fig.3. where an

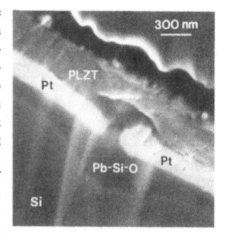

Fig.3. SEM micrograph of cross-section of the PLZT/Pt/Si(100) sample, fired at 800°C, 5 min.

SEM micrograph of the cross-section of a sample on Si/Pt substrate, fired at 800°C for 5 min is shown, a defect in the Pt layer and an amorphous Pb-Si-O layer could be noticed.

4. CONCLUSIONS

During the firing of sol-gel derived PLZT thin films on various substrates, monocrystalline (100) textured perovskite rosettes consisting of slightly misoriented subgrains, surrounded by fine-grained Pb-deficient pyrochlore grains were formed. Due to defects in the PLZT film and Pt coating, a Pb-Si-O amorphous reaction layer at the film/substrate interface was found in samples on Si-based substrates.

5. ACKNOWLEDGEMENT

Financial support from the Ministry of Science and Technology of Slovenia is gratefully acknowledged.

6. REFERENCES

Beltram T, Kosec M, Stavber S 1991 Proceedings of 2nd ECERS, Augsburg 1991
    to be published
Huffman M 1990 Thin Solid Films, 193/194, pp1017-1022
Myers S A and Chapin L N 1990 Mat. Res. Soc. Symp. Proc. Vol. 200, pp231-236
Tuttle B A, Doughty D H and Martinez S L 1991 J. Am. Ceram. Soc. 74 (6)
    pp1455-1458

Inst. Phys. Conf. Ser. No 130: Chapter 3
Paper presented at Int. Congr. X-ray Optics and Microanalysis, Manchester, 1992

319

Structure/property relationships in carbon fibres

Yanling Huang and Robert J. Young

Manchester Materials Science Centre, UMIST, Manchester M1 7HS

ABSTRACT: Raman spectroscopic studies on the deformation of carbon fibres showed that the rates of Raman shift per unit strain for both PAN and mesophase pitch based fibres, increase linearly with Young's modulus with different slopes. Both types of fibre have been examined using electron microscopy and Raman spectroscopy. The results demonstrate that there is a profound skin-core difference in PAN based fibres. The skin of the PAN based fibre has a higher modulus, which gives rise to a larger rate of Raman shift.

1. INTRODUCTION

There are two major types of carbon fibre based on their precuror materials, one is PAN based and the other is mesophase pitch based. Most of PAN based fibres have high strength, while the mesophase pitch based fibres have high modulus.

Structural heterogeneities in carbon fibres have also been reported in previous studies. They are inherited from the processing stages and vary from fibre to fibre and different precursors (1, 2). Two-phase structure has been observed (3,4) with almost perfectly oriented material close to the surface and the poorly oriented material in the centre of PAN based fibres.

Raman spectroscopy has been proved to be a very useful technique for studying the surface structure of carbon fibres (5). In this work, the differences between skin and core can be studied by this technique and compared with the results from transmission electron microscopy.

2. EXPERIMENTAL

Materials used in this study were PAN based HMS4 from Hercules with a modulus of 345GPa and an average diameter of 7.78μm and mesophase pitch based P75 from Amoco with a modulus of 517GPa and a diameter of 10.74μm.

Uncoated fracture surfaces have been examined under a field emission SEM (JSM 6300F) to show the fine detail of the transverse structure.

TEM specimens were ion beam thinned to electron transparency for observation. The selected area diffraction patterns through the whole fibre were recorded by re-positioning the selective aperture so that orientation distribution could be determined from the spread of the (002) arc.

Raman spectra of carbon fibres were obtained via a Raman microprobe system with the 488.0 nm line of an Argon ion laser. The details have been introduced elsewhere (6).

3. RESULTS AND DISCUSSION
3.1 Fibre Deformation

The effect of deformation on the Raman spectra for carbon fibres and graphites has been studied before (6, 7, 8) and it was shown that Raman bands shift under applied stress. It is found that the rate of Raman shift per unit strain increases linearly with Young's modulus for both types of fibre, but they have different slopes. This difference in slope is due to the difference in the structure of two types of fibre.

3.2 Fibre Structure

Optical micrographs of polished longitudinal sections in Figure 1 show that fibre P75 has a highly anisotropic phase along the fibre axis but, there is no difference through the fibre. In HMS4, there is a dark core surrounded by a brighter surface and the dark core occupies about one third of the whole fibre, suggesting the existence of a two-phase structure.

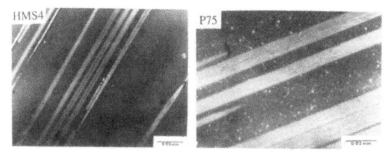

Figure 1 Optical micrographs of longitudinal sections of HMS4 and P75.

High resolution scanning electron micrographs in Figure 2 demonstrate that both HMS4 and P75 have sheet-like structures although the graphite sheets in HMS4 are not well defined. In HMS4, the graphite sheets are radially oriented in the skin and random in the core. In fibre P75, the graphite sheets in the skin region are also radially oriented, then they become circumferentially oriented and random in the core.

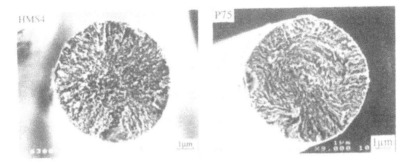

Figure 2 High resolution scanning electron micrographs showing transverse
structure of HMS4 and P75.

SAD patterns have been found to be very effective in giving a semi-quantitative indication of the degree of molecular orientation with respect to fibre axis (9). The degree of orientation can be determined from the spread of the (002) arc. The SAD patterns from skin and core regions for fibre HMS4 have been shown in Figure 3. The orientation distributions from skin to core are shown in Figure 4. The orientation angle changes from 14.7 to 17.0° for PAN based fibre HMS4. The degree of orientation in P75, however, is almost the same through skin to core, with an average value of 10.0°.

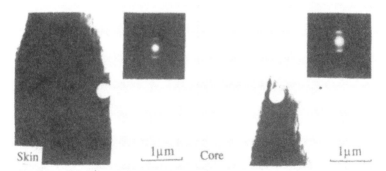

Figure 3 Transmission electron micrographs and corresponding SAD patterns
from skin and core regions for HMS4.

Raman spectra in the region 1200 to 1700 cm^{-1} have been recorded through the skin to core. The ratio of intensities of two bands, I_{1360}/I_{1580}, is related to the crystal size, L_a, (5) and the band width reflects the orientation of graphite with respect to the fibre axis (10). The differences of crystal size and orientation across the fibre can be observed from the changes in Raman spectra. Figure 5 (a) and (b) are the ratio of intensities and the band width against x/R, here x is the distance from the centre of the fibre and R is the radius of the fibre. The crystal size decreases from skin to core gradually in HMS4, while is almost the same in P75 except for the surface. The band width increases from skin to core in HMS4, whereas, in P75 the band width is almost constant throughout the fibre. The orientation distributions determined from the transmission electron microscope are in accordance with the Raman results.

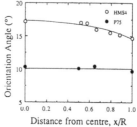

Figure 4 The orientation distribution across the fibre determined from
        SAD patterns.

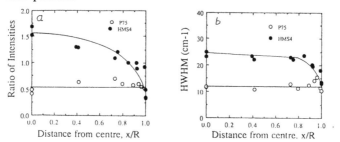

Figure 5 (a)The ratio of intensities and (b) the band width, against the distance
        from the centre of the fibre to the skin.

4. CONCLUSION

A significant skin-core effect has been observed in PAN based fibres, which tend to have more highly oriented skin than core. The skin of PAN based has a higher modulus due to higher orientation, which gives rise to a larger Raman shift rate.

5. REFERENCES
1.   S. M. Zeng, Y. Korai, J. Mochida, T. Hino and H. Toshima, *Bull.
        Chem. Soci. Jpn.*, 63, (1990) p2083
2.   J. B. Barr, S. Chwastiak, R. Didchenko, I. C. Lewis and R. T. Lewis and
        L. S. Singer, *Appl. Polym. Symp.*, No.29, (1976) p161
3   B. J. Wicks and R.A. Coyle, *J. Mater. Sci.*, 11, (1976) p376
4.   M. Guigon, A. Oberlin and Degart, *Fibre Sci. Tech.* , 20 (1984) p177
5.   F. Tuinstra and J. L. Koenig, *J. Comp. Mater.*, 4 (1970) p492
6.   I. M. Robinson, M. Zakikhani, R.J. Day, R.J. Young and C. Galiotis, *J.
        Mater. Sci., Lett.* 6 (1987) p1212
7.   H. Sakata, G. Dresselhaus and M. Endo, *Carbon Conf.* XVIII (1987) p18
8.   C. Galiotis and D. N. Batchelder, *J. Mater. Sci., Lett.*7 (1988) p545
9.   R. J. Young, D. Lu and R. J. Day, (submitted for publication)
10.   G. Katagiri, H. Ishida and A. Ishitani, *Carbon* 26 (1988) p565

Inst. Phys. Conf. Ser. No 130: Chapter 3
Paper presented at Int. Congr. X-ray Optics and Microanalysis, Manchester, 1992

323

Diffuse scattering in nickel-chromium alloys

A C Rainford, G W Lorimer, R Pilkington and A Marucco*

Manchester Materials Science Centre, Grosvenor St, Manchester M1 7HS
* Instituto per la Technologia dei Materiali Metallici non Tradizionali, del Consiglio
Nazionale delle Ricerche, Via Bassini 15, 20133 Milano

ABSTRACT: Short-range ordering (SRO) has been investigated in Inconel 182, a Ni-Cr
weld filler metal. Specimens were aged to a short-range ordered state, as indicated by
electrical resistivity measurements and electron diffraction patterns were then examined for
evidence of diffuse scattering, which has previously been associated with SRO. A sample
of pure nickel (99.99+Wt%) was also examined. It was found that both Inconel 182 and
pure nickel exhibited diffuse scattering which can be interpreted as double diffraction from
NiO rings.

1. INTRODUCTION

The identification of diffuse scattering in an electron diffraction pattern has been widely
considered to be the most reliable method of identifying short-range order (SRO) within a
material. A number of different authors have offered ideas concerning the origin of such
scattering (Clapp et al. 1969, Moss 1969, DeRidder 1975); although no one hypothesis
adequately describes the phenomenon. Inconel 182 is a nickel chromium based weld filler
metal used in the manufacture of dissimilar metal transition joints. Alloys with similar Ni:Cr
ratios to Inconel 182 have been shown to form SRO of the type Ni_2Cr, stable up to $590°C$
(Marucco et.al. 1987); the present work was carried out to determine whether Inconel 182
undergoes a similar transformation.

2. EXPERIMENTAL

Specimens of Inconel 182 (Table 1), were solution treated ($1250°C$ 1h), water quenched and aged at $500°C$ for times up to 25h. Resistivity measurements were then performed on these specimens using a comparative voltage drop technique. Samples were then examined in the transmission electron microscope (TEM). Electron diffraction patterns were taken with the beam in the $<031>$, $<112>$ and $<111>$ directions, as diffuse scattering effects had been reported to be most pronounced in these directions (Nath 1990). Pure nickel (Table 2) was examined as a comparison.

3. RESULTS AND DISCUSSION

The resistivity of Inconel 182 showed an increase with ageing time at $500°C$ (Fig.1). This is due to SRO scattering conduction electrons. The diffraction patterns examined exhibited low intensity diffuse scattering of the type seen in Figure 2. For the diffuse scattering to be maximised in the diffraction pattern, it was necessary to de-focus the beam slightly (Tanaka et al. 1986). The diffraction patterns of pure nickel examined also displayed the diffuse scattering associated with SRO (Fig.2).

The position of the diffuse intensity maxima coincided with the intersection of higher intensity double diffracted nickel oxide (NiO) rings. This can be seen when compared to the computer generated schematics (Fig.2). The oxide appears in the diffraction pattern as diffuse maxima rather than sets of distinct rings because of two simultaneous effects: at such low intensities the positions at which the rings overlap appear as indistinct points, and these points expand when the beam is slightly defocused.

It was found that the visibility of diffuse maxima increased as the thickness of the foils decreased, an effect similar to that observed by other authors (Stobbs et al 1978 and Thomas et.al.1977). This can be explained by considering the oxide-to-foil thickness ratio. In a wedge shaped foil with a surface oxide of constant thickness, the ratio of oxide to foil will rise as the foil thickness decreases. This causes a relative increase in the oxide scattering, thus the diffuse maxima appear more intense.

It is concluded on the basis of resistivity measurements that Inconel 182 forms a SRO structure when aged at 500°C. The use of diffuse electron scattering to confirm the presence of SRO in Ni-Cr alloys is difficult because of the double diffraction of NiO rings.

The authors would like to thank National Power (Dr.B.Nath) and SERC for supporting this project.

4. REFERENCES

Clapp P C and Moss S C, 1966, Phys. Rev., 171, p754.

DeRidder R, 1976, Acta Cryst., A32, p216.

Marucco A and Nath B, 1987, Proc. Conf. Phase Transformations,   ed G W Lorimer, London, p588.

Moss S C, 1969, Phys. Rev. Lett., 22, p1108.

Nath B, 1990, Private Communication.

Nordheim R, 1953-54, J. Inst. Met., 82, p440.

Stobbs W M and Chevalier J-P A A, 1978, Acta Met., 26, p233.

Tanaka N and Cowley J M, 1987, Acta Cryst., A43, p337.

Thomas G and Sinclair R, 1977, Acta Met., 25, p231.

Ni	Cr	Fe	Mn	Nb	Ti	Si	C
Balance	14.4	8.74	7.14	1.75	0.45	0.61	0.034

Table 1. The composition of Inconel 182 (Wt%) used in this study.

Ni	Fe	Si	Zr	Ti	Na	Ca
Balance	20	10	10	8	4	0.3

Table 2. The purity of the nickel used in this study (impurity levels are in ppm).

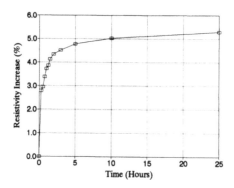

Figure 1. Resistivity as a function of temperature of In 182 aged at 500°C

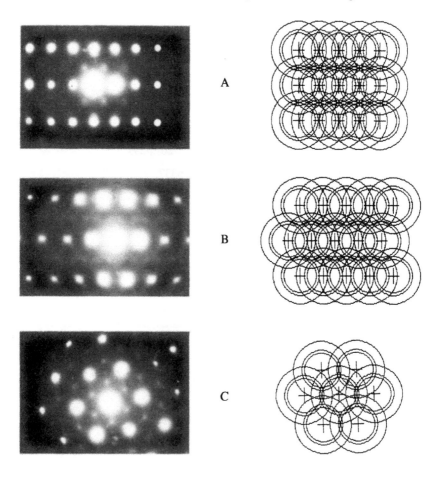

Figure 2. Electron diffraction patterns and simulated computer images of NiO rings, taken from pure nickel in the (A) <112>, (B) <031> and <111> directions.

Inst. Phys. Conf. Ser. No 130: Chapter 3
Paper presented at Int. Congr. X-ray Optics and Microanalysis, Manchester, 1992

Determination of the processing window and structural characteristics of an austempered compacted cast iron

A.L.Rimmer and R.Elliott
Manchester Materials Science Centre, University of Manchester, Grosvenor Street, Manchester M1 7HS, England.

ABSTRACT:

X-ray diffraction (XRD), transmission electron microscopy (TEM) and quantitative optical microscopy have been used to define the heat treatment processing window and identify the microstructural changes in austempered compacted cast iron (ACI). The width of the processing window gradually decreases as the austempering temperature is increased. Carbides forming in intercellular areas are identified.

INTRODUCTION.

The austempering of ductile irons (ADI) has been studied extensively over the past decade by authors including Darwish et al (1992), Dorazil (1991), Elliott (1988), Johannson (1987) and Voigt (1989). This heat treatment has created a new branch in the cast iron family. These irons are characterized by high strength and ductility. Compacted irons, in which graphite is present in a compacted flake form in contrast to fine flake in ordinary grey irons and the spheroidal form in ductile iron, have properties intermediate of the other two forms of iron. In particular, they are an attractive alternative to flake irons in applications such as cylinder heads and engine blocks. To date there have been few studies of the austempering of compacted irons (Guilemany et al (1990) and Riposan et al (1991)), and any enhancement of properties has not been documented.

Austempering is achieved by austenitizing at 850 to 950°C, rapidly cooling to an austempering temperature in the range 250 to 450°C, holding at this temperature for 1 to 4 hours followed by air cooling to room temperature. The matrix microstructural changes occur in two stages. The austenite produced during austenitizing transforms into a metastable structure of bainite ferrite and high C austenite during the first stage. Coarse plates of upper bainitic ferrite form if the austempering temperature is above 330°C, thinner plates of lower bainitic ferrite if below 330°C. The high C austenite decomposes into a stable structure of ferrite and carbide in the second stage. The high Si content of the iron delays the second stage creating a time interval between the two stages. Little microstructural change occurs in this interval and an iron air cooled from this condition shows optimum properties. Consequently it is important to define this time

interval or heat treatment processing window. The beginning and end of the
window, times t_1 and t_2, are defined as the times at which an acceptable
quantity of property damaging microstructural constituent is present. This
is 1% of untransformed austenite (UAV) in stage I. t_2 is defined as the time
at which carbide forms. This corresponds with a fall in the volume fraction
of retained high C austenite. It is determined using the definition

$$\ln t_2 = (\ln t_x + \ln t_y)/2$$

where t_x and t_y are the times at which a line drawn at 90% of the maximum in
the retained austenite volume fraction curve intersects the curve.

It is necessary to make alloying additions to increase hardenability.
However, most alloying additions segregate to areas between the graphite
particles. They may reduce properties by forming carbides or delay the
transformation kinetics so that the processing window is closed.

EXPERIMENTAL.

An unalloyed compacted cast iron of 3.5%C-2.25/2.3%Si-0.023%Mn and 1%Cu
was produced by the Sintercast Process. Small specimens were austenitized at
890°C for 90 minutes and then austempered at 400,375,350,300 and 250°C for
times between 1 and 1440 minutes. The specimens were polished to 1μm and used
for XRD. CuKα radiation was used at a voltage of 40kV and a current of 30mA.
Each specimen was scanned over the range 2θ = 40° to 90° to provide a general
picture and then slowly over the ranges 47.8° to 52.8° and 80.5° to 85.5°.
The relative intensity of austenite and ferrite peaks was measured from which
the volume fraction of retained austenite was calculated. UAV values were
measured by point counting on polished and etched microsections. A minimum
of 3000 counts was made. Two stage carbon replicas and TEM/EDAX analysis was
used to identify carbides in irons containing 0.2%Mo and 0.2%Mn.

EXPERIMENTAL RESULTS AND DISCUSSION.

Figure 1 shows a typical x-ray scan for an ACI.

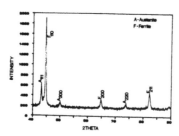

Figure 2 shows how the volume fraction of retained austenite (VRA) varies with time at the different austempering temperatures. As the temperature is reduced the maximum VRA decreases and the time t_y increases. Consequently, the time t_2 is increased and stage II is delayed. Carbide precipitation occurs during stage I in the lower bainite structure leading to a marked reduction in the VRA at low austempering temperatures.

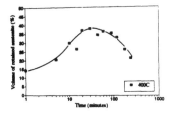

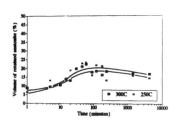

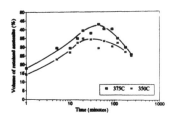

Figure 2: Variation in VRA with time and austempering temperatures.

Figure 3 shows the variation in UAV for the corresponding austempering times and temperatures. Decreasing the austempering temperature results in a greater driving force for the stage I process. Hence the time t_1 decreases.

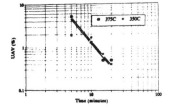

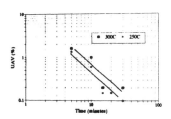

Figure 3: Variation in the UAV with time and austempering temperatures.

These relationships are shown in figure 4 which shows how the width of the processing window decreases with increasing austempering temperature until it closes at approximately 400°C. Figure 5 identifies a Mo rich preciptate, and figure 6 one rich in Mn. Ti and Cr rich precipitates have also been identified in areas between graphite particles in alloyed irons.

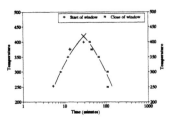

Figure 4: Variation in the processing
window with temperature.

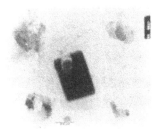

Figure 5:   Figure 6:

CONCLUSIONS.
   1. Reducing the austempering temperature increases the window width.
   2. Various carbides have been identified when the iron is lightly alloyed.

REFERENCES.
   1. Darwish N and Elliott R to be published.
   2. Dorazil E 1991 High Strength Austempered Ductile Cast Iron (Ellis Horwood Ltd).
   3. Elliott R 1988 Cast Iron Technology (Butterworths).
   4. Guilemany J.M and and Llorca-Isern N 1990 Mat. Sci and Eng. A130 pp 241-6.
   5. Johannson M 1987 AFS Trans. Vol 85 pp 117-22.
   6. Riposan I and Chisamera M 1991 Austempered Compacted Graphite Cast Iron Polytechnic Institute of Bucharest.
   7. Voigt R.C 1989 Cast Metals 2 pp.71-93.

Inst. Phys. Conf. Ser. No 130: Chapter 3
Paper presented at Int. Congr. X-ray Optics and Microanalysis, Manchester, 1992

331

Electron microscopy and XRD analysis of austempered and plasma nitrided S. G. Iron

S Korichi[*] and R Priestner[+]

[*]University of Constantine, Institute of Physics, Algeria.
[+]University of Manchester/UMIST Materials Science Centre, Grosvenor St., Manchester M1 7HS

ABSRACT: The austempering of S.G. iron and its subsequent plasma nitriding was investigated using electron microscopy supplemented with quantitative XRD analysis. The optimally austempered microstructure of bainitic ferrite and high-carbon retained austenite yielded the shallowest nitrided compound layer, because layer growth was impeded by coarse carbide particles precipitated from the austenite at the nitriding temperature. The finer dispersions of carbide in under- and over-austempered iron were more readily nitrided and subsumed into the ϵ/γ' compound layer.

1. INTRODUCTION

The ideal microstructure of austempered S.G. iron is a mixture of low-carbon ferrite and high-carbon austenite phases, brought about by an austempering heat treatment cycle such as shown schematically in Fig.1. During isothermal austempering a first bainitic ferrite forms by rejecting carbon into the austenite with very little precipitation of carbides. C-enrichment stabilises the retained austenite against the martensite transformation on subsequent cooling. At longer austempering times a second stage of bainite formation is accompanied by the precipitation of carbide. In normal processing compositional heterogeneities are inherited from segregation during solidification, and high-carbon martensite and carbides coexist with ferrite and austenite.

Austempered S.G. castings are frequently surface-engineered by plasma nitriding, in order to improve corrosion resistance and to reduce friction and wear. However, the austempered microstructure is unstable, and the constitutions of nitrided layers are sensitive to the way in which it decomposes at the nitriding temperature.

Electron microscopy was found to be useful for investigating morphological changes in the matrix during austempering and in the matrix and the compound surface layers during plasma nitriding. However, significant information on phase compositions and volume fractions was unobtainable by electron optical methods. In this paper, therefore, we show how electron microscopical information was supplemented with quantitative XRD analysis in order to reveal details of the role of the austempered microstructure in determining the character of plasma nitrided layers.

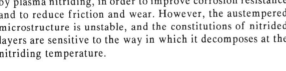

Fig.1. Schematic temperature cycles for austempering and plasma nitriding.

2. EXPERIMENTAL DETAILS

The composition of the cast iron was; 4.05%C, 2.1%Si, 0.19%Mn, 0.01%P, 1.14%Ni. The complete range of heat treatments investigated are shown schematically in Fig.1; those referred to in this paper are represented by bold lines. Austenitising times were 1/2hour, during which the carbon content equilibrated with respect to the graphite. Austempering times were from 15s to 120 hours, but only selected results are presented here. For austempering times from approx 15-120 minutes, austempering at 300°C produced an under-austempered microstructure of ferrite, austenite and high-carbon martensite, at 380°C an optimum microstructure of ferrite and retained austenite, and at 440°C the over-austempered austenite is completely decomposed to a mixture of first and second stage bainites. Plasma nitriding was done at 570°C for 4 hours in a 50/50 N_2/H_2 atmosphere. Scanning electron micrographs were obtained from metallographically polished sections etched in nital. Two-stage plastic-carbon replicas were also examined in the TEM; only the finer fractions of the carbide particles that were present were extracted. XRD spectra were obtained from plane polished sections using a focussing spectrometer and either Cu or Cr targets. Carbon contents of austenite were derived from its lattice parameter (Roberts) after correction for the effect of silicon (Ruhl and Cohen). Volume fractions were determined from relative integrated peak intensities.

RESULTS AND DISCUSSION

Fig.2 illustrates the morphology of first stage bainite formed at 300°C. At optical magnifications the bainite is in the form of dark-etching needles which partition the austenite into small volumes. At high magnification the bainite needles are seen to consist of ferrite particles interleaved by austenite. XRD showed the austenite present to be high in carbon and, therefore, stable with respect to transformation to martensite. However, some martensite was also present. The first stage of transformation to bainite was complete at two hours, and the second stage had not started at 24 hours.

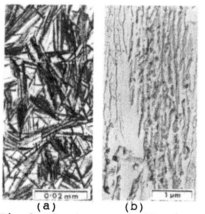

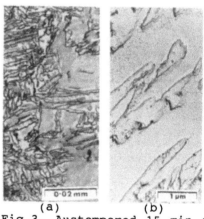

(a)   (b)
Fig.2. Austempered 15 min at 300°C, (a) optical (b) TEM-replica.

(a)   (b)
Fig.3. Austempered 15 min at 380°C, (a) optical (b) TEM-replica.

At 380°C, Fig.3, the bainitic ferrite is much coarser and the austenite phase is less finely partitioned and more continuous. This microstructure is the ideal one, conferring high strength and toughness; it remains stable at 380°C for two hours, when the slower transformation to second stage, carbide-containing bainite begins. At 440°C the first stage bainite is yet coarser, and the second stage bainite transformation is virtually complete at two hours.

Fig.4 illustrates the shift in the $(220)_\gamma$ peak due to progressive enrichment of the austenite with carbon during austempering. Fig.5 shows the strong carbon enrichment of austenite at 380°C, to a plateau at 2 hours. The retained austenite increases to peak value at 1-2 hours; at shorter times its carbon content is still low enough for part of it to transform to martensite on cooling, beyond the peak the second stage bainite forms with the precipitation of bainitic carbide.

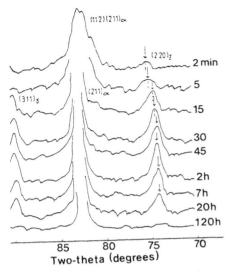

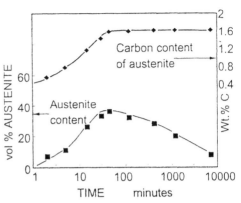

Fig.4. XRD spectra after austempering at 380°C.

Fig.5. C-enrichment of retained austenite at 380°C

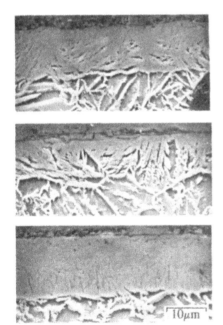

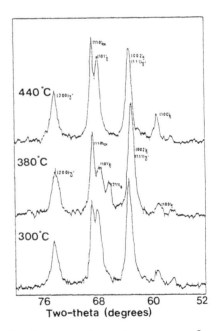

Fig.6. Compound layers formed on material austempered 2hrs at (top)300°C (middle)380°C (bottom)440°C

Fig.7. XRD spectra from compound layers in Fig.6.

During heating to the plasma nitriding temperature of 570°C these base microstructures temper completely, any martensite and retained austenite decomposing to ferrite and cementitic carbide. A Fe(NC) compound layer also forms on the surface, from nitrogen in the plasma and internal carbon present as carbide. Fig.6 illustrates compound layers formed under identical conditions on characteristically under- and over-austempered iron and from the ideal microstructure. It is plain that the correctly austempered microstructure produced the least satisfactory (shallowest) compound layer. Fig.7 illustrates XRD spectra from these compound layers. The main constituents of all the layers were ε and γ' nitrocarbides. However, cementite (Θ-phase) was also present in the layer formed on the correctly austempered microstructure, but absent from the others.

Fig.8 shows the distribution of carbides in replicas taken from specimens austempered at 300°C and 380°C and then tempered at 570°C. Carbides are both replicated and extracted in these pictures. The under-austempered microstructure (300°C) yielded a much finer and more uniform distribution of carbides than the correctly austempered microstructure. This difference in tempered carbide morphology results directly from the different fineness of the original microstructures. The austenite present after austempering at 300°C was finely divided by bainite needles and contained some high-C martensite, whereas that at 380°C was coarser, more connected, and virtually free of martensite (Figs.2 and 3). The latter resulted in the coarse carbides on tempering at the nitriding temperature, which inhibited the growth of the ε/γ' compound layer. It is proposed that the formation of the compound layer requires nitrogen to diffuse into cementite particles and convert them to C-rich ε; the excess carbon released from those particles then diffuses to the ferrite/compound layer interface to form N-rich

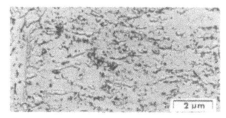

Fig.8. Carbide distributions after 2hrs at 570°C following austempering at (a)300°C and (b)380°C.

γ'. Shorter diffusion paths due to the presence of fine carbide particles should then enhance the kinetics of layer formation. On the other hand, large carbide particles and large interparticle spacing should inhibit layer formation. Large carbide particles may be expected to persist in the compound layer for some time before being converted to ε, as was observed. Over austempering to complete the second stage of the bainite reaction also produced a finer distribution of bainitic carbides prior to nitriding, and resulted in an improved compound layer, Fig.6.

CONCLUSIONS

Morphological data obtained from electron microscopy were supplemented by quantitative XRD data to show that in properly austempered S.G. iron the retained austenite was relatively coarse and highly enriched in carbon rejected from first-stage bainitic ferrite. At the nitriding temperature this austenite tempered to coarse cementite and ferrite, and the compound layer was unsatisfactory because the large cementite particles inhibited its growth. Lower and higher austempering temperatures yielded finer dispersions of carbide on tempering, which resulted in thicker and more uniform compound layers after nitriding.

ACKNOWLEDGEMENTS

The authors are grateful to BCIRA for providing the material, and to P. Kenway, G. Cliff and I. Brough for technical assistance.

REFERENCES

C.S.Roberts; J.Metals, 1953, vol.5, p.203.
R.C.Ruhl and M.Cohen; 1969, Trans. Met. Soc. AIME, vol. 245, p. 241.

Inst. Phys. Conf. Ser. No 130: Chapter 3
Paper presented at Int. Congr. X-ray Optics and Microanalysis, Manchester, 1992

335

Silicide formation in Mo thin films and Si single-crystal interactions

A. BOUABELLOU, R. HALIMI and A. BECHIRI

Unite de Recherche de Physique des Materiaux et Applications, Universite de Constantine, Ain El Bey, 25000, Algerie

ABSTRACT: Silicide formation resulting from molybdenum thin films and silicon single-crystal interactions is investigated by x-ray diffraction (XRD) and transmission electron microscopy (TEM). Samples of Mo-Si configurations are prepared by e-gun deposition of Mo (1500 A) onto Si wafers. The samples are annealed in vacuum at temperatures in the range 500-1000°C. It is established that the tetragonal disilicide phase $MoSi_2$ forms and grows in the considered temperature range. However, the TEM detects two silicide phases: tetragonal $MoSi_2$ and Mo_5Si_3.

1. INTRODUCTION

The formation of a silicide film by the thermal treatment of a thin metal film on a silicon substrate is extensively studied because of its technological importance and scientific interest (Murarka 1983).

Molybdenum silicide is one refractory metal silicide which is being actively investigated as highly conductive gate and interconnection material for applications in the semiconductor industry (Stearns et al 1990 and Slaughter et al 1991).

In the present work we studied the formation of molybdenum silicides by conventional thermal annealing of 1500 A of Mo on Si substrate in the temperture range of 500-1000°C. Both the structure and the phase identification of these silicides were investigated by use of XRD and TEM.

2. EXPERIMENTAL PROCEDURE

Single-crystal Si (100) and Si(111) wafers (p-type) with resistivities of 60-80 Ω.cm and 5-8 Ω.cm respectively were used as substrates. Half of the samples were implanted with antimony at 130 keV to doses of 5×10^{14} and 5×10^{15} atoms/cm^2 while the rest were not implanted. The implantation damage was removed by annealing the wafers for 30 min at 900°C. Si wafers were first cleaned using standard chemical procedures. In order to eliminate as far as possible the residual layer of silicon oxide, the samples were etched in buffered HF solution (HF = 1 : 50) for 2 min before loading into an electron gun deposition system. Molybdenum thin films, 1500 A in thickness, were deposited onto Si substrates maintained at 100°C. The depositions were performed at a rate of 2 A/s in vacuum better than 5x10⁻⁶ Torr. The samples were annealed in vacuum in a quartz tube furnace in the temperature range of 500-1000°C.

X-ray diffractometry was done on a Siemens D500 automated diffractometer using CuKα radiation and nickel filter. Transmission electron microscopy was done on a Philips EM300 at 100 keV.

3. RESULTS AND DISCUSSION

In the temperature range of 500-1000°C the Mo/Si phase diagram (Hawkins et al 1973) indicates three intermetallic compounds may be formed: Mo_3Si, Mo_5Si_3, and $MoSi_2$. The disilicide ($MoSi_2$) occurs in two different crystalline forms, tetragonal (t-$MoSi_2$) and hexagonal (h-$MoSi_2$).

Figue 1 gives the two typical diffractograms of the unannealed Mo/Si(100) and Mo/Si(111) systems. In these samples, the x-ray diffraction spectra show the characteristic peaks of polycrystalline Mo films, Si(100) (Fig. 1a) and Si(111) (Fig. 1b) substrates. In addition to observed reflections corresponding to Mo and Si constituents, the x-ray diffraction pattern indicates the presence of MoO_3 peaks. The 1500 A thick Mo films deposited at 100°C on Si(100) and Si(111) substrates possess a polycrystalline disorientated structure. The important feature revealed in

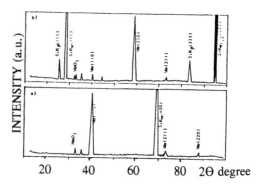

Fig. 1. X-ray diffraction spectra from a as-deposited 1500 A Mo film on single crystal Si(100) (a) and Si(111) (b) substrates.

figure 1 is the prominent (200) peak in the Mo/Si(111) diffractogram. However, according to ASTM powder diffraction data card No . 4-0809, the strong peak which would be expected is Mo(110). This means that the grains of the Mo films deposited onto Si(111) are preferentially oriented.

In figure 2, x-ray diffraction patterns of Mo/Si(100) samples annealed in vacuum at 650, 800, and 900°C are shown. Analysis of these patterns reveals unambiguously the appearance only of polycrystalline tetragonal disilicide $MoSi_2$. Then the interaction of polycrystalline Mo film with Si substrate leads to the formation and the growth of t-MoSi$_2$ phase, and there is no direct evidence on the presence of other silicides. The formation of t-MoSi$_2$ phase in the 500-1000°C range is in agreement with the published studies of most recent experiments (Van Ommen et al 1988 and Doland et al 1990). It is important to note that MoO$_3$ peaks disappear after annealing at 800-1000°C. This fact suggests probably the volatilization of MoO$_3$ at these temperatures. This result agrees with that reported by Ho et al (1992).

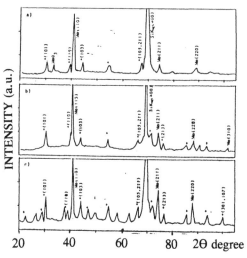

Fig. 2. X-ray diffraction spectra from annealed Mo/Si(100) samples at 650°C for 1 hr (a), 800°C for 30 min (b) and 900°C for 30 min (c). (*: t-MoSi$_2$)

Silicide phases formed at the Mo-Si interface were also investigated by transmission electron microscopy.

TEM microdiffraction analysis reveals coexistence of both t-MoSi$_2$ and Mo$_5$Si$_3$ silicide compounds. Thus the Mo$_5$Si$_3$ phase, not seen in x-ray diffraction spectra, is detected. Mixed Mo$_5$Si$_3$ and t-MoSi$_2$ silicide phases are detected in the considered temperature range independently of silicon crystal orientations and Sb-implanted doses. However, it is observed that the reaction is retarded when the dose implantation increases. Figure 3 shows electron diffraction patterns of Mo/Si(100) samples as-deposited and annealed at 500°C, and 650°C for 1 hour. For the unannealed sample, only electron diffraction rings characteristic of polycrystalline Mo appear (Fig. 3a). Annealing of samples leads to the formation of t-MoSi$_2$ and Mo$_5$Si$_3$ silicide phases as shown in Fig. 3b and 3c.

The coexistence of Mo$_3$Si, Mo$_5$Si$_3$ and MoSi$_2$ silicide phases in the considered temperature range has not been previously reported. Several studies have observed the formation of MoSi$_2$ and Mo$_3$Si (Guivarc'h et al 1978), MoSi$_2$ and Mo$_5$Si$_3$ (Stearns et al 1988) or MoSi$_2$ (Oertel et al 1976).

X-ray and electron diffraction methods confirm that a transformation of Mo film to tetragonal MoSi$_2$ and Mo$_5$Si$_3$ silicides is obtained at temperatures higher than 500°C. But, an amount of Mo is always remained even for a long period of anneal at 1000°C. This fact indicates that, for these deposition conditions, the Mo layer has not been completely transformed into polycrystalline mixtures of the two detected silicides Mo$_5$Si$_3$ and t-MoSi$_2$. This suggests that during annealing at temperature required for silicide formation oxygen, probably incorporated initially in the deposited metal layer, is expected to move towards the silicon interface and block any further reaction. The influence of oxygen on Mo silicides formation has been reported in previous studies (Ottaviani 1979,

Wakita et al 1984 and Bomchil et al 1986).

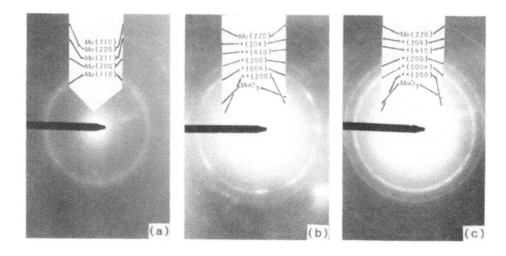

Fig. 3. Electron diffraction pattern (100 keV) of a 1500 A Mo/Si(100) samples: before annealing (a), and annealed for 1 hr at 500°C (b) and 650°C (c). (*: t-MoSi$_2$, **: Mo$_5$Si$_3$)

REFERENCES

Bomchil G, Goeltz G and Torres J 1986 Thin Solid Films 140 59
Doland C M and Nemanich R J 1990 J. Mater. Res. 5 2854
Guivarc'h A, Auvray P, Berthou L, Le Cun M, Boulet J P, Henoc P, Pelous G and Martinez A 1978 J. Appl. Phys. 49 233
Hawkins D T and Hultgren R 1973 (in Metals Handbook 8th ed (ASM, Metals Park, OH) 8 pp 321- 368
Ho C H, Prakash S, Doerr H J, Deshpandey C V and Bunshah R F 1992 Thin Solid Films 207 294
Murarka S P 1983 Silicides for VLSI Applications (New York: Academic Press)
Oertel B and Sperling R 1976 Thin Solid Films 37 185
Ottaviani G 1979 J. Vac. Sci. Technol. 16 1112
Slaughter J M, Shapiro A, Kearney P A and Falco C M 1991 Phys. Rev. B 44 3854
Stearns D G, Stearns M B, Cheng Y, Stith J H and Ceglio N M 1990 J. Appl. Phys. 67 2415
Van Ommen A H, Reader A H and de Vries J W C 1988 J. Appl. Phys. 64 3574
Wakita A S, Sigmon T W and Gibbons J F 1984 Appl. Phys. Lett. 45 140

Inst. Phys. Conf. Ser. No 130: Chapter 4
Paper presented at Int. Congr. X-ray Optics and Microanalysis, Manchester, 1992

Imaging surfaces with scanning tunnelling and scanning force microscopes

H J Butt

Max-Planck-Institut für Biophysik, Kennedyallee 70, 6000 Frankfurt a.M. 70, Germany

ABSTRACT: Scanning tunnelling and scanning force microscopes can provide atomic-resolution images of samples in ultra-high vacuum, air and liquids including water. This makes it an interesting tool for science and industry. After giving examples of applications, technique and theory of the scanning tunnelling and the scanning force microscope are discussed.

1. INTRODUCTION

The invention of the scanning tunnelling microscope (STM) in 1982 by Nobel Prize winners Gerd Binnig and Heinrich Rohrer (Binnig et al. 1982) has led to the invention of a host of other scanning probe microscopes. The scanning force microscope (SFM), invented by Binnig, Quate and Gerber in 1986 is one of the most successful of these new devices. These new scanning probe microscopes provide complementary structural information not accessible by other methods such as electron microscopy and X-ray diffraction. Scanning probe micrographs are generated by moving a nanometre sized probe laterally over the surface and writing the signal to a storage device. Depending on the sensor and the mode of operation, micrographs provide information of the electronic structure, the topography and mechanical properties of the sample.

Excellent reviews and monographs about scanning probe microscopy have appeared (Wickramasinghe 1989, Rugar and Hansma 1990, Hamann and Hietschold 1991, Güntherodt and Wiesendanger 1992). This paper focuses on the STM and the SFM - the two most important scanning probe microscopes. After giving a few examples of applications, instrument design and theoretical background are described.

2. THE SCANNING TUNNELLING MICROSCOPE

To be imaged with an STM, the sample must be a conductor, a semiconductor or a thin ($<$ 1 nm) molecule adsorbed to a conducting surface. This is the main limitation in scanning tunnelling microscopy. Within this limit the STM is an unsurpassed tool to image the topography and electronic structure of surfaces. STMs operating in UHV have yielded detailed information on semiconductor surface reconstruction (Binnig et al. 1983, Tromp et al. 1986, Feenstra et al. 1987), metal surface reconstructions (Kuk 1992), the condensation of metals on semiconductors (Brodde et al. 1991) and organic (Rabe 1992) and inorganic (Dujardin et al. 1992, Winterlin and Behm 1992) adsorbates on conducting surfaces. It turned out that STMs can operate not only in UHV, but also in air or liquids (Liu et al. 1986, Sonnenfeld and Hansma 1986). Hence, it is possible to investigate electrochemical processes (Cataldi et al. 1990). The combination of high resolution without the necessity of UHV made the STM an interesting tool to study biological objects (Engel 1991, Guckenberger et al. 1992). Besides the ability to image surfaces, STMs were used to manipulate surfaces in the nanometre range. Eigler and Schweizer (1990) could even position single xenon atoms on a nickel surface.

2.1 Design

The principles and operation of an STM are surprisingly simple (Figure 1). The conducting sample is brought close enough to a sharp tip that, at a convenient voltage (2 mV - 2 V), a tunnelling current (typically 0.1-10 nA) is measurable. Then the sample is raster-scanned underneath the tip in x,y direction, while the tunnelling current I is sensed. A feedback network changes the height z of the sample to keep the current constant. The tunnel current is sensitive to the distance between sample and tip. Hence, keeping the tunnel current constant also the distance remains nearly constant. An image of the sample is obtained by displaying the height z versus the lateral position x, y of the sample.

Figure 1.
Schematics of an
STM and an SFM
with beam deflection.

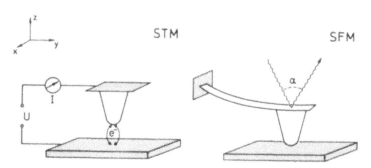

To move the sample it is mounted to an xyz translator, which is usually a single-tube piezoelectric crystal. By applying appropriate voltages to the piezo, the sample can be moved in three dimensions. Scan ranges are up to 100 μm laterally and 10 μm vertically.

Mainly two types of tips are used: Pt-Ir (8:2) wires which have been cut at an oblique angle or tungsten tips which have been electrochemically etched in NaOH or KOH. These tips often produce atomic resolution images with no further treatment. Electron microscope studies of STM tips showed that at the end these tips have a radius of curvature of about 10 nm - 100 nm (Garnaes et al. 1990). This indicates that there was at least one atomically sharp tip projecting out from the (relatively dull) end. Often the tips are sharpened in situ by applying a relatively high voltage (typically 5V) for few microseconds or even by collision with sample. Tips suitable for imaging in electrolyte solutions are glass- or epoxy-coated Pt-Ir wires or are made from etched metal wires coated with wax. While scanning the tip sometimes changes its shape. Particularly in air tip alterations are likely to be the rule rather than the exception.

The operation mode described above is called the *constant current mode* because the current is kept constant while scanning. Alternatively, in the *constant height mode* the sample is scanned at nearly constant height while the current is monitored. In this case the feedback circuit responds slowly and only keeps the average current constant. An image displays the current I versus x, y. In the constant current mode image interpretation is simpler than in the constant height mode. The advantage of the constant height mode is a higher possible scan speed, which is important when processes are imaged.

2.2 Image Interpretation

Early analysis of the tunnelling current in STMs was based on the analogy with the well known one-dimensional tunnelling problem. The situation is illustrated in figure 2a for two pieces of a metal, separated in vacuum by a gap d. At zero temperature electrons occupy states up to the Fermi energy E_F, and states above the Fermi energy are empty. If an external voltage U is applied, electrons tunnel from occupied states of the tip (left) to empty states in the sample (right). With this simple one-dimensional model the influence of important parameters can already be understood. For low voltages an electron coming from the left encounters a rectangular potential barrier with a height given by the work function of the metal Φ. The transmission coefficient for this process, and hence the tunnelling current I, is proportional to

$$ exp\left(-\frac{d}{\hbar}\sqrt{8m\Phi}\right), \tag{1} $$

where m is the electron mass. For a typical work function of 5 eV the tunnelling current decreases tenfold when the gap is increased by only 0.1 nm. This extreme sensitivity of the tunnelling current is the basis of the STM.

Figure 2. (a) tunnelling between a metallic tip and a metallic sample. If an external voltage is applied electrons can tunnel from the tip (left) into empty states above the Fermi level E_F of the sample. (b) tunnelling between a metal tip and a semiconductor. The voltage must be high enough that electrons can either tunnel from the tip into empty states of the conduction band of the sample (hatched) at positive sample potential, or that electrons can tunnel from the valence band of the semiconductor (dotted) into empty states above the Fermi level of the tip at negative sample potentials. For low voltages tunnelling is impossible.

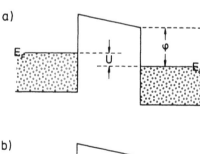

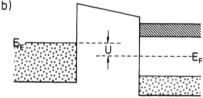

With the first atomic resolution images it became clear that a more thorough understanding of the tunnelling process was needed. Therefore the full three-dimensional tunnelling problem was considered by many researchers and various methods. The most widely used theory of STM is that of Tersoff and Hamann (1983, 1985), based on Bardeen's (1961) first-order perturbation approach for the tunnelling current between weakly coupled electrodes. Tersoff and Hamann found that for low voltages the tunnel current is proportional to the *sample local density of states at the Fermi level*. Under certain reasonable assumptions, the STM image would even be similar to the total charge density. Their finding was confirmed by Lang (1986) who calculated the tunnelling current between two planar metal electrodes, on each of which is an adsorbed atom.

Images of noble metals predicted with Tersoff and Hamann's theory agreed nicely with experimental results. Their theory also helped to interpret STM images of semiconductor surfaces like the Si(111) 7 × 7 reconstruction. Another striking example is the GaAs(110) surface. In GaAs(110) occupied states are concentrated preferentially on the As atoms and empty states on the Ga atoms (Tersoff and Hamann 1985). Thus, it was theoretically predicted that STM images of the occupied states should reveal the position of the As atoms and images of empty states should reveal the position of the Ga atoms. This prediction was experimentally confirmed by Feenstra et al. (1987). At a sample voltage of +1.9 V electrons tunnel from the tip into empty states of the Ga atoms, at -1.9 V electrons tunnel from occupied states of the As atoms into the tip.

To obtain more detailed information about the electronic structure of a surface, spatially resolved tunnelling spectroscopy can be done (Tromp et al. 1986). In this approach, one varies the tunnelling voltage at a fixed height in order to obtain current-voltage spectra from specific points on the surface. Peaks in the derivatives of these spectra correspond to energies of high local density of surface states or resonances. Alternatively, images may be taken at specific voltages to highlight the distribution of particular states.

Despite the success of Tersoff and Hamann's theory on GaAs(110) and Si(111) in many cases of semi-metallic or semiconducting surfaces STM images deviated from the local density of states; one such case is graphite. Some of these discrepancies could still be accounted for in the framework of Bardeen's approximation by a rigourous treatment of the tip influence (Chen 1990, Tersoff 1990). Recently, however, some authors have stressed the need to go beyond Bardeen's first-order perturbation approach (Doyen et al. 1988, Chen 1991, Sacks and Noguera 1991) because the assumption of sufficiently spaced, nearly independent electrodes is often not valid: Direct measurements of the force between tip and sample under normal STM operation yielded values between 10 nN and 1 μN (Düring et al. 1988, Mate et al. 1989), which can not be considered weak coupling. Furthermore, Kuk and Silverman (1990) showed that for distances smaller than 0.5 nm the local barrier height is drastically reduced. Tekman and Ciraci (1991) studied the effects of tip-sample-interaction based on an empirical tight-binding approximation. They distinguished three regimes. For large distances between tip and sample coupling is weak and Tersoff and Hamann's theory can be applied. For intermediate distances, and this is the normal tunnelling case, the tip influences the electronic structure of the sample. In this case *tip-induced localized states* rather than the *sample local density of states* is imaged. Contrast can be significantly enhanced due to these tip-induced local states. For small distances chemical bonding between sample and tip atoms becomes important. Ciraci et al. (1990) calculated that for graphite scanned by a single-atom aluminium tip at a typical distance of 0.4 nm, binding energies are between 0.33 eV and 0.61 eV, depending on the exact location of the aluminium tip. These calculations show that generally no simple interpretation of STM images can be given.

3. THE SCANNING FORCE MICROSCOPE

With the SFM conducting and insulating samples can be imaged. Albrecht and Quate (1987) were the first who imaged an insulator, boron nitride, at atomic resolution. In addition, SFMs can be operated in vacuum, air or liquids including water. This opened a wide range of applications. Polymers (Magonov and Cantow 1992) and biological objects (Engel 1991, Guckenberger et al. 1992) have been studied. Electrochemical processes (Chen et al. 1992) or the dissolution of crystals like quartz and calcite could be imaged with atomic resolution (Gratz et al. 1991, Hillner et al. 1992). Barrett and Quate (1991) were able to deposit and read electric charges on a nitride-oxide-silicon system with a modified SFM. This stored charge can be interpreted as a digital memory. Barrett and Quate could read and write bit sizes as small as 100 nm at a rate of 100 kHz.

3.1 Design

In the SFM (figure 1) the sample is scanned by a tip which is mounted to a cantilever spring. In operation the tip actually touches the sample surface, much like the stylus of a record player. For the SFM, however, the tip is sharper and the tracking force is smaller: typical forces are 0.1 nN to 100 nN. To obtain these small forces the spring constant of cantilevers used must be small. Typical spring constants are 1 N/m. Hence, if an atom while being scanned deflects such a cantilever by 0.1 nm, a force of only 0.1 nN is required. This is low compared with binding forces of atoms.

Like in the STM the sample is scanned by a piezoelectric translator. While scanning the deflection of the cantilever is sensed and a feedback loop keeps the deflection, and hence the force, constant. A topographic image of the sample is obtained by plotting the height of the sample versus x and y. To measure the deflection of the cantilever two optical methods are currently used: interferometry and beam deflection. Both methods are capable of measuring deflections on the order of 0.01 nm. In a typical beam-deflection SFM, light from a laser diode is focused onto the back of a mirror like cantilever surface. The beam is reflected and the direction of the reflected light is sensed by a position sensitive photodetector. Beam deflection is usually easier to implement than the interferometric system. For this reason commercial SFMs use the beam deflection technique.

Tip and cantilever are microfabricated and made of silicon nitride, silicon oxide or pure silicon (Akamine et al. 1990, Albrecht et al. 1990, Buser et al. 1991, Wolter et al. 1991). When beam deflection is used the back of the cantilevers is coated with gold to get a high reflection for light. Beside the small spring constant cantilevers should have a high resonant frequency. The higher the resonant frequency, the less sensitive the cantilever is to external vibrations. Typical resonant frequencies are 10-100 kHz. Tips should have a small radius of curvature at the end to obtain a good lateral resolution. A typical radius of curvature is 20 nm, but new fabrication methods promise smaller radii (e.g. Keller and Chung 1992, Kado et al. 1992).

3.2 Image Interpretation

The SFM is based on the repulsive force which occurs when electron orbitals of tip atoms overlap with orbitals of sample atoms. This short range repulsion is caused by Pauli's exclusion principle for electrons. The strong distance dependence of this ionic force has made atomic resolution possible.

To understand atomic-resolution images microscopic calculations were done. There the sample is supposed to consist of periodically arranged atoms which are scanned by a single-atom (or few-atom) tip. With microscopic calculations experimental images of graphite could be reproduced (Gould et al. 1989, Ciraci et al. 1990). Asymmetric images were explained by tips ending in more than one atom. The maximum allowed force was calculated to be in the order of 1 nN for graphite or silicon samples (Abraham et al. 1988, Abraham and Batra 1989, Landman et al. 1989, Zhong et al. 1991). Imaging with a force above 1 nN should lead to rearrangements of tip and sample atoms. However, in most experiments on crystal surfaces much higher forces were used. This indicates that normally the effective contact area is larger than a single atom. Another observation points in the same direction. Usually only perfect two-dimensional crystals were observed with atomic resolution (Albrecht et al. 1987, Binnig et al. 1987, Manne et al. 1990). In contrast to the STM, atomic defects or single atoms adsorbed to a substrate were rarely imaged with the SFM. While it is plausible that a periodic corrugation can be reproduced with a tip consisting of ten or more atoms at the extreme end, defects should be smoothed out. Indeed, in order to image steps and defects (and not only a two-dimensional lattice) with atomic resolution, Giessibl and Binnig (1992) found that on KBr(001) the force had to be kept below a few nanonewtons.

In contrast to the microscopic approach, which helped to understand atomic resolution, continuum theory was used to understand which long range forces act between tip and sample. One such interaction is the attractive van der Waals (vdW) force. The vdW force depends on the medium between tip and sample (Hartmann 1991). In air or vacuum it is typically 1 nN. It can be reduced by imaging in water or ethanol to 0.1 nN or even 0.01 nN, respectively. The vdW force decays like $1/d^n$, with n between 1 or 2, depending on the tip shape. Hence, if the distance is doubled the vdW force decreases by a factor 2 to 4. This relatively weak decay makes it a long range force. An implication is that more than the few atoms at the extreme end of the tip contribute to the vdW force. Also those parts of the tip which are not directly in contact with the sample contribute to the total force. This has an important practical consequence: the long range attraction has to be compensated by the few atoms at the extreme end of the tip and the corresponding atoms in the sample. If the force is high, the tip or the sample might be destroyed.

When imaging in air the *meniscus* or *capillary force* is present. This attractive force is due to capillary condensation of water around the contact sites between tip and sample. The Laplace pressure in the water pulls the tip down onto the sample. The meniscus force is typically 10-100 nN. Fortunately, it is easy to eliminate capillary forces by imaging in an aqueous medium (Weisenhorn et al. 1989). For this reason many SFM studies have been done in electrolyte solutions, which is the natural medium for many applications like imaging electrochemical processes or biological objects. In an electrolyte other forces such as the electrostatic and the hydration forces may contribute significantly to the total force (Ducker et al. 1991, Butt 1991).

Experimentally, forces in scanning force microscopy are 10-100 nN in air, around 1 nN in UHV and few 0.1 nN in water. In liquids like ethanol the force can further be reduced down to the order of 0.01 nN due to the low vdW interaction.

Forces between tip and sample have been studied mainly for two reasons. First, knowledge about the different components is important to minimize the total force and thus prevent a possible deformation or even destruction of the sample. Second, the force might be used to obtain additional information about local surface properties. SFMs with magnetic tips have been used to image magnetization patterns on storage media (Martin and Wickramasinghe 1987, Mamin et al. 1988). Terris et al. (1990) were able to create and image electric charges with a nickel tip on insulators. Local surface charge densities of biological samples could be measured in electrolyte solutions (Butt 1992).

Figure 3. For a periodically corrugated sample (of wavelength λ) scanned by a tip with a spherical end it can be seen that the observed profile depends on the tip size. If the tip size is small, the observed profile is similar to the true surface corrugation. With increasing radius of the tip R the observed image becomes more and more distorted. Even if the tip shape is known, the true surface can not totally be reconstructed by image processing because the observed image contains no information about the structure of depressions.

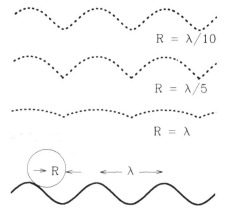

A severe problem in interpreting SFM (and STM) images is the influence of the tip shape. For small tips imaging large but flat objects the interpretation is relatively simple: the tip behaves as a point like probe and the measured topography is authentic. Unfortunately this is often not the case and tip shape effects have to be considered. This is demonstrated in Figure 3, where a periodically corrugated sample is scanned by a tip with a spherical end. When the tip size becomes comparable to the wavelength of the surface corrugation, depressions appear smaller while protrusions appear broader. For large tips depressions are observed as sharp features which might lead to an over-estimated resolution.

4. ACKNOLEDGEMENTS

I would like to thank M. Stedman for critically reading the manuscript and E. Bamberg for his help. This work was supported by the Deutsche Forschungsgemeinschaft SFB 169.

5. REFERENCES

Abraham F F, Batra I P and Ciraci S 1988 Phys. Rev. Lett. 60 1314
Abraham F F and Batra I P 1989 Surf. Sci. 209 L125
Akamine S, Barrett R C and Quate C F 1990 Appl. Phys. Lett. 57 316
Albrecht T R, Akamine S, Carver T E and C F Quate 1990 J. Vac. Sci. Technol. A8 3386
Albrecht T R and Quate C F 1987 J. Appl. Phys. 62 2599

Bardeen J 1961 Phys. Rev. Lett. 6 57
Barrett R C and Quate C F 1991 J. Appl. Phys. 70 2725
Binnig G, Rohrer H, Gerber C and Weibel E 1982 Phys. Rev. Lett. 49 57
Binnig G, Rohrer H, Gerber C and Weibel E 1983 Phys. Rev. Lett. 50 120
Binnig G, Quate C F and Gerber C 1986 Phys. Rev. Lett. 56 930
Binnig G, Gerber C, Stoll E, Albrecht T R and Quate C F 1987 Europhys. Lett. 3 1281
Brodde A, Harazim E, Kliese R, Röttger B and Neddermeyer H 1991 Ber. Bunsenges. Phys. Chem. 95 1434
Buser R A, Brugger J and de Rooij N F 1991 Microelectronic Engineering 15 407
Butt H J 1991 Biophys. J. 60 1438
Butt H J 1992 Biophys. J. 63 578
Cataldi T R I, Blackham I G, Briggs G A D, Pethica J B and Hill H A O 1990 J. Electroanal. Chem. 290 1
Ciraci S, Baratoff A and I P Batra 1990 Phys. Rev. B 41 2763
Chen C J 1990 Phys. Rev. Lett. 65 448
Chen C J 1991 Modern Phys. Lett. B 5 107
Chen C, Vesecky S M and Gewirth A A 1992 J. Am. Chem. Soc. 114 451
Doyen G, Drakova D, Kopatzki E and Behm R J 1988 J. Vac. Sci. Technol. A6 327
Ducker W A, Senden T J and Pashley R M 1991 Nature 353 239
Düring U, Züger O and Pohl D W 1988 J. Microscopy 152 259
Dujardin G, Walkup R E and Avouris P 1992 Science 255 1232
Eigler D M and Schweizer E K 1990 Nature 344 524
Engel A 1991 Ann. Rev. Biophys. Biophys. Chem. 20 79
Feenstra R M, Stroscio J A, Tersoff J and Fein A P 1987 Phys. Rev. Lett. 58 1192
Garnaes J, Kragh F, Mørch, K A and Thölén A R 1990 J. Vac. Sci. Technol. A8 441
Giessibl F J and Binnig G 1992 Phys. Rev. Lett., in press
Gould S A C, Burke K and Hansma P K 1989 Phys. Rev. B 40 5363
Gratz A J, Manne S and Hansma P K 1991 Science 251 1343
Guckenberger R, Hartmann T, Wiegräbe W, Baumeister W 1992, in Güntherodt and Wiesendanger 1992, II 51
Güntherodt H J and Wiesendanger R 1992 Scanning tunnelling Microscopy I-III (New York: Springer)
Hamann C and Hietschold M 1991 Raster-Tunnel-Mikroskopie (Berlin: Akademie)
Hartmann U 1991 Phys. Rev. B 43 2404
Hillner P E, Gratz A J, Manne S and Hansma P K 1992 Geology 359
Kado H, Yokoyama K and Tohda T 1992 Rev. Sci. Instr. 63 3330
Keller D J and Chung C C 1992 Surf. Sci., in press
Kuk Y 1992, in Güntherodt and Wiesendanger 1992, I 17
Kuk Y and Silverman P J 1990 J. Vac. Sci. Technol. A8 289
Landman U, Luedtke W D and Nitzan A 1989 Surf. Sci. 210 L177
Lang N D 1986 Phys. Rev. Lett. 56 1164
Liu H Y, Fan F R, Lin C W and Bard A J 1986 J. Am. Chem. Soc. 108 3838
Magonov S N and Cantow H J 1992 J. Appl. Polymer Sci., in press
Mamin H J, Rugar D, Stern J E, Terris B D and Lambert S E 1988 Appl. Phys. Lett. 53 1563
Manne S, Butt H J, Gould S A C and Hansma P K 1990 Appl. Phys. Lett. 56 1758
Martin Y and Wickramasinghe H K 1987 Appl. Phys. Lett. 50 1455
Mate C M, Erlandsson R, McClelland G M and Chiang S 1989 Surf. Sci. 208 473
Rabe J P 1992 Ultramicroscopy, in press
Rugar D and Hansma P K 1990 Physics Today Oct. 23
Sacks W and Noguera C 1991 Phys. Rev. B 43 11612
Sonnenfeld R and Hansma P K 1986 Science 232 211
Tekman E and Ciraci S 1991 Phys. Rev. B 43 7145
Terris B D, Stern J E, Rugar D and Mamin H J 1990 J. vac. Sci. Technol. A8 374
Tersoff J and Hamann D R 1983 Phys. Rev. Lett. 50 1998
Tersoff J and Hamann D R 1985 Phys. Rev. B 31 805
Tersoff J 1990 Phys. Rev. B 41 1235
Tromp R M, Hamers R J and Demuth J E 1986 Science 234 304
Weisenhorn A L, Hansma P K, Albrecht T R and Quate C F 1989 Appl. Phys. Lett. 54 2651
Wickramasinghe H K 1989 Sci. Am. 260, 10 98
Wintterlin J and Behm R J 1992 in Güntherodt and Wiesendanger 1992, I 39
Wolter O, Bayer T and Greschner J 1991 J. Vac. Sci. Technol. B9 1353
Zhong W, Overney G and Tománek D 1991 Europhys. Lett. 15 49

Inst. Phys. Conf. Ser. No 130: Chapter 4
Paper presented at Int. Congr. X-ray Optics and Microanalysis, Manchester, 1992

347

The performance and limits of scanning probe microscopes

M Stedman

National Physical Laboratory, Teddington, Middlesex TW11 0LW

ABSTRACT: A wide range of surface sensing probes of high spatial resolution were devised in the 1980s, and developed into scanning probe microscopes (SPMs). The principles of several types of SPM are examined. A theoretical comparison of performances is made by considering the ability of an SPM to measure sinusoidal surfaces of varying amplitude and wavelength. SPMs are essentially miniature co-ordinate measuring machines, and validation of dimensional data requires a suitable calibration procedure. The requirements for the traceable calibration of SPMs are discussed, and progress towards the development of calibration artefacts presented.

1. INTRODUCTION

The development over the last decade of scanning probe microscopes having very high and even atomic resolution, is part of a wider development of surface measuring systems aimed at improved performance with respect to resolution, speed, lack of specimen contact and tolerance to poor environmental conditions. It also reflects an increased awareness of the relevance of surface texture to surface function. To quantify the performance and limits of these instruments, one must model them along with the surfaces they are to measure. The modelling process emphasises the metrological nature of the instruments, and that validation of performance requires calibration procedures and artefacts.

2. THE SCANNING PROBE MICROSCOPE

In classical optical and electron microscopy a magnified image is produced simultaneously over the whole of the field of view. In scanning microscopy a probe is scanned sequentially in a raster pattern over the specimen, and the magnified image generated by a corresponding but larger raster scan in the viewing device. The concept of raster scanning to encode and regenerate an image was embodied in the iconoscope and the television transmission experiments of John Logie Baird in the 1920s. The scanning principle has been applied to electron, acoustic and optical probes to produce versatile and well known microscope systems. However, in all cases the lateral resolution of the scanning microscope (equally as in the case of the classical microscope) is limited by the wavelength of the probing radiation.

The last decade has seen the birth and development of a new generation of scanning microscopes based on the use of probes which sense the specimen surface only at very close proximity, and thereby have the potential for very high lateral as well as vertical resolution. Such probes are known as local or proximal probes, and have opened up new vistas in the study of surfaces at nanometric resolutions. The proximal sensitivity of these probes contrasts with that of the electron probe which has little ability to sense depth. Indeed, it is the very depth of focus of the electron beam that leads to the visual impact of the scanning electron microscope (SEM) image. It is depth sensitivity that distinguishes SPMs from electron microscopes and has led to their being called complementary microscopes. The characteristics of a range of proximal probes is examined in the next section.

3. PRINCIPLES OF SOME PROXIMAL PROBES

The essential quality of a proximal probe is that the interaction or property sensed is strongly dependent on the gap between the probe and the specimen surface. Some probe interactions that have been made the basis of an SPM include quantum tunnelling current, interatomic and intermolecular forces, photon tunnelling intensity and capacitance. The variations of these interactions with probe-surface gaps are shown in Figure 1 (after Pohl).

The archetypal SPM is the scanning tunnelling microscope (STM) invented by Binnig and Rohrer (1982), for which they later received a Nobel Prize. A quantum tunnelling current is established by a small potential between specimen and tip, the current and gap being of the order of nanoamperes and nanometres respectively. If the tip is scanned at a constant level over the specimen, then the exponential dependence of the current on gap size ensures good image contrast, but the non-linearity of the dependence makes height interpretation difficult. For topographic application the preferred mode of use is for the probe to be servo driven in height to keep the current and hence the gap constant. Thus, the tip effectively traces the topography of the surface. The large rate of change of current with height ensures a high sensitivity and hence vertical resolution for the probe.

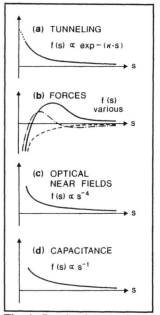

Fig. 1 Proximal probe interactions versus gap (s) (courtesy of D Pohl)

The sub-nanometric lateral resolution also derives from the current-height characteristic which confines the effective volume of interaction to a small area surrounding the part of the tip closest to the surface. In practice this may be as small as a single atomic diameter. However, on specimens with significant changes of local slope, the point of interaction on the tip may change leading to discontinuities in the image. However, this is only an expression of the unavoidable convolution of surface and tip shapes that affect all SPMs.

The limitation of the STM to conductive specimens was overcome by the development of the atomic force microscope (AFM) by Binnig *et al* (1986). In the AFM the interaction sensed is the force between a sharp tip attached to a cantilever spring and the specimen surface. The compliance of the cantilever is such that deflection caused by a force of 1 nN can be readily detected and used to servo drive the cantilever and tip to keep a constant repulsive force. Thus the AFM acts as a conventional stylus profilometer, though at a far lower contact force. More recently, attractive mode AFMs have been designed to work at a gap which takes the tip beyond the regime of repulsive forces into that of the attractive van der Waals forces.

The possibility of obtaining optical resolution below the Rayleigh limit was foreseen by O'Keefe (1956), who proposed that an optical near-field would penetrate a small aperture to a depth of the order of its diameter. Thus, in scanning near-field optical microscopy (SNOM) an illuminated aperture (which could conveniently be an optical fibre drawn to a fine tip) is scanned proximally over the specimen so that its surface interrupts the near field. The evanescent wave at the low refractive index side of a total internal reflection (TIR) also has near field characteristics and can be made the basis of an SPM. If the TIR is frustrated by a probe scanning within the evanescent field, then the intensity lost from the total reflection or picked up by the probe provides the interaction, and the probe will sense the topography of the surface and any transparent specimens (eg biological) thereon. The decay rate of the evanescent field with distance is exponential, and the analogue with the quantum tunnelling current has led to the acronym SPTM for scanning photon tunnelling microscope.

Finally, the scanning capacitance microscope (SCM) will be mentioned. Here, a conductive tip and substrate are required, though not necessarily a conductive specimen. The capacitance between the proximal tip and substrate forms the interaction, but change with gap size is less than with probes described earlier, even when the tip is oscillated to improve sensitivity. Although the resolution of the SCM is poorer than for other SPMs, the system of Bugg and King (1988) has been developed into a commercial instrument with specialist applications.

4. A MODEL OF SURFACE TEXTURE

How do the various new SPMs compare in their ability to measure topography, and how do they compare with the classical profilometers? A sinusoidal model of surface topography has been developed by Stedman (1987) as a basis for comparing the performance of surface measuring instruments. A sinusoidal profile has height, surface wavelength, slope and curvature, and these are the principal topographic features of a surface which determine its functional characteristics. Comparison of these parameters with critical parameters of the SPM - such as the radius (R_s) of the tip and slope (S_s) of the flanks of the stylus, the vertical range (R_v) and

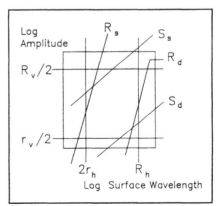

Fig. 2 Limit lines for instrument parameters plotted in AW space

resolution (r_v), the horizontal range (R_h) and resolution (r_h), and the slope (S_d) and radius of curvature (R_d) errors of the datum - leads to a series of limit lines which can be plotted in amplitude-wavelength (AW) space to define a zone within which the SPM is able to measure surface texture modelled as sine waves (Figure 2).

5. PERFORMANCE OF SPMs IN AW SPACE

Since SPMs usually use similar PZT actuators their scan ranges and resolutions are likely to be similar. The main differences in their AW charts are likely to arise from limits associated with probe geometry. However, in the current state-of-the-art sharp STM tips cannot be made with reproducible geometry; AFM tips made by micromachining methods vary considerably in tip radius even within one wafer batch. Therefore, the AW chart shown in Figure 3 (Stedman and Lindsey 1988) should be regarded as that of a generic SPM with typical values of tip or probe parameters. The shaded area represents the collective limits for commercial stylus instruments; also shown in the figure is the AW chart for Nanosurf-2, an NPL-built instrument

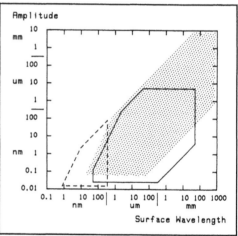

Fig. 3 AW charts of Nanosurf-2 (solid line) and a generic SPM (dashed line)

which approaches the practical limits for measurement by contact stylus. The limits of performance of STMs and AFMs have been discussed in AW terms by Stedman (1988a); charts for a range of optical instruments have been given by Stedman (1988b).

6. CALIBRATION PROCEDURES

The increasing use of scanning probe microscopes (SPMs) for making small scale dimensional measurements emphasises the need for developing appropriate calibration methods and standards. In many ways the SPM can be considered as a miniature co-ordinate measuring machine (CMM), since the probe is guided and positioned in 3D space and is able to detect proximity to the specimen surface thereby acting similarly to a CMM touch trigger probe. Various procedures have been developed for calibrating CMMs, from the calibration of each of the 21 independent sources of error to the repeated measurement of an array artefact (such as a ball plate) at various positions and orientations within the measuring volume. However, the latter approach has proved suitable only for performance verification, rather than formal calibration, and although such approaches could be applied in principle to the SPM, they are not useful in practice.

A strategy for successful calibration must also take account of the non-linear and scan-rate dependent behaviour of the SPM. The probe geometry and dimensions also play a crucial role in the process of imaging the surface, and the image obtained is always a convolution of the topography of both specimen and probe. The most effective approach is through the use of transfer artefacts which have been calibrated at a standards laboratory. Although the calibration of instrument scales can be made to be reasonably independent of probe characteristics, it is also advisable to characterise or 'calibrate' the probe. Although electron microscopy has proved useful for assessing STM and AFM tips, artefacts which interact strongly with the probe are also required for probe calibration.

7. CALIBRATION ARTEFACTS

A variety of prototype calibration artefacts of periodicities of 4 μm and higher is being developed at NPL; a series of artefacts with smaller periods has been developed in collaboration with the University of Southampton, using electron beam lithography and micromachining (Ensell et al 1992). The artefacts provide for a wide range of topographic dimension and the characteristics of different SPMs. The

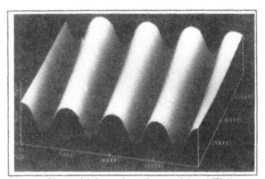

Fig. 4 Sinusoidal grating (imaged by AFM)

artefacts take the form of gratings having a variety of profiles or arrays of features such as pits and domes. The preferred grating profile is sinusoidal, since it enables an SPM to be calibrated in a particular zone of AW space. The AW chart of an instrument typically covers several decades in both axes, and since performance, or calibration, is unlikely to be constant over such a large area, it is advisable to calibrate at a number of zones within the AW chart. Only sinusoidal gratings provide the harmonic purity required for calibration relevant to a single AW zone. The sine wave is one of the calibration specimen profiles recommended for the calibration of stylus instruments in BS 6393:1987 and ISO 5436-1985. NPL artefacts conform to these standards. An example of a 20 μm pitch sinusoidal grating is shown in Figure 4 as an AFM image.

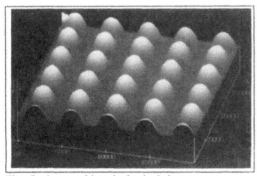

Fig. 5 Array of hemispherical domes

In addition, a series of array artefacts has also been developed. The arrays are of either pyramidal pits with Si (111) faces or domes of reflowed phosphosilicate glass (Figure 5). Both types could be used as analogues of ball plates used in CMM calibration. The domes can take the form of hemispheres of excellent sphericity and surface smoothness. SPM imaging of such domes reveals information on the spatial characteristics of the probe, since a hemisphere presents all possible local slopes during a scan. AFM images of hemispherical domes encode not only the effective topography of the tip under scanning conditions, but also dynamic effects such as the onset of vibration (note the zone in Figure 6 showing induced tip vibration).

8. CONCLUSIONS

The rapid development of a wide range of different types of SPM over the last decade has been accompanied by a great expansion in the breadth of applications. However, the impressive imaging abilities of SPMs has tended to obscure the deficiencies of their capabilities for dimensional measurement. In addition, the problems of allowing for the

Fig. 6 A single hemisphere imaged by AFM (spacing of simulated fringes is 210 nm)

effects of probe geometry are much more severe for SPMs than for the traditional stylus profilometers and CMMs, simply because of the ratio of probe size to the scale of measurement. A wide range of calibration artefacts is being developed to allow traceable calibration of SPMs by users under normal operating conditions. Specialised instrumentation to calibrate the artefacts is being developed at NPL.

REFERENCES

Binnig G, Quate C F and Gerber C 1986 Phys. Rev. Letts. 56 930-933
Binnig G and Rohrer H 1982 Helv. Phys. Acta 55 726-735
Bugg C D and King P J 1988 J. Phys. E 21 147-151
Ensell G, Evans A, Farooqui M, Haycocks J A and Stedman M 1992 Micromech. and
    Microeng. (in the press)
O'Keefe J A 1956 J. Opt. Soc. Amer. 46 359-360
Stedman M 1987 Prec. Eng. 9 149-152
Stedman M 1988a J. Microscopy 152 611-618
Stedman M 1988b Proc. SPIE 1009 62-67
Stedman M and Lindsey K 1988 Proc. SPIE 1009 56-61

Inst. Phys. Conf. Ser. No 130: Chapter 4
Paper presented at Int. Congr. X-ray Optics and Microanalysis, Manchester, 1992

Aspects of 3-D imaging, display and measurement in light and scanning electron microscopy

Alan Boyde.

Dept of Anatomy and Developmental Biology, University College London, London WC1E 6BT.

ABSTRACT. Methods for enhancing the understanding of complex structures, including stereo-image display (static stereo-parallax), pseudo-holograms (multiple stereo viewpoints), and continuous motion parallax from rotational displays are common to all the 3-D microscopies. Measurement and reconstruction of relative co-ordinates is necessary for the exact understanding of a structure or topography. Methods for the acquisition, analysis and measurement of 3-D images in methods ancillary to x-ray microscopy (SEM, confocal and conventional LM) are considered.

1. INTRODUCTION.

Methods using two or more projections in the microscopy of 3-D surfaces or thick objects avoid misinterpretation and create the ability to measure. Procedures for recording, studying, projecting and measuring stereo-pair images have been described often enough in the literature: for example, for x-ray microscopy as such by Cosslett and Nixon (1952, 1960), Latham (1966) and Buckland-Wright (1989); and in the ancillary methods of light microscopy by Osborn et al (1978), Inoue (1986) and Wolf (1989); transmission electron microscopy by Helmcke (1954, 1955); and SEM by Boyde and Ross (1975), Boyde et al (1986) and Boyde and Jones (1992).

2. STEREOSCOPY AND STEREO MEASUREMENT IN SEM.

Most specimens examined in SEMs are imaged using secondary electrons (SE). The SE image is often difficult to interpret correctly and guesses about the 3-D relationships of features in an object will often be wrong. If there are overlap cues, relative depths of the features concerned may be obvious. However, slopes across a rough field and absolute distances can not be inferred from a single SE image. Another important cause of difficulty in interpreting 3-D relationships from SE images is simply that they may have such a large depth of field. In the real world, we obtain much depth information from dynamic focussing when we look at real objects, and from foreground and background blur in photographs of close objects. With backscattered electron (BSE) images acquired with directional detectors, shadows due to parts of the surface obscuring others from the direct line of sight to the detector give depth sense.

The correct relative positions of features in the specimen surface can be understood and measured if they are represented in sets of images taken from different viewpoints: in the SEM, this will usually mean that the specimen was tilted to allow a different projection, but the electron beam may be tilted to change the effective origin of the raster scanning the specimen (Boyde et al 1972; Thong and Breton 1992). Two projections will suffice if all relevant features are represented in two views which are sufficiently similar that they can be viewed with the two eyes as a stereoscopic pair: both requirements are assumed to be satisfied in the following. However, more projections from more defined angles will make it possible to determine 3-D locations of more features with more accuracy.

The following contains some brief reminders of procedures and definitions for convenient SEM work, defining Y as the direction towards a directional BSE and SE detector, the direction at right angles to this as X and the electron optical axis as Z.   It will be convenient if X and Y are parallel with the line and frame directions of the SEM monitors. If they are not, then adjust the scan rotation control. If the SEM does not have a scan rotation unit, change the working distance to find one at which X movement parallels the line axis. By choosing to tilt around the Y axis, the same mean signal level will be shown in both members of the stereo-pair and there will be the same degree of overall foreshortening of the field of view. With a ring detector, it does not matter which axis is used for the tilt, and it will be best to tilt symmetrically about mean normal beam incidence. If the plane of interest is not coaxial with the tilt mechanism, the field must move in Y and go out of focus on tilting. To retain the same magnification and rotation of the field of view after tilting, it is necessary to refocus using the stage Z control after recentering the field of view. The processed photographs must be rotated 90° for viewing if the tilt was around the X-axis.

The tilt angle difference between the two projections of the specimen may vary from around $5°$ for a very rough specimen surface and an unskilled observer, to around $7°$ to $8°$ for a roughly correct impression of depth, to $10°$ for the minimal tilt angle difference which can be obtained with any degree of precision with the majority of manufactured stages, to $20°$ to obtain greater parallaxes and greater apparent depth for flatter surfaces. Higher tilt angle differences should be symmetrical about the perpendicular to the specimen surface.

To view 68mm wide stereo pairs unaided, the eyes are allowed to diverge to a parallel gaze, but focussed at say 25cm. Simple viewing aids are: a pair of lenses in a frame held at their focal length from the stereo pair: two mirror Wheatstone stereoscopes: two pairs of mirrors (±lenses) to allow increased separation distance between larger photos and to correct the mirror inversions: one or two prisms:- a prismatic System Nesh viewer uses the stereo-pair mounted one above the other. Stereo slides may be projected, one overlapped on the other, using two projectors with $45°$ and $135°$ oriented polarising filters and a silvered screen (e.g. aluminium spray paint on smooth hardboard), or by using a red filter in one and a blue or green filter in the other projector: in each case the observer wears appropriate spectacles to decode the images to the correct eyes. The latter anaglyph display method is commonly used with a colour TV monitor to display real time stereo images in SEMs, the two projections achieved by deflecting the electron beam axis right and left, rather than mechanical tilting of the object in the SEM (Boyde et al 1972; Thong and Breton, 1992).

Exact knowledge of the tilt angle difference, α, is important if measurements of parallax, p, are to be made in order to calculate height differences:

$$Z_c = p/2(\sin(\alpha/2))$$

where Z_c is a height measured with respect to the plane of an image which might have been taken in the central position, halfway between the two positions in which the left and right photos were actually recorded, and assuming parallel projection. For heights, Z_l perpendicular to the plane of the less tilted, left photo or Z_r measured with respect to the plane of the right photo [both of which exist and may be used as maps],

$$Z_l = (X_l\cos(\alpha) - X_r)/\sin(\alpha) = X_l/\tan(\alpha) - X_r/\sin(\alpha)$$
$$Z_r = (X_l - X_r\cos(\alpha))/\sin(\alpha) = X_l/\sin(\alpha) - X_r/\tan(\alpha),$$

where X_l and X_r are the X distances between pairs of features measured in the left and right photos respectively, and the distances, parallaxes are all measured and the heights calculated at the photo scale and to be corrected by the overall instrumental and photographic magnification. Thus Z co-ordinate errors depend on the accuracies in measuring the tilt angle difference, parallaxes and distances. Relative errors in parallaxes are reduced by increasing the tilt angle difference. Measurements of parallax can be made with great precision employing dual marks, one introduced into each photo in stereo viewing, so that the two components fuse to form a floating mark which is adjusted to appear to lie in the plane of the surface of the object (Boyde and Ross 1975; Boyde et al 1986; Boyde and Jones 1992a).

3. CONFOCAL SCANNING OPTICAL MICROSCOPY (CSLM): 3-D IMAGING

In confocal microscopy, the illumination in the intermediate image plane of the objective is apertured so that real or virtual apertures are imaged in the focal plane in the specimen to illuminate, efficiently and brightly, only corresponding points (Wilson and Sheppard 1984). The imaging system only examines the same point, by collecting the signal via a "confocal" aperture in the intermediate image plane of the objective on the image forming side of the microscope. Thus only a small volume element in the plane of the specimen surface (or within a translucent object) is probed. There is a sharp maximum at focus for an object being translated through this one point in space so that the signal intensity can be used as a measure of focus. To turn this principle into an image forming system requires scanning: for example, the specimen, the lens, or mirrors or acousto-optic deflection devices. There are now many different types of confocal microscope available commercially. When we began our work in this field, we had to commission the construction of a microscope and we chose the so called Tandem scanning microscope (TSM) (Petran et al 1985). In this type of microscope, a field of apertures in the intermediate image plane of the objective is scanned to provide the scanning of points of illumination in the sample. The scanning of multiple detectors is provided by the same aperture array.

Laser CSLMs have become very popular and were soon recognised for their potential in the precise measurement of surfaces (Sheppard et al 1983).

However, in the laser based instruments only one spot is illuminated at one moment and their frame speeds have been too slow for practical survey work. Recent commercial instruments using acousto-optic beam deflection have been able to achieve video rate confocal performance. These instruments perform well in the practical measurement field for solid objects (Jones et al 1992) and for high temporal and spatial resolution observation in both reflection and fluorescence of events in rapidly moving live cells (Boyde and Jones 1992b).

Specimen scanning confocal microscopy for surface measurement and mapping was introduced by the Oxford group (Sheppard et al 1983; Wilson and Sheppard 1984). We developed a system for rapid mapping plus area, depth and volume measurement based upon the TSM (Boyde et al 1990). More recently, we have employed video rate laser CSLM (Jones et al 1992a). In the latest equipment, 256 focus levels are sampled in less than 10 seconds to generate a 3-D map in which the value at each pixel is the value of focus at which the highest signal level was returned. Confocal microscopic mapping allows the extraction of derived 3-dimensional information, such as volume and shape, from surfaces. There are problems at steep edges where no light is retro-reflected, but if enough is known about the surface, the missing parts can be filled in by simple image processing. One of our major areas of application remains in the determination of the volume of osteoclastic resorption lacunae (Boyde et al 1990; Boyde and Jones 1992a; Jones et al 1992). By subtracting a smoothed version of a topographic map image from the original map one obtains the true value of the local surface relief (Boyde and Fortelius 1991).

Another method for confocal surface mapping may eventually displace methods which depend on mechanical focus and measurement of signal intensity. This will exploit the fact that TSMs illuminated with uniform white sources generate images in which the apparent colour of a flat reflected specimen surface depends upon the focal distance. For specimens with surface roughness, the local colour reflects the local height. Very rapid methods for the determination of the height may then be based upon the measurement of the colour generated by the longitudinal chromatic dispersion of the objective lens (Browne et al 1992).

CSLMs produce images of narrow focal planes but the depth of field may be increased by adding the components of a stack of through-focus images. An interesting possibility arises with very fast scan rate CSLMs of rapidly cycling the depth interval to generate an instantaneous view with increased depth of field (Frosch and Korth 1974), something which we have instituted in a TSM using piezo-electric oscillation of the objective lens.   Using frame store memory, Kalman averaging whilst focusing provides the same end result in slow scan CSLMs, but with an according time delay.

For most computer based, laser CSLMs, the procedure used for generating stereo pairs is first to record a stack of separate images of separate focal planes, after which oblique views through the stack are generated. In the simplest possible case this is achieved by shifting each successive image by a whole number of pixels before adding it to the Kalman sum of the stack. This may be done in opposite senses for the synthesised right and left views, or it may be done for only one projection, the other being left as the straight sum of the images in the original line of projection. More elaborate schemes bias the signal level as a function of depth into the object or take the maximum intensity at each pixel in the reconstructed line of sight. More simply, however, the specimen can be moved on two axes simultaneously whilst focusing to simulate an oblique through focus direction corresponding to a new line of sight. Done twice, on two inclined axes, and two Kalman averaged images will produce an excellent stereo pair in much less time than is used in the usual current practice (Boyde 1990). The simplest high resolution permanent image capture medium remains the photographic emulsion, and with a direct view (real image) CSLM one can simply record an extended focus image by photographing whilst through focusing to sum the components in a stack. Again, if done twice on two inclined axes, one produces beautiful high resolution stereo images which may also be in natural reflection or fluorescence colours (Boyde, 1990). In a system which we employ, the objective lens of the TSM is controlled by two piezo electric activators giving movements of up to 50µm in X (a horizontal plane) and Z (along the optical axis) which are combined to give skew axes for stereo (and integram - see later) imaging.

We have recently developed means by which high magnification and high resolution confocal stereo images may be viewed in real-time by looking through a binocular head as in a conventional binocular microscope (Maly and Boyde, in preparation). This is achieved with a combination of newly designed objective lenses with large, linear, longitudinal chromatic spread (see also Courtney-Pratt and Gregory 1973 and Courtney-Pratt 1976) with a

TSM illuminated with a xenon arc source. The system focuses at different
depths in different colours. By viewing the coloured image with opposite
handed split prism blocks one achieves instant stereo, still colour coded
for depth.   3-D images are also simply obtained by tilting an object
between the recording of two photographs with these extended depth of field
objectives. A similar system for a slit confocal microscope was proposed by
Baer (1970).

4. MULTIPLE VIEWPOINTS: INTEGRAMS AND ROTATING DISPLAYS

Series of deep focus views using small differences in viewpoint can be
processed into integrams or pseudoholograms.   The many images of the
sequence are printed in sequence through a lenticular screen overlay,
before the photographic paper is processed and eventually cemented behind
the lenticular screen.   These images are permanent 3-D displays of great
beauty. With the material which we have used, the angle of view into the
structure changes with the position of the observer over intervals of 24°
For TSM (and other CSLM) range images, we have to take account of the fact
that in summing flat optical sections to generate oblique views there is no
foreshortening due to tilt. Each TSM image has therefore to be
rephotographed at the appropriate tilt angle (computer CSLM images can be
deformed by computer image scaling) in order to generate the integram
series. An alternative system can be used with samples in which the depth
and the width of the field of view is much larger: the tilt series can be
obtained simply by tilting the object in small increments (for which we
have used a stepper motor) prior to recording Kalman (or photographically)
averaged CSLM images whilst focusing through the required range.

For integrams from SEM images, series of from 18 to 36 very well-registered
views of the same point on the sample separated by small tilt angle
differences are recorded. We have used a precision eucentric goniometer
stage in which the tilt angle difference could be read reliably to 0.1°,
the series recorded with symmetrical tilts about normal beam incidence. A
total tilt angle difference of 24° in increments up to one degree (to a
total of 36 steps) produced the best results, matching the angles of view
reproduced in the lenticular screen.

Nowadays, many expensive 3-D reconstructions from serial physical or optical sections (or rotational tomographic reconstructions) use rotational displays to convey complex 3-D relationships, and there can be no doubt that the resultant continuous motion parallax is a powerful means in aiding interpretation. In some cases, however, it may be possible to record the rotating image of the real structure by direct means: it would be very welcome if this were done for certain classes of structure using synchotron x-radiation. So far, we are only aware of it having been done by video recording in light optical systems, but it could also be done by TV rate SEM. Large depth of field optics for the lower range of light microscopic magnifications are now becoming commonplace, and fibre optic coupled video camera based microscope systems particularly lend themselves well to this excellent, cheap method for obtaining and conveying 3-D information.

Ackowledgements. I thank Peter Howell, Mojmir Petran, Milan Hadravsky, Charles E Dillon, and Mirek Maly for their several forms of assistance. Work developing 3-D SEM and confocal imaging techniques has been supported by grants from the MRC, the SERC and The Wellcome Trust.

5. REFERENCES

Baer SC 1970 US Patent No. 3,547,512
Boyde A 1985 Science 230 1270-72
Boyde A 1990 The Handbook of Biological Confocal Microscopy ed JB Pawley (New York: Plenum) pp 163-168
Boyde A, Cook D and Morgan JE 1972 UK Patent No. 1,393,881
Boyde A, Dillon CE and Jones SJ 1990 J. Microsc. 158 261-5
Boyde A, Fortelius M 1991 Scanning 13 429-30
Boyde A, Howell PGT, Franc F 1986 J. Microsc. 143 257-64
Boyde A and Jones SJ 1992a Photogrammetric Record 14 59-84
Boyde A and Jones SJ 1992b Binary 4 119-123
Boyde A, Ross HF 1975 Photogrammetric Record 8 408-57
Browne MA, Akinyemi O, Boyde A 1992 Scanning 14 145-53A 387 171-86
Buckland-Wright JC 1989 Brit J Radiol 62 201-8
Cosslett VE and Nixon WC 1952 Proc. Roy. Soc. B 140 422-31
Cosslett VE and Nixon WC 1960 X-ray Microscopy (Cambridge: Cambridge University Press)
Courtney-Pratt JS 1976 Optical Engineering 15 379-83
Courtney-Pratt JS and Gregory RL 1973 Applied Optics 12 2509-19
Frosch A, Korth HE 1975 US Patent No. 3,926,500
Helmcke JG 1954 Optik 11 201-25
Helmcke JG 1955 Optik 12 253-73
Inoué S 1986 Video Microscopy (New York: Plenum Press)
Jones SJ, Boyde A, Piper K and Komiya S 1992 Microsc. & Anal. July 18-20
Latham RV 1966 J. Royal Microscopical Soc. 85 255-82
Osborn M, Born T, Koitsch H-J and Weber K 1978 Cell 14 477-88
Petran M, Hadravsky M and Boyde A 1985 Scanning 7 97-108
Sheppard CJR, Hamilton DK and Cox IJ 1983 Proc. Roy. Soc. A 387 171-86
Thong JTL and Breton BC 1992 Scanning 14 65-72
Wilson T and Sheppard CJR 1984 Theory and Practice of Scanning Optical Microscopy (London: Academic Press)
Wolf R 1989 J. Microsc. 153 181-6

Inst. Phys. Conf. Ser. No 130: Chapter 4
Paper presented at Int. Congr. X-ray Optics and Microanalysis, Manchester, 1992

361

Determining the form of atomic force microscope tips

P Siedle[a], H-J Butt[a], E Bamberg[a], D N Wang[b], W Kühlbrandt[b], J Zach[b] and
M Haider[b]

[a] Max-Planck-Institut für Biophysik, Kennedyallee 70, 6000 Frankfurt a. M. 70, Germany
[b] European Molecular Biology Laboratory, Meyerhofstrasse 1, 6900 Heidelberg, Germany

ABSTRACT: To obtain the form of atomic force microscope (AFM) tips, cubic crystals of
MgO and NaCl were imaged. Low voltage scanning electron microscopy showed the shape
of the crystals. The edges of the crystals appeared rounded in the AFM image, because at
sharp edges the AFM image reflects the form of the tip rather than the crystal. Hence, the
tipform and radius of curvature could be determined from AFM images of crystals. Also
transmission electron microscope micrographs of tips were made after coating the back of
the cantilever and tip, but not the front of the tip, with carbon. In this way charging
effects were reduced, and the micrographs showed the shadow of the tip.

1. INTRODUCTION

In the atomic force microscope (AFM) (Binnig et al. 1986, Rugar and Hansma 1990), the tip
of a flexible force sensing cantilever is scanned over the suface of a sample. Surface features
cause deflections of the cantilever which are detected and displayed as an image. AFM images
are influenced by the form of the tip, and in general the image is a convolution (not in the
mathematical sense) of tip and sample (Horie and Miyazaki 1987, Stedman 1988, Stemmer
and Engel 1990). Once the form of the tip is known, AFM images can be partly corrected
(Keller 1991).

Various attempts have been made to reveal the form of the tip: Coating non–conductive
AFM tips with metal and taking scanning electron microscope (SEM) images is a standard
procedure (see e. g. Albrecht et al. and Wolter et al. 1991), but the resolution is limited by the
coating. The tip structure of single atomic tips was examined by Paik et al. (1991). Hellemans
et al. (1991) and Möller et al. (1990) imaged one AFM tip by another tip, but the problem
remained to deconvolute the tips.

In this paper we present a method to determine the form of AFM tips by reversing the role
of tip and sample: AFM tips are imaged by edges of cubic crystals (figure 1). We used MgO
and NaCl crystals in a size up to about 2 μm. In addition, we imaged the shadow of AFM
tips with a transmission electron microscope (TEM).

Figure 1: *A schematic drawing of the imaging of a cubic crystal by an AFM tip. A sharp edge cannot be seen in the image; instead the edge images the tip.*

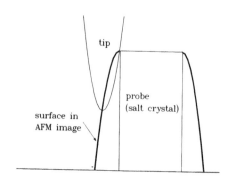

tip

surface in
AFM image

probe
(salt crystal)

2. MATERIAL AND METHODS

We used microfabricated cantilevers with integrated tips made of Si_3N_4 (Nanoprobes, Digital Instruments) and a commercially available AFM (Nanoscope II, Digital Instruments). The tips were pyramidal formed with one face of the pyramid in scan direction.

To obtain MgO crystals, 0.4 g magnesium powder was mixed with 0.03 g KNO_3 and heaped on a copper plate. KNO_3 served as oxidizing agent and kindled the magnesium while the copper plate was heated by a gas flame. About 20 cm above the copper plate a glass microscope cover slide caught the ascending white smoke of MgO crystals.

We produced NaCl crystals by placing 10 ml of a saturated solution of NaCl in an ultrasonic bath and adding the same quantity of ethanol as precipitant. The precipitate was filtered through a 0.22 μm Millipore filter. The filter was put into organic solvent (ethanol or chloro-

a)

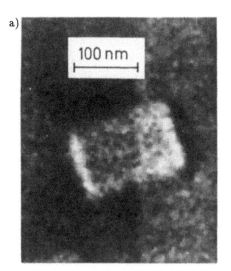

100 nm

b)

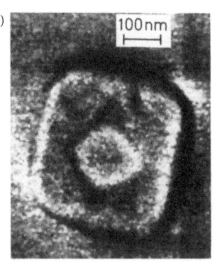

100 nm

Figure 2: *Low voltage SEM micrographs of a MgO crystal (a), and a NaCl crystal (b). The hole in the center of the crystal was characteristic for many NaCl crystals.*

a)

b)

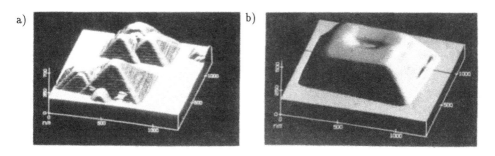

Figure 3: *AFM images of a few MgO crystals (a). The crystals are so small that the image is nearly a sole image of the tip. (b) AFM image of a NaCl crystal.*

form) and sonified to detach the crystals from the filter. In this way we obtained a water-free suspension of crystals in organic solvent, which we evaporated on mica.

The uncoated crystals were imaged with a low voltage SEM at 1 kV electron energy.

After use in the AFM, the tips were prepared for the TEM: we attached the cantilever on a TEM grid with epoxy glue. Then we coated the back of the cantilever and tip, but not the front of the tip, with carbon to reduce charging effects. Thus the shadow of the uncoated tip could be imaged.

a)   b)

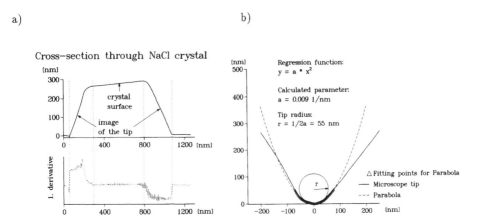

Figure 4: *(a) The figure shows a cross section of the NaCl crystal shown in figure 3 b). The edges of the crystal image the tip. The tip image is identified by inspecting the changes in the first derivative of the cross section (see dotted vertical lines). In our AFM the apex of the pyramid is slightly pointing to the right. Hence the slope at the left side of the crystal is steeper than the slope at the right side. (b) The form of the tip was reconstructed from a) by removing the crystal surface at the top. The triangles at the lower part of the tip denote the points where we fitted a parabola. The tip radius was obtained by equating the curvature of parabola and circle at the origin leading to $r = 1/(2a)$.*

Figure 5: *TEM micrograph of an AFM tip. We coated the back of the cantilever and tip with carbon to reduce charging. But charging effects still blur the image.*

3. RESULTS AND DISCUSSION

Figure 2 shows the shape of typical crystals. Crystals were imaged with the AFM (see Figure 3) and the image cross–sectioned. Figure 4 describes the procedure for obtaining the tip radius from AFM images. Parabolas approximated the tip form well, while hyperbolas did not. To test if the tip scrapes off the edge of the crystal, we doubled the force to 80 nN, and scanned a single crystal for several hours. No effect could be seen. Also the scan direction had no effect on the measured tip radius.

The accuracy of the tip radius depended on the ability to identify the crystal surface in the cross sections. It was limited by the form of the tip, irregularities of the crystals, and on possible roundness of the crystal edges. The eight examined tips had tip radii between 25 nm and 60 nm. The standard deviation which was obtained by averaging over up to five images of different crystals was 5 - 10 nm. We compared these tip radii to results obtained from TEM micrographs of tips after use in the AFM (see Figure 5). A typical value is (46 ± 10)nm for a tip radius measured in the AFM, whereas in TEM micrographs of the same tip we measured (38 ± 5)nm.

The presented method to measure tip radii relies only on the AFM itself, and is independent of other apparatus like TEM's. In addition, it allows one to rule out bad tips.

4. ACKNOWLEDGEMENTS

We would like to thank B. Kelety, T. Schauer and H. Wartig for helpful dicussions, M. Stedman for critical reading of the manuscript.

5. REFERENCES

Albrecht T R, Akamine S, Carver T E and Quate C F 1990 J. Vac. Sci. Technol. A 8 3386
Binnig G, Quate C F and Gerber Ch 1986 Phys. Rev. Lett. 56 930
Hellemans L, Koen W, Hennau F, Stockman L, Heyvaert I and Van Haesendonck C 1991
    J. Vac. Sci. Technol. B 9 1309
Horie C and Miyazaki H 1987 Jpn. J. Appl. Phys. 26 L995
Keller D 1991 Surf. Sci. 253 353
Möller R, Esslinger A and Rauscher M 1990 J. Vac. Sci. Technol. A 8 434
Paik S M, Kim S and Schuller I K 1991 Phys. Rev. B 44 3272
Rugar D and Hansma P K 1990 Physics Today Oct. 23
Stedman M 1988 Journal of Microscopy 152 611
Stemmer A and Engel A 1990 Ultramicroscopy 34 129
Wolter O, Bayer Th and Greschner J 1991 J. Vac. Sci. Technol. B 9 1353

Inst. Phys. Conf. Ser. No 130: Chapter 5
Paper presented at Int. Congr. X-ray Optics and Microanalysis, Manchester, 1992

365

Developments in image processing

P W Hawkes

CEMES/Laboratoire d'Optique Electronique du CNRS, B.P. 4347 F - 31055 Toulouse Cedex

ABSTRACT: Some of the most spectacular developments in digital image processing have been made in recent years by theoreticians. The impact of such findings on the processing of images from scanning instruments is discussed, with particular reference to median and related nonlinear filters and to image algebra.

1. INTRODUCTION

Scanning instruments lend themselves to digital image processing, since the transfer of the image to the computer, pixel by pixel, is so straightforward. The development of suitable equipment for this transfer and of algorithms for subsequent treatment of various kinds was reviewed in past ICXOMs (Hawkes, 1984, 1989). Image processing came into being in response to a need initially to improve images sent back from space, and rapidly spread into a host of very different fields: forensic science, astronomy, microscopy, medicine, agriculture and fisheries, and this is only a small sample. Over the past few years, however, we have seen a change from the applications-driven developments of the early days: image processing now exists as an autonomous activity, with research effort not always directed towards a specific problem. It is this aspect of the subject that we shall examine here though of course we concentrate on topics of present or potential interest in scanning microscopy. This is, we believe, timely for major advances have been made on certain theoretical questions in image processing in the past few years and these will no doubt soon be impinging on the microscope world.

Before turning to these themes, we briefly recall the major divisions of the subject, for subsequent reference. Despite its somewhat shapeless appearance, due perhaps to the fact that radiographers, pig-farmers, electron microscopists, fishermen, astronomers, carpet-manufacturers and policemen (and many others) have all had a hand in fashioning it, some pattern can be recognized in - or some would say, imposed on - image processing. There are four broad chapters: (1) acquisition, sampling and coding; (2) enhancement; (3) restoration; and (4) analysis including description and pattern recognition. In chapter 2, enhancement, we put all the methods of improving images that do not make use of prior knowledge of the image-forming process : the image is poor in some way and we attempt to do something to improve it. Conversely restoration of an image (chapter 3) involves using all available knowledge to render the image (or set of images) more informative. Among the many ways of enhancing an image, we distinguish in particular those that use the grey-level histogram to guide subsequent processing, linear operations, essentially convolutions, and nonlinear operations, such as the median filter and the morphological procedures. In restoration, we might mention the inverse problem (phase from amplitude), three-dimensional reconstruction, Wiener and related filtering. The first and fourth chapters are reasonably self-explanatory and image analysis is probably the most important single activity in scanning microscope image processing.

2. NONLINEAR FILTERS FOR IMAGE ENHANCEMENT: MEDIAN AND RELATED FILTERING

One of the most important operations of image enhancement involves filtering of some kind to sharpen the raw image, often with a view to accentuating fine detail or shallow contrast steps. It has long been believed that the (nonlinear) median filter is capable of giving better results than the (linear) convolutional filters although the bias in favour of Laplace, Fourier and "Convolute" (Gabbouj *et al*, 1992) has not yet been fully overcome. At first, the improvement was claimed on purely subjective grounds but the efforts of the theoreticians in recent years have not only justified this belief but also generated a host of related nonlinear filters. We draw attention to these in the remainder of this section.

A median filter is an operator that replaces each pixel-value in an image by the *median* of that grey-level and those of the nearest neighbours; the neighbourhood may be defined in various ways, 4-neighbours, 8-neighbours or more extended neighbourhoods. In a simple extension, the grey-level values may be weighted before taking the median, to emphasize the values closest to the pixel in question if the neighbourhood is large, for example. This is clearly quite different from the convolutional filters, in which each pixel-value is replaced by a weighted sum of the grey-level values at that pixel and at its nearest neighbours, again defined in one of several ways.

Over the years, our understanding of median filtering has developed extensively. It was shown that ordinary median filters have roots or eigenvalues, which any image reaches if the filter is applied repeatedly but that recursive median filters, in which the new pixel-values are used as the filter explores the image, reach their roots immediately: further application produces no new changes. Filters in which some other member of the set of pixel-values in the chosen neighbourhood than the median is adopted (k-th rank-order filters, when the k-th largest sample is chosen) were explored. Hybrid filters in which averaging and selection are both employed were introduced, notably the alpha-trimmed filter in which pixel-values in the neighbourhood are retained or rejected, depending on their closeness to the value at the pixel in question (nonlinear operation) after which the survivors are averaged (linear).

Another important step was the recognition of the utility of threshold decomposition, which shows that these filters exhibit a kind of weak superposition, not as strong as that of linear operations but present nonetheless: if we threshold the image at each of its levels and apply a binary median filter of the same size as the original median neighbourhood, we obtain a stack of ones with zeros above at each pixel; adding these yields the correct median filter of the original grey-level image. This observation led to research into *stack filters*, that is, filters other than the elementary median or rank-order filters that possess this threshold decomposition-stacking property. Once the class of such filters had been established, optimal filtering over this class could be investigated.

Finally, in this brief glance at these filters, we mention *vector* median (and related) filters. Electron microscopy is one of the fields in which several signals are commonly associated with a single pixel: in everyday scanning microscopy, in analytical electron microscopy and of course in STEM. These signals may be grey-level values, as in the case of signals from different SEM detectors, but may also be labels or more abstract structures. Ordinary median filtering could of course be applied to each of the individual images separately but this would neglect any relation that might exist between thse images. It was for such situations as this that vector median filters were devised.

This account gives no more than the flavour of this approach to image enhancement. For further details, a convenient reference is the survey by Gabbouj *et al.* (1992) already cited or the book by Pitas and Venetsanopoulos (1990).

3. IMAGE ALGEBRA: A FERTILIZER OF THE IMAGE TERRAIN

A drawback of the enthusiasm with which scientists in so many diverse fields have taken to image processing is that the subject has become inchoate, the same algorithms being used in different fields under different names and expressed in different notation. In order to remedy this, or to set a good example at least, a mathematical structure has been devised, simple enough to be grasped rapidly but general enough to encompass most, ideally all, image processing methods. This is known as *image algebra* and although several versions have been proposed, we confine this account to the algebra of Ritter and colleagues, fully described in Ritter *et al.* (1990) and Ritter (1991), though new growths continue to spring out vigorously from the trunk. The idea is to create a mathematical scheme with as few operations and operands as possible, in terms of which all image processing techniques can be expressed compactly and, in the best of cases, elegantly. One very important point to be borne in mind is that the algebra works with images, not individual pixels though of course these are there in the background.

In fact, we need define only one operand, the *image*, though in practice we give a special name to a particular type of image and tend to regard it as a second type of operand, the *template*. The operators are the everyday ones of addition, multiplication and max (or sup), to which it is natural to add their duals, subtraction, division and min (inf). Images are assumed to have been rendered discrete and suitably quantized (that is, a grey-level has been ascribed to each pixel) so that an image is characterized by a set of position coordinates and the corresponding grey-level values - in short, it is an array but the latter need not necessarily be arranged as a square or rectangular mesh and the grey-level values need not be integers or even real numbers. The obvious example is an image that is the Fourier transform of another, in which case the grey-level values may be complex. More generally, the "grey-level" value might itself be an array or even a matrix or it might be heterogeneous, containing both numerical and logical elements, or labels. The set of data generated by each pixel in a SEM by the various detectors is again the obvious example, to which a label, the chemical element present at that pixel for example, might be added. We then have a multivalued image with several strata.

All these possibilities (and others) are included if we define an image **a** to be the following set :

$$a = \{ (x, a(x)) \mid x \text{ in } X \} \quad a(x) \text{ in } F$$

in which **X** tells us which coordinates are being used and **F** denotes some value set, for example the real or complex numbers or a subset of the integers. For multivalued images, **F** will be the product of the value sets of the individual strata ; if there are n strata, then $F = F_1 \times F_2 \times ... F_i ...\times F_n$, F_i corresponding to the i-th stratum.

The operators act on the same pixel in different images when these are conformable ; if **a** and **b** are images, then

$$c: = a + b = \{ (x, c(x)) \mid c(x) = a(x) + b(x) \}$$

and likewise for the other operators.

Let us now consider an image **t**, defined on coordinates **Y**, so that $t = \{ y, t(y) \}$, each of whose values t(y) is itself an image, defined on **X**, with pixel-values (t(y))(x). Thus **t** might be a 64x64 image at each pixel of which is situated a 3x3 image or even an image defined on a polar or hexagonal grid, whereas the original (**Y**) may have been square. It is usual to lighten the notation by writing t_y instead of **t(y)** so that (t(y))(x) can be replaced without ambiguity by $t_y(x)$. Such images are so important that they have been given a separate name and are called *templates*. Being images, we need say no more about operations between templates but operations between images and templates do need further consideration. Here the most useful operations are compound: they involve *two* of the basic operations, addition and multiplication for example; this particular choice yields convolution:

$$b: = a\,(+)\,t\ = \{\ (y, b(y))\ |\ b(y) = {}_{xinX}\ a(x).\,t_y\,(x), y\ in\ Y\ \}$$

summation   multiplication

The other operations, max in particular, may be used to generate other ways of combining an image and a template. Care must be tahen to specify the appropriate value sets (**F**) correctly.

4. ALGEBRAIC REPRESENTATION OF SOME IMAGE PROCESSING TECHNIQUES

Although the process of expressing the numerous image processing algorithms in terms of this algebra is by no means complete, the results available show how very powerful it is. In the case of enhancement, for example, we expect to find histogram modification, linear and nonlinear filtering represented. In restoration, the Wiener-like filters, three-dimensional reconstruction and the Gerchberg-Saxton algorithm and its descendants are among the candidates. In image analysis, apart from mathematical morphology, we expect to find image description, pattern recognition in general and techniques for measurement of areas, perimeters and the like. We can give only a few examples here; for further details see the work of Ritter already cited, Ritter and Gader (1987), Davidson (1992) and the annual SPIE meetings on this theme (Gader, 1990; Gader and Dougherty, 1991; Gader *et al*, 1992).

Histogram: In order to express the histogram **h** of an image **a**, we need the following template, $t_x\,(y)$:

$$t_x\,(y) = \begin{cases} 1 & \text{if } y = a(x) \\ 0 & \text{otherwise} \end{cases}$$

where **y** in **Y** and $Y = \{\ 0, 1, ...\ G\ \}$ is the set of grey-levels (e.g. 0 - 255) in the image. Then $h = t\,(+)\,1$, where **1** is a uniform image, all of whose grey-levels are unity, since $h(g) = {}_{xinY}\ t_x(g)$ is the number of times an **a(x)** has the grey-level g (example taken from Ritter *et al.*, 1990).

Convolutional filtering. We have already seen that such filters are represented by expressions of the form $a\,(+)\,t$ but we note that the extension to more complicated cases in which the convolution function **t** depends on the position of the pixel in question is easily made: instead of being translation-invariant, the template **t** varies with position.

Median and related filters: Selection of the neighbourhood is easily made by multiplying the image with a template with unit weights in the window area. The operator max then finds the highest grey-level value and the corresponding pixel is deleted from the window; successive applications of max and deletions give us the sequence of grey-level values in the window in decreasing order after which the median (or any other member if we are interested in rank-order filtering) is simply found as the $(n+1)/2$ -th member for a window admitting n pixels. We have not introduced the notion of recursive templates here (fully described in Li, 1992) but these permit a very neat formulation of the median and rank-order filters.

SEM and other multivalued images: A whole chapter of image algebra is devoted to images for which several grey-level or other values are associated with each pixel. Operators are available that allow us easily to combine like strata in different images (Ritter *et al.*, 1990) or different strata in the same or different images (Hawkes, 1992). We have seen that vector median filtering is attractive for such images and these operators should make it easy to incorporate such filters in the image algebra. Work is currently in progress on this.

Analysis. Extensive studies on the algebraic representation of the operations of mathematical morphology have been made. See Davidson (1992) in particular. Structural description has been explored only more recently (Shi and Ritter, 1992).

5. CONCLUDING REMARKS

We hope to have shown that image processing has now become an autonomous science and that this has been beneficial for the development of the subject. So far as image algebra is concerned, we wish to stress that it has proved remarkably fertile. Once algorithms are expressed in its compact notation, it is natural to enquire what will happen if a different template or set of operators is introduced.

REFERENCES

Davidson J L Adv. Electron. Electron Phys. 84 61
Gabbouj M, Coyle E J and Gallagher N C 1992 Circuits Systems Signal Process. 11 7
Gader P D 1990 Proc. SPIE 1350
Gader P D and Dougherty E R 1991 Proc. SPIE 1568
Gader P D, Dougherty E R and Serra J 1992 Proc. SPIE 1769
Hawkes P W 1984 J. Phys. 45 C2 : 195
Hawkes PW 1989 Proc. 12 ICXOM ed S Jasienska and L J Maksymowicz (Cracow: Acad. Mining. Metallurgy Printing House) pp. 351-356
Hawkes P W 1992 J. Math. Imaging & Vision 1 (to be published)
Li D 1992 J. Math Imaging & Vision 1 23
Pitas I and Venetsanopoulos A N *Nonlinear Digital Filters* (Boston and Dordrecht: Kluwer)
Ritter G W 1991 Adv. Electron. Electron Phys. 80 243
Ritter G X and Gader P D 1987 J. Parallel Distrib. Comput. 4 7
Ritter G X, Wilson J N and Davidson J L 1990 Comput. Vision Graphics & Im. Proc. 49 297
Shi H and Ritter G X 1992 Proc. SPIE 1769

Inst. Phys. Conf. Ser. No 130: Chapter 5
Paper presented at Int. Congr. X-ray Optics and Microanalysis, Manchester, 1992

Image analysis of clay scanning electron micrographs

Xiaoling Leng and Peter Smart
Civil Engineering Department, Glasgow University, Glasgow G12 8QQ

ABSTRACT: This paper summarizes a whole set of new image analysis methods for studying individual and grouped particles and global microstructure of soil scanning electron micrographs.

1. INTRODUCTION

Scannning electron microscopes are widely applied in studies of natural and artificial materials. Human examination of these micrographs is far slower and much more empirical than research requires. Image analysis offers a quick and accurate automatic way of analyse these micrographs.

The images used in this paper are back-scattered scanning electron micrographs of clay. The cross-sections of particles are seen as bright linear features and the pores as dark background. In order to cover a sufficiently large field of view and to show details, scanning electron micrographs have to be kept in a certain range of magnification, normally around 2k in our research. In this case, the small clay particles can just be seen, and the whole image is full of particles of different brightness and sizes with blurred edges (Figure 1).

Normal image segmention methods are unsuitable for separating particles from background in this type of micrograph. Simple binary thresholding will ignore the darker particles, and adaptive thresholding will be difficult to apply because the filter size required will depend on particles size. A whole set of new image processing methods has therefore been developed, to processes grey-level image at the pixl level.

2. A NEW TOOL FOR PARTICLE ANALYSIS

The image analysis of individual particle includes the measurement of particle size, shape, size distribution, orientation and contact angle. In all these measurements skeletons (centre axis of particles) play an important role. There have been numerous attempts to obtain accurate skeletons. However the usual binary morphology transformations are not

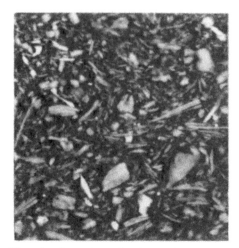

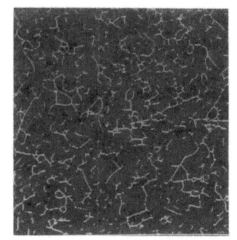

Fig. 1. Scanning electron microscopy of soil with 2k magnification.

Fig. 2. Skeletonized result
of Fig. 1.

suitable for such images as Figure 1, since the smaller and darker
particles need to be considered as well as the largeer and brighter ones.
We have developed a new grey-scale morphological method that is particulaly
well suited to images such as Figure 1, in where many size and grey-level
ranges are present. Difficult step of converting into a binary image is
thereby avoid. A non-linear structure function, namely the local intensity
gradient, is used. Each transformation involves a contrast correction based
on a cubic curve. The erosion formula (for details see Leng 1992) is:

$$I'_e(\imath,\jmath)=\frac{2}{m^2}\left[I(\imath,\jmath)-\frac{1}{2}U(\imath,\jmath)\right]^3-\frac{5}{2m}\left[I(\imath,\jmath)-\frac{1}{2}U(\imath,\jmath)\right]^2+\left[I(\imath,\jmath)-\frac{1}{2}U(\imath,\jmath)\right]$$

where: $I(\imath,\jmath)$, $U(\imath,\jmath)$, and $I'_e(\imath,\jmath)$ are the values of intensity, intensity
gradient, and eroded result at pixel $(\imath,\jmath)$ respectively; m=maxgrey (255).

Figure 2 shows the successful skeletonized image of Figure 1. Although
there are join-points and a few one pixel gaps on the skeletons, they can
be very easily improved by using Hough transform (Costa et al 1991).

3. A NEW TOOL FOR DOMAIN ANALYSIS

Clay particles are often grouped together in sub-parallel areas with
very strong local preferred orientation called domains, or occur in random
clusters. In order to study their behaviour during different mechanical
processes, two image segmentation procedures have been developed to map
these domains and random clusters automatically. Both are based on the
intensity gradient, we define:

$$\mathbf{U} = \text{grad } I, \qquad U = \text{mod } \mathbf{U}, \qquad A = \text{arg } \mathbf{U}$$

thus I, **U**, U, and A are the intensity, intensity gradient vector, strength, and angle of intensity gradient respectively. The '20-14 formula' of Smart & Tovey (1988) was used to calculate the intensity gradient (briefly, this formula uses a neighbourhood of area 20 pixels and is based on calculating th partial derivatives of intensity with respect to x and y up to order 4, there being 14 these derivatives).

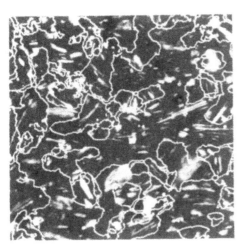

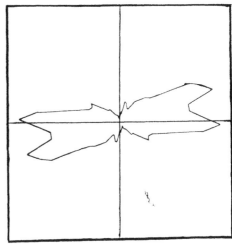

Fig. 3. Segmented domains and random clusters of Fig. 1.

Fig. 4. Polar histogram of enhanced preferred orientation of Fig. 1.

The first procedure encodes angle A into a limited number of directions (Leng & Smart 1991), typically 4, 8, 12 being used. These angles are then smoothed by using a large circular uniform filter . Guided by the semi-variogram, a 20 pixel radius filter is used (Smart et al 1992).

The second procedure first sums the doubled angle of intensity gradient within the filter, then smooths and halves the angle to obtain the local preferred orientation:

$$X = \sum U_i \cos 2A_i, \quad Y = \sum U_i \sin 2A_i, \quad R = \sqrt{X^2 + Y^2}, \quad T = 1/2 \ \arctan(Y/X)$$

where U_i, A_i are the strength and angle of intensity gradient at pixel i, and R and T are the strength and angle of the summed vector; Figure 3 shows the segmented domains of Figure 1; sub-horizontal, down-left, sub-vertical, down-right, and random clusters, are visible.

4. ANISTROPIC ANALYSIS

Platy particles lead to anistropic stress-strain behaviour and permeability. Unitt (1975) first using the polar histogram of angles of intensity gradient to calculate the anistropic index of the image. Later, many improvements have been made, to eliminate the spikes and provide accurate results (Smart & Tovey 1988). However these histograms were still elliptical. The aim of the author's work in this area is to enhance the directional information, thereby smoothing out the noise of the orientation distribution and revealing a more meaningful histogram of direction.

The summed vector of intensity gradient within the filter replaces the intensity gradient at each pixel, giving the local preferred orientation. The weighted polar histogram is seen to have a bow tie shape in Figure 4 rather than an elliptical form, which indicates that the most common directions in the image are between 85° and 110°. Other directions have been chopped off. Within the main tendency, there are two sub-peaks at 90° and 105° giving more detailed information about particles that lie parallel and packed very close to each other.

5. CONCLUSIONS

Pixel-by-pixel based digital image processing provides a whole set of tools for analysing individual and grouped particles and global microstructure of scanning electron micrographs. These new algorithms avoid the first and very difficult step of normal image analysis, binary conversion, and can therefore be applied to some very awkward images. Thus more accurate measurements can be obtained.

ACKNOWLEDEGEMENTS

US AFOSR; SERC; Drs N K Tovey, M W Hounslow and J E S MacLeod.

REFERENCES

Costa L, Leng X, Smart P and M Sandler 1991 *Analysis of clay microstructure by transputer*, Proc. 3rd Int. Conf. on Applications of Transputer, (Amsterdam: IOS Press) pp 329-344.

Leng X 1992 *Analysis of some textured images by transputer*, Ph D Thesis to be submitted to Glasgow University.

Leng X and Smart P 1991, *Improved methods of textural analysis*, Proc. 3rd Int. Conf. Applications of Transputer, (Amsterdam: IOS Press), pp 323-328.

Smart P and Tovey N K 1988 *Theoretical aspects of intensity gradient analysis*, Scanning 10, 115.

Smart P , Leng X, and Bai X 1992, *Image analysis of soil microstructure*, Proc. Int. Conf. on Geotechnics and Computer, Sep.

Unitt B M 1975, *A digital computer method for revealing directional information in images*, J Phys E Scientific Instru., 8 423.

Inst. Phys. Conf. Ser. No 130: Chapter 5
Paper presented at Int. Congr. X-ray Optics and Microanalysis, Manchester, 1992

375

Prospects for a deconvolution of the electron probe size in transmission X-ray microanalysis

E. VAN CAPPELLEN

Laboratoire de Structures et Propriétés de l'Etat Solide; URA 234, bât. C6 : Univ. des Sciences & Technologies de Lille. F-59655 VILLENEUVE D'ASCQ, CEDEX, FRANCE.

ABSTRACT : Spatial resolution of X-ray microanalysis in the (S)TEM is limited by the initial electron probe size and by subsequent beam broadening in the foil. A finite probe size will thus inevitably deteriorate resolution when compared to a hypothetical point source. The adverse effects on X-ray maps and concentration profiles, introduced by the use of finite probes can be attenuated via Fourier deconvolution. A mathematical algorithm susceptible to retrieve hidden information from profiles and maps is presented.

1. INTRODUCTION :

At relatively high magnifications, X-ray maps are blurred as soon as the size of the electron intensity distribution in the specimen becomes visible (i.e. becomes of the order of the sampling distance on the specimen, which is the pixel size of the map divided by the magnification). In the same way a concentration profile across an interphase boundary also suffers broadening because of the finite dimensions of the primary excitation volume. Both blurring and broadening effects can be attenuated via appropriate processing (which in the case of an X-ray map would be called image processing). Instead of using classical sharpening algorithms with their obvious dangers, a model based on Fourier deconvolution of the electron density distribution is investigated. The X-ray signal used to record a one-dimensional profile or a two-dimensional map is basically proportional to the convolution of the real concentration distribution function of the specimen and the electron intensity distribution function; henceforth the latter one will be called the probe function. On condition that this probe function can be modeled with sufficient accuracy, a deconvolution in Fourier space becomes feasible. For reasons mentioned elsewhere (Van Cappellen (1992)), the multiple scattering theory of Doig et al. (1981) is used to describe beam broadening in the specimen and similar considerations will be made here to obtain a probe function depending on microscope and material parameters.

2. ELECTRON INTENSITY DISTRIBUTION :

According to Doig et al. (1981), the intensity distribution in a thin foil generated by a point source is given by the following normalized Gaussian:

$$b(r,z) = \frac{1}{\pi \beta z^3} \exp[\frac{-r^2}{\beta z^3}] \qquad (1)$$

r is the radial coordinate normal to the incident beam and z is the distance from the foil top surface in a direction parallel to the incident beam. The β parameter is given by :

$$\beta = 500 \left[\frac{4Z}{E_0}\right]^2 \frac{\rho}{A} \qquad (2)$$

where E_0 is the electron accelerating voltage in volts, ρ is the specific gravity, Z is the atomic number and A is the atomic weight of the material. In general, the incident intensity in a (S)TEM probe is Gaussian and yields a Gaussian probe function j, normalized for a total incident electron intensity I_e :

$$j(r,z) = \frac{I_e}{2\pi(\sigma^2 + \frac{1}{2}\beta z^3)} \exp[\frac{-r^2}{2(\sigma^2 + \frac{1}{2}\beta z^3)}] \qquad (3)$$

where σ is the variance of the incident beam.

3. DECONVOLUTION IN FOURIER SPACE :

The intensity of the X-ray signal obtained with the beam at position (x_0, y_0) is denoted $s(x_0, y_0)$ and is the product of the real concentration function $c(x,y,z)$ and the probe intensity function $j_{x_0,y_0}(x,y,z)$, integrated over the whole specimen volume (Fig. 1.).

$$s(x_0,y_0) = \int_{-\infty}^{+\infty} \int_{-\infty}^{+\infty} \int_{Z_1(x,y)}^{Z_2(x,y)} c(x,y,z) \, j_{x_0,y_0}(x,y,z) \, a(x,y,z) \, dx \, dy \, dz \qquad (4)$$

where $Z_1(x,y)$ and $Z_2(x,y)$ are the entry and exit surfaces of the foil. $a(x,y,z)$ accounts for X-ray absorption and equals $\exp[-\mu_x \rho \, d(x,y,z)]$, if $d(x,y,z)$ is the distance the photons have to propagate in the specimen before leaving towards the detector and μ_x is the mass-absorption coefficient (in cm^2/g) of the considered X-ray line. The general problem to be solved is : given $s(x_0,y_0)$, how can $c(x,y,z)$ be retrieved ? In its most general shape, this problem can not be solved and a number of hypotheses are needed.

A.   It is assumed that the real concentration function is independent of z, the depth coordinate so that : $c(x,y,z) = c(x,y)$.   This means that all concentration gradients are perpendicular to the beam. For a thin foil this hypothesis is quite realistic, nevertheless in some cases interphase boundaries will have to be oriented edge on to meet this condition.

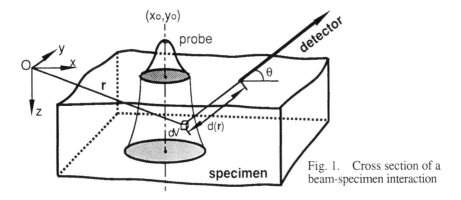

Fig. 1. Cross section of a beam-specimen interaction

B. The entry and exit surfaces of the specimen are assumed to be very slowly varying functions when compared to the lateral dimensions of the probe. This condition is certainly met since the probe rarely exceeds 50nm, broadening included and generally over such distances the foil can be considered uniformly thick. This enables one to replace the boundaries $Z_1(x,y)$ and $Z_2(x,y)$ of the integral over z (depth) by values independent of the coordinates x and y : $\int_{Z_1(x,y)}^{Z_2(x,y)}$ becomes $\int_{0}^{T(x_0,y_0)}$ where $T(x_0,y_0)$ is the foil thickness of the specimen in (x_0,y_0). As a consequence the order of the integrals over x,y and z can be changed. Another consequence of this second hypothesis is that the absorption function becomes independent of x and y : $a(x,y,z) = a(z) = \exp[-\mu_x \, \rho \, z \, \mathrm{cosec}(\theta)]$ where θ is the take-off angle of the X-rays.

With hypotheses A. and B. the expression for the X-ray signal in position (x_0,y_0) becomes :

$$s(x_0,y_0) = \int_{-\infty}^{+\infty} \int_{-\infty}^{+\infty} c(x,y) \int_{0}^{T} j(x-x_0,y-y_0,z)\, a(z)\, dz \quad dx\, dy \qquad (5)$$

In order to simplify notations the integral over z of the probe function j times the absorption function a will be referred to as the generating function g :

$$g(x-x_0,y-y_0) = \int_{0}^{T} j(x-x_0,y-y_0,z)\, a(z)\, dz \qquad (6)$$

In other words $s(x_0,y_0)$ is a two-dimensional convolution of the concentration function $c(x,y)$ and the generating function $g(x,y)$:

$$s(x_0,y_0) = c(x_0,y_0) * g(x_0,y_0) \qquad (7)$$

If the Fourier transforms of the functions s, c and g are denoted with their respective capitals $S(u,v) = \mathcal{F}[s(x,y)]$; $C(u,v) = \mathcal{F}[c(x,y)]$ and $G(u,v) = \mathcal{F}[g(x,y)]$, then :

$$S(u,v) = C(u,v) \cdot G(u,v) \qquad (8)$$

The real concentration function $c(x,y)$ is simply obtained via an inverse Fourier transform :

$$c(x,y) = \mathcal{F}^{-1}[S(u,v)/G(u,v)] \qquad (9)$$

The ratio $S(u,v)/G(u,v)$ becomes dangerous when $G(u,v)$ is of the order of the noise and a Norbert Wiener filter has to be introduced in eq. (9) :

$$c(x,y) = \mathcal{F}^{-1}\left[\frac{S(u,v) \cdot G(u,v)}{|G(u,v)|^2 + \varepsilon}\right] \qquad (10)$$

When $|G(u,v)|^2 \gg \varepsilon$, there is no difference between (9) and (10) and when $|G(u,v)|^2 \ll \varepsilon$ the inverse Fourier transform of $[(S \cdot G)/\varepsilon]$ rapidly decreases. The value of ε has to be adjusted so as to minimise fringe patterns in the final concentration function $c(x,y)$.

4. FOURIER TRANSFORM OF THE GENERATING FUNCTION

$$G(u,v) = \int_{-\infty}^{+\infty}\int_{-\infty}^{+\infty}\int_{0}^{T} \frac{I_e\, a(z)}{2\pi(\sigma^2 + \frac{1}{2}\beta z^3)} \exp\left[\frac{-r^2}{2(\sigma^2 + \frac{1}{2}\beta z^3)}\right] \exp[-i(ux+vy)] \; dz \; dx \; dy \qquad (11)$$

According to hypothesis B. the order of the integrals can be inverted and the fact that the Fourier transform of a Gaussian is a Gaussian leads to :

$$G(u,v) = I_e \exp[-\frac{1}{2}(u^2+v^2)\sigma^2] \int_{0}^{T}\exp[-\beta(u^2+v^2)z^3] \exp[-\mu_x\, \rho\, z\, cosec(\theta)] \; dz \qquad (12)$$

The integral in (12) is denoted $I(u,v)$ and equals :

$$I(u,v) = \int_{0}^{T} \exp(-az^3-bz) \; dz \qquad \text{with :} \quad a = \beta(u^2+v^2) \text{ and } b = \mu_x\, \rho\, cosec(\theta) \qquad (13)$$

The integrand of (13) is expanded in T instead of 0 in order to avoid an alternating series and subsequently integrated :

$$I(u,v) = \sum_{n} (-1)^n\, f_{uv}^{(n)}(T)\, \frac{T^{n+1}}{(n+1)!} \qquad (14)$$

with $f_{uv}^{(n)}(T)$ being the nth derivative of $f(z) = \exp(-az^3-bz)$ in T :

$$f_{uv}^{(n)}(T) = -[(3aT^2+b)f^{(n-1)}(T) + 6aT(n-1)f^{(n-2)}(T) + 3a(n-1)(n-2)f^{(n-3)}(T)] \qquad (15)$$

Series (15) is not an alternating one since the successive derivatives of $f(z)$ have alternating signs. With the two first derivatives of $f(z)$, the recurrent formula (15) enables one to calculate $G(u,v)$ and together with the Fourier transform $S(u,v)$ of the X-ray map, eq. (10) yields the deconvoluted map.

References :

Doig P., Lonsdale D. and Flewitt P.E.J. 1981 Quantitative Microanalysis with High Spatial Resolution (The Metals Society London book 277) pp 41.
Van Cappellen E. and Schmitz A. 1992 Ultramicroscopy **41** pp 193-199.

Inst. Phys. Conf. Ser. No 130: Chapter 5
Paper presented at Int. Congr. X-ray Optics and Microanalysis, Manchester, 1992

379

A STEM study of core-shell structures in an X7R-type dielectric ceramic

R. Al-Saffar, F. Azough and R Freer

Materials Science Centre, University of Manchester/UMIST, Grosvenor Street, Manchester
M1 7HS, U.K.

ABSTRACT: STEM studies of a commercial X7R dielectric, of grain size 0.5 -
2.0 μm, revealed distinct core-shell structures within individual grains. The cores
contained a number of ferroelectric domains whilst the shells were free of domains.
EDS analyses showed that the cores were almost pure barium titanate, whilst the
shells were rich in Nb, Bi and Pb plus Ba and Ti.

1. INTRODUCTION

Ferroelectric Barium Titanate, $BaTiO_3$, is used extensively as a ceramic dielectric in
monolithic and multilayer capacitors. Within the ferroelectric regime (approximately 0-
130°C) undoped $BaTiO_3$ has a relative permittivity which varies in a non-linear manner
with temperature, being ~ 1200 at room temperature (and 1 kHz), but increasing rapidly
to 8000-12000 in the vicinity of the ferroelectric Curie temperature (~ 130°C). A variety
of additives are therefore used to flatten the dielectric peak, to move it to lower
temperatures and to control the change of relative permittivity (ϵ_r) with temperature
(Moulson and Herbert, 1990; Kahn et al, 1988). For so-called "X7R" materials, which
typically contain 90-95% $BaTiO_3$, ϵ_r must not change by more than $\pm 15\%$ from the value
at 25°C over the range -55 to 125°C. Additives used to give X7R behaviour include
$CdBi_2Nb_2O_9$ (Hennings and Rosenstein 1984), $Bi_4Ti_3O_{12}$ (Rawal et al 1981), ZrO_2 (Lu et
al 1990; Hennings et al 1982), and Bi and Nb (Chiang et al 1987). Microstructures of such
ceramics are characterised by heterogeneities in the form of core-shell structures (Daniels
et al 1976; Kahn et al 1988). Using electron diffraction techniques Hennings and
Rosenstein (1984) were able to demonstrate that the cores and shells were not different
phases, but were coherent structures, with paraelectric shells growing epitaxially on
ferroelectric cores.

In this paper we report a STEM investigation of the core-shell structures in an X7R-type
dielectric ceramic.

2. EXPERIMENTAL

Samples were supplied in the form of sheets of fired ceramic approximately 1 mm thick by
a commercial manufacturer of multilayer ceramic capacitors. Phases present in the samples
were identified by X-ray diffractometry, using Cu k_α radiation. Dielectric properties were

determined at 1 kHz over the temperature range 20-150°C using a Wayne-Kerr LCR bridge in conjunction with a dedicated sample cell.

For TEM studies 3 mm diameter discs were machined from the stock samples. These were mechanically thinned to 120 μm and ion-milled to the final form. A Philips EM430 Transmission Electron Microscope operating at 300 keV was used for microstructural investigations.

3.   RESULTS AND DISCUSSION

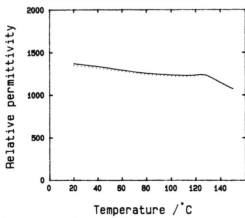

Fig. 1.   Relative permittivity as a function of temperature for X7R dielectric.

The samples were mainly single phase perovskite, of high density (i.e. at least 98 theoretical density on the basis of pure $BaTiO_3$) with a grain size of 1-2 μm.

Figure 1 shows the relative permittivity as a function of temperature. The dielectric peak at the Curie temperature was very small, and the room temperature ϵ_r value was ~ 1450. The temperature dependence of ϵ_r is in accordance with X7R characteristics.

Figure 2 shows the typical duplex structure of X7R-type dielectrics exhibiting core-shell formations. Individual grains comprise an inner core of ferroelectric domains, surrounded by a domain-free shell. Most of the grains had one core, but occasionally two individual cores could be identified (Figure 3).

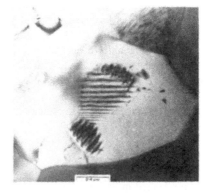

Fig. 2. Typical TEM micrograph showing core-shell structure of grains in X7R type ceramics

Fig. 3. TEM micrograph of a single grain containing two cores.

Chemical analysis (EDS) of the grains showed that the cores were almost pure BaTiO₃, whereas the shells contained small amounts of Nb, Bi and Pb (Figure 4 and Figure 5).

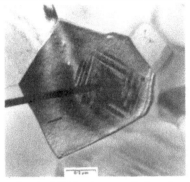

Fig. 4. TEM micrograph showing the domain formation in a grain exhibiting core-shell structure.

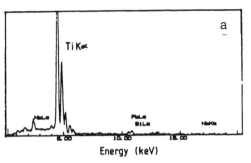

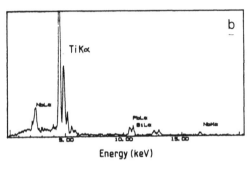

Fig. 5. (a) EDS spectrum of the core of the grain shown in Fig. 4; (b) EDS spectrum of the shell region of the grain shown in Fig. 4.

Figure 6 shows a Scanning Transmission Electron Microscope (STEM) X-ray map for two adjacent grains in the ceramic dielectric. The TEM image of the grains, for which the elemental maps were obtained, is shown in the top left hand corner of the figure. The bottom row shows the original (unprocessed) X-ray data, the middle row shows the processed X-ray data, and the image in the top right hand corner shows an overlay of the processed data. The X-ray map clearly indicates that the additive elements (Nb, Bi and Pb) are mainly located in the shells of the grains.

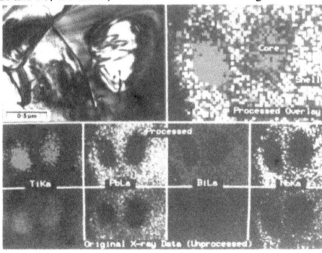

Fig. 6. STEM X-ray map of the X7R ceramic (see text for details).

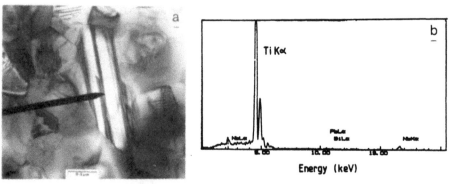

Fig. 7. (a) TEM micrograph of second phase in X7R material; (b) EDS spectrum of second phase.

The X7R formulation was found to be predominantly single phase, but occasionally a minor second phase, containing Nb, Ti and Ba was observed (Figure 7). A comparison of the EDS spectrum of this phase (Figure 7b) with that for the core $BaTiO_3$ (Figure 5a) indicated that the minor phase was rich in Ti, and could be associated with compounds in the Ti-rich region of the $BaO-TiO_2$ binary system.

4.   CONCLUSIONS

The STEM investigation showed that the X7R dielectric ceramic exhibited a core-shell structure in which the grain cores were almost pure $BaTiO_3$, whilst the shells also contained Bi, Nb and Pb.

REFERENCES

Chiang S K, Lee W E and Readey D W Am Ceram Soc Bull 66 (1987) 1230
Daniels J and Härdtl K Philips Res Repts 31 (1976) 489
Hennings D and Rosenstein G J Am Cer Soc 67 (1984) 249
Hennings D, Schnell A and Simon G J Am Ceram Soc 65 (1982) 539
Kahn M, Burks D, Burn I and Schulze W in Electronic Ceramics: Properties, Devices and Applications, Ed Levinson L M Marcel Dekker 1988 p 191
Lu H, Bow J and Deng W J Am Ceram Soc 73 (1990) 3562
Moulson A J and Herbert J M, Electroceramics: Materials, Properties and Applications, Chapman & Hall, London, 1990
Rawal B S, Kahn M and Buessem W R in Advances in Ceramics Vol 1 Ed Levinson Am Cer Soc Ohio 1981 172

Inst. Phys. Conf. Ser. No 130: Chapter 6
Paper presented at Int. Congr. X-ray Optics and Microanalysis, Manchester, 1992

383

Never mind the baby, how about the bath water? Insights from the secondary electron background in electron spectroscopy

J A D Matthew, W C C Ross and M M El Gomati

University of York, Heslington, York YO1 5DD

ABSTRACT: Measurements of the secondary electron background in the range 200-2000 eV are reviewed with an emphasis on the need for a good spectrometer transmission function over a wide energy range. It is shown that the background is characterised by a Sickafus form AE^{-m}, with m of the order of unity. The origin of the Sickafus equation is investigated using a model in which fast secondary electrons are created according to a modified Mott cross section. Monte Carlo studies are supplemented by simple analytic models, which give m values reasonably well in accord with experimental information.

1. Introduction

It was Austin and Starke (1902) who discovered the phenomenon of secondary electron emission in a study of the reflection of cathode rays by metals: they observed that it was possible for more electrons to be emitted than were incident. The bonus of "electrons for nothing" is a bargain that microscopists and spectroscopists have been trying to cash in on for the rest of this century.

An important milestone on the way to Davisson and Germer's classic discovery of the wave nature of the electron was Davisson and Kunsman's earlier measurement of the energy distribution N(E) of secondary electrons emitted from solid metal surfaces by applying a retarding potential to the emitted electrons. They were the first to identify an elastically scattered component in the spectrum, and it was a study of the angular distribution of this feature that led them by something of a random walk to the phenomenon of electron diffraction - see Gehrenbek (1978). Between then and around 1960 ultra high vacuum remained an elusive goal achieved by only a few dedicated workers, and the vast majority of experiments on secondary electron emission were carried out under high vacuum. The general picture of what controlled the energy distribution steadily emerged, for example in the work of Rudberg (1930,1934): below the elastically scattered component characteristic loss peaks could be identified with multiple losses merging into a continuous background associated with rediffused primary electrons which finally emerge from the surface (Figure 1). Below a certain energy, which depends on primary energy and atomic number, N(E) falls with decreasing energy. These rediffused electrons lose energy either by exciting plasmons or by single electron excitations to one electron states above the vacuum level. The excited electrons in turn lose energy by the same mechanisms and so generate a secondary electron cascade distribution (the true secondary distribution) which peaks a few eV above the vacuum level and falls rapidly with increasing energy. The low energy secondary signal originates from up to 100 Å below the surface and is highly topography dependent - an effect exploited in the scanning electron microscope. Above this there is a "no man's land" range of

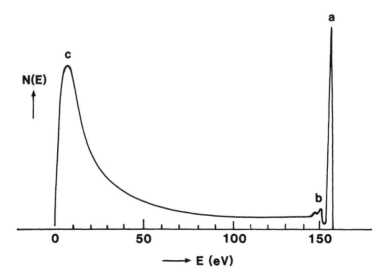

Fig. 1. The secondary electron distribution as observed by Rudberg for Ag distinguishing (a) the elastic peak, (b) characteristic losses, (c) the secondary cascade.

energy where both the true secondaries and the rediffused primaries may contribute significantly to N(E) - typically between 200 eV and 2 keV for a primary energy of 20 keV.

As time went on people became enthusiastic about small bumps in N(E): Lander (1953) and Harrower (1956) correctly interpreted features whose energies were independent of primary energy as Auger electrons characteristic of the target atoms; they were aware of the surface sensitivity of their observations, but the time was not ripe for surface analysis to take off. In the 1960s ultra high vacuum became routinely available and improved electronic techniques accompanied by signal enhancement by acceleration following retardation in LEED systems improved reproducibility. Harris (1968) provided the next key to progress: using a dispersive 127° analyser of the type employed by Lander and Harrower he became irritated by the smallness of the Auger signal, and, prompted by colleagues at General Electric, he differentiated the spectrum with respect to energy using a simple rc time differentiating circuit, a linear sweep of analyser voltage and a phase sensitive detector. He shared his idea with Weber and Peria (1967) who quickly saw that *double* differentiation of the signal on their LEED retarding grid system would similarly yield dN/dE and emphasise Auger peaks. Paradoxically the referees gave Harris a hard time so that Weber and Peria were first to publish. It was like the bursting of a dam: there were now many LEED systems round the globe and semi instantly they could all now do surface analysis; but there was a price to pay in that the differentiation process removed most of the information on the smooth secondary electron background on which the Auger peaks were superimposed. Surface scientists obsessed with their new baby were glad to throw out the bath water.

2. Instrument Evolution

The need to improve signal to noise ratios in experiments led to the gradual replacement of retarding grid analysers by dispersive analysers (cylindrical mirror analysers (CMA) or concentric hemispherical analysers (CHA)): using counting techniques it was convenient to measure N(E) once more, but unlike retarding grid analysers whose transmission function T and detection efficiency D varies with energy in a straightforward way, the more sophisticated spectrometers had more complex T(E) and D(E) functions, ie although the bath water was still there it was adulterated by the measuring instrument! Paradoxically the advent of good highly reproducible surface environments did not lead to better measurements of N(E): most experimenters were glad to concentrate either on the Auger peak to peak heights in their dN/dE or the peak profiles N(E) obtained after carefully subtracting out the local background, which they regarded merely as an unfortunate nuisance associated with the technique. Their interest was no longer in the secondary electron spectrum as a whole.

As the field matured quantification of Auger surface analysis evolved to keep managers happy on the one hand and to keep obsessive physicists fruitfully employed. It was realised how important the electron background was as a secondary source of inner shell ionisation, but also recognised that the measured spectrum should be converted into a true emission spectrum taking account of the transmission function of the instrument and the detector sensitivity. As discussed by Seah and Smith (1990) this is essential if all instruments are to give consistent results, if theory is to be used as an aid to quantification, and if reliable Auger data banks are to be established. Furthermore it is the key to getting the true N(E).

Paradoxically progress in describing the secondary electron background came from application of the earlier instrument technology (the retarding grid analyser) used in N(E) mode. Sickafus (1977a,b), following the ideas of Seah (1969), examined plots of log N(E) v log E to reveal linear sections obeying the simple relation

$$N(E) = AE^{-m}$$

where A and m are constant for a material at a particular primary energy E_p. Typically, m ~ 1 with m modified below intense Auger peaks. Most workers were happy to ignore this information, but gradually in the 1980s the electron background crept back into prominence. Following the first demonstration of Auger imaging by MacDonald and Waldrop (1971) it was realised that the Auger signal from a particular pixel was strongly dependent on topography as well as local elemental concentration, and various schemes were conceived to correct for this using background ratioing - see for example El Gomati *et al* (1992). To quantify you really have to understand what makes the background tick.

In parallel progress was made in understanding how electron spectrometers really worked. Erickson and Powell (1986) and Peacock *et al* (1984) studied how spectrometer field of view varied with electron energy, and Seah and Smith (1990) developed a metrology spectrometer which sets a standard for the removal of spectrometer transmission function T(E) and detector sensitivity D(E) from measured spectra. Figure 2 shows a schematic of the metrology spectrometer, a modified VG Scientific ESCALABII in which the channel electron multiplier is exchangeable for a Faraday cup, the very device that Davisson and Germer (1927) used to make quantum mechanics respectable. Why this reversion to a former previous detector era? Insensitive though Faraday cups are, their response is exceptionally weakly dependent on the energy of the electrons to be analysed, unlike the channel multiplier where D(E) varies strongly with energy. By using a well understood spectrometer in a regime where T(E) ~ E coupled to a detector whose performance can be calibrated by a Faraday cup, Seah and Smith (1990) are able to present standard reference AES spectra in EN(E) form, which, for a particular primary energy, will be characteristic of the material, but

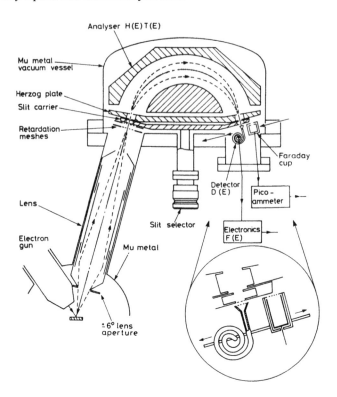

Fig. 2. The metrology spectrometer of Seah and Smith (1990) incorporating a Faraday cup.

the background is just as standard as the Auger spectrum! The way is now open to more discriminating tests of the Sickafus profile.

3. Modelling N(E)

There are two basic approaches to modelling the secondary electron background (1) transport theories, (2) Monte Carlo modelling. The huge amount of work carried out in the field has, however, concentrated on two energy regions: (a) $0 < E < 50$ eV the true secondary region, where a complicated electron cascade of creation and decay governs the distribution. This is particularly important for the understanding of the scanning electron microscope, but controversy still rages over the details of what is going on, see for example Luo and Joy (1990); (b) at the other end of the energy scale huge effort has gone in to modelling the energy distribution of rediffused primaries because of their importance in quantifying electron microprobe analysis and the Auger backscattering factor (Shimizu and Ding (1992)). The "Sickafus" region, typically 200 eV - 2 keV is in a kind of no mans land between the two regimes: in the lower part of the range it will be dominated by the high energy tail of the secondary electron distribution, while the lower energy tail of the rediffused primaries will make contributions at higher energy.

In some respects the true secondary tail is much simpler than the lower energy cascade region, which is particularly complex due to the diversity of electron excitation mechanisms and the low energy of the emerging electrons. Electrons excited to energies above 300 eV simply by one electron excitations of valence electrons and core electrons of binding energy much less than that of the exciting electrons, ie the rediffused primaries. The cross section for scattering a single free electron of zero binding energy to an energy in the range W to W + dW is given by Mott (1930),

$$\frac{d\sigma}{dW} = \frac{4\pi}{E_o} \left[\frac{1}{W^2} - \frac{1}{W(E_o\text{-}W)} + \frac{1}{(E_o\text{-}W)^2} \right] \text{au}^2 \text{ Ryd}^{-1} \; ,$$

where E_o is the energy of the exciting electron.

The formula takes account of exchange with the exciting electron, and is a remarkably good approximation even at quite low values of E_o. Note the symmetry in the expression between "loss" electrons and "excited" electrons due to the fact that they cannot be distinguished. In addition it is possible to generalise the formula to take account of electrons with finite binding energy W_i simply by replacing W by $W + W_i$ ($E_o \gg W_i$).

It is usual to separate the two halves of the curve rather arbitrarily, assigning the top half to the rediffused primary camp, the bottom to the secondaries. It is also important to recognise that generation of fast secondaries is a relatively rare process. The mean free path for excitation of secondaries of energy greater than 300 eV in Al is of order 3×10^4 Å for a 20 keV electron so that few electrons will be created sufficiently close to the surface to escape with energy above 300 eV. The further implication of this is that fast secondaries will themselves seldom create the fast secondaries: this implies that the high energy tail is not a genuine cascade, and, to a first approximation, models which take account of fast secondary generation by rediffused primaries and subsequent transport to the surface will describe the main features of the background.

This was the approach suggested by Matthew *et al* (1988) in some simple quasi analytic models. For $W \ll E_p$ the incident beam will act as an electron generator with approximately W^{-2} weighting, and these electrons will move to the surface suffering both elastic scattering and electron loss. In the simplest case (the straight line approximation, no elastic scattering), where electrons are assumed to have a mean free path independent of energy over the relevant energy range, δ function sources of energy W with a uniform depth density lead to a constant energy distribution at the surface, for $0 < E < W$, as discussed by Dwyer and Matthew (1984). Then the total fast secondary signal is obtained by summing contributions of all sources at energies $W > E$, which leads to

$$J(E) \sim E^{-1} \qquad\qquad (E \ll E_p)$$

ie of Sickafus form with m = -1. Elastic scattering, variation of mean free path with energy, and generation of fast secondaries from states of finite binding energy can individually be shown to modify m somewhat, but in real materials all these and other effects are operating simultaneously. It is a situation ripe for the Monte Carlo approach.

The model of Ross *et al* (1992) adapts the single scattering Monte Carlo method of Joy (1988) and El Gomati *et al* (1991): the primary electron undergoes elastic scattering (screened Rutherford, single scattering model) and energy loss (Bethe (1930)), Rao-Sahib and Wittry (1974)) in the conventional way. On each straight line path between elastic scatterings there is a small probability of fast secondary generation. Random numbers determine both the position and energy of the event, which is given a probability weighting proportional both to $d\sigma/dW$ and the path length between elastic events. The calculations consider generalised Mott inelastic scattering of all electrons (core and valence) that can be excited to energies greater than 300 eV. Indeed the contributions of core electron ionisation to the high energy secondary tail is very significant. The excitation energy is *not* debited to the rediffusing primary: it is assumed to have a normal continuum loss profile, so that a rare fast secondary can, with appropriate low probability weighting, be "created" on each leg of the path, whilst the rediffused primary electron follows a "typical" path history. For an electron with energy $E_o(< E_p)$ the secondaries span the energy range $0 < W < E_o/2$. The loss in energy of the primary electron in creating the secondary electron is not taken account of explicitly, but over many scatterings continuous loss will allow for this loss mechanism. The secondaries are assumed to be emitted in a direction randomly perpendicular to the direction of the primary electron, a good approximation for $E_o \gg W$. Now each fast secondary (originating energy W) is followed according to the same model as the primary electrons, and, when it emerges from the surface with energy $E(< W)$, its original creation probability is added to the appropriate energy bin. Over 30000 trajectories reasonably good statistics are established for electrons in the range 300 eV $< E < E_p/2$. The lower limit is chosen somewhat

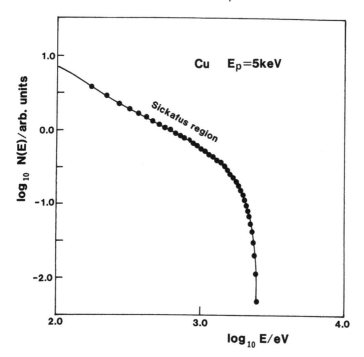

Fig. 3. The energy distribution of fast secondaries emitted from Cu for Ep = 5 keV at normal incidence in log N(E) versus log E form (Monte Carlo model of Ross *et al* (1992))

arbitrarily both as the lowest energy of generation considered and the lowest bin energy.

Below 300 eV the continuous energy loss model becomes increasingly suspect; however, in practice this is a reasonable limiting energy for comparison with experiment. Figure 3 shows the resulting energy distributions for fast secondaries generated in Cu at normal incidence with $E_p = 5$ keV. The yield of secondaries at their maximum possible energy (2.5 keV), but between 300 eV and 1500 eV there is a good linear section in the log N(E) versus log E plot

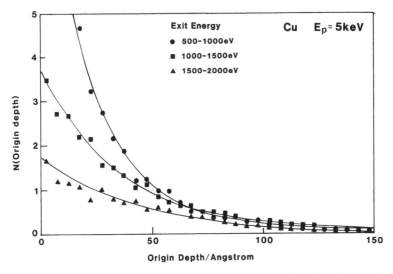

Fig. 4. The depth distribution of fast secondary electron creation for exit energies in the ranges 500-1000 eV, 1000-1500 eV, and 1500-2000 eV in Cu (E_p = 5 keV).

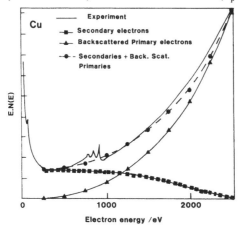

Fig. 5. Comparison of the EN(E) reference spectrum of Seah and Smith (1990) with a Monte Carlo simulation of secondaries and rediffused primaries.

with Sickafus parameter m = 1.0. As E_p varies, modifications of the rediffused primary profile leads to modest variation in m; atomic number and angle of incidence also influences the Sickafus gradient - see Ross *et al* (1992) for a detailed study. It is of interest to know where the fast secondaries that emerge from the surface originate: Figure 4 considers emitted electrons in various energy ranges (500-1000 eV, 1000-1500 eV, 1500-2000 eV) and plots the spatial distribution of their creation depth. Most originated within 100Å of the surface, a result not dissimilar to that for slow secondaries.

Having shown that the model produces Sickafus behaviour it is of interest to see if we can simulate the standard spectra of Seah and Smith (1990). Figure 5 models both the fast secondaries and the rediffused primaries in Cu at $E_p = 5$ keV: note the substantial overlap region where both contribute significantly to $EN(E)$, and the comparison of theory with experiment. The rediffused primary signal is fitted to experiment (scale arbitrary) at 2.5 keV where there is no true secondary contribution, and the relative weight of the fast secondary signal is adjusted to give an optimum fit. The agreement between theory and experiment is reasonable considering the complexity of both the rediffused primaries and fast secondaries.

4. Discussion and Summary

The Monte Carlo model presented here appears to contain the basic ingredients of the Sickafus secondary electron background profile, but it can be improved in a number of ways - a better description of elastic scattering particularly in the lower energy regime, improved treatment of energy loss, and more realistic angular distributions for fast secondary creation. The emergence of sound spectrometer characterisation procedures paves the way for some reliable experimental specifications of the secondary electron distribution $N(E)$ - at last. The York group of Martin Prutton is carrying out a comprehensive study of elemental solids at E_p ~ 20 keV, but investigations at lower primary energies are urgently required. When experimental concensus blends with computer simulation and theoretical insight for homogeneous systems, the backgrounds of more complex materials will come under scrutiny. Until we know more about the bath water we are not going to be able to weigh the baby properly.

The work was partly supported by DTI under contract NPL 82/0476. The authors would like to thank Professor D Joy for providing a single scattering Monte Carlo backscattering code that provided the initial base for the present program.

References
Austin L and Starke H, 1902, Ann Physik 9, 271
Bethe H, 1930, Ann Phys 5, 325
Dwyer V M and Matthew J A D, 1984, Surf Sci 193, 549
El Gomati M M, Barkshire I, Greenwood, J C, Kenny, P G, Roberts R and Prutton M,
  Microscopy: The Key Research Tool March 1992, p29
El Gomati M M, Ross W C C and Matthew J A D, 1991, Surf Interface Anal 17, 183
Erickson N E and Powell C J, 1986, Surf Interface Anal 9, 111
Gehrenbeck R K, 1978, Physics Today, Jan, p34
Harris L A, 1974, J Vac Sci Technol, 11, 23
Harrower G A, 1956, Phys Rev 102, 340
Joy D C, 1988, Private Communication
Lander J J, 1953, Phys Rev 91, 1382
Luo S and Joy D C, 1990, Scanning Microscopy Supp 4, p127
MacDonald N C and Waldrop J C, 1971, App Phys Letts 19, 315
Matthew J A D, Prutton M, El Gomati M M and Peacock D C, 1988, Surf Interface Anal
  11, 173
Mott N F, 1930, Proc Roy Soc London A126, 259
Peacock D C, Prutton M and Roberts R H, 1984, Vacuum TAIP 34, 497
Rao-Sahib T S and Wittry D B, 1974, J App Phys 45, 5060
Ross W C C, El Gomati M M and Matthew J A D, 1992, Surf Interface Anal
  (to be presented for publication)
Rudberg E, 1930, Proc Roy Soc London) A127, 111
          1934, Phys Rev 4, 764
Seah M P, 1969, Surf Sci 17, 132
Seah M P and Smith G C, 1990, Surf Interface Anal 15, 751
Shimizu R and Ding Z-J, 1992, Rep Prog Phys, 487
Sickafus E N, 1977, Phys Rev B16, 1436
Sickafus E N, 1977, Phys Rev B16, 1448
Weber R E and Peria W T, 1967, J Appl Phys 38, 4355

Inst. Phys. Conf. Ser. No 130: Chapter 6
Paper presented at Int. Congr. X-ray Optics and Microanalysis, Manchester, 1992

391

Expressions for $\varnothing(0)$ and the Auger backscattering factor at normal and oblique incidences

H. Benhayoune[1], O. Jbara[1], C. Merlet[2], J. Cazaux[1]

[1] LASSI, Faculté des Sciences, BP 347, 51062 Reims Cedex, France

[2] CGG CNRS, Université de Montpellier II, Place E. Bataillon 34095 Montpellier Cedex 5

Abstract: Two different approaches have been used for finding analytical expressions for the ionisation function at the surface $\phi(0)$ and the Auger backscattering factor. These calculations have been compared to experimental results obtained at normal and oblique incidences by using the tracer method and including the fluorescence corrections (even at oblique incidence).

INTRODUCTION

Most of the quantification procedures in Electron Probe Microanalysis (EPMA) require the knowledge of the $\phi(\rho z)$ and its surface value $\phi(0)$. The same requirements hold for the Auger electron Spectroscopy (AES) with the backscattering factor, r. The definition of $\phi(0)$ in EPMA and of r in AES, is the same, it is the ratio of the X-ray photons (or the Auger electrons) generated at the surface of the specimen (element A, level c) by both primary (dn_p) and backscattered (dn_s) electrons to primary electrons alone. $\phi(0)$ (or r) is given by:

$$\phi(0) = \frac{dn_p + dn_s}{dn_p} = 1 + \frac{\cos(\alpha)}{Q(E_o)} \int_{E_c}^{E_0} \int_{2\pi} \frac{d^2\eta}{dEd\Omega} \frac{Q(E)}{\sin\theta} . dEd\Omega. \tag{1}$$

where E_0 and E_c are respectively the primary energy and the ionisation energy of the element A of interest, α is the angle of incidence with respect to the normal, $Q(E)$ the ionisation cross-section and $d^2\eta/dEd\Omega$ is the angular energy distribution of backscattered electrons, i.e., the number of backscattered electrons that leave the specimen with an energy E and are emitted into the solid angle $d\Omega$, θ is the corresponding angle of emission with respect to the surface. Some empirical or semi-empirical models of $\phi(0)$ at normal incidence have been established, unfortunately there are only few analytical expressions able to give $\phi(0)$ at oblique incidence α: Shimizu (for only $\alpha=30°$ and $\alpha=45°$) and Pouchou et al. (1989). The goal of the present paper is to compare on the one hand the results of two recent approaches developed independently by two of us (Cazaux 1992b, and Merlet 1992c) which can be applied at any angle of incidence and the other hand to new experimental results obtained at normal and oblique incidence by using the tracer method.

THE MODELS

The success of expressions proposed by Cazaux (1992a) and Merlet (1992a, 1992b), at normal incidence allows us to develop, on the same bases, expressions for oblique incidence.

a) The expression of Cazaux (1992b) is :

$$\phi(0)_\alpha = 1 + \frac{\eta(\alpha)}{1+\eta(\alpha)}(3\cos\alpha+1)\left(1-\frac{1}{U_0}\right)\left(1+\ln\left(\frac{1+\eta(\alpha)}{2}\right)/\ln U_0\right) \qquad (2)$$

where $U_0 = Eo/Ec$ is the overvoltage of the element of interest, and $\eta(\alpha)$ is the backscattering coefficient at angle α. To obtain $\phi(0)$, this formulation does not require the knowledge of the substrate composition if the backscattering factor $\eta(\alpha)$ is deduced from the simultaneous measurement of the specimen current (the specimen holder being correctly polarized). If $\eta(\alpha)$ is not measured, it can be calculated by using an equation derived from the work of Arnal et al. (1969): $\eta(\alpha) = \eta(0)[2/(1+\cos(\alpha))]^P$ with $p = -\ln(\eta(0)/\ln(2)$, and with the formulation of Hunger and Kuchler (1979) for $\eta(0)$:

$$\eta(0) = (0.1904 - 0.2236\ln(Z) + 0.1292[\ln(Z)]^2 - 0.01491[\ln(Z)]^3)E_0^{0.1382-0.9221/\sqrt{Z}}.$$

b) The expression of Merlet (1992c) is:

$$\phi(0)_\alpha = 1 + 1.85\cos(\alpha)a.[0.27b_1 + f(\alpha)^q(1.1+5/Z)(b_2-1.1b_3)] \qquad (3)$$

with $b_i = d_i^{-1}\left[1-\left(1-1/U_o^{d_i}\right)/d_i\ln(U_o)\right]$, $f(\alpha) = (2/(1+\cos\alpha))^{15/\sqrt{Z}}$, $q = 1.58 + 0.007Z$,

$d_1 = 0.02Z - m + 1$, $d_2 = 0.1Zf(\alpha) - m + 1$, $d_3 = 0.4Zf(\alpha) - m + 1$,

$$a = Z(-0.00391\ln(Z+1) + 0.00721(\ln(Z+1))^2 - 0.001067(\ln(Z+1))^3)(E_0/30)^{-0.25+0.0787Z^{0.3}}$$

where Z is the mean atomic number.

The two models use the same basis:

i) a simplified Bethe cross-section: $Q(U) = cte.\ln(U)/(E_c^2.U^m)$, with m=1 for K and L levels and m=0.8 for M levels for exp. 3.

ii) The energy distribution of backscattered electrons, $d\eta(\alpha)/dE$ is expressed by δ-function at $\overline{E} = (1+\eta(\alpha))E_o/2$ (4) for exp. 2 while exp. 3 uses a complex functional expression: $d\eta(\alpha)/dW = a[0.27(W)^{0.02Z} + f(\alpha)^p(1.1+5/Z)(W^{0.1Zf(\alpha)} - 1.1W^{0.4Zf(\alpha)})]$ (5) where W=E/Eo. Fig. 1 illustrates the results obtained by using exp. 4 and exp. 5 and shows that exp. 5 represents with accuracy the experimental and Monte Carlo data.

iii) The angular distribution of backscattered electrons $d\eta(\alpha)/d\theta$ is expressed by $(3\cos(\alpha)+1)/4$ in eq. 2 and by $f(\alpha)^{0.2}$ in the second term of the eq. 3.

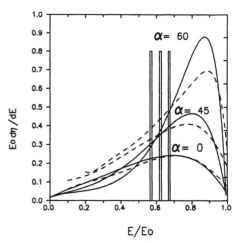

Fig 1. Energy spectra for Al at 20keV. Broken curve experiments of Matsukawa et al. (1974), δ-function calculated by eq. 4, full curve energy distribution calculated by eq. 5.

The two expressions take into account the fact that the backscattered electrons are composed of a diffuse fraction and a single scattering fraction. At normal incidence they lead to a Lambert cosine distribution (Jablonski 1979) and to a factor 2 when integrated over all backscattering angles $\int_0^{\pi/2}(d\eta/d\theta)(1/\cos\theta)d\theta = 2$.

RESULTS

The $\phi(0)$ experimental values (reported in table 1, table 2) have been determined by using the tracer method of EPMA (Castaing and Henoc 1965). The results reported on table 1 have been also submitted to characteristic and continuous fluorescence corrections, by using Monte Carlo simulations at normal and oblique incidence.

Table 1 Comparison of experimental $\phi(0)$ values obtained in this study by the tracer method at normal and oblique incidence with the values calculated by the proposed models (C for eq. 2, M for eq. 3).

System	Eo	0°			30°			45°			60°		
		tracer	C	M	tracer	C	M	tracer	C	M	tracer	C	M
Al/Cu	15	1.71	1.68	1.71	1.63	1.68	1.69	1.62	1.67	1.65	1.57	1.65	1.58
K	20	1.79	1.71	1.74	1.65	1.71	1.71	1.66	1.70	1.67	1.62	1.68	1.59
1.6 keV	25	1.82	1.73	1.76	1.74	1.73	1.73	1.68	1.72	1.68	1.66	1.69	1.60
	30	1.85	1.74	1.77	1.75	1.74	1.74	1.72	1.73	1.69	1.69	1.70	1.60
Al/Au	15	2.01	2.01	2.00	1.82	1.96	1.92	1.79	1.90	1.83			
K	20	2.12	2.06	2.04	1.89	2.01	1.96	1.87	1.95	1.85			
1.6 keV	25	2.13	2.10	2.06	1.94	2.05	1.98	1.94	1.98	1.87			
	30	2.13	2.12	2.08	2.04	2.07	2.00	1.94	2.00	1.89			
Au/Al	15	1.29	1.33	1.34	1.33	1.35	1.38	1.38	1.38	1.42			
M	20	1.34	1.35	1.37	1.37	1.38	1.40	1.40	1.41	1.44			
2.2 keV	25	1.36	1.36	1.38	1.48	1.39	1.42	1.52	1.42	1.45			
	30	1.46	1.37	1.39	1.50	1.40	1.42	1.54	1.43	1.45			
Au/Cu	15	1.45	1.62	1.66	1.50	1.62	1.65	1.49	1.62	1.62	1.47	1.60	1.55
M	20	1.58	1.66	1.70	1.57	1.66	1.68	1.52	1.66	1.64	1.51	1.64	1.67
2.2 keV	25	1.67	1.69	1.72	1.58	1.69	1.70	1.54	1.69	1.66	1.53	1.66	1.58
	30	1.75	1.71	1.74	1.69	1.71	1.72	1.69	1.70	1.67	1.61	1.68	1.58
Au/Ag	15				1.70	1.77	1.80	1.62	1.74	1.73	1.61	1.69	1.62
M	20				1.88	1.83	1.83	1.68	1.79	1.76	1.65	1.73	1.64
2.2 keV	25				1.93	1.86	1.86	1.77	1.82	1.78	1.71	1.76	1.65
	30				1.99	1.89	1.87	1.84	1.85	1.79	1.74	1.78	1.66

Table 2 Same as table 1 but with experimental $\phi(0)$ values obtained by Sewell et al. (1987).

System	Eo	0°			30°			50°		
		tracer	C	M	tracer	C	M	tracer	C	M
Al/C	5	1.11	1.10	1.08	1.07	1.11	1.12	1.01	1.15	1.28
K	10	1.13	1.14	1.12	1.13	1.17	1.18	1.25	1.22	1.34
1.6 keV	15	1.11	1.15	1.14	1.16	1.18	1.21	1.16	1.24	1.35
Si/Al	10	1.20	1.31	1.32	1.26	1.34	1.36	1.32	1.38	1.43
K	15	1.26	1.35	1.36	1.34	1.38	1.40	1.40	1.42	1.46
1.8 keV	20	1.34	1.37	1.39	1.33	1.39	1.42	1.40	1.44	1.46
Bi/Au	20	1.40	1.11	1.50	1.35	1.13	1.48	1.48	1.16	1.45
L	25	1.56	1.32	1.65	1.54	1.32	1.62	1.55	1.32	1.55
13.4 keV	29	1.60	1.44	1.72	1.53	1.46	1.68	1.58	1.44	1.60

As shown on tables 1 and 2, the proposed
equations can be applied with a reasonable
degree of confidence in regard to the
precision of the experimental results (up to
10% errors may be expected, due to the
difficulties encountered in the measurements
in particular for the high α). To show the
behaviour of the two models when α is
changed, the calculated values have been
compared to the results of transport
calculation obtained by Batchelor et al.
(1983). Fig. 2 shows that the results are very
similar for the two models in the angular
range 0-70°. For α > 70° the two models have
the same downward trend, but with a different
limit at 90°.

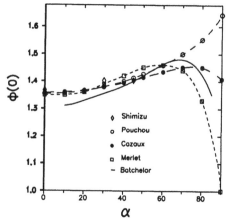

Fig. 2 φ(0) calculations for Si Kα as a function of
the angle of incidence.

CONCLUSION

Despite some restricted discrepancies, it is surprising to observe the good agreements between
the two proposed approaches and the experimental results. In fact we believe that the expression
proposed by Merlet is more accurate than that of Cazaux for reduced energies Uo in the 1-5
interval and its field of application is the accurate point analysis of EPMA. The advantage of
the Cazaux's expression is that it does not require the knowledge of the substrate composition
(if η(α) is measured in parallel). Its field of application concerns the microanalysis of
heterogeneous materials by characteristic x-y imaging (EPMA and AES) and z profiling by
AES.

REFERENCES

F. Arnal, P. Verdier, and P.D. Vincensini 1969 C.R. Acad. Sci 268, 1526
R.D. Batchelor, P. Rez, D.J. Fathers and J.A. Venables 1988, Surf. Interf. Anal. 13 193
R. Castaing and J. Henoc 1965, Proc. ICXOM 4, R. Castaing, P. Deschamps, J. Philibert,
    Hermann Paris, 120
J. Cazaux 1992a, Mikrochim. Acta, suppl. 12, 37
J. Cazaux 1992b, Micros. Microanal. Microstruct. 3, 1
H. J. Hunger and L. Küchler 1979, Phys. Stat. Sol. (a) 56, K45
A. Jablonski 1979, surf. Inter. Anal. 1, 122
H. Kulenkampff and W. Spyra 1954, Z. Phys. 137, 416
T. Matsukawa, R. Shimizu and H. Hashimoto 1974, J. Phys. D: Appl. Phys. 7, 695
C. Merlet 1992a, X-Ray Spectrom. (in press).
C. Merlet,O. Jbara, S. Rondot and J. Cazaux 1992b, Surf. Int. Ana. 19, 192
C. Merlet 1992c, X-Ray Spectrom. (submitted).
J.L. Pouchou, F. Pichoir and D. Boivin 1989, Proc. ICXOM 12, S. Jasienska and LJ.
    Maksymovicz Eds. Vol 1, 52
D.A. Sewell, G. Love and V.D. Scott 1987, J. Phys. D: Appl. Phys., 20,1567
R. Shimizu 1983, Jpn J. Appl. Phys. 22, 1631

Inst. Phys. Conf. Ser. No 130: Chapter 6
Paper presented at Int. Congr. X-ray Optics and Microanalysis, Manchester, 1992

395

The use of integrated circuits for position-sensitive detection

J V Hatfield,[a] J Comer,[b] P J Hicks,[a] A A Wills[b] and T A York[a]

a) Department of Electrical Engineering and Electronics, UMIST, Manchester M60 1QD, UK

b) Physics Department, The University, Manchester M13 9PL, UK

ABSTRACT: An integrated circuit has been developed which, when fitted with microchannel plates, can be used for position-sensitive detection of charged particles, ultra violet radiation and X rays. The integrated circuit consists of an array of charge-sensing electrodes with a corresponding array of amplifiers and counters all integrated on a single chip and giving parallel detection. This has significant advantages over existing position-sensitive detection systems.

1. INTRODUCTION

There are many analytical instruments in which charged particles or photons are detected and spectra are often obtained by spatially dispersing the particles and scanning the image across a slit. It is becoming increasingly common to improve the sensitivity of these systems by employing position-sensitive detectors to allow the simultaneous detection of particles from an extended section of the spectrum. Such systems are subject to various limitations which are largely overcome by using integrated position-sensitive detection. Figure 1 shows an example of an instrument which is a photoelectron spectrometer. A hemispherical deflector disperses the electrons to give an extended image which is linear in energy and can be recorded with a position-sensitive detector.

2. EXISTING POSITION-SENSITIVE DETECTOR SYSTEMS

Most position sensitive detector systems employ microchannel plates (Wiza, 1979) as their first stage. These consist of an array of multiplier channels spaced by about 10-20μm, with a hole area of about 50% and producing a pulse of about 10^6 electrons when a particle is registered at the input. This device preserves the spatial information about the incident particles. It is then necessary to record the arrival and position of this pulse, and techniques for achieving this have been reviewed, for example, by Richter and Ho (1986). These methods can be divided into three classes discussed below.

2.1 Charge Division

The most common example of this technique uses a resistive anode (Moak et al, 1975). The charge pulse from the microchannel plates falls on the resistive anode and is picked off at its edges. Measurements of the resulting pulses are used to deduce the arrival position of the initial particle. The events are processed one at a time and the interval needed for this operation represents a dead time and limits the overall count rate of the detector.

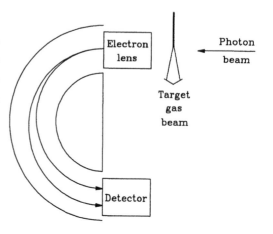

Fig. 1.   Photoelectron spectrometer

2.2 Optical Conversion

It is possible to convert the electron image at the output of the microchannel plates into an optical image by accelerating the electrons onto a phosphor screen. The optical image can then be projected onto a position-sensitive optical sensor using lenses or fibre optics. Specific devices have used vidicon systems (Gelius et al, 1973) and charge coupled imaging devices (CCD's) (Hicks et al, 1980). Such systems can detect simultaneous events at different points and are therefore true parallel detectors capable of operating at high signal levels. However they have significant problems, which arise from the fact that they are not counting detectors but are analog in operation. This can lead to significant problems with intensity non-linearity.

2.3 Discrete Detector Arrays

This technique is an extension of the single detector system. Typically microchannel plates are followed by an array of metal electrodes fabricated on a substrate and placed close to the exit of the microchannel plates. In its simplest form, each electrode is independent with its own discrete set of amplifier, discriminator and counting electronics (Timothy and Bybee, 1975). Such a system is capable of detecting simultaneous events, and operating at high count rates. However, the large number of components places a limit on the number of detector elements that can be conveniently used. Cross talk between adjacent electrodes can also be a major problem. This results from interchannel capacitance; the reduction that can be achieved in the value of this capacitance is limited when discrete components are used.

3. THE INTEGRATED MULTICHANNEL DETECTOR

3.1 Operation

The integrated multichannel detector is based on the discrete detector array concept but it overcomes the disadvantages associated with that technique by integrating all the necessary circuitry, as well as the detector electrodes, onto a single chip (Hatfield et al, 1989/92). The detector is illustrated in figure 2. A pair of microchannel plates is fitted to the integrated circuit, and a particle impinging on them gives a charge pulse of typically 10^6 electrons and 1ns duration. Such charge pulses are collected on an

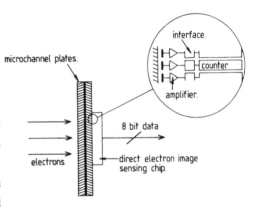

Fig. 2. Schematic diagram of integrated position-sensitive detector

array of electrodes, fabricated as an extra metallization layer on top of an insulating polyimide film over the surface of the chip. Each electrode is connected to a separate, and novel, on-chip preamplifier/comparator whose output controls an 8-bit counter. Silicon area is saved, and the interconnections simplified, by fabricating each channel of the detector directly under its own electrode. The sensing electrodes are on a 160 µm pitch and make a direct contact through the polyimide film to a single stage amplifier. This gain stage is an inverter which is dynamically biased into its cross-over region. When a periodic bias signal is applied to the analog switch, the inverter output is shorted to its input and the charge-capture electrode is automatically discharged to the cross-over voltage. At the same time the comparator samples the potential on the holding capacitor. For each channel a comparison is made between the output of the gain stage and a threshold voltage, and the counter may be incremented, depending on the result of the comparison. A scanning shift register is responsible for periodically multiplexing the data from the counters onto the data bus.

The integrated circuit detector is a 29 channel device, based on a commercial $2\mu m$, double-layer metal, n-well, CMOS process. In practice the detector chips have been extended as hybrid circuits. A 116 channel detector employs four detector chips mounted on a custom-built ceramic substrate. A programmable gate array chip is mounted on the same substrate. Communication between the external world and the detector is via this gate array, whose function is to make the sensing array appear to be a monolithic device and also to provide control signals for the detector chips.

During operation of the integrated multichannel detector the counters are read out at a pixel period which is typically $50\mu s$. In order to achieve this without interrupting data accumulation each counter is disabled for one sample period during readout onto the data bus. the period during which all the counters are read out, and therefore the period during which data is collected in each counter before readout, is the frame time.

3.2 Dead Time

The system dead time is determined by the readout frequencies. Two contributions are significant. The time taken to discharge the electrodes after sampling is 25ns and the electrodes are not sensitive during this time. In addition the counters are read out during one sample period for each frame and they are not sensitive during this time. The total system dead time is therefore given by:

((sample time/frame time)+(discharge time/sample time))x100%

Taking a worst-case example with an electrode sampling rate of 1MHz and with very high count rates up to 10^8 counts/cm^2 sec expected, the frame time cannot exceed $256\mu s$ to avoid counter overflow. Under these conditions the total system dead time is 3%. If lower data rates are expected, the frame and sample time can be extended, with a corresponding decrease in dead time. It should be noted that having fixed the operating frequencies, the dead time is fixed and calculable, and does not then depend on the actual count rate, so that this does not lead to any intensity non-linearity.

3.3 Count Rates

The maximum count rate for the integrated multichannel detector system may be determined by the microchannel plates or by the rate at which data can be transferred to the computer. When operating a system in which the pixels are read directly into the computer the minimum pixel time is limited to about 64μs for a typical 386PC. This corresponds to a frame time of about 8ms. Since the counters on the detector each have a capacity of 256, this leads to a maximum count rate of 3×10^4 counts/pixel sec. The pixels each have an area of $3 \times 10^{-3} cm^2$ so the maximum count rate is then 10^7 counts/cm^2 sec.

In order to increase this count rate capability a hardware buffer has been constructed. This system has two blocks of memory. Up to 256 frames of data are accumulated in one block and accumulation is then switched to the second block while the first set of data is read out. This process then continues, and effectively the higher data rate is made possible because up to 256 frames of 8 bit pixels are converted in hardware into one frame of 16 bit pixels. The use of this buffer between the detector and the computer extends the maximum count rate to 10^6 counts/pixel or 3×10^8 counts/cm^2. In order to achieve such high count rates the detector has been fitted with high output microchannel plates and has then been operated successfully at count rates of 3×10^8 counts/cm^2 sec.

The electrodes and amplifiers of the integrated detector present a very low input capacitance making them sensitive to small electron pulses, down to about 10^4 electrons. This in turn makes the system very forgiving of variations in microchannel plate gain which in other systems leads to intensity non-linearities at high count rates.

3.4 Interchannel Cross Coupling

Cross coupling is potentially a severe problem with discrete anode detectors and can result in spurious signals being recorded particularly on pixels which are adjacent to one that is registering a genuine signal. This can result in degradation of resolution and other distorsions in the spectrum. The effect is a consequence of the wide pulse height distribution from the microchannel plates. It is a serious problem if there is only about an order of magnitude difference between a cross coupled pulse and a signal pulse. It implies that a discriminator threshold that rejects all cross coupled pulses will also reject a significant number of real signal pulses from the lower end of the pulse height distribution.

The geometry of the integrated detector helps to reduce cross coupling by maintaining a very low inter-electrode capacitance. The microchannel plates provide an effective ground plane 10μm above the electrodes. The final surface layer of each detector chip in the array consists of an insulating spacing layer, aluminized so that it acts as an electrical terminal to the back face of the microchannel plates. Windows cut through this layer expose the electrodes to the incoming electrons. There is an additional ground plane, on the silicon substrate, about 10μm below the electrodes. The electrode array is thus effectively surrounded by a ground plane which reduces interchannel coupling to less than 1% between adjacent electrodes.

This conclusion has been experimentally verified. In preliminary work, an array was fabricated with electrodes all connected to amplifiers, but with selected electrodes covered with an insulating layer. Even when the device was operated with a strong signal registering on the exposed electrodes no signal was observed on the covered electrodes.

4. APPLICATIONS

Since the first stage of the integrated multichannel detector is a pair of microchannel plates, the detector is sensitive to any particles for which microchannel plates are suitable. These include UV, X rays, electrons, positive and negative ions and energetic neutral particles.

One example of a specific application is in UV photoelectron spectroscopy. A 116 channel integrated multichannel detector has been fitted to a hemispherical analyser in the photoelectron spectrometer illustrated in figure 1. This has been used at the Daresbury Synchrotron Radiation Source to study a variety of gas phase systems. In hydrogen chloride, for example, a number of interesting effects are observed. The use of multichannel detection has allowed detailed spectra to be taken as a function of both photon energy and photoelectron kinetic energy. This is done by measuring a conventional photoelectron spectrum at one photon energy and then incrementing the photon energy by a step, typically 10meV, before measuring the next photoelectron spectrum. This process is repeated over the energy range of interest in order to produce a two dimensional spectrum.

An example of such a spectrum is shown in figure 3. The electron yield is plotted on a gray scale with intensity increasing from light to dark. The

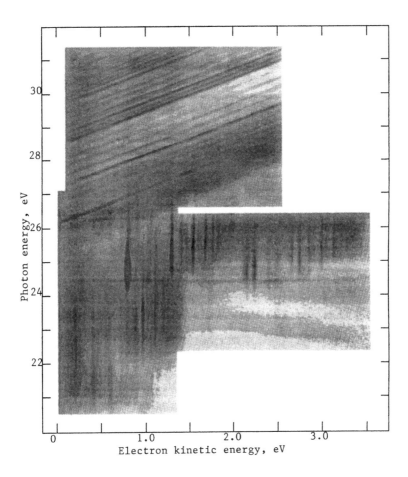

Fig.3. Photoelectron spectrum of hydrogen chloride

diagonal bands in the upper part of the figure have electron kinetic energy increasing with photon energy, and are due to molecular photoionization into both bound and dissociative states of HCl^+.

$$h\nu + HCl \rightarrow HCl^+ + e$$

The sharp lines can be assigned as vibrational bands and represent the first vibrationally resolved measurements of inner valence states of HCl^+.

A significant advantage of the technique of two dimensional spectroscopy is that a range of lines are measured in the same experiment, together with backgrounds, so that normalization is more reliable. A further advantage is that the technique often reveals other processes which may be difficult to identify in conventional spectra. This is illustrated in the lower half of figure 3. This energy region is dominated by bands of constant electron kinetic energy. These are due to neutral photodissociation followed by fragment autoionization.

$$h\nu + HC\ell \rightarrow H + C\ell^* \rightarrow H + C\ell^+ + e$$

These measurements provide a technique for obtaining detailed information about free radicals which may be difficult to obtain by more conventional techniques. The interpretation of the data on HCℓ has been discussed in some detail by Wills et al (1992).

ACKNOWLEDGMENTS

The authors would like to thank the Science and Engineering Research Council for initial funding of this project, Alan Gundlach and Staff of the Edinburgh Microfabrication Facility for their assistance with extra layer processing and Hybritek (Oldham) Ltd., for providing wafer slicing and bonding facilities. They would also like to thank D Cubric, M Ukai, F Currell and T Reddish for producing the results on HCℓ shown in section 4.

REFERENCES

Gelius U, Basilier E, Svensson S, Bergmark T and Siegbahn K 1973 Uppsala University, Institute of Physics, Report No 817
Hatfield J V, York T A, Comer J and Hicks P J 1989 IEEE J. Solid State Circuits SC-24 704
Hatfield J V, Burke S A, Comer J, Currell F, Goldfinch J, York T A and Hicks P J 1992 Rev. Sci. Inst. 63 235
Hicks P J, Daviel S, Wallbank B and Comer J 1980 J. Phys. E: Sci. Instrum. 13 713
Moak C D, Datz S, Garcia Santibanez F and Carlson T A 1975 J. Electron. Spectrosc. Relat. Phenom. 6 151
Richter L J and Ho W 1986 Rev. Sci. Inst. 57 1469
Timothy J G and Bybee R L 1975 Appl. Opt. 14 1632
Wills A A, Cubric D, Ukai M, Currell F, Goodwin B J, Reddish T and Comer J 1992 to be published.
Wiza J L 1979 Nucl. Instrum. Methods 162 587.

Inst. Phys. Conf. Ser. No 130: Chapter 6
Paper presented at Int. Congr. X-ray Optics and Microanalysis, Manchester, 1992

The use of the Rutherford cross section for elastic scattering of electrons at low beam energies and high atomic number targets

A M Assa'd, M M El Gomati and J Robinson[*]
Electronics Department, * Computing Service, University of York, Heslington York YO1 5DD, UK

ABSTRACT: A modified version of the Rutherford elastic scattering cross section containing a universal relationship of the screening parameter as a function of electron energy, the target atomic number and density is proposed. The modified formula has been used to predict the backscattering coefficient η, the Auger backscattering factor and its spectral background in the energy range 200-2500 eV. The results obtained compare well with the experimental values.

1. Introduction:

The use Monte Carlo calculation in electron probe microanalysis has been one of the most important methods for understanding electron solid interaction. Several models have been proposed over the years varying in the degree of sophistication and scattering details described. The availability of faster computers has seen a gradual shift from the popular and simpler Rutherford scattering cross section to describe elastic scattering to using the more accurate Mott cross section. This has particularly been the case for electrons of energies below 10 keV and elements of atomic number (Z) greater than Cu where the former expression is less accurate. Tabulated results using the partial wave expansion (PWE) method of the Mott cross section for a wide range of elements and energies has been published by Czyzewski et al (1990) and Browning (1991) has suggested an empirical fit to one such set of data. While the use of the Mott cross section has been necessary in some cases, a wide range of microprobe applications can be adequately described by a less detailed treatment.

The approach adopted in the present work has concentrated on modifications to the screening parameter used in the Rutherford formula in a single scattering Monte Carlo model (Joy 1988) to yield the experimentally measured η values. We propose here an expression for this parameter as a function of the incident electron energy, E_P; the target atomic number and density, Z and ρ respectively. A comparison of the differential cross section of the Rutherford formula using the present expression for the screening parameter and another suggested by Bishop (1966) and with the Mott cross section given by Czyzewski et al (1990) is made and the justification of the present approach is given. This expression has been included in our Monte Carlo model and used in several microprobe applications. The results obtained are quite encouraging.

2. Elastic scattering:

The screened Rutherford cross section for elastic scattering of electrons through an angle θ in a solid angle $d\Omega$ can be written as

$$d\sigma(\theta) = \frac{Z^2 e^4}{16(4\pi\epsilon E)^2} \frac{d\Omega}{[\sin^2(\theta/2) + (\theta_o^2/4)]^2} \qquad (1)$$

where

$$\theta = \lambda/2\pi R \qquad (2)$$

λ = h/mv is the de-Broglie wavelength, where h is the Planck's constant, m and v are the mass and velocity of the incident electron, respectively. R is the screening radius of the atom and is given by

$$R = j \, a_H \, Z^{-1/3} \qquad (3)$$

where a_H is the Bohr radius and j is a numerical constant. In equation 1, $\theta^2/4$ is called the screening parameter, α, that can be derived from eqs 2 and 3 as:

$$\alpha = K \, Z^{0.67}/E \qquad \text{(E in eV)} \qquad (4)$$

Several authors have used different values of K and j to fit the calculated to the measured η values:

K = 3.43	at	j = 1	(Bishop, 1966)
K = 4.711	at	j = 0.790	(Murata et al, 1971)
K = 4.34	at	j = 0.885	(Werner and Heydenreich, 1984)

In the present study K was adjusted in a single scattering Monte Carlo model (Joy 1988) so as to obtain the experimental η values measured by Reimer (1985). The latter was chosen because it constituted a more complete and consistent set of data. K values obtained for 12 elements at incident energies between 2 and 20 keV were calculated. A numerical fit of K as a function of E_p (in kev), the density ρ (in g/cm^3) and Z was generated. This has the following form:

$$K = a + e^{bE_p + c} \qquad (5)$$

where
$$a = \ 3.706 + 8.784 \times 10^{-3} \, Z + 0.2613 \, \rho,$$

$$b = -0.283 - 6.820 \times 10^{-4} \, Z + 7.191 \times 10^{-3} \, \rho,$$

$$c = \ 1.641 + 3.595 \times 10^{-2} \, Z - 2.776 \times 10^{-3} \, \rho,$$

3. Results:

A comparison of the experimental η values of Reimer (+) and those calculated using K of equation (5) solid lines, K = 3.43 dashed lines and the Mott cross section (o) (Czyzewski et al) is given in figure 1. As can be seen, the agreement between the experiment and the modified formula is generally very good and particularly at the low energy region and heavy elements in comparison to the other formulae. These results can be explained by inspecting the form of the differential scattering cross section of the formulae used, figure 2. As expected, the agreement is quite good for the higher energy data between all three formulae while at lower energies the discrepancies are much larger. For example, Au at 2 keV the new formula acts to lower the cross section and hence reduce η. However, although the reduction at lower angles is particularly large, this is not significant because small angle scattering has little influence in the backscattering coefficient and the Auger backscattering factor, and this region is normally dominated by inelastic scattering. A comparison of the angular region where elastic scattering is dominant, (i.e between 20° and 120°) shows that the present modification averages that of the Mott cross section. The net effect is that the behavior of both the present expression and the Mott cross section at these angles and for low energies are similar.

Further, the use of equation 5 has not greatly altered the shape of the energy distribution of the backscattered electrons except at the lower energy region. In a separate study to simulate the spectral background in Auger electron spectroscopy (Ross et al 1992) it has been found that by including the part of the background due to backscattered electrons calculated using the present formula, the simulated and measured backgrounds agree rather better than if the backscattered part with a K value of 3.43 is used.

In addition, the Auger backscattering factor was also calculated at different energies to compare with recent measurements by Cazaux et al (1989). Again the agreement with the experiment is rather good and a comparison with the data of Ichimura et al (1983) is given in figure 3.

4. Conclusion:

The modification to the screening parameter proposed here greatly extends the application of the Rutherford elastic scattering cross section to lower energies than has been previously possible. In particular, it has been successfully used in two separate investigations involving the magnitude and energy distribution of Auger electrons and the spectral backgrounds in Auger spectroscopy. The addition of these modifications will further extend the application of the Rutherford cross section in a single scattering Monte Carlo model to layered samples and those featuring buried and topographically raised structures.

The present work was partly supported by DTI under the NPL 82/0476 contract and is duly acknowledged. The authors would also like to thank Prof D. Joy for a copy of the tabulated Mott cross section data.

REFERENCES:

Bishop H. (1966), Ph.D Thesis Univ. of Cambridge.
Browning R, (1991), Appl. Phys. Lett. $\underline{58}$ (24).
Cazaux J, Jbara O, Nassiopoulos A G and Valamontes E, (1989), Paper presented at 12th Int. Congr. on X-ray Optics and Microanalysis, Krakow.
Czyzewski Z, MacCallum D O, Romig A, and Joy D C, (1990) J. Appl. Phys. $\underline{68}$ (7).
Ichimura S, Shimizu R, and Langeron J P (1983) Surf. Sci. $\underline{124}$, L49.
Joy D C (1988), private communication.
Murata K, and Matsukawa T, and Shimitzu R, (1971) Jap. J Appl. Phys, $\underline{10}$.
Reimer L (1985), Scanning Electron Microscopy, Springer-Verlag.
Ross W C, El Gomati M and Matthew J A D (1992), to be published.
Wentzal G, (1927), Z. Phys., $\underline{40}$.
Werner U, Heydenreich J (1984), Ultramicroscopy $\underline{15}$.

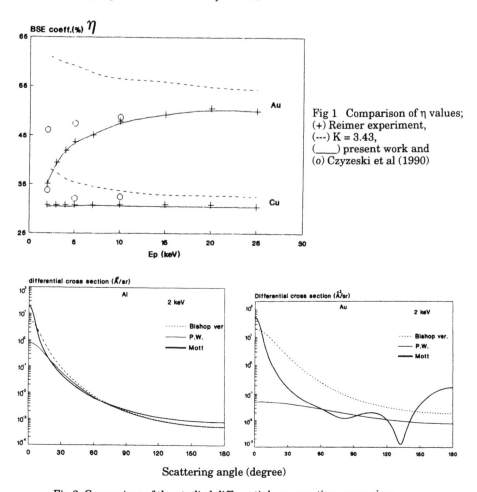

Fig 1 Comparison of η values;
(+) Reimer experiment,
(---) K = 3.43,
(___) present work and
(o) Czyzeski et al (1990)

Scattering angle (degree)

Fig 2 Comparison of the studied differential cross section expressions

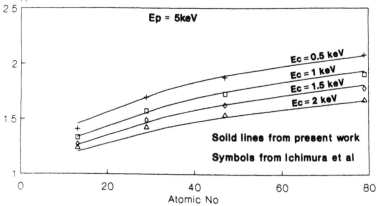

Fig 3 Auger backscattering factor R as a function of Z for different ionization
energy Ec and electron beam energy Ep = 5 keV at normal incidence

Inst. Phys. Conf. Ser. No 130: Chapter 6
Paper presented at Int. Congr. X-ray Optics and Microanalysis, Manchester, 1992

A simple method of testing the cleanliness of ion bombarded surfaces in Auger microanalysis

L Frank

Institute of Scientific Instruments, Czech. Acad. Sci., Královopolská 147, 612 64 Brno, Czechoslovakia

ABSTRACT: Contrast differences between line-scans in the energy filtered total electron emission taken respectively at several tens of eV and at several hundreds of eV were found to show up remarkably small differences in surface chemical composition, i.e. local differences in the work function on a partially cleaned surface, so that comparison of such scans can be used as a test for the cleanliness of a rough ex-situ prepared surface subject to ion beam cleaning. The comparison can be quantified by a single numerical quantity which allows one to monitor the cleaning process in a highly effective way.

1. INTRODUCTION

For a rough specimen surface, being prepared by ion beam cleaning for a surface microanalysis, the question always arises whether the surface, including any cavities present, is sufficiently cleaned. The result is, as a rule, checked by measurement of spectra at several points selected from the secondary electron image; this is relatively time consuming and there is some uncertainty about the selection of the analyzed points. This paper suggests a novel method of monitoring the contamination layer removal, suitable for the case described.

2. PRINCIPLES OF THE METHOD

In the most frequent practical case, the "contamination layer" is composed of oxygen and carbon in various types of physical and chemical coexistence with the specimen material below the layer. Oxygen and carbon (but not only they) change, as a rule, the work function; oxidation increases it in most cases while the carbidation has usually an opposite influence. If an ex-situ prepared and environment exposed, i.e. homogeneously contaminated specimen is being ion beam cleaned, then, with an exception of an ideally smooth surface, the sputtering-off rate locally varies with the surface relief. Consequently, the contamination layer removal varies from point to point and local work function differences arise. After some bombardment, the contamination layer is fully removed and the work function stops changing. It remains either laterally constant for a homogeneous specimen or some true work function contrast of the clean specimen is preserved.

In other words, the cleaning process is characterized by work function changes in time. For not perfectly smooth surfaces, these changes are transformed into variations accross the surface plane so that the work function value statistics within the field of view is in movement during the cleaning process. This phenomenon can be utilized for making the self-calibrating measurement, dependent on relative differences only.

The second principle of the proposed method is the way of detecting the local work function differences. The work function influences the total rate of the secondary electron emission, i.e. the balance between SE and BSE parts of the overall spectrum. In the $E \times N(E)$ type of spectra (e.g. from CMA), this ratio determines the overall slope of the inelastic BSE spectrum background growing from the absolute minimum somewhere around or below 100 eV. Work function mapping is, therefore, equivalent to mapping of this slope which can be very simply done by comparison of two energy filtered images taken at sufficiently distant energies within the slope, say at some tens of eV (near the absolute minimum of the $E \times N(E)$ curve) and at some hundreds of eV. The differences between these images, mainly major ones as are the contrast reversals between selected pairs of pixels, signalize the work function differences. The absolute value of the image signal difference between both mappings taken as vectors could be a measure of the work function inhomogeneity within the field of view. Registering this quantity during the ion beam cleaning of the contaminated surface, we can expect a random virgin situation that is probably difficult to observe owing to measuring beam damage. Then, mostly some increase followed by a sequential decrease takes place and the difference approaches some final "true" value indicating the "end-of-cleaning". Instead of the full mappings, two single line-scans accross a field of view or a region dubious from the point of view of cleanliness have proved to be sufficient.

The method proposed can be applied when there are similar topographic components in the images under comparison or when at least the differences are small with respect to the work function effects observed. The experiments with various specimens have shown that in the range between the absolute minimum of $E \times N(E)$ and the energy of approximately 600 to 1000 eV (for the primary energy of 3 keV), i.e. in the range of the overlap of SE and BSE regions, the topographic contrast exhibits no drastical variations.

3. EXPERIMENTAL VERIFICATION

The experiments were made in the Perkin Elmer PHI 595 Auger microprobe using 3 keV primary electrons. The standard multi-specimen carousel with no rotation facility was used, so that a fixed ion beam/surface geometry was available only by using the 4 keV Ar^+ gun. Various metal sheets were used as the specimens exhibiting a more or less regular topography produced by parallel grinding or by sand blasting and glass ball blasting. A "perfectly" cleaned surface was produced by sufficiently long ion beam bombardment when the one-directional grinding marks were oriented as exactly as possible parallel to the ion beam projected onto the surface, and a "partially" cleaned surface was produced when the grinding marks were in another orientation.

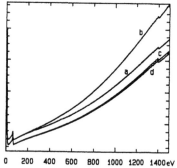

Fig. 1. demonstrates the difference in the inelastic background slope (for a $E \times n(E)$ curve) connected with the work function variations. The curve couples are mutually scaled to fit in the absolute minima for a better comparison but, in fact, it is the lower energy part that changes owing to cleaning.

Fig. 1. Spectra taken on the opposite slopes of a relief top on the Al ground surface; a, b: partially cleaned, work function differences, c, d: well cleaned surface.

Having the well cleaned Al surface imaged at various emission energies (see Fig. 2.), we can compare the topographic contrast components for these energies; within the region 100 eV to 600 or 1000 eV the contrast changes are obviously not large, they surely do not prevail over the work function effects. This conclusion was verified with more various metal specimens.

The measurement energies actually used should be selected from the outside of any Auger peaks in order to simplify the interpretation. A true measurement series shown in Fig. 3. clearly demonstrates the sequential mutual approach of the both scans when an increasing ion dose is applied (up to 5.6×10^2 Ccm^{-2}). Some stepwise removed oxidized sites are observable as local minima along the 90 eV virgin curve. In principle, the 90 eV curve approaches the 600 eV one, in accordance with the expectation. The rest contrast difference of the last couple of curves is well negligible with respect to the virgin one.

The first three figures present examples of the experimental results verifying qualitatively the assumptions used in the previous section. Attempting at quantification of the method, first of all an experiment was made using a highly artificial geometry, namely a cavity in a Cu surface cratered by using a steel ball (see Fig. 4.). In the course of cleaning, the correlation was searched between the dose dependences of the contaminant element percentages and of the ratio of contrasts

$$\varepsilon = S_1(E_x) S_2(E_y) / S_1(E_y) S_2(E_x) \qquad (1)$$

where 1 and 2 characterize the surface points, and x and y are the energies selected (100 and 600 eV, respectively). Point 1 was located in the directly bombarded region but point 2 (inside the shadowed region the boundary of which is visible in the micrograph) was impacted only by primary, reflected or sputtered particles rediffused in the residual atmosphere or by ions reflected at an angle larger than 125° from the mirrored ray. This gives a sputtering rates ratio of about 1:30 and causes consequently large differences in the cleanliness and in local work function values within an extended interval of ion doses. Just within this interval the ε value changes significantly, reaching a pronounced maximum fitted to the maximum difference in, e.g., Cu contents, and then again approaches unity after prolonged bombardment sufficient even for cleaning the indirectly bombarded area. These results clearly demonstrate the sensitive and unambiguous reaction of the contrast difference to the uniformity of the surface cleanliness.

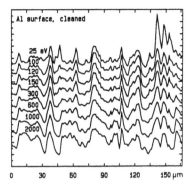

Fig. 2. Line-scans taken at various emission energies (see curve labels) across the grinding marks on a well cleaned Al surface.

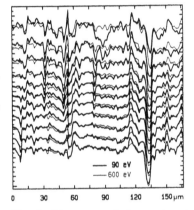

Fig. 3. Line-scans of the sand and glass ball blasted Cu surface for ion doses growing from top to bottom.

Quantification of the contrast difference between both line-scans was developed in the form of a simple algorithm easily implementable for routine measurements. Let the scans be represented by sets of pixel intensities $\{X_a\}$, $\{X_b\}$. The items X_b are transformed into $X'_b = A X_b + B$ so that $E(X'_b) = E(X_a)$ and $D(X'_b) = D(X_a)$ where $E(X)$ denotes the mean value and $D(X)$ the variance of X, respectively; this compensates for any possible measurement differences and drifts. Having the scans mutually fitted in this way, the contrast difference, i.e. the separation of the curves, can be measured in terms of the quantity

$$\Delta = \left[\frac{E\left[(X_a - X_b')^2\right]}{D\,(X_a)} \right]^{1/2}$$

(2)

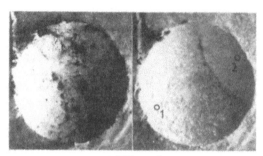

|—150 μm—|

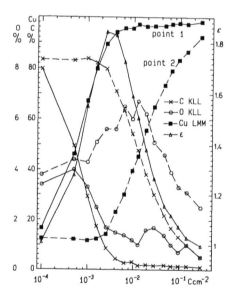

Fig. 4. Ball cratered cavity in Cu surface before (left) and after (right) cleaning; bottom : ion dose dependences of the point analyses and the contrast ε for the points indicated.

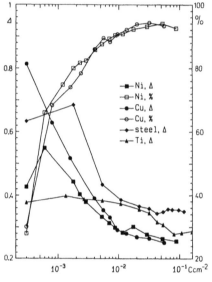

Fig. 5. The course of cleaning several metal surfaces measured with the help of Δ, for Ni and Cu compared with the development of the substrate material contents.

Fig. 5 presents examples of the practical use of the proposed method for monitoring the cleaning process for some sand and glass ball blasted metal surfaces. From the Δ(D) curves, namely from the location of their knees, the ion doses sufficient for the surface cleaning can be determined in the range of 10^{-2} to 10^{-1} Ccm^{-2}.

Regular measurement of two line-scans with a low energy resolution and high signal-to-noise ratio (because no background is subtracted) and their evaluation and monitoring in the form of the single quantity seem to be a highly effective and easily implementable procedure suitable for the routine check of the surface cleanliness during the ion beam bombardment.

ACKNOWLEDGEMENTS: The author is deeply indebted to Professor E. Bauer and Dr. A. Pavlovska for having been given the opportunity to carry out all the measurements in the Institute of Physics, Technical University Clausthal.

Inst. Phys. Conf. Ser. No 130: Chapter 6
Paper presented at Int. Congr. X-ray Optics and Microanalysis, Manchester, 1992

411

A new ion optical system for SIMS and SNMS

S M Smith, R C Wilson*, J C Vickerman.

Chemistry Department, UMIST, Manchester, M60 1QD.
*VSW, Warwick Rd, Old Trafford, Manchester.

We have designed and constructed an ioniser and ion optical system for the performance of Secondary Ion Mass Spectrometry (SIMS) and Sputtered Neutral Mass Spectrometry (SNMS). SNMS can be performed using an electron beam postioniser or by a novel thermionically maintained discharge. Both modes of operation are significantly more efficient than existing instrumentation. These improvements have been made without degradation of the SIMS efficiency. The basic principles and calibration of the instrument are described and an application to the analysis of noble metal alloys is presented.

1. INTRODUCTION

Although SIMS is an excellent tool for chemical and physical characterisation of surfaces (Vickerman 1988) it is unable to provide data which can be easily related to the concentration of surface species. This is due to severe matrix effects which result in the ionisation probability of some species varying by as much as three orders of magnitude depending on the chemical environment (Benninghoven 1987). Quantification of SIMS data is thus only possible when the chemical nature of the sample matrix is known and matrix matched standards are available. This obviously prevents the application of SIMS to the quantification of genuine "unknown" samples. However, when materials are sputtered the vast majority of sputtered species (>99% in most cases) leave the sample in an uncharged state. These sputtered neutrals will not be affected by any electrostatic forces originating from the sputtered surface and their yield should be far less matrix dependent. Before the sputtered neutrals can be mass analysed they need to be ionised, it is this postionisation step which up until comparatively recently has prevented the analytical application of sputtered neutral mass spectrometry (Reuter 1985).

Postionisation by electron impact offers a simple and mechanically compact solution to the application of SNMS to surface analysis. The simplest electron impact postionisers employ an electron beam which intersects the plume of sputtered neutral particles (Wilson 1989). A second, more complicated, application of electron impact ionisation uses a radio frequency induced plasma (Oechsner 1977). Although the plasma provides a more efficient form of postionisation due to the negation of space charge and larger ionising volume, the useful yields of the two techniques are approximately the same. This is because of the geometry constraints placed on the r.f. plasma technique by the accommodation of large r.f. coils. As the transmission of neutrals into the postionising region is purely a function of geometry a short sample to ioniser distance is essential for high transmission. Our new instrument employs the close geometry of an electron beam postioniser but also allows the use of a non-self sustaining d.c. argon discharge/plasma. Using this unique combination we have increased the useful yield of electron impact postionisation by an order of magnitude.

2. INSTRUMENTATION.

A VSW electron beam transport system incorporating a 90 degree electrostatic energy analyser was used as the basis of the ion optical system. To one end of this a VG12-12 800 mass unit range quadrupole mass spectrometer with an off axis pulse counting detector was fitted. To the other end the ion source was fitted on a telescopic mount so that optimum sample to ioniser distance could be obtained. A schematic of the ioniser is shown in figure 1. The targets were introduced on a rotating XYZ manipulator into a UHV vacuum system with a base pressure of 1×10^{-9} Torr. A Leybold

Figure 1- Ioniser Schematic.

(Mica washers & ceramics not shown)

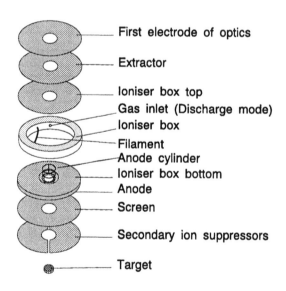

- First electrode of optics
- Extractor
- Ioniser box top
- Gas inlet (Discharge mode)
- Ioniser box
- Filament
- Anode cylinder
- Ioniser box bottom
- Anode
- Screen
- Secondary ion suppressors
- Target

IQE 12-38 with Wien filter was used as the primary source of 5keV Argon ions. When operating the gun the pressure in the analysis region was typically 5×10^{-7} Torr and during discharge mode experiments pressures of 1×10^{-5} to 1×10^{-4} Torr were experienced.

3. CHARACTERISATION.

Extensive characterisation experiments were performed on both the electron beam and discharge modes of the ioniser. The characterisation of the electron beam mode centered mainly on the energy spread of ions leaving the ioniser. The energy distribution of particles leaving the ioniser is the property by which parasitic signals originating from the ionising region (residual gas (RG) and electron stimulated desorbed species(ESD)) are suppressed. It is therefore important to have well defined potentials in the ioniser region. Computer modeling of the potential spread in the ioniser allowed the geometry of this region to be optimised.

The final design had a band width of 20eV and allowed stable electron emission currents of 0-25 mA to be generated. It has been conventional in electron beam SNMS to use emission currents of between 3-5mA. Our characterisation experiments have shown that as the emission current was increased above 10mA the signals from the RG and ESD species were attenuated. This suppression is likely to be due to space charge attraction which has a greater influence on the lower energy RG and ESD species. The primary method of parasitic suppression at low emission currents involves the use of potential barriers, these barriers also reduce the amount of sputtered neutral signal. The extra suppression obtained at high emission currents in our ioniser allowed these barriers to be lowered and consequently the sputtered neutral signal to increase. The accepted benchmark for assessment of SNMS instruments is the signal produced from polycrystalline silver foil. The signal from both isotopes of silver is summed and expressed as counts/μC of primary beam current. Our instrument is capable of 4×10^5 counts/μC. This is significantly better than the electron beam SNMS specifications of commercially available instruments (1×10^5-2×10^5 counts/μC). This assessment was carried out using a 100nA 5keV argon primary beam and 12mA of emission current; RG, ESD and secondary ion species were totally absent from the spectrum.

Characterisation of the discharge/plasma mode of operation has also been performed. The stable operation of the discharge is dependent on more factors than the electron beam mode. The characterisation experiments were separated into those relating to discharge current stability and those relating to the nature of the species emitted from the discharge. Optimisation of these two qualities by altering potentials, gas pressure and filament material and shape has been carried out. The final configuration operates at a pressure of 8×10^{-5} Torr, using a thoriated tungsten filament and can deliver stable currents of up to 120mA. Operating under these optimised conditions the discharge mode gives a figure of 4×10^6 counts/μC in the silver benchmark. The spectra produced in discharge mode show signals from argon and a small constant secondary ion signal from the stainless steel body of the ioniser. It is hoped that the secondary ion signal originating from the stainless steel can be eliminated by modification of the ioniser construction, as at present it prevents the determination of Fe, Cr and Mn at levels below 0.01 Atomic%.

4.APPLICATIONS.

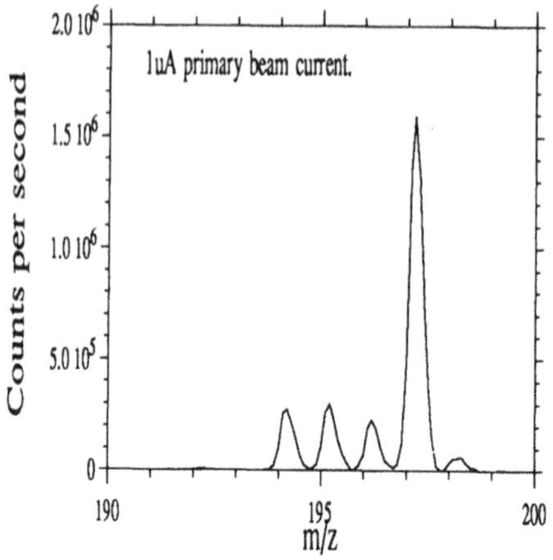

Before SNMS can be applied to the quantification of materials, relative elemental sensitivity factors must be determined. These have been obtained for approximately twenty elements relative to Fe from a variety of materials. Sensitivity factors determined for the same element sputtered from different matrices show no significant matrix dependency and vary from 2.5 (Mo) to 0.5 (Pb). Although sensitivity factors for the electron beam and discharge

Figure 2. SNMS spectrum of Au/Pt alloy.

modes are not numerically identical due to the difference in electron energy (80eV and 120eV respectively) they show the similar range and standard deviations.

A variety of noble metal alloys have been examined using SNMS. Such materials are difficult to analyse using SIMS due to the poor ion yields of elements such as Au. Figure 2 shows part of a discharge mode spectrum of a Pt/Au alloy. In +ve ion SIMS platinum and Au give low yields (Pt $6x10^5$ and Au $5x10^4$ counts/μC). In discharge SNMS both give yields of $3x10^6$ counts/μC, quantification of such alloys by SNMS is simply achieved by comparison of peak areas. Detection of Au and Pt concentrations lower than 0.001 Atomic% can be achieved with an error of 5%.

5. REFERENCES.

Benninghoven A, Rudenauer F G, Werner H W, 1987, SIMS, basic concepts, instrumental aspects, applications and trends (New York: Wiley) pp224.

Gnaser H, Fleischhauer J, Hofer O, 1985, Appl. Phys. A 37 211-220.

Oechsner H, Stumpe E, 1977, Appl Phys 14 43

Reuter W, 1985, SIMS V, (Berlin: Springer) pp94-101.

Vickerman J, 1988, in SIMS its principles and applications, ed Vickerman, Read, Brown (Oxford: Oxford Univ. Press)

Wilson R, Van den Berg J, Vickerman J, 1989, SIA 14 393-400.

The authors would also like to acknowledge the help of the SERC in the funding of this work.

Surface studies in UHV–SEM and STEM

J.A. Venables[*][+], G.G. Hembree[+], J. Liu[+] and J.S. Drucker[+]

M. Huang[*], F.L. Metcalfe[*], C.J. Harland[*] and R.H. Milne[*]

[+]Department of Physics, Arizona State University, Tempe, Arizona 85287, USA

[*]School of Mathematical and Physical Sciences, University of Sussex, Brighton.

ABSTRACT: Secondary and Auger electron spectroscopy and imaging has been further developed in the 'MIDAS' FEG-STEM instrument in Arizona. Secondary electron imaging at 1nm resolution, and Auger image resolution at the 2-5nm level has been demonstrated on selected samples. Applications to Ge/Si(100), Ag/Si(100) and Ag/Al$_2$O$_3$ are described. A new UHV-SEM has been constructed at Sussex around a high collection angle Cylindrical Mirror Analyser (CMA), for angle resolved electron spectroscopy at energies up to about 10keV, with parallel recording of electrons emitted over a wide range of angles. The design is described, and initial work on Ag/Si(111) and Ge(111) discussed.

1. INTRODUCTION

Both secondary and Auger electrons are surface sensitive. Secondary electrons form the most widely used signal in the Scanning Electron Microscope (SEM), which has a very high signal strength (typically 0.1-1 electrons per incident electron). In an ultra-high vacuum (UHV) environment, secondary electrons are sensitive to changes in work function, and to electric fields generated, either inside or outside the sample (Venables *et al* 1986). Auger electron spectroscopy (AES) finds wide application because of its chemical specificity, but it also has low signal strength (10^{-5} - 10^{-3} electrons per incident electron). Exploiting the advantages of AES at spatial resolutions less than 50nm has proved difficult. The minimum requirements for high spatial resolution AES, and the imaging equivalent, scanning Auger microscopy (SAM) are: 1) an all-UHV specimen environment; 2) a high brightness source, typified by a Field Emission Gun (FEG), and 3) an efficient low energy spectrometer (Venables and Hembree 1991). As described here in sects. 2-4, these conditions are fulfilled in the Microscope for Imaging, Diffraction and Analysis of Surfaces (codenamed MIDAS) at the ASU-NSF Facility for HREM at Arizona State University. Recent work, which has improved the resolution limit to 2-5nm, and the detection limit below 20 atoms on selected samples, is described.

Information is also contained in the angular distribution of emitted electrons. A new UHV-SEM has been constructed around a large Cylindrical Mirror Analyser (CMA) at the University of Sussex. This is being developed for Angular Resolved (AR) spectroscopy for electrons with energies up to about 10keV, using a parallel recording detector (Huang *et al* 1991,1992). This work is described in section 5.

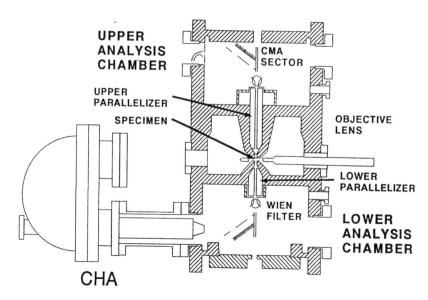

Figure 1. Scale cross-sectional drawing of the objective lens region of MIDAS, showing the parallelizer coils, Wien filter, CMA and CHA. See text for discussion.

2. ELECTRON EMISSION SPECTROSCOPY IN MIDAS

The principles behind, and progress of, the MIDAS project have been described in some detail (Kruit and Venables 1988; Venables and Hembree 1991); the bulk of this material will not be repeated here. In outline, the small probe and high collection efficiency are both achieved with an immersion lens with 'through the lens' detection of secondary and Auger electrons, with a 'magnetic parallelizer' to control the cone angle of emerging electrons. In principle, 100% collection efficiency into a cone of semiangle around 6° can be achieved up to a certain cut-off energy; this cone is then interfaced to a commercial concentric hemisperical analyser (CHA). In this paper we demonstrate the achievement of high energy resolution AES from small regions, and high spatial resolution secondary and Auger imaging. The examples shown are taken from published work on the deposition systems Ag/Si(100) (Luo *et al* 1991; Hembree and Venables 1992), Ge/Si(100) (Krishnamurthy *et al* 1990, 1991) and Ag/Al$_2$O$_3$ (Liu *et al* 1992).

Figure 1 shows the region on the MIDAS column concerned with the collection and detection of secondary (SE) and Auger (AE) electrons. The analysis chambers below and above the specimen allow collection of SE and AE either from both sides of electron transparent samples, or from the input side of bulk samples. The cone of SE is coaxial with the primary beam at the exit aperture of the parallelizer, where a Wien (**ExB**) deflector provides a 15° deflection off axis. The polarity of the deflection fields enables selection of either SE, into a Everhart-Thornley detector, or AE into the spectrometer system. For spectroscopy, the Wien filter is followed by a gridless cylindrical mirror analyser (CMA) sector, which deflects the emitted electrons a further 75°.

This CMA sector produces a simple spectrometer, illustrated in the upper analysis chamber, which produces about 3% energy resolution. For higher energy resolution, this focus is interfaced to a commercial 100mm radius analyser (VSW HA100). By operating the spectrometer appropriately, we can maximize the signal, while retaining an energy resolution better than 1%. Typically the sample is biased to about -500V, in order to acquire biased (b)-SE images, and to optimise the collection efficiency of the objective lens- parallelizer combination.

Examples of high energy resolution Auger spectra are shown in figure 2, for Si(KLL) and Ag(MNN) (Hembree *et al* 1990,1991). The Si spectrum, from a bulk Si(100) substrate, cleaned in-situ, is comparable in quality to high resolution X-ray excited AES, and shows all the minor, and plasma loss, peaks expected. The energy resolution on the main peak is 0.3% FWHM; the high signal to background ratio, caused by working at high probe energy (100 keV) is noteworthy. The Ag spectrum was taken from a 100nm wide Ag island grown on Si(100). Here the doublet structure can be clearly seen, as well as the strong signal to background ratio. The peaks can still be seen for locations between the islands, where we know, from detailed studies of the Auger amplitudes as a function of coverage, that the layer contains only 0.3 - 0.5 monolayer (ML) (Luo *et al* 1991, Hembree and Venables, 1992). This system is therefore a good choice for exploring sensitivity limits of AES and SAM.

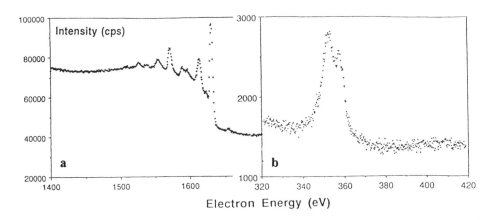

Figure 2: a) Si KLL Auger spectrum from bulk Si(100) sample, bias -500V; b) Ag MNN spectrum from 100nm sized Ag island on Si(100), each acquired in around 10 min at 1.5 nA at 100keV probe energy. See text for discussion.

3. BIASSED SECONDARY AND AUGER ELECTRON IMAGING

Energy filtered images, collected with the spectrometer at (A) and above (B) the Auger peak energy, can be converted into an elemental map by image subtraction and normalization procedures developed previously (Harland and Venables, 1985). For the images shown in figures 3 and 4, the quasi-logarithmic algorithm (A-B)/(A+B) was used; this works well because the high energy spectral background is very flat. Figure 3 shows the data, the processed image and the corresponding b-SE image of a Ag island. There is topographic contrast in the individual energy selected images, due to the strongly sloping (111) faces on the (100)-oriented crystal, but this has been largely removed by the processing algorithm, which is known to reduce topographic effects substantially. Similar Auger maps, and extracted linescans, from facetted Ge islands, grown on top of 3 ML of Ge on Si(100), have been published (Hembree *et al* 1991, Krishnamurty *et al* 1991). Using the 20-80% edge resolution criterion, these figures show around 5nm resolution. Resolution is of course a complex topic, and we need to distinguish carefully between 'image' and 'analytical' resolution. As argued elsewhere in more detail (Hembree and Venables 1992), these figures show essentially the 'image' resolution, which is determined by the probe diameter, estimated at 2-3nm, the Auger mean free path (0.8 nm for Ag MNN, and 1.5-2nm for Si KLL and Ge LMM), and sample topography. The Auger images are also, and most noticeably, limited by the signal to noise ratio.

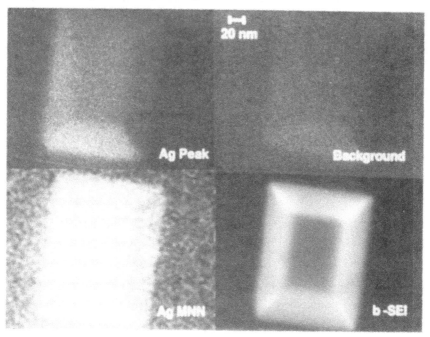

Figure 3. Energy selected, Auger and b-SE images of a silver crystal grown on Si(100) at 500 °C (Hembree *et al* 1990, 1991). The energy selected images were acquired using the CHA in 20 min. with 1.5nA beam current. The Auger image was produced by image processing off-line, by correlating, and then ratioing, the two energy selected images. The b-SE image, by contrast, took 1 min with 0.3nA.

4. MULTIPLE SIGNALS IN STEM

The secondary and Auger electron detectors on MIDAS are additional to the more standard STEM detectors, such as Bright Field BF and Annular Dark Field ADF; there is also a diffraction camera and an EELS spectrometer. Thus for thin samples, there is considerable scope for correlating these different images, and thereby relating 'bulk' and surface information.

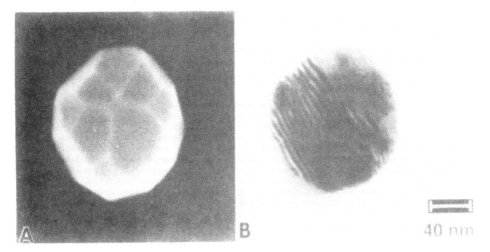

Figure 4. Simultaneously collected images of the same germanium island, produced during deposition of 6 ML Ge on Si(100) at 375 °C (Krishnamurthy *et al* 1991). a) biassed SE image from deposited (beam entrance) surface showing well developed facets; b) BF image showing moire fringes. See text for discussion.

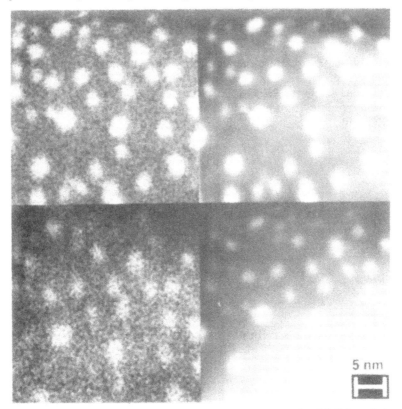

Figure 5. Images of Ag/Al2O3 from Liu *et al* (1992); a) Ag MNN peak image; b) secondary electron image; c) Ag MNN background at 20eV higher energy; d) high angle annular dark field image. Images a, b and d were obtained simultaneously.

An example from recent work on Ge/Si(100) is shown in figure 4, which compares the b-SE and BF images of a Ge island (Krishnamurthy *et al* 1991). The SE image shows the outer surface facets very clearly, while the BF image shows the 4 nm moire, caused by the 4% misfit between Ge and Si lattices. Islands smaller than 10nm diam do not show these moires, which strongly suggests that the small islands are coherent with the substrate. This has been correlated with detailed size distributions obtained for islands as small as 2nm diam, constructed from many b-SE images at a wide range of magnifications.

A second example is provided by the work of Liu *et al* (1992), who have studied the model catalyst system Ag/Al_2O_3. Figure 5 compares the Ag Auger peak image, the ADF and secondary images, and demonstrates a resolution of 2-3nm. Further, particles which are smaller than this are clearly visible. By analysing the intensity of these particles, they concluded that particles of <1nm diameter, containing less than 20 Ag atoms, could be detected in the Auger peak image. The signal to background ratio is so high in these images that dividing by the background image merely serves to intensify the noise. The more appropriate measure is simply (A-B), or (A-B)/<B>, where <B> is the average background for the whole picture. The local background does however contain extra information, and one can envisage using this information in more detailed schemes in future.

5. DEVELOPMENT OF AN ANGLE-RESOLVING CMA

A complementary development of an angle-resolving cylindrical mirror analyser is being undertaken in Sussex. Here the aim is to collect the maximum signal, in the SEM geometry with the sample outside the lens, and to preserve the emission angle information. Angle resolved (AR) spectroscopies are widely used in surface science, giving specific information about electron states and surface atomic geometry. Recently, the possibility of direct visualisation of atomic positions via

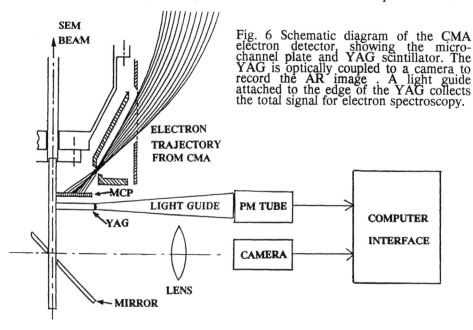

Fig. 6 Schematic diagram of the CMA electron detector, showing the micro-channel plate and YAG scintillator. The YAG is optically coupled to a camera to record the AR image . A light guide attached to the edge of the YAG collects the total signal for electron spectroscopy.

'electron holography' in a variety of forms has been explored (Harp *et al* 1990, Riley and Leckey 1992). The interest here is to see whether a parallel recording version, suitable for application with a low current SEM beam, is feasible. Serial recording of Auger Electron Diffraction patterns in an SEM has been reported by Nakamura et al (1991), but naturally data collection times are very long.

The construction and initial performance of this CMA has been reported previously (Huang *et al* 1991), and the recently developed dual detector arrangement is shown in Fig. 6. Here a channel plate/ YAG screen combination is used to record both the electron spectrum, and the angular distribution of emitted electrons at the analysed energy at the same time. The AR image, seen by a low-light level TV camera carries information about the environment of the emitting atom. Thus the initial aim is to see if AR Auger images formed from manipulating such AR images according to algorithms such as (A-B)/(A+B) contain useful information on surface structure. Our initial test systems are Ag/Si(111) and Ag/Ge(111), since crystal growth and surface diffusion have aleady been studied in these systems (Metcalfe *et al* 1991,1992).

So far, we have obtained SEM pictures of the islands, Auger spectra from and between the islands, and initial AR images from both positions. Comparisons have been made with Ag deposits on amorphous Si and Ge (Huang *et al* 1992). This work has established that Auger spectra can be obtained which are shot-noise limited, and that differences exist in the Angular Auger images, which have an angular resolution of 3-5o. However, some technical improvements are needed before these results can be considered definitive. First, the low light level SIT TV camera has insufficient integration time, so that the AR images tend to be noisy. This camera is about to be replaced by a CCD camera which can integrate internally for many minutes. Second, there is a fixed pattern background on these images, due to the grids across the exit slot in the inner cylinder of the CMA. We have designed a new gridless version of the CMA, which has no such grids. The diverging lens of the slot is compensated by extra electrodes, which additionally can steer the analysed electrons so as to produce the optimum focus. This new design has been developed using the SIMION electron-optics package; the details are described by Huang *et al* (1992) and in future publications.

6. DISCUSSION AND CONCLUSIONS

The results presented in this paper indicate that a certain stage in the development of this high resolution secondary and Auger imaging project has been reached: we have demonstrated 1nm b-SE and 3-5nm Auger image resolution, using reasonable probe currents and acquisition times. Small particles, smaller than the resolution limit, can be detected, down to below 20 atoms. We are, however, approaching the point where we can begin to ask questions about fundamental limits to resolution imposed by finite escape depths (Auger inelastic mean free paths of order 0.5-2nm) and localisation of the Auger process itself. As discussed elsewhere (Hembree and Venables 1992) we need to be clear about distinctions between 'image' and 'analytical' resolution, the latter including the effects of backscattering. We also need to select test samples carefully to minimise topographic effects; the Auger images are currently limited by signal to noise ratio considerations.

A second development project, to explore the usefulness of angular resolved electron spectroscopy and imaging, is at an earlier stage. In this case, further development of a gridless CMA is in progress, and parallel recording of the angle resolved, energy selected image looks promising. In addition, we continue to find the b-SE signal very useful, since it has high spatial resolution, and is surface sensitive at the sub- monolayer level. Further efforts, aimed at understanding the origin of b-SE contrast, and quantifying both SE and AE signals are in progress.

ACKNOWLEDGEMENTS

Support in Arizona was provided by the NSF-DMR, ONR, Shell Development Corporation, and Arizona State University. Support in Sussex came from SERC, the British Council, and the University of Sussex.

REFERENCES

Harland CJ and Venables JA (1985) *Ultramicroscopy* **17** 9

Harp GR, Saldin DK and Tonner BP (1990) *Phys. Rev. Lett.* **65** 1012; *Phys. Rev.* **B42** 9199

Hembree GG and Venables JA (1992) *Ultramicroscopy* in press.

Hembree GG, Luo FCH and Venables JA, (1990) *Microbeam Analysis -1990* ed J R Michael and P Ingram (San Francisco Press) pp 249-252; *Proc. 12th Intl. Cong. on Electron Microscopy (Seattle)* **2** 378

Hembree GG, Drucker JS, Luo FCH, Krishnamurthy M and Venables JA, (1991) *Appl. Phys. Lett.* **58** 1890

Huang M, Harland CJ and Venables JA (1991) *Inst. Phys. Conf. Ser. (EMAG 91)* **119** 147; (1992) *Surf. Interface Analyis*, submitted.

Krishnamurthy M, Drucker JS and Venables JA (1990) *Proc. 12th Intl. Cong. on Electron Microscopy (Seattle)* 1 308; *Materials Res. Soc. Symp.* **198** 409

Krishnamurthy M, Drucker JS and Venables JA (1991) *Materials Res. Soc. Symp.* **202** 77; J. Appl. Phys. **69** 6461

Kruit P and Venables JA (1988)*Ultramicroscopy* **25** 183

Liu J, Hembree GG, Spinnler GE and Venables JA (1992) Surf. Sci **262** L111; Catalysis Letters **15** 133

Luo FCH, Hembree GG and Venables JA (1991) *Materials Res. Soc. Symp.* **202** 49

Metcalfe F, Raynerd G and Venables JA (1991)*Inst. Phys. Conf. Ser. (EMAG 91)* **119** 177; (1992) *Surf. Sci.*, in preparation.

Nakamura N, Anno K and Kono S (1991) *Surf. Sci,* **256** 129

Riley JD and Leckey R (1992) *Springer Mod. Phys.* in press.

Venables JA and Hembree GG (1991) *Inst. Phys. Conf. Ser. (EMAG 91)* **119** 33

Venables JA, Batchelor DR, Hanbucken M, Harland CJ and Jones GW (1986) *Phil. Trans. Roy. Soc.* **A318** 243

Inst. Phys. Conf. Ser. No 130: Chapter 6
Paper presented at Int. Congr. X-ray Optics and Microanalysis, Manchester, 1992

423

Auger-profiling of multilayer structures: new approach

S V Kuchaev, F F Balakirev and A F Plotnikov

P N Lebedev Phisical Institute, Leninsky prosp., 53, 117924 Moscow, Russia

Abstract. A new approach is suggested to solve Auger-profiling deconvolution problem. The approach is based on the computer calculation of loop-like dependence of the Auger-signal amplitude of the chemical element under investigation versus the amplitude of another Auger-peak with the energy and electron escape depth different from the first ones. The comparison of the experimental data with the dependences calculated in frames of various model assumptions allows one to define the element depth distribution and artifacts.

1. Introduction

High depth resolution is very important to investigate a chemical element depth distribution in different multilayer structures with thin layers. Amorphous and semiconductors superlattices, x-ray optics, neutron filtres and so on can be mentioned among such objects. Auger-profiling is one of the best analytical methods from the point of view to receive the higher depth resolution. Nevertheless, the Auger-profiling depth resolution value measured by 0.01 and 0.99 levels of signals does not exceed 20–30 Å in the best case. Moreover, the last value can be estimated only in any assumption about etching rate dependence on sample composition. So, when the layer thicknesses are in the range of 10–100 Å it is impossible to answer the main questions of interest which are: (1) Are the experimental data caused by the sample nature only or are they distorted by the analysis procedure? (2) What is the element concentration in the layer? (3) What are the layers thicknesses? (4) Where are the interfaces situated? What is the interface sharpness? The impossibility to answer these questions caused by the number of processes and effects concerned with the interaction of both electrons and ions with the sample. The main difficulty is that there is not adequate theoretical nor empirical descriptions of the most effects mentioned above for the resulting Auger-profile form near interfaces. This situation is completely presented in review [1]. That is why the main purpose of Auger-profiling—to define the real concentration depth distribution from Auger profile—is not achieved up today. The new approach to overcome these problems is presented in this paper. In spite of its evidence this approach allows one to receive more information about the sample nature and artifacts.

2. The calculation model and experience

Let A-atom has two characteristic Auger-peaks in the spectrum with the energy values E_1 and E_2 and corresponding electron escape depth λ_1 and λ_2 for definitiveness $\lambda_2 > \lambda_1$. Consider the simplest model of Auger-profiling of B-A-B-A-... multilayer structure in frames of the next assumptions.

(1) Both A and B layers are thick compared with the thicknesses of layers.

(2) Escape depths λ_1 and λ_2 are independent on sample composition.

(3) Neglect electron backscattering and put the calibration $(1 + r_i)I_i^0 = 1$ ($i = 1$, 2) for both Auger-peaks, where I_i^0 are ith Auger-peak amplitudes of pure element, r_i are backscattering coefficients. Let the B-A and A-B interfaces of the first A-layer are situated at z_1 and z_2 correspondingly. It is evident from well known integral expression for I [2] that the Auger-profiles $I_i(y)$ are defined by equations

$$\begin{cases} I_i(y) = e^{-(z_1-y)/\lambda_1} & y \to z_1 \text{ (signals growth)} \\ I_i(y) = 1 - e^{-(z_2-y)/\lambda_1} & y \to z_2 \text{ (signals drop)} \end{cases} \quad (1a,b)$$

Consequently, I_2 depends on I_1 according to equations

$$\begin{cases} I_2(y) = I_1(y)^{\lambda_1/\lambda_2} & y \to z_1 \text{ (signals growth)} \\ I_2(y) = 1 - (1 - I_1(y))^{\lambda_1/\lambda_2} & y \to z_2 \text{ (signals drop)} \end{cases} \quad (2a,b)$$

So, the A layer is characterized by two branches $(2a)$ and $(2b)$ of the $I_2(I_1)$ dependence at the first (B-A) and at the second (A-B) interfaces. This result is shown in figure 1 for the ratio $\lambda_2/\lambda_1 = 3$. When $y = 0$ the amplitudes I_i are equal to zero because we have assumed $z_1 \gg \lambda_i$. Then the point in figure 1 moves along the upper branch according to $(2a)$ due to ion etching. When B-layer is sputtered completely it arrives at the [1,1] position and remains in it. Then $I_1(y)$ and $I_2(y)$ decrease according to $(1b)$ and the point $I_2(I_1)$ in figure 1 moves to zero along the lower branch according to $(2b)$. Each consequent A layer would be characterized by just the same $I(I)$ loop-like dependence. Figure 2 shows $I_2(I_1)$ dependences calculated analogously for various layers thicknesses d_A and d_B in the periodical multilayer structures B-A-B-A... with ideally sharp interfaces (solid line). The calculations for each small square picture are made for d_A and d_B values corresponding to square centre, λ_2/λ_1 ratio is equal to 3. Figure 2 shows also the data calculated for the concentration profiles $c_A(z)$ changing linearly in the intermediate regions near the interfaces. The internal loops correspond to different values G of this regions thicknesses: G = 10 Å - dashed line, G = 20 Å - dotted line and G = 30 Å - dashed-dotted line. It is seen from figure 2 that each loop is individual enough and could be used to define the set of d_A, d_B and G values. One of the most important features of loops mentioned above is that they are independent on etching rate variations caused by either ion beam instability or sample composition changes since time of sputtering is excluded as a parameter from equations (1). The I_1 and I_2 amplitudes are produced by atoms with just the same 'momentary' depth distribution for every sputtered thickness value y. The only difference is that of λ_1 and λ_2.

Let us consider the experimental data. The samples under investigation are different carbon-nickel multilayer structures produced by either laser ablation or RF-sputtering techniques. In view of paper volume deficiency the laser ablation data are not presented. The structures produced by RF-sputtering are preferred as experimental samples because they have more sharp interfaces. The samples are shown in figure 3(a) (the left and the right pictures). The corresponding experimental $I_2(I_1)$ dependences

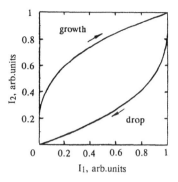

Figure 1. The $I_2(I_1)$ dependence for idealized model.

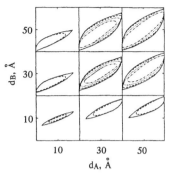

Figure 2. The $I_2(I_1)$ calculated data for various d_A and d_B values. G = 0, ———, G = 10 Å, ----, G = 20 Å, ·······, G = 30 Å, — · —.

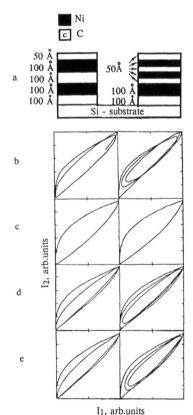

Figure 3. The structures (a) experimental $I_2(I_1)$ dependences, (b) and $I_2(I_1)$ dependences calculated in various assumptions (c − e).

(i.e. 844 eV Ni Auger-peak amplitude versus 61 eV one) are presented in figure 3(b) under the samples picture. They have some interesting features: (1) they are not symmetric unlike the data of figure 2; and (2) they are spiral. It is evident that the calculation model resulting in figure 2 data is very imperfect.

Let us consider the next effects in the calculation model. The upper branch of loops mentioned above corresponds to Ni-electron escape through carbon layer (C-index) and the lower one—through nickel layer (Ni-index). Consequently, the values of λ_1 and λ_2 should be different in Ni and C layers as is seen from figure 3(b). Figure 3(c) shows the data calculated for structures shown in figure 3(a) and escape depths $\lambda_{1C} = 5$ Å, $\lambda_{2C} = 15$ Å, $\lambda_{1Ni} = 8$ Å and $\lambda_{2Ni} = 12$ Å. The spiral form of $I_2(I_1)$ dependences could be achieved be either increase of G or the surface roughening increase during the ion etching. In fact, figure 3(d) shows the data calculated for Gaussian roughening value Δy depending on y linearly $\Delta y = 0.4y$.

At last the Ni into C 'knock-on' effect and atom mixing caused by ion bombardment should be considered. This can be done by step-by-step calculation of I_i amplitudes for the $c_{k+1}(z) = c_k(z) + \Delta c_k(z)$ distributions, where k is the number of etching steps, $c_k(z)$—A-element depth distribution after kth step, $\Delta c_k(z)$—concentration change due to ion bombardment during this step of sputtering. Different assumptions (but the same for C-Ni structures) can be done about $\Delta c_k(z)$. Exponential decrease of $\Delta c_k(z)$ is suitable for 'knock-on' effect description, while the atom mixing result in the mean concentration value in the surface layer of some thickness. Figure 3(e) shows the data calculated in frames of the next model: (1) λ_{iC}, λ_{iNi} and the roughening amplitude dependence $\Delta y = 0.4y$ are the same as in the case of figure 3(d); (2) when one thickness step h is sputtered the part $P = 0.8$ of Ni-atoms amount in this layer is implanted into a sample and $\Delta c_k(z)$ is proportional to $\exp(-(z-y)/\Lambda)$, where $\Lambda = 3$ Å. These dependences are similar to experimental ones in figure 3(b). We have no intention of comparing the experimental and calculated data now. This is not because the calculation model is imperfect, but in view of insufficient accuracy of experimental data. Nevertheless, the approach allows one to explain some features of experimental data and to estimate some effects already.

3. Conclusions

There are many parameters in the model to be defined. So, the number of 'experimental equations' should be sufficient to define the algorithm which should be unique for all Ni-C structures. There are some ideas for how to overcome this problem and improve the accuracy of experimental data. Both equipment modification and new equipment generation can be developed based on the approach mentioned above to receive a new possibilities of analysis.

References

[1] Powell C J and Seah M P 1990 *J. Vac. Sci. Technol.* A **8** 735
[2] Briggs D and Seah M P (ed) 1983 *Practical Surface Analysis by Auger and X-ray Photoelectron Spectroscopy* (Wiley)

Inst. Phys. Conf. Ser. No 130: Chapter 6
Paper presented at Int. Congr. X-ray Optics and Microanalysis, Manchester, 1992

427

Electron spectroscopy at high spatial resolution

P. Kruit

Delft University of Technology, Department of Applied Physics, Lorentzweg 1, 2628 CJ Delft, The Netherlands

ABSTRACT:

X-ray optics is pushing photo electron spectroscopy to sub μm resolution, while new electron optics is pushing Auger electron spectroscopy to nm resolution. Coincidence detection of the emitted electrons and the transmitted electrons in an electron microscope allows virtual-photo electron spectroscopy also at nm resolution. The expected signal per nm^2 in the electron microscope is of the same order as in modern synchrotrons.

1. INTRODUCTION

The disturbance of the electron configuration of a solid is a popular way of forcing the material to give information on elemental composition and chemical properties. Quite often it is for the physical process not important what causes the disturbance: the effect of an x-ray absorption is identical to the effect of an inelastic scattering event of a fast electron. This implies that the information that can be obtained in x-ray based analysis is very similar to that which can be obtained in electron beam analysis. The differences between these methods are related to secondary effects like background levels, instrumental possibilities, etc. For electron spectroscopy of emitted electrons, the pro's and contra's of both excitation methods have led to a situation in which Auger electron spectroscopy is usually performed with electron beam excitation and "electron spectro-scopy for chemical analysis" (ESCA) almost exclusively with x-ray excitation, which makes that the latter also listens to the name x-ray photo electron spectroscopy (XPS). It is well known that Auger spectroscopy can also be performed with x-ray excitation, leading to a better signal to background ratio but at a loss of spatial resolution. I shall argue in this paper that XPS can also be performed with electron excitation, yielding unprecedented spatial resolution, but at the expense of more complicated instrumentation and not applicable to all kinds of sample. The clue , of course, to using electron excitation is that it must be known how much energy the electron has deposited in the material. This can only be done in a thin, transmitting sample where the energy loss of the fast electron is analyzed and the occurrence of the emitted electrons is measured in coincidence with the excitation.

2. INSTRUMENTATION FOR COINCIDENCE ELECTRON SPECTROSCOPY

Coincidence experiments with a high energy primary beam which scatters under a small angle to ensure a dipole excitation have been performed on gases (Van der Wiel and Brion 1972/1973, Haak et al. 1984, Ungier and Thomas 1985). To obtain spatial resolution we have modified a transmission electron microscope in such a way that the emitted electrons from the sample can be analyzed in a concentric hemispherical analyzer. For high spatial resolution the sample should stay in the magnetic field of the objective lens. One of the modifications is the change of the magnetic field form by reshaping the pole pieces and adding an extra coil, such that the field of a magnetic monopole is created (Bleeker and Kruit 1991). The straight flux lines guide the emitted electrons to the entrance of a 90^0 deflector, at which point the field suddenly stops. The electron trajectories are "parallelized" by the action of the magnetic field. In the adiabatic approximation, the angles of the trajectories are reduced by a factor $\sqrt{B_f/B_i}$, where B_i and B_f are the magnetic field strength at the sample and at the exit (Kruit 1991, Bleeker and Kruit 1990). The 90^0 deflector is not meant as a dispersive element, which can be realized by super imposing electrostatic and magnetic fields with opposite dispersion. An energy range between 300 and 800 eV can pass without a change of field strength, and thus without a change in deflection of the primary beam. After this separation of the secondary beam from the microscope axis, an electrostatic transfer lens system brings the electrons towards a 140 mm 180^0 hemispherical analyzer. The energy analysis of the transmitted electrons is performed with a standard magnetic sector. In Delft we have first build a prototype based on a Philips EM 400 for testing the electron optics and performing preliminary coincidence measurements in a non UHV environment (Kruit and Venables 1988). Presently a fully UHV compatible instrument is under construction based on a Philips EM 430. In a similar project at the Arizona State University a microscope has been build in conjunction with VG Microscopes Ltd (Kruit and Venables 1988, Hembree et al.1988).

Coincidence experiments in which both the energy loss of the primary electron and the energy of the emitted electron were measured, have been performed in the low-loss region, where there is an abundant signal (Pijper and Kruit 1991). The results (Figure 1) show a clear peak where the difference between the energy loss and the energy of the emitted electron is constant, this peak can only be interpreted as equivalent to a peak in a UPS spectrum. Other factors in the comparison are specimen damage, favouring x ray excitation, availability of the instruments, etc.

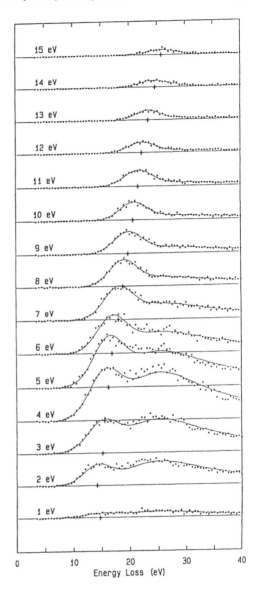

FIG. 1. Energy loss coincidence spectra (shown as dots) for 15 consecutive values of the secondary-electron energy. Spectra have been vertically displaced over an equivalent of 500 counts. False coincidence counts have been subtracted.

3. SPATIAL RESOLUTION IN ELECTRON BEAM SYSTEMS

Commercially available scanning Auger instruments usually work with beam voltages of 5-20 kV. Typical spatial resolution is 1 μm, going down to 50 nm for modern instruments with field emission guns. At 100 kV in a STEM equipped with a 150^0 spherical analyzer which can collect electrons from a sample placed outside the objective lens, Cazaux et al (1988) were able to demonstrate a line resolution of 8 nm from a GaAs/Si edge. The high spatial resolution obtainable with the sample in the immersion lens has now been demonstrated in several studies in the Arizona instrument. Hembree et al (1991) used Ag deposits on Si as a test system showing a line resolution of 3 nm using the Ag MNN lines. Islands of Ag on Al_2O_3 containing approximately 15 atoms could be distinguished in an element map (12). A probe size of 2 - 3 Å is possible in a 100 keV system with immersion lens, so in principle, for special cases, this could also be the resolution for the electron spectroscopic mode. The coincidence detection does not influence the probe forming, so the resolution argument stays valid. In practice, the amount of signal available, or the non locality of the physical process, might well put other limitations on the resolution.

4. SPATIAL RESOLUTION IN X-RAY SYSTEMS

It is much more difficult to focus a beam of x-rays than it is to focus an electron beam. This paper is not meant to give an overview of the most recent situation, for that the reader is referred to other papers in these proceedings. Using laboratory sources and curved crystals for focusing, typical spot sizes obtainable at the moment are 50 - 150 μm (Pella et al 1992, Wittry et al. 1989). Using total reflection capillary optics, probes of about 5 μm might be possible (York 1992). An alternative for using x-ray optics is to create the x-rays in a small spot close to the sample under investigation by focusing an electron beam on a foil. Cazaux (1984) reports a resolution of 20-30 μm. Using synchrotron radiation and zone plates, resolutions of 30 - 50 nm are reported in scanning x-ray microscopy (Bradshaw 1990, Michette 1992), but not yet in combination with XPS. Similar resolutions might be expected from laser plasma sources.

For x-ray generated electron spectroscopy there is an alternative to probe forming for obtaining spatial resolution: an image of the sample can be produced using the emitted electrons. This image can either be energy filtered entirely, or a spectrum can be obtained from a part of the image, selected with an aperture. With the sample in a field free area, a resolution of a few μm's has been obtained in routine operation. If the sample is placed in a strong electrostatic field, as in a cathode lens, a high resolution can be obtained. The limit is set by the combination of diffraction and spherical aberration of the acceleration. Veneklasen (1991) has recently analyzed the potential of this system and predicts resolutions of 3 nm and up, or 30 nm and up for reasonable detection efficiencies. An alternative imaging system is a diverging magnetic field, where the electrons are guided by the fluxlines and thus keep the information on their starting position to within the diameter of the Larmor motion. King et al (1991) report on studies with such an instrument at a synchrotron, obtaining a resolution of 1 - 30 μm, depending on the energy of the electrons.

5. SIGNAL STRENGTH AND OTHER FACTORS

Although x-rays and electrons can produce the same excitation, the cross-section for doing so is quite different. But also the obtainable current density is different and it is the combination which determines the signal strength. A good comparison of these two factors was made by Isaacson and Utlaut in 1978, at that time primarily directed at comparing the EELS technique with x-ray absorption. For 100 keV primary electrons, they found an approximate relation between the photo absorption cross-section $\sigma_x(E)$ and the differential electron energy loss cross-section $\partial\sigma/\partial E$:

$$\frac{\sigma_x(E)}{\frac{d\sigma}{dE}(E)} \approx \frac{170E}{\ln\left(\frac{4x10^5}{E}\right)}$$

a factor of about 10^2 for 10 eV excitations going up to about 3×10^4 for 1000 eV excitations. Current densities obtainable with electron beams are about 10^{24} el/cm^{2}sec (1nA in a probe of 1 nm). A comparable number for x-ray beams is not easy to find: the usual information contains the brightness of the source expressed in photons per sec cm^2 mrad^2 within 0.1 or 0.01 % of the bandwidth. Roughly estimating for 1000 eV photons the aperture angle coming from a zone plate 20 mrad and the efficiency of the monochromator 10%, then a .synchrotron source like BESSY can have 10^{17} photons/cm²sec eV. Newer sources with wigglers or undulators can give a factor 10^6 more brightness. For the comparison of signal strength there is one other factor that comes in for electron beams: the EELS spectrum can be detected with a parallel detector. If the coincidence electronics can accommodate this, it improves the detection efficiency by a large factor, maybe 10^2. With that, the number of emitted electrons per cm² using electron beam excitation would be of the same order of magnitude as what can be expected using the most modern photon sources.

6. CONCLUSION

In the near future we expect significant advances in electron spectroscopy at high spatial resolution, both in x-ray based systems and in electron beam systems. Since similar information can be obtained, using x-rays or electrons for excitation, the technology of the instrumentation decides what will be easier. At the moment electron beams are preferred for high spatial resolution and x-ray beams for high energy resolution or easier interpretation of chemical effects. However, the two techniques are starting to overlap now that x-ray optics is pushing to higher resolution and electron optics is giving electron microscopes new analytical power.

REFERENCES

Batchelor D R, Bishop H E and Venables J A 1989 Surface and Interface Analysis 14 709-716.
Bauer H E and Seiler H 1986 Proc. XIth Int. Cong. on Electron Microscopy, Kyoto, 1986.
Bleeker A J and Kruit P 1990 Nuclear Instruments and Methods in Physics Research A298 pp. 269-278.
Bleeker A J and Kruit P 1991 Rev. Sci. Instrum. Vol. 62 No. 2 pp. 350-356.
Bradshaw A M 1990 Proc. XIIth Int.Congr. Electron Micr. pp. 388, San Francisco Press
Cazaux J 1984 Ultramicroscopy 12, pp. 321.
Cazaux J, Chazelas J, Charasse M N and Hirtz J P 1988 Ultramicroscopy 25 pp. 31-34.
Chazelas J, Friederich A and Cazaux J 1988 Surface and Interface Analysis, Vol. 11 pp. 36-39.
Della P A, Lankosz M and Holynska B 1992 Proc. 50th EMSA meeting 1992 pp. 1754.

San Francisco Press.

Glezos N M and Nassiopoulos A G 1991 Surface Science 254 pp. 309-319.

Haak H W, Sawatzky C A, Ungier L, Gimzewski J K and Thomas T D 1984 Rev.Sci.Instrum 55, pp. 696.

Hembree G G, Luo F C H, Bennett P A and Venables J A 1988 Proc. 46 the Annual Meeting of EMSA, Milwaukee. San Francisco Press.

Hembree G G, Luo F C H and Venables J A 1990 Microbeam Analysis, pp. 249-252. San Francisco Press.

Hembree G G, Drucker J S, Luo F C H, Krishnamurthy M and Venables JA 1991 Appl.Phys. Lett. 58 (17), pp. 1890-1892.

Isaacson M and Utlaut M 1978 Optik 50 pp. 213.

King P L, Borg A, Kim C, Yoshikawa S A, Pianetta P, Lindau I 1991 Ultramicroscopy 36 pp. 117.

Krishnamurthy M, Drucker J S and Venables J A 1990 Proc. of the XIIth Int. Congr. for Electron Microscopy, Seattle, pp. 308-309.

Kruit P 1986 Proc. XIth Int. Cong. on Electron Microscopy, Kyoto, pp. 593-594.

Kruit P 1991 Advances in Optical and Electron Microscopy, Vol. 12, pp. 93-137 (Academic Press Limited).

Kruit P, Bleeker A J and Pijper F J 1988 Inst. Phys. Conf. Ser. No. 93: Vol. 1, Chapter 5, pp.249-254. Proc. EUREM 88, York.

Kruit P and Pijper F J 1990 Inst. Phys. Conf. Ser. No. 98: Chapter 7, pp. 271-276. Proc. EMAG-MICRO 89, London.

Kruit P and Venables J A 1988 Ultramicroscopy 25 pp. 183-194.

Liu J, Hembree G G, Spinnler G E and Venables J A 1991 Proc. of the 49th Annual Meeting of the Electron Microscopy Society of America pp. 690-691. San Francisco Press.

Liu J, Hembree G G, Spinnler G E and Venables J A 1992 Proc. of the 50th Annual Meeting of the Electron Microscopy Society of America pp. 308. San Francisco Press.

Michette A 1992 European Microscopy and Analysis july pp. 25.

Müllejans H, Bleloch A L, Howie A and McMullan D 1991 Inst. Phys. Conf. Ser. No. 119: section 3, pp. 117-120. Proc. EMAG 91.

Pijper F J .... (unpublished).

Pijper F J and Kruit P 1991 Phys. Rev. B, Vol. 44, Nr. 17 pp. 9192-9200.

Powell C J 1976 Rev. Mod. Phys. 48 pp. 33.

Ungier L and Thomas T D 1985 J. Chem. Phys. 82 pp. 3146.

Venables J A, Cowley J M and von Harrach H S 1988 Electron Microscopy and Analysis 1987 Inst.Phys.Conf. Ser 90 pp. 85.

Veneklasen L H 1991 Ultramicroscopy 36 pp. 76.

Wiel M.J. van der and Brion C E 1972/1973 J. electron Spectrosc. Ralt. Phenom 1, pp. 439.

Wittry D B, Golijanin D M and Sun S 1989 Proc. XIIth Intern. Conf. on x-ray Optics and Microanalysis, Cracow, pp. 31.

York B R 1992 Proc. 50th EMSA meeting 1992, ppp. 1764. San Francisco Press.

Sputter-induced concentration profiles in binary alloys

GN van Wyk[1], WD Roos[1], J du Plessis[1] and E Taglauer[2]

[1]University of the Orange Free State, ZA-9300 Bloemfontein, Republic of South Africa.
[2]Max Planck-Institut für Plasmaphysik, W-8046 Garching bei München, Germany.

ABSTRACT: Auger electron spectroscopy and ion scattering spectroscopy were used to determine sputter-induced concentration profiles in Pt-Pd, Ni-Cu and Cu-Ti alloys. It was found that Pd in Pt-Pd, Cu in Ni-Cu and Cu in Cu-Ti were the segregating elements. A depletion of the second and consecutive layers, which is a manifestation of the segregation process, was found and the results were used to determine segregation energies and diffusion constants.

1. INTRODUCTION

Sputter-induced altered layers in alloys cause a major complication in the interpretation of experimental results from surface analytical techniques such as AES, XPS and ISS. In the case of AES, the signal originates from several atomic layers, depending on the energy of the transition. However, if the composition of these layers was altered by sputtering, no meaningful interpretation of the AES signal can be given. On the contrary, AES measurements are often used to determine the altered layer. It is therefore clear that iterative modelling is necessary to get a clear picture of the situation.

It became apparent during the last 10-15 years that the altered layer can not be explained by preferential sputtering alone. Therefore radiation enhanced processes at room temperature were suggested to explain results for Cu-Ni (Watanabe et al 1976), Ag-Au (Overbury and Somorjai 1976) and Al-Cu (Chu et al 1976). This led to a model (Ho 1978) introducing radiation enhanced diffusion into the sputtering model of Patterson and Shirn (1967). Segregation effects were included in later models (Swartzfager et al 1981, Lam and Wiedersich, 1987). Excellent reviews on the role of segregation in bombardment induced compositional changes have been published recently by Kelly (1989) and Lam (1990).

In an ongoing programme, the effect of sputtering was investigated by the authors for Pt-Pd (du Plessis et al 1989), amorphous Cu-Ti (van Wyk et al 1989,1991) and Ni-Cu (Roos et al 1992). A model which contains sputtering aspects as discussed by Swartzfager (1981) and diffusion effects as developed by du Plessis and van Wyk (1988) was used to analyze the results. In this paper the latest results for Ni-Cu is discussed in context with the previous findings within the framework of a theoretical model which was used by the authors to explain the results.

2. THEORETICAL CONSIDERATIONS

A one-dimensional model is used in which the sample is divided into a row of n cells, which is a distance a apart. The cells are then allowed to exchange atoms by means of three flux processes, whereafter the equilibrium concentration values in the cells are used to construct a concentration profile.

The flux terms are the diffusion flux $J_1^{D(i+1,i)}$ of component 1 from cell $i+1$ to cell i and the sputter flux J_1^S of component 1 from the surface. The recession flux, $J_1^{R(i+1,i)}$, is merely a mathematical transformation in order to keep the surface at $x = 0$. The detailed expressions of the flux terms are

$$J_1^{D(i+1),i} = M_1 \frac{\Delta\mu^{(i+1,i)}}{a} C_1^{(i+1)} \tag{1}$$

$$J_1^S = I^+ \left(Y_1 \frac{N}{N_1} \right) X_1^{(1)} \tag{2}$$

$$J_1^{R(i+1,i)} = I^+ \left(Y_1 \frac{N}{N_1} X_1^{(1)} + Y_2 \frac{N}{N_2} X_2^1 \right) X_1^{(i+1)} \tag{3}$$

In these equations, M_1 is the mobility of species 1 related to the diffusion coefficient by the equation $D = M_1 RT$, where R is the universal gas constant and T the temperature. The term $\Delta\mu_{(i+1,i)}$ is a function of the chemical potentials of the species in the cells and can be related to the surface free energy ΔG. $C_1^{(i)}$ is the concentration and $X_1^{(i)}$ the fractional concentration of species 1 in the ith cell. The layer atom density is indicated by N, N_1 and N_2 for the alloy and elemental densities of species 1 and 2 respectively. The ion flux is denoted by I^+ while the elemental sputter yields and fractional concentrations for elements 1 and 2 are given by Y_1, Y_2 and $X_1^{(1)}, X_2^{(2)}$ respectively (van Wyk et al 1991).

A system of coupled nonlinear differential equations can now be set up using the relation between the rate of concentration change in each cell and the various fluxes, with the provision that sputtering occurs in the first cell only. All the parameters except the diffusion coefficient D and segregation energy ΔG are known from the experimental setup. It is therefore possible to solve the system of differential equations numerically while fitting D and ΔG values in such a way that the calculated intensities correspond to the experimentally observed AES intensities.

3. EXPERIMENTAL METHOD

Binary policrystalline alloys which form homogeneous solid solutions (Cu-Ni, Pt-Pd) as well as Cu-Ti in the amorphous state were investigated. In each case the specimens were mounted in an UHV chamber (10^{-8} Pa) and bombarded with 2 keV Ar$^+$-ions with a current density of 40 μA.cm^{-2} for 30 minutes until a steady state was reached. Immediately after bombardment the AES and ISS analyses were done using 3 keV electrons and 1 keV Ne$^+$-ions respectively. For the Pt-Pd and Cu-Ti systems only one Auger transition per element was used, which was 418 eV for Ti, 920 eV for Cu, 64 eV for Pt and 330 eV for Pd.

In the case of Ni-Cu, the difference in escape depth of Auger electrons from different transitions was used to construct the concentration profile. Because the Ni and Cu peaks overlapped in the energy range studied, special precautions were taken to prevent variations in the measured current due to changes in secondary electron emission from different surfaces. The data from the lockin-amplifier was fed via a 12-bit analog to digital converter to a computer, while the scan rates and energy ranges were controlled by the computer using a 12-bit digital to analog converter. All spectra were collected under identical experimental conditions.

4. RESULTS AND DISCUSSION

The elemental Auger spectra for Ni and Cu are shown in Figure 1 for the energy range 650-950 eV. The measurements and fitting procedures were done for the entire energy range up to 950 eV but these results were omitted due to a lack of space. In Figure 2 the Ni-Cu alloy spectrum for the Ni70at% alloy in the same energy range is shown. Spectra were also recorded for Ni50at% and Ni30at% alloys.

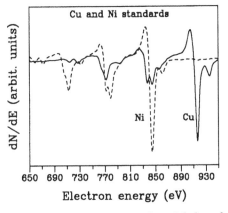
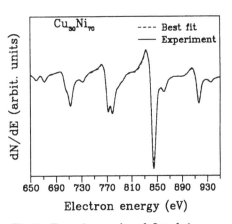

Fig 1. Auger spectra for nickel and copper standards.

Fig 2. Experimental and fitted Auger spectra for the Ni70at% alloy.

The fitting procedure is to use an equation (van Wyk 1991) for the elemental quantities I_i which contains the layer concentrations X_i^n. The sum of these quantities is then compared with the alloy spectrum. The concentrations are then adjusted and the procedure is repeated until the best fit is obtained. The fitting results are presented in fig. 2 and it can be seen that excellent agreement was obtained. This agreement is valid over the entire energy range and it can therefore be concluded that the assumed values for the layer concentrations are the correct ones. The layer concentration values used to obtain the fitted spectrum are presented as depth profiles in Figure 3. It turns out to be an enriched first layer (which is also preferentially sputtered in this case), a depleted second layer and subsequent increase to the bulk concentration.

For the Pt-Pd and Cu-Ti investigations, ISS and AES results were used to determine concentration profiles. The results were similar to that of Ni-Cu and can be explained in terms of Pd and Cu segregation to the respective surfaces of the Pt-Pd and Cu-Ti alloys where it is preferentially sputtered. In this case segregation energies and diffusion coefficients of Pd and Cu were determined using the model described in section 2. These results are fully described in the references given above and only summarized results will be given here. For Pd in Pt-Pd, diffusion coefficients in the range $(1.2 - 8.0) \times 10^{-20}$ m^2s^{-1} and segregation energies in the range $(2000 - 8000)$ J/mol were obtained. The dependence of the segregation energy on concentration for Cu-Ti is shown in Figure 4. In this case diffusion coefficients in the range $(2.0 - 2.5) \times 10^{-20} m^2s^{-1}$ were obtained. These values are larger than the 10^{-21} limit set by Pickering (1976) for radiation enhanced diffusion to occur.

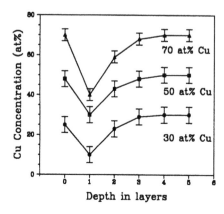

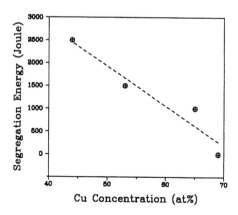

Fig 3. Bombardment induced depth profiles for the Cu-Ni alloys.

Fig 4. Segregation energy of copper as a function of concentration.

In conclusion, it was shown that concentration profiles in sputter induced altered layers can be successfully determined by AES and ISS. Furthermore, diffusion constants and segregation energies can be determined by applying the appropriate model.

REFERENCES

Chu WK, Howard JK and Lever RF 1976 J. Appl. Phys. **47** 4500.
du Plessis J and van Wyk GN 1988 J. Phys. Chem. Solids **49** 1441.
du Plessis J and van Wyk GN 1989 J. Phys. Chem. Solids **50** 237.
du Plessis J, van Wyk GN and Taglauer E 1989 Surf. Sci. **220** 381.
Ho PS 1978 Surf. Sci. **72** 253.
Lam NQ 1990 Scanning Microscopy Supplement **4** 311.
Lam NQ and Wiedersich H 1987 Nucl. Instrum. Methods B **18** 471.
Overbury SH and Somorjai GA 1976 Surf. Sci. **24** 104.
Patterson WL and Shirn GA J. Vac. Sci. Technol. **4** 343.
Pickering HW 1976 J. Vac. Sci. Technol. **13** 618.
Roos WD, van Wyk GN and du Plessis J 1992 Surf. Interf. Anal. (in press).
Swartzfager DG, Ziemecki SB and Kelley MJ 1981 J. Vac. Sci. Tech. **19** 185.
van Wyk GN, du Plessis J, Zhang B and Taglauer E 1989 S. Afr. J. Phys. **12** 58.
van Wyk GN, du Plessis J and Taglauer E 1991 Surf. Sci. **254** 73.
Watanabe K, Hashiba M and Yamashina T 1975 Surf. Sci. **28** 104.

Inst. Phys. Conf. Ser. No 130: Chapter 6
Paper presented at Int. Congr. X-ray Optics and Microanalysis, Manchester, 1992

439

Weathering of wood measured by XPS and imaging XMA

P.J. de Lange, A.K. de Kreek, A. van Linden (1), N.J. Coenjaarts (1)

Akzo Research Laboratories Arnhem, Analytical and Environmental Chemistry Department, PO Box 9300, 6800 SB Arnhem, The Netherlands
(1) Akzo Coatings bv, Research Centre Sassenheim, PO Box 3, 2170 BA Sassenheim, The Netherlands

ABSTRACT: X-ray Photoelecton Spectroscopy (XPS) and X-ray Micro-analysis (XMA) were used to study the upper layer of wood during weathering. With XPS an increase of the oxygen to carbon ratio and a change of the shape of the carbon (1s) signal was found, pointing to the formation of a cellulose rich weathered layer. The thickness of the weathered layer was determined by imaging the oxygen to carbon ratio of a cross section of the wood with XMA.

1. INTRODUCTION

Interactions of photons, moisture and fungi with wood lead to changes in colour and erosion of the wood surface. A very effective protection against this weathering is a surface treatment with aqueous chromium trioxide or with other solutions containing hexavalent chromium. But the toxicity of these protection agents is very high. A better understanding of the degradation processes may lead to the development of non-toxic alternative treatments. Surface analytical techniques, like XPS, and micro analytical techniques, like XMA, may be helpful tools in studying these degradation processes.

2. XPS OF WOOD

The principal chemical constituents of the bulk of dry wood are (hemi) cellulose, lignin and extractives. The XPS spectrum of wood therefore consists of an O1s and a C1s peak (Dorris 1978). The C1s peak is made up of three components, corresponding to carbon atoms bound only to carbon and hydrogen (C1), carbon atoms bound to one non-ketonic oxygen (C2) and carbon atoms bound to a ketonic oxygen or to two non-ketonic oxygens (C3)

In order to understand results of XPS measurements on wood, it is useful also to study the principal components of wood, i.e. cellulose and lignin. Results from XPS measurements on filter paper, used as model compound for cellulose, and lignin are given in Table 1. It is clear from these results that cellulose can be distinguished from lignin with XPS. Cellulose has a relatively high O/C ratio and the maximum of its C1s peak on the C2 position. Lignin has a much lower O/C ratio and the maximum of its C1s peak on the C1 position.

Using the measured carbon and oxygen percentages of cellulose and lignin, together with the expected data of a representative extractive, e.g. abietic acid, it is possible to calculate a weighted average of the three (De Lange 1992). The same can be done for the C1:C2:C3 ratio (see Table 1). If the surface of the wood were to consist of a homogeneous mixture of the three components, these weighted averages would be the expected percentages to measure a wood sample. As can be seen in Table 1, the deviation in a real measurement is very large. The O/C ratio of a freshly split wood surface is much lower than expected and the C1 contribution is too high. This can be explained by the cellular structure of wood. The cell walls consist largely of cellulose, while the lignin is found in-between the cell walls in order to bind them together. Wood split parallel to the fibres will contain a relatively large amount of lignin on the upper surface. Together with the presence of extractives at the surface this will result in a low O/C ratio and a high C1 contribution. A more homogeneous mixture of cellulose, lignin and extractives, made by filing wood and measuring the powder, shows a much higher O/C ratio and a lower C1 contribution (Table 1). Now the percentages are much more in agreement with the calculated values of a mixture.

Table 1: XPS results of cellulose, lignin and wood

	abundance in wood (%)	atomic % C	atomic % O	ratio O/C	percentage C1	C2	C3
cellulose	65	57.0	43.0	0.75	17	64	19
lignin	30	77.4	22.6	0.29	68	29	3
extractives	5	91.0	9.0	0.10	95	0	5
wood (calculated)	100	64.8	35.2	0.54	36	50	14
wood surface (measured)		77.1	22.9	0.30	65	29	7
wood powder (measured)		67.7	32.2	0.48	47	44	9

3. WEATHERING OF WOOD

Chemical changes in the upper layer of wood during artificial weathering
were studied with XPS. The results are given in Table 2. The O/C ratio
increases from 0.3 to a value of >0.7. The shape of the C1s signal also
changes considerably, which is shown in Figure 1. The increase of the C2
contribution is very clear and the shape shows much resemblance with the
carbon peak of cellulose. This is in agreement with the current theories
on degradation of wood (Hon 1984, Feist 1984). Lignin is more susceptible
to the absorption of UV radiation than cellulose, which results in
preferential lignin degradation. The formation of a cellulose rich
weathered layer increases the O/C ratio and the C2 component.

Table 2: Weathering of wood

Hours of weathering	atomic % C	atomic % O	ratio O/C	percentage C1	C2	C3
0	77.1	22.9	0.30	65	29	7
50	69.4	30.6	0.44	49	40	11
100	66.4	33.6	0.51	42	45	13
200	61.8	38.2	0.62	30	56	14
400	60.2	39.8	0.66	27	58	15
600	58.5	41.5	0.71	20	63	17
800	58.3	41.7	0.72	20	63	17

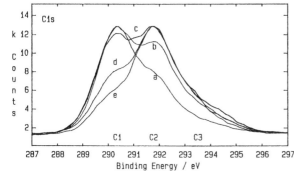

Fig.1: Carbon 1s signal of
wood after 0 (a), 50 (b),
100 (c), 200 (d) and 800
(e) hours of weathering.
Binding energies are not
corrected for charging.

To study the thickness of the weathered layer a cross section of a
weathered wood sample (800 hours WOM) was made. With a windowless X-ray
Microanalysis detector a carbon and an oxygen image was made. By dividing
(for every pixel) the oxygen intensity through the carbon intensity an
O/C image was obtained (see Figure 2). The result is in agreement with
the XPS measurements, a higher O/C ratio near the surface than in the
bulk. But now the thickness of the weathered layer can be determined,

which turns out to be about 200 um. A cross section of unweathered wood
was imaged as reference, for this sample a homogeneous O/C ratio was
found.

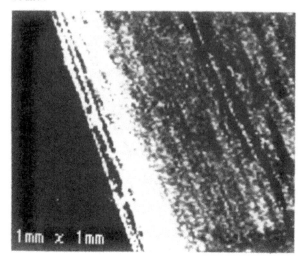

Fig. 2: O/C ratio of a
cross section of a
weathered wood sample
(800 hours). A brighter
intensity corresponds to
a higher O/C value.

4. CONCLUSIONS

XPS and imaging XMA can be nicely combined in studying the degradation of
wood during weathering. XPS shows an enrichment of lignin on the upper
surface of freshly split wood. This is due to the cellular structure of
wood. Weathering of wood starts with the degradation of the lignin. The
formation of a cellulose rich layer on the surface increases the O/C
ratio and the C2 component of the carbon signal.

By imaging the O/C ratio of a cross section of the wood with XMA it is
possible to determine the thickness of the cellulose rich weathered
layer. After 800 hours of weathering this thickness turns out to be about
200 um. This combination of XPS and XMA measurements may have many other
applications where degradation of materials is studied.

5. REFERENCES

De Lange P J, De Kreek A K, Van Linden A and Coenjaarts N J 1992 Surf.
    Interface Anal. 19 397
Dorris G M and Gray D G 1978 Cellul. Chem. Technol. 12 9
Feist W C and Hon D N S 1984 The Chemistry of Solid Wood (Washington DC:
    American Chemical Society) Chapt. 11
Hon D N S 1984 J. Appl. Polym. Sci. 29 2777

Inst. Phys. Conf. Ser. No 130: Chapter 6
Paper presented at Int. Congr. X-ray Optics and Microanalysis, Manchester, 1992

443

The performance of the photoelectron spectromicroscope at low and high energies

D.W. Turner and I.R. Plummer

Physical Chemistry Laboratory, Oxford OX1 3QZ

ABSTRACT: Images and spectra utilising the information content of low and high energy electrons derived from both high energy primary electron sources and low energy electron sources are presented. These were obtained from a new photoelectron spectromicroscope equipped with a monochromated Al Kα X-ray source operating at 450W.

1. INTRODUCTION

Electron spectroscopy and imaging devices are generally optimised for the elementally characteristic high energy region ($<\sim 1500eV$ in XPS). Many primary excitations however yield low energy (<10eV) electrons, amongst these are neutral and ion bombardment, UVPES and XPS near a band edge. Additionally there are processes which liberate low energy electrons without requiring auxiliary sources such as thermionic emission and exoelectron emission. High energy electrons yield a long low energy tail in the emitted electron flux due to material dependent energy loss processes. In conventional electron microscopies these electrons are often a nuisance.

In addition image contrast will depend on some (often complex) way on the depth distribution of the materials of interest. The depth of information is a function of the penetration depth of the ionising particle and the electron escape depth, the latter being a function of energy. Buried high yield objects may produce an image if they emit electrons with large escape depths (e.g. with <5eV or >500eV). If the object is a strong emitter then the energy loss processes will cause the energy of moderate energy electrons to be degraded to energies at which they have long path lengths and hence generate a diffuse image.

2. THE PHOTOELECTRON SPECTROMICROSCOPE

The photoelectron spectromicroscope (PESM) (Turner et al. 1986a) involves no lenses and hence is not subject to the same optimisation restrictions as conventional spectrometers and microscopes. The main considerations are i). the collection efficiency (solid angle) and ii). the analyser optimisation. In the PESM all of the emitted electrons are guided to the image band pass filter (IBPF) (Turner et al. 1986b, 1987, 1989) which extracts an image comprising of electrons lying between two defined energies. The operating parameters of the analyser are totally independent of the electron energy. Once the bandwidth has been defined the analysis energy is achieved by applying a retardation to the whole analyser.

Energy analysis of low energy electrons is difficult because of their extreme sensitivity to electrostatic and magnetic fields. Surface and contact electrostatic micropotentials and mild magnetic fields readily deflect the slow moving electrons. They can also cause an electron to be returned to the surface at a different place to be re-adsorbed or emitted after further energy loss. To transport successfully the low energy electrons to an electron optical system it is normal to accelerate those electrons to higher energy to make them less susceptible to patch fields in the apparatus. This can however make the measurement of the true energy of the emitted electron difficult since the momentum of the electron will be altered depending on its initial angle to the accelerating field.

The PESM overcomes these problems by the immersing the sample in a strong (7T) magnetic field. At the sample the deflection on the electron due to surface electrostatic fields is now dominated by the magnetic field. There exists now an ExB force on the electron which causes it to leave the surface with a tight helical trajectory. Once the electron is emitted it is guided by the magnetic flux to an imaging screen lying in a low magnetic field. Consequently there is no need to accelerate the electrons to avoid patch fields. Finally the magnetic field serves to re-orient the electron's momentum so that it becomes aligned with the magnetic axis (Beamson et al. 1981). This greatly facilitates electron energy measurement.

A new PESM has been built in collaboration with Kratos Surface Analysis Division. It consists of a 7T superconducting solenoid surrounding a sample in UHV. The sample can be heated and cooled (77 - 900K) and is mounted on a sample manipulator. Five sources currently converge upon the sample: an Al Kα X-ray source with a 1m quartz crystal monochromator, a He(I) resonance lamp, CO_2 IR laser, He-Cd laser and a fast atom source. The image band-pass filter is preceded by a 'skimmer' electrode and the IBPF has been modified to improve the low-pass filter and rotation correction.

The magnetic configuration is controlled by 19 external coaxial solenoids powered by computer-controlled power supplies. These serve to produce a homogeneous magnetic field region of ~0.015T over the extent of the IBPF and then control the flux divergence to the image screen and hence the magnification. Coils are also provided to position the image in the horizontal and vertical directions.

Image Acquisition. The image electrons are guided to a Galileo microchannel plate and phosphor on a fibreoptic screen. The screen is viewed by a low light CCD TV camera connected to a hardwired image processing unit and then to a frame grabber in an IBM PC clone. Images can be averaged and integrated at TV rate in the image processor then quantified in the IBM computer. Once in the computer images can be differenced, colour coded etc.

Energy Analysis. The IBPF is controlled by both fixed and computer-controlled voltage supplies. These allow complete computer control of the IBPF. The IBPF has been modified since the last published design. The first low-pass filter, referred to as the super dump (Turner and Plummer 1989), has been modified to eliminate electron background from high energy electrons (<1000eV) scattering from an internal surface in the analyser whilst working at low retardation voltages.

To obtain energy spectra a region is defined on the image and the intensity within that region is measured as a function of the retardation applied to the whole analyser, and hence of the selected energy of the electrons. The retardation is under computer control and can be stepped over two ranges; 0 to -50V in 0.76meV steps and 0 to -1500 in 23meV steps. Once acquired the data can be smoothed, background subtracted etc.

Spatial and Energy Resolution Enhancement. Electrons are emitted in all directions at the sample and hence have components of both axial and transverse momentum. The axial momentum causes them to drift out of the high magnetic field region towards the analyser and detector, whilst the transverse momentum causes a cyclotron orbit perpendicular to the magnetic field. This circular orbit is responsible for limiting the spatial resolution as it generates a circle of confusion about an emitting point equivalent to four times the cyclotron orbit radius r_c. Since the electrons always return to the point of emission irrespective of the initial direction of emission, the probability of finding an electron is always greatest about the emission point and hence $4.r_c$ represents the worst case. The spatial resolution can be improved by removing (skimming) electrons which have large components of transverse momentum from the electron flux. In the extreme if all but totally axially oriented electrons were removed then there would be no loss of resolution. In practice however this extreme would reduce the signal level greatly.

Removing electrons with large transverse momenta also improves the energy resolution. The momentum interconversion from transverse to axial when moving from a high magnetic field to a low one would only approach completeness if the ratio of the high to low fields was near infinity. At finite ratios there is always some transverse momentum remaining and hence the full momentum is not represented by the axial momentum. This causes a small spread in the measured energies. In the Kratos PESM, for an electron emitted near perpendicular to the magnetic axis in a 7T field, it would present 99.85% of its momentum in the axial direction in a 0.02T field. A skimmer electrode positioned between the sample and the analyser allows retarding field analysis before the full electron collimation caused by the change in magnetic field. This serves to reject electrons with small axial momentum components, and consequently large transverse components, from those reaching the IBPF.

Charge neutralisation. Only electrons higher in energy than the retardation potential applied to the IBPF and within the skimming angle finally leave the sample. Those of lower energy and lower axial momentum are returned to the sample along the magnetic flux lines, to a point within the $4.r_c$ diameter and most probably to the point of origin. This limits the electron loss from the sample and resistive samples are hence more amenable to study.

3. PERFORMANCE AT DIFFERENT ENERGIES

3.1 Spatial and Energy Resolution Enhancement

The skimmer electrode lies at an intermediate magnetic field (~0.03T) between the sample and the IBPF. By applying a negative bias to the skimmer slightly smaller than that applied to the IBPF, electrons with transverse momenta are turned back towards the sample. The degree of skimming is defined as the half cone angle of electrons leaving the sample. Figures 1b and 1c show that skimming is capable of producing chemical mapping to a few microns resolution in the PESM even at these high energies. If one calculates the Worst Case spatial resolution based on $4.r_c$ then for a 1keV electron in a 7T field this would be ~60µm. Skimming has improved on this by at least a factor of 10. The X-ray spectra and images in Figure 1 were obtained using a 1m diameter Rowland circle monochromator with an Al Kα source. This is designed to operate at 1500W but in preliminary test it was operated at 450W and unoptimised focus. Figure 1a demonstrates a wide scan XPS with inset an expansion of the Ag 3d peaks at ~1120eV kinetic energy. The half widths of these peaks are 1.99eV and 2.07eV before skimming and 0.78eV and 0.66eV after. Figure1b illustrates spatial resolution enhancement on 20 micron diameter tungsten wires over copper substrate imaged on the W 4f band at 1453eV, skimming 20°, 20 minutes acquisition at 450W Al Kα   1c imaged at 1485eV, valence and Cu 3d band. Note the shadow on the copper cast by the tungsten wire.

a   b   c

Figure 1. a) Selected area XPS of the Ag 3d $_{5/2,3/2}$ peaks. Upper trace with no skimming, lower with 20° acceptance angle at the sample. b), c) Selected energy images of Tungsten mesh over Copper at 1453eV and at 1485eV with 20° acceptance angle. Image dia ca. 230 μm.

3.2 Sample Charging

Major problems are often encountered in applying electron spectroscopic methods to analysis of insulators. As an example of the possibilities with the method indicated above Figure 2 shows the He(I) image and X-ray spectrum and image of a cross section of a decorative glazed ceramic tile. We first heated the specimen noting the decrease of charging, which effects the He(I) image, as the temperature rose to 350° C where there was no further change. Al Kα images and spectra were then recorded at successively lower temperatures and finally at 98° C. At this temperature the shift to lower energy of 3V was apparent on all identifiable peaks from O(1s) at 955eV to F(2s) at 1456eV.

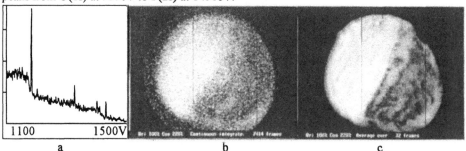

a   b   c

Figure 2. Ceramic tile glaze layer cross section. a). Selected area of glaze interface XPS at 96° C, 1100-1500eV. Glaze-tile interface at 350° C; b). Al Kα image using O(1s) electrons. c). He(I) image ca. 1eV electrons. Image dia ca. 230 μm

Other peaks confirm the presence of Lead and Silicon and show the absence of Tin in this glaze. The effect of the 'bottling" of the electrons between the sample and the retardation field at the analyser is most effective in XPS at the higher energies. Almost all electrons leaving the sample are of low axial momentum and return to the sample, therefore flood gun techniques are not needed.

3.3 Low energies: Scattered Electrons Derived from He(I) and X-ray Irradiation

The band structure of graphite has been extensively studied by X-ray valence electrons and UVPES. The experimental density of states is in good agreement with the calculated band structure (Tatar and Rabii 1982). Graphite has a low photoelectron yield for He(I) and its spectrum has a narrow band at 3eV kinetic energy. This narrow band has been attributed to the trapping of electrons following inelastic collisions within the crystal collecting in a sigma band which shows little dispersion in momentum space and is hence almost flat (Law et al. 1983). We have found that photoelectron images of larger crystals show marked face to face differences in their UVPES especially at the lowest kinetic energy end of the energy spectrum. This range is difficult for most conventional instruments to investigate.

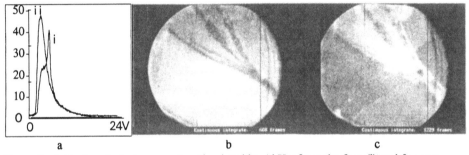

a   b   c

Figure 3. a) Scattered electron spectra stimulated by Al Kα from the face (i) and fracture regions (ii) of a graphite flake, and selected energy images at 3eV kinetic energy, of graphite flake, b) and at 9eV kinetic energy, c). Image dia ca 230 μm

Figure 3 shows the low energy scattered electron spectra obtained from the face and fractured edges of a graphite flake under Al Kα irradiation. These electrons result mainly from inelastic processes involving high energy primary photoelectrons. The 3eV peak is absent on spectra from the disrupted edge, both in these scattered electron spectra and He(I), He(II) and X-ray excited spectra. The overall emission from the smooth crystal faces of the graphite is much lower than from the edges. The corresponding He(I) PESM images in the figure show that the band at 3eV is associated only with the least emissive region, whilst the fractures are bright. This relatively intense 3eV emission indicates that electron transport is facilitated along the six-fold axis in the absence of defects and slip planes. The band structure indicates that the sigma band about 7eV above the Fermi level (i.e. producing 3eV electrons) is a sink for electrons created in the bulk, and is accessible in the K_z direction. At the edges there is still a concentration in the sigma band but it has a broader energy spread which is reflected in the UVPES.

3.4 Exoelectron Emission

Exoelectron emission is the release of electrons from substances subject to a small triggering stimulus. It is a source of the lowest energy of all electron images. There is a wide variety of stimulation e.g. heating from sub zero temperatures to a few hundred Celsius, exposure to light of an energy less than the work function, radio frequency field, mechanical stress, etc.

Exoelectron emission may occur in ambient atmospheric conditions and a vacuum is not essential. For a detailed study however, and in particular to allow emission images to be recorded, conventional high vacuum spectrometers and microscopes have been used (Yamamoto et al. 1985, Veerman 1969).

We have used PESM to study post-stimulus low energy electron emission (afterglow exoelectron emission) following pulses of U.V. light. Crystals of K_2SO_4 are exposed to brief pulses (e.g. 10 msec) of He(I) irradiation, and dependent on temperature, subsequently emit electrons for up to many minutes. A cyclic system was devised to irradiate the crystal with it held at a bias of +30V, to avoid photoelectron loss and hence cause charging, then bias it to -30V to encourage electron emission by inducing a strong extraction field at the surface. The chopping of irradiation and emission periods was controlled by pulse generators. At high pulse rates an apparently continuous image could be displayed arising solely from the exoelectron emission.

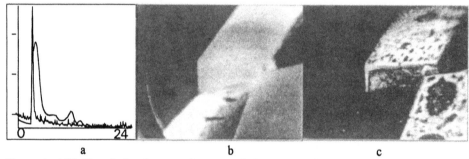

a   b   c

Figure 4. a) He(I) and afterglow exoelectron emission spectra from K_2SO_4 at ~170°C.
b) He(I) all-electron image of K_2SO_4 crystals on a Copper support mesh.
c) Afterglow all-electron image using a chopping sequence. Image dia ca 300 μm

Figure 4 shows the He(I) spectrum and the corresponding images, of K_2SO_4 crystals grown on a Copper support mesh from hot saturated aqueous solution. The He(I) spectrum shows the characteristic low energy intense peak which we have come to associate with exoelectron activity. This region is seldom investigated in conventional spectrometers. The low energy peak remains in the exoelectron emission spectrum. The afterglow image shows the non-uniformity of emission over the crystal reflecting a number of factors. The intensity is a function of the local defect concentration and escape conditions. If the surface barrier to electron escape is high, either by surface contamination or surface charging then the emission is reduced or absent.

3.5 Caesium Film Growth on Silicon

We have also used PESM to follow the growth of Caesium films from sub-monolayers to multilayers on p-type (100) Silicon (48-62 Ωcm) via its low energy electron emission.This allowed us to simultaneously perform energy analysis and imaging and record the dynamic changes in both the extent of coverage and the changes in electronic structure.

Silicon was etched in 40% hydrofluoric acid and washed in distilled water. It was heated to 300°C and exposed to argon FAB for 1 hour. This procedure generated a Silicon surface whose UVPES did not evolve further. The surface thus produced contains a considerable defect concentration. Caesium vapour was generated in a Knudsen cell 2mm below the He(I) source so that the Caesium impinged on the sample at the same angle of incidence as the He(I) beam. In the prototype PESM in which this experiment was performed the He(I) source is at 90° to the magnetic axis. The vacuum within the prototype PESM is poor and there is an appreciable partial pressure of Oxygen. Moderately fast deposition rates were used to limit the Oxygen absorption. The Caesium coverage as monitored by the height of the Cs $5p_{3/2}$ peak in the He(I) reached saturation after ~60 minutes. Other workers have interpreted this as indicating a monolayer coverage on defect free samples. The formation of the Caesium film was monitored by the change in the work function (Böttcher et al. 1991, Souziz et al. 1989).

During the caesiation the all-electron image was viewed either under He(I), He-Cd(2.8eV) or by thermally induced self-emission (Martinelli 1974). The main features of the image changes were an increase in photoemission under He-Cd illumination during 60 minutes exposure then reaching a steady level. After about 5 minutes into the exposure the sample, in the absence of external sources, self-emitted thermal electrons. This peaked after 30 minutes having increased in intensity by a factor of 3×10^5 times. This emission, showing a negative electron affinity, continued when the sample was rotated out of the Caesium beam. The sample was then left exposed to the Caesium source after the self-emission intensity had decayed. Between 60 and 80 minutes a very weak self-emission started which was dependent on exposure to Caesium. If the sample was rotated out of the beam the self-emission stopped and thus was Caesium induced exoelectron emission.

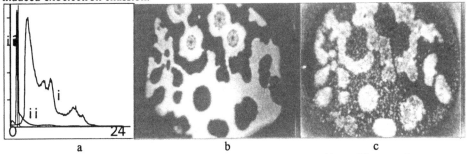

Figure 5. a) He(I) spectrum of Caesium coating of Silicon (100), (i) to (iii) succesive increases of caesium. b) He-Cd laser stimulated all-electron photoemission images of heavily caesiated Silicon. c) Spontaneously emitted electrons after heavy caesiation. Image diameter is ~400µm. Sample at ~300°C

During this evolution the electron images of the surface displayed complex growth phenomena. During the increase and decay of the thermal self-emission signal small black areas originating in edges, scratches and similar defects appeared on the surface, grew, some coalesced and they developed growth rings. The form of these patterns is displayed in figures 5b and 5c. Note particularly a contrast reversal between 5b and 5c.

We interpret these observations as the formation of multilayer films on the surface of defective Si-(100). The ring patterns we infer as being due to the edges of layers of Caesium on the surface, and as these layers grow the extent of the patterns change. Finally there is a thick layer of Caesium on the surface which, within the escape depth of the He(I) generated electrons, appears as bulk Caesium. The self-emission is due to the release of energy on the formation of Cs-Cs bonds when the vapour condenses on the surface and attacked to the edge of a layer.

3.6 Fast Atom Bombardment as an Imaging Source.

Fast atom bombardment results in the copious release of electrons from solids. The electrons lie in the energy range of 0-5eV and the resulting low energy images are extremely intense. The image contrast obtained in FAB images is almost entirely due to the variation in total electron yield. The energy spectra from different materials are all very similar - a Maxwell-Boltzmann form of distribution with energy about 2eV. There are shifts in the spectra relating to the differences in the work function.

No characteristic high energy electron emission has been detected under FAB bombardment, consequently it is feasible to depth profile simultaneously with X-ray or He(I) data acquisition.

6. CONCLUSION

The photoelectron spectromicroscope has been shown to be capable of accurate and sensitive response from near thermal to 1.5keV electrons. By using skimming both energy and spatial resolution enhancement have been demonstrated showing the utility of the PESM for elemental mapping at the micron level. In addition skimming proves beneficial in eliminating specimen charging in XPS. Novel low energy emission processes have been investigated including afterglow exoelectron emission from K_2SO_4, thermal stimulated exoelectron emission from Cs/Si and exoelectron emission on condensation and bond formation of Caesium on a Silicon surface. The complex growth of Caesium films on defective (100) Silicon has been revealed by images of both stimulated and electron self-emission.

The authors are grateful to Kratos Analytical and the U.K. Science and Engineering Research Council for Support.

REFERENCES

Beamson G, Porter H Q and Turner D W 1981 Nature 290 556
Böttcher A, Grobecker R, Imbeck R, Morgante A and Ertl G 1991 J. Chem. Phys. 95 3756
Ernst H and Yu M L 1990 Phys. Rev. B 41 12953
Hasselkamp D, Hippler S and Scharmann A 1987 Nucl. Instr. and Methods In Physics
    Research B18 561
Law A R, Barry J J and Hughes H P 1983 Phys. Rev. B 28 5332
Martinelli R U 1974 J. Appl. Phys. 45 1183
Souzis A E , Seidl M, Carr W E and Huang H 1989 J. Vac. Sci. Technol. A7 720
Tatar R C and Rabii S 1982 Phys. Rev. B 25 4126
Turner D W and Plummer I R 1989 J. Phys. E: Sci Instrum. 22 593
Turner D W, Plummer I R and Porter H Q 1986a Phil. Trans. R. Soc. Lond. A318, 219
Turner D W, Porter H Q and Plummer I R 1986b Rev. Sci. Instrum. 57 1494
Turner D W, Plummer I R and Porter H Q 1987 Rev. Sci. Instrum. 59 45
Veerman C C 1969 Mater. Sci. Eng. 4 329
Yamamoto S, Yokogawa H and Hashimoto H 1985 Japan J. Appl. Phys. 24 277

Inst. Phys. Conf. Ser. No 130: Chapter 6
Paper presented at Int. Congr. X-ray Optics and Microanalysis, Manchester, 1992

451

Magnetic lens technology in X-ray photoelectron spectroscopy (XPS): facts and fallacies

K S Robinson, F J Street, I W Drummond

Kratos Analytical Ltd. Barton Dock Road, Urmston, Manchester M31 2LD, UK

ABSTRACT: An XPS spectrometer has been built, incorporating a magnetic immersion lens. The low aberration coefficient of this lens enables analysis of small areas, with increased sensitivity, compared to conventional electrostatic lenses. The use of a magnetic lens obviously raises questions about the ability to analyse "difficult" specimens, including magnetic and insulating specimens. Analysis of such specimens is discussed with practical examples, showing that excellent results may be obtained.

1. SPECTROMETER DESIGN

Figure 1 shows a schematic diagram of the new spectrometer (Drummond 1992, Walker 1987). It consists of a conventional 180 degree, 127mm radius hemispherical analyser with three channeltrons positioned across the energy dispersive plane. A magnetic immersion lens of the 'snorkel' type (Mulvey 1982) is positioned below the sample and focusses photolectrons onto the spot size aperture. A set of electrostatic deflection plates enables the position of the analysis spot to be moved and also to be scanned to produce images. There is an electrostatic probe lens which can be used instead of the magnetic lens or in combination with it. Without the magnetic lens, the minimum useful analysis size possible is about 80µm diameter. The magnetic lens with its higher magnification and lower aberrations reduces the the analysis area and also increases the acceptance angle of the lens. This larger angle significantly increases the sensitivity, particularly at smaller analysis areas. An analysis diameter of less than 30µm (16-84% edge resolution) can now yield an intensity greater than five times that previously obtained at 100µm with a purely electrostatic small area XPS analysis system (Drummond et. al. 1990)

2. MAGNETIC SPECIMENS

Manufacturers and users of electron spectrometers have always made great efforts to remove all magnetic fields from inside the spectrometer. This includes the use of magnetically permeable "mu" metal to prevent all stray magnetic fields, including the earth's magnetic field, from entering the spectrometer. Magnetic fields greater than a few milliGauss are thus eliminated. A magnetic field will deflect electrons and the degree of deflection is determined by the electrons kinetic energy. If the field remains constant, then the position on the specimen from which the photoelectrons are analysed will vary with the kinetic energy of the electrons. It may therefore seem counterproductive to introduce a magnetic field with a peak value of several hundred Gauss into the spectrometer. The large magnitude of this field actually tends to swamp all other fields. The field is also well defined and controlled such that the optical properties are constant with energy. The field is also localised and careful design of magnetic shielding eliminates the field from the remainder of the spectrometer. Before discussing the effect of magnetic specimens in the magnetic lens, it is important to remember that such specimens may also cause analysis difficulties in a conventional spectrometer for the reasons given above.

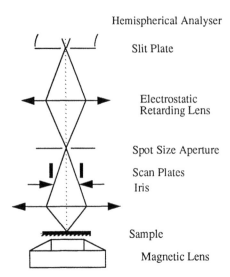

Hemispherical Analyser

Slit Plate

Electrostatic
Retarding Lens

Spot Size Aperture

Scan Plates

Iris

Sample

Magnetic Lens

Figure 1 - Schematic Diagram of AXIS HS
Spectrometer with Magnetic Lens.

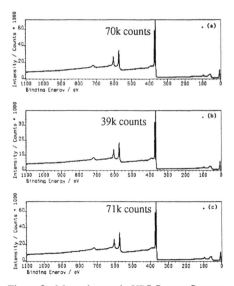

Figure 2 - Monochromatic XPS Survey Spectra
from Silver: (a) No Fe; (b) 10mm Fe, magnetic
lens normal; (c) 10mm Fe, magnetic lens
adjusted.

Figure 2 shows survey spectra obtained from
sputter cleaned silver, using monochromatic Al
X-radiation. Figure 2(a) was obtained from a
sample mounted over a hole in the sample stub. In
Figure 2(b), the hole was filled with a 6mm
diameter x 10mm length piece of iron. Figure 2(c)
shows the spectrum from the same sample after
adjusting the magnetic field to optimise the
sensitivity, and shows very similar intensity to
that obtained without the iron. Note that the
overall spectral shape is also identical, indicating
that there is no change in the transmission
function of the spectrometer.. In this case, the iron
has acted as an additional element of the
magnetic lens. The field lines appear to have
followed parallel paths through the iron and
where they emerge into the vacuum, they diverge
again, as they normally do in the magnetic lens.
The specimen therefore experiences a field
similar to that normally found closer to the
magnet. Hence, the magnetic lens has to be
refocussed to ensure the photoelectrons are
focussed into the electrostatic lens.

Imaging with magnetic samples is also possible
with few problems. Figure 3 shows an image of a
gold coated electron microscope grid with bar
width of 10μm and pitch of 100μm, mounted
above a 1.5mm thick piece of iron.

Figure 3 - Au4f XPS Image of TEM grid,
mounted above 1.5mm thick iron.

3. CHARGE NEUTRALISATION

When insulators are irradiated with X-rays, the emitted photoelectrons give rise to a positive charge on the surface of the insulator. Charging will continue and may eventually prevent the emission of further photoelectrons, unless a balancing flux of electrons is directed onto the surface. Charging is particularly severe when a monochromatic X-ray source is used. The usual method for charge neutralisation involves directing a beam of focussed low energy electrons onto the sample. However, this can cause differential charging if the relative flux density of electrons is not well matched to the X-ray flux density. Changing the relative angle of electrons, X-rays, sample and analysis area, as for example in angle-dependent XPS studies, can require refocussing of the electron source.

When a magnetic lens is used, it is not possible to direct low energy electrons onto the sample, particularly from the usual incidence angles of 30 to 60 degrees, since the low energy electrons cannot cross the magnetic field without being deflected and few electrons reach the sample surface. An alternative patented method therefore has to be used. A reflector electrode is fitted below the entrance to the electrostatic lens. Electrons (photoelectrons and secondary electrons) which are not focussed through the spot size aperture, by the magnetic lens may strike this plate. The application of a negative potential reflects low energy electrons back into the magnetic field. They are then constrained by the magnetic field and follow their initial direction of motion, back onto the sample surface. Since the same field also controls the analysis area, there is a good charge balance across the analysis area. Figure 4 shows typical C1s spectra from two polymers showing excellent energy resolution. Recently, modified polymer surfaces have been studied and characterised with particular reference to their relatively low intensity valence bands (Wells et al 1992). Also the high sensitivity and small analysis area has allowed polymer lithography to be investigated (Robinson et al 1992)

Optimisation of neutralisation parameters is straightforward. The net peak height (or Full Width - Half Maximum (FWHM)) is monitored whilst adjusting the potential on the reflector electrode. As the reflected electron flux is directed onto the sample, the peak height maximises and

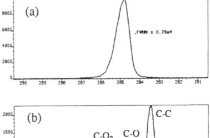

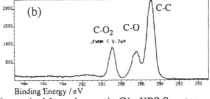

Figure 4 - Monochromatic C1s XPS Spectra (a) Polyethylene (FWHM=0.79eV) (b) PET showing FWHM <0.70eV on C-O$_2$ peak.

FWHM minimises. Similar neutralisation conditions are found for many materials, and further adjustment on individual samples is not usually necessary. The system is also partially self regulating as the X-ray power is changed. As the X-ray flux is increased, the outgoing flux of photoelectrons also increases, but this is compensated by an increase in the number of secondary reflected electrons.

Figure 5 shows valence band spectra from two similar hydrocarbon polymers (polyethylene and polypropylene). Both polymers exhibit similar C1s spectra and only the valence band shows significant differences.

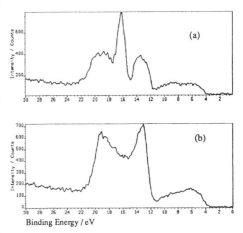

Figure 5 XPS valence band spectra (a) Polypropylene (b) Polyethylene.

Figure 6 shows the C1s and valence band maps obtained from a sample consisting of polyethylene and polypropylene sheets mounted close together. The C1s map shows a relatively uniform distribution. The valence band maps were obtained from the two characteristic energies, 13eV for polyethylene and 16eV for polypropylene.

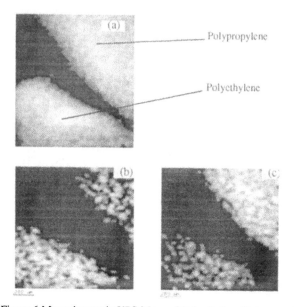

Figure 6 Monochromatic XPS Maps of Polyethylene/Polypropylene
(a) C1s (b) Valence band BE 13 eV (c) Valence band BE 16eV

4. CONCLUSIONS

The use of a magnetic lens in an XPS spectrometer has been shown, including the improved performance compared to conventional spectrometers. Its suitability for analysing difficult samples such as magnetic specimens and insulators, in both spectroscopic and imaging modes has also been demonstrated.

REFERENCES

Drummond I W, Street F J, Ogden L P, and Surman D J 1990 Scanning 13 149

Drummond I W 1992 Microscopy and Analysis March 1992 29

Mulvey T 1982 In "Magnetic Electron Lenses"; Ed P W Hawkes pp 359 -412, (Berlin, Springer-Verlag)

Robinson K S, Street F J, Drummond I W 1992 Paper 2.3.8 of these Proceedings

Walker A R 1987 American Patent No. 4,810,879

Wells R K, Drummond I W, Robinson K S, Street F J, Badyal J P S 1992 J Adhes. Sci. Tech. (In Press)

Inst. Phys. Conf. Ser. No 130: Chapter 6
Paper presented at Int. Congr. X-ray Optics and Microanalysis, Manchester, 1992

455

The role of transfer film chemistry in automotive friction couples

Allan Wirth* and Robert Whitaker +

*Materials Research Institute, School of Engineering,

Sheffield City Polytechnic, Pond Street, Sheffield S1 1WB UK

+ Mintex Don, Cleckheaton, PO Box 18, West Yorkshire BD19 3UJ UK

ABSTRACT: During automotive braking a friction heat affected layer is introduced onto the surface of the two components of the friction couple together with the transfer of material between rubbing faces to form a transfer film (TF) or third body layer (TBL).

In this investigation energy dispersive X-ray analysis (EDX) and X-ray photo electron spectroscopy (XPS) have been used to study the chemistry of transfer films formed on the grey cast iron

The results show a clear connection between transfer film chemistry, asperity temperature and the composition of the friction material, especially lubricant additions. Evidence is presented which relates the friction characteristics of the couple to transfer film chemistry.

1. INTRODUCTION

There is a general consensus that the tribological characteristics of the transfer film (TF) formed at the friction interface controls the performance of automotive braking systems. Liu and Rhee(1976) have shown both (TF) formation and friction characteristics are extremely temperature sensitive. Recent work by Wirth and Whitaker(1992) has shown that temperature induced compositional changes within the (TF) can also influence the friction coefficient of the couple.

In this paper the use of energy dispersive X-ray analysis (EDX) and X-ray photo-electron spectroscopy (XPS) for the analysis of transfer films formed on cast iron is described. The use of imaging (XPS) to provide spatially resolved chemical information is also discussed. The results obtained extends our initial understanding of the influence of transfer film chemistry upon friction performance.

Thus the possibility of designing friction materials from a knowledge of transfer film chemistry may one day become a reality.

2. MATERIALS AND EXPERIMENTAL PROCEDURE

Two semi-metallic friction materials were selected for this work. The materials differed only in respect of an antimony sulphide addition of 10 vol%. Constant torque and constant friction test methods were used to generate transfer films on the grey cast iron discs. After each run samples were removed from the disc and transferred to a dessicator.

3. RESULTS AND DISCUSSION

Figure 1 shows two typical dynamometer traces illustrating the effect of a 10 vol% antimony sulphide addition to the friction material. There is an immediate increase, visible during the first stop, in the overall coefficient of friction associated with the antimony sulphide addition. The presence of sulphide also provides a greater degree of stability during braking although some fade was evident during each stop.(i.e since the experiments were carried out at constant torque any increase in applied pressure during braking was an indication of a decrease in friction coefficient). The inclusion of antimony sulphide was also responsible for a decrease in the low temperature wear rate of the friction material, 110.88mm^3 /MJ compared with 126.53mm^3 /MJ for the base material.

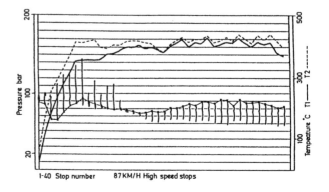

Figure 1.1 Base Material

Previous work by Liu et al (1980) has shown differences in the wear rates of asbestos based friction materials can be related to the degree of transfer film coverage. It could be argued that the evidence presented in figure 2.2 supports this view since antimony sulphide promotes continuous transfer film coverage. Differences in transfer film coverage alone cannot be used to explain the improved friction characteristics when antimony sulphide is added to the friction material since the frictional force acting between two rigid bodies is generally considered to be independent of the contact area. Recent work by Wirth

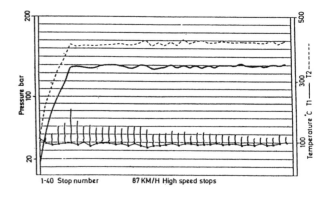

Figure 1.2 Base Material + Antimony Sulphide

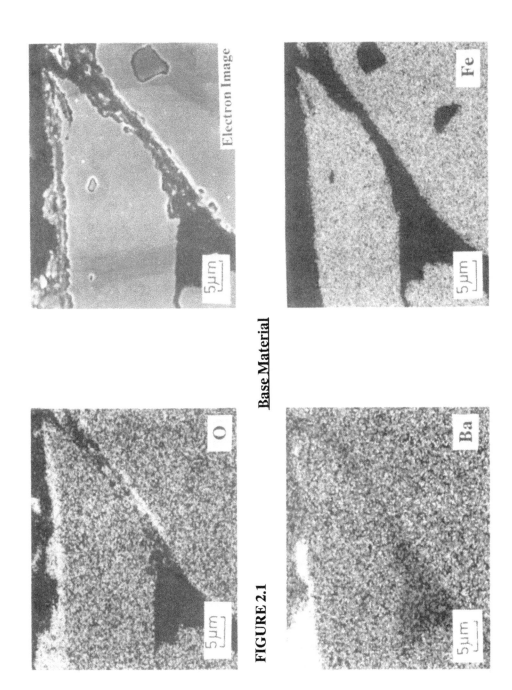

Base Material

FIGURE 2.1

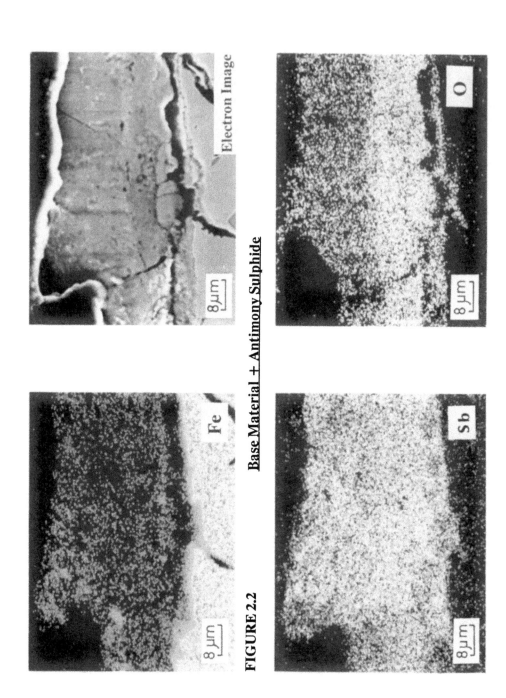

Base Material + Antimony Sulphide

FIGURE 2.2

and Whitaker(1992) has demonstrated the importance of transfer film composition in friction coefficient studies of automotive friction materials.

Energy dispersive X-ray (EDX) and X-ray photo electron spectroscopical XPS analysis of the transfer films confirmed appreciable compositional differences between the two transfer films shown in figures 2.1 and 2.2. Comparison of the EDX data presented in figures 3.1 and 3.2 highlights the presence of significant quantities of antimony in the transfer film after rubbing against the antimony sulphide bearing friction material. The low energy portions of the two spectra, notably the carbon oxygen and iron levels, provide further evidence of differences in the transfer film chemistry when antimony sulphide is added to the friction material. These results suggest the antimony sulphide is acting as a solid lubricant. It is suggested, from evidence presented in the X-ray elemental maps illustrated in figure 2.1 (intensity plotted on a thermal image scale) that the friction debris partially destroys the transfer film by a ploughing action in the absence of antimony sulphide.

Total oxygen coverage of the surface (see map in figure 2.1) does not conflict with this view since the surface of the disc, after losing a portion of transfer film by a ploughing action, will naturally oxidise on cooling from approximately 450^0 C. Confirmation of an antimony presence throughout the transfer film is shown in figure 2.2. Unfortunately the X-ray mapping technique did not provide any information regarding the chemical state of the antimony.

X-ray photo-electron spectroscopical data presented in figure 4 confirms a sulphide contribution to the S2p peak indicating some if not all of the antimony sulphide in the friction material transfers and remains in the film as the sulphide form. Imaging the transfer film surface using the sulphide S2p peak at 168eV clarified the situation (see figure 5). The central white band indicates a region of high sulphide concentration, surrounded by darker bands with a lower sulphide level. A similar effect

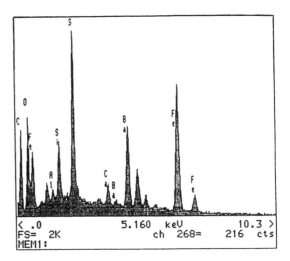

Figure 3.1 Base Material

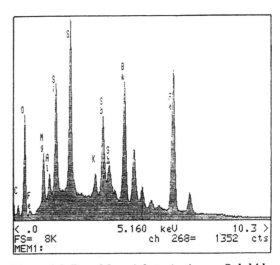

Figure 3.2 Base Material + Antimony Sulphide

has been observed by Wirth and Whitaker (1992) with molybdenum sulphide additions to friction materials, albeit at much lower interface temperatures. They argued that bands of sulphide retained at the interface, when the asperity temperature was below the molybdenum sulphide decomposition temperature, was responsible for a decrease in the friction coefficient of the couple.

Similarly the evidence obtained in this investigation supports the view that retention of antimony sulphide at the interface during braking produces fade, (i.e. at constant torque additional pedal pressure was required as demonstrated in the dynamometer data shown in figure 1).

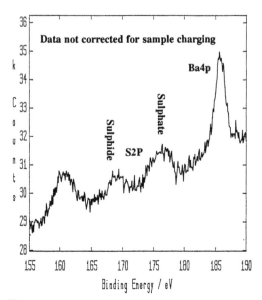

Figure 4.

4. CONCLUSIONS

(i) There was a marked increase in overall friction coefficient of an automotive braking system when antimony sulphide was added to an asbestos free friction material.

(ii) The fade observed during braking with an asbestos free friction material containing antimony sulphide was due to the retention of sulphide at the interface during the brake application.

(iii) The wear rate of an asbestos free friction material was reduced by the addition of antimony sulphide.

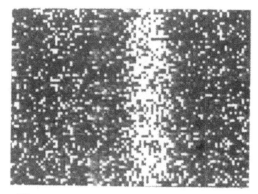

Figure 5.

REFERENCES

Liu T and Rhee S K 1976 Wear 37 291-7

Liu T, Rhee S K and Lawson K L   1980 Wear 60 1-12

Wirth A and Whitaker R 1992   J.App Phys D 25 38-43

Inst. Phys. Conf. Ser. No 130: Chapter 6
Paper presented at Int. Congr. X-ray Optics and Microanalysis, Manchester, 1992

461

Auger image cross correlation with secondary image

P. MORIN, E. VICARIO
Département de Physique des Matériaux Université LYON I
43 Bd du 11 novembre 1918 - 69622 VILLEURBANNE - FRANCE

ABSTRACT: Auger image cross correlation with the secondary emission image leads to an increase in the signal to noise ratio. First the two-dimensional Fourier transform (FT) of the Auger and secondary image is performed. The product of these transforms gives the inter-spectrum; its root-square is the FT modulus of the expected image, and the FT phase of this image is that of the Auger image FT. Results are obtained on a aluminium-silicon alloy containing other elements at low concentration: Fe, Cu, Ni, Mg, O, C.

1. INTRODUCTION

Poor signal to noise ratio of Auger emission requires long acquisition times and high beam current for Auger images. In contrast the collection of all secondary electrons gives good images quickly. As chemical contrast is present in these images, we propose to use them to improve the signal to noise ratio of Auger images (Vicario 1992).

It is well known that the principal contrast in the secondary electron images is due to the topography of the specimen, but it can be reduced by surface polishing, and the acquisition conditions of Auger images are favourable to exhibit chemical contrasts: clean surfaces in ultra high vacuum, low energy beam (10keV).

2. PROCESSING

The two-bidimensional Fourier transforms are calculated for the secondary and Auger images. The product gives the inter-spectrum. We suppose that the inter-spectrum is the power spectral density of the expected Auger image (Max 1981). This image is thus given by the inverse Fourier transform in which the modulus is the root square of the inter-spectrum and the phase is that of the initial Auger image.

The secondary images being not noisy, the noise of the FT modulus of the expected image will be lower than the noise of the initial Auger image.

Before processing the images are made symmetrical in the X and Y directions by adding mirror symmetry images (Vicario 1989). The symmetrical Fourier transform is real, the value of the phase is 0 or π. Thus the noise affects the phase only near the zero values of modulus.

This process is not linear, it enhances contrasts at low level of images. This nonlinearity is partially corrected by shift of the cross correlated image histogram. A value is added to each pixel to fit the histograms of the initial and cross correlated images, and, we give zero value at all pixels that

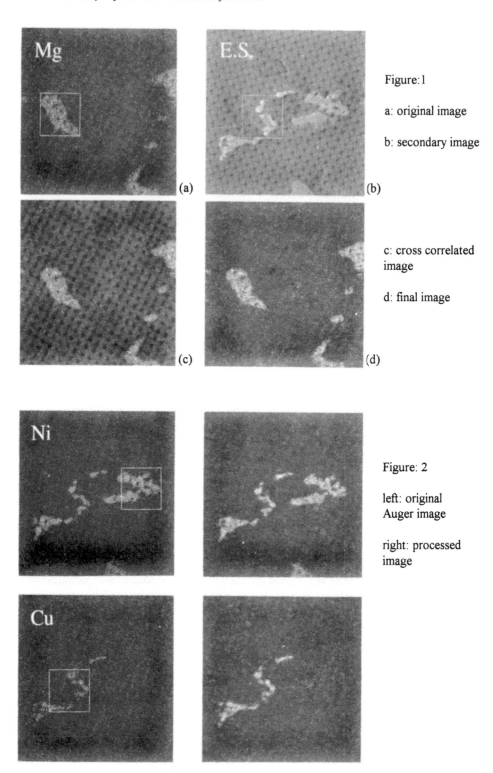

Figure:1

a: original image

b: secondary image

c: cross correlated image

d: final image

Figure: 2

left: original Auger image

right: processed image

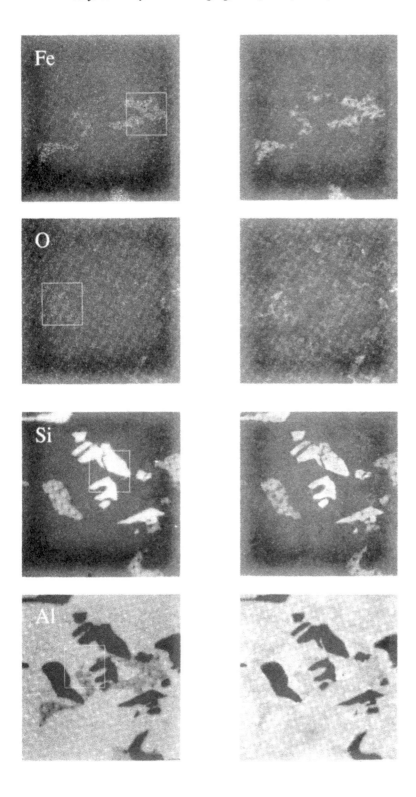

are null in the initial Auger image, these zero are supposed significant.

3. RESULTS

The specimen is an aluminium-silicon alloy containing other elements with low concentration: Fe, Cu, Ni, Mg, O, C. The surface is mechanically polished and cleaned by ionic sputtering.
The observations are performed with an Auger microprobe: Nanoscan 100 CAMECA. Auger and secondary images are accumulated at TV rate with a 256*256 pixel format.
Experimental settings for Auger acquisition are given in the table. The beam parameters are the same for Auger and secondary images (Ibeam = 15 nA, Ebeam = 10 keV). The acquisition time for secondary images is 10 s. All the micrographs are taken on the same area. The image width is 30 μm.
Figure 1a and 1b, show respectively the initial Auger image of Mg and the secondary image. Figure 1c is the cross correlation of 1a and 1b. The final image is obtained Figure 1d after level correction.
Figure 2 gives the original and treated Auger images for the other elements: Al, Si, Fe, Ni, Cu, O.
The measure of the signal to noise ratio (S/N) is not easy. It is estimated by calculation of the FT averaged over a significant part of the image; then we suppose that the noise is white, thus the high frequency components of this FT give the noise amplitude. The results in the table have been obtained for areas marked by white borders in figure 1 and 2. The S/N gain of treated image is lower than for cross correlated image : the level correction introduces signal discontinuities that fail the S/N result. The S/N gain related to the treated images and given in the table is under-estimated. The gain is more important for the high noise images; a gain of +7 dB corresponds to a 5 time longer accumulation.
In spite of complicated image structure the process introduces low additional contrast only in images with good S/N (Al, Si in figure 2). The correction of the nonlinearity is not yet satisfying for these images. But for low S/N images this process allows us to improve sensitivity to give qualitative identification of elements at very low concentration with very high spatial resolution.

Element	Auger pic (eV)	Acquisition time (s)	Auger image S/N	Cross correl. image S/N	Treated image S/N	Gain S/N
O	503	600	-9dB	11dB	-2dB	+7dB
Mg	1184	600	-4dB	11dB	4dB	+8dB
Al	64	180	11dB	20dB	15dB	+4dB
Si	89	180	12dB	17dB	13dB	+1dB
Fe	700	1200	-8dB	1dB	-1dB	+7dB
Ni	844	600	0dB	12dB	7dB	+7dB
Cu	916	1200	-8dB	10dB	0dB	+8dB

Table: experimental setting and S/N results

Max J 1981 Méthodes et techniques de traitement du signal - tome I (Paris: Masson et Cie).
Vicario E, Dogmane N, Tholomier M 1989 J. Microsc. Spectrosc. Electron. 14 pp 95-17.
Vicario E, Morin P, Chanel J 1992 Microscopy Microanalysis Microstructure to be published

Inst. Phys. Conf. Ser. No 130: Chapter 6
Paper presented at Int. Congr. X-ray Optics and Microanalysis, Manchester, 1992

465

Photoemission microscopy—A novel technique for failure analysis of VLSI silicon integrated circuits

Danilo Golijanin

Fisons Instruments, X-Ray Division, Valencia, California 91355, USA

ABSTRACT: The radiative electron-hole recombination in silicon gives rise to the emission of visible light. Photoemission microscopy and photoemission spectrocopy have recently become important techniques in the failure analysis of functionally-failing complex VLSI silicon devices. A variety of semiconductor failure modes can be imaged and characterized. The spectral distribution of the emitted light can be used for the identification of an actual failure mechanism.

1. INTRODUCTION

When electron-hole pairs recombine in a semiconductor material, light can be generated. If the exited carriers are generated by the injection of current into the material, the ensuing luminescence (photoemission) is called the electroluminescence. Visible light emission from silicon in both forward and reverse biased p-n junctions has been known since the dawn of the semiconductor research. A faint yellowish-white light was first seen by Newman (1955) and Chynoweth and McKay (1956) at the edge of a p-n junction under the reverse breakdown condition. The brightness and the location of the light spots varied with current intensity. The spectral distribution was also characterized by a quartz monochromator. As time progressed the weak light emission was observed from a forward-biased p-n junction, and from a silicon dioxide dielectric in an integrated gate-oxide capacitor.

The mechanism for photon generation in a forward-biased silicon p-n junction is the radiative electron-hole recombination. If the junction is reverse-biased into an avalanche breakdown, photons are generated due to the intraband relaxations and the hot electron-hole recombinations. Khurana and Chiang (1986) found that in the dielectric breakdown of the gate-oxide capacitor, photons are generated due to the recombination of majority carriers that cross the oxide barrier through Fowler-Nordheim tunneling. The difference in carrier energies, for each of these recombination mechanisms, gives rise to a specific photon wavelength (energy) distribution in the visible range and thus creates a possibility of performing photoemission spectroscopy. Additionally, all photoemitting events are found, by Ong et al (1983), to be characterized by a very low light intensity level, due to the low quantum efficiency of 10^{-5} to 10^{-4} photons per electron-hole recombination.

2. PHOTOEMISSION MICROSCOPY

The first practical light-emission microscope (LEM) was made by Khurana and Chiang (1986). They took the advantage of the development of night vision technology and used it for imaging the faint ("invisible") light coming from various structures on a silicon integrated circuit (IC). This invention created a flurry of activity in the area of VLSI device failure analysis, testing and equipment manufacturing.

A typical photoemission microscope consists of an x-y-z stage with a device holder, an optical microscope, and a light sensitive camera, all set within a light-tight enclosure and a computer system for image acquisition and processing. If spectral analysis of the emitted light is desired, a holder for band-pass filters is inserted in the optical path.

Since the intensity of the emitted light is very low (a few photons per pixel per second) it is necessary to employ an image amplification scheme. There are currently two different approaches to image acquisition and processing: (i) an image intensifier (e.g. a microchannel plate) directly coupled to a solid-state CCD camera, and (ii) a cooled CCD camera for long exposure times.

The system, based on an image intensifier, acquires images in the real time (TV rate) and the image processor performs the integration over a period of time. The light amplification is limited to a factor of 10^5 (Khurana and Chiang, 1986), due to the quantum and multiplication noise limitations. In the system based on a thermoelectrically-cooled solid-state camera, the intensifier-generated noise is eliminated, but the exposure time is still limited by the dark-current leakage associated with the CCD elements. The image acquisition time is typically several minutes long in either system.

3. APPLICATION OF PHOTOEMISSION MICROSCOPY IN FAILURE ANALYSIS

The initial work, related to light emission from various semiconductor structures, was concentrated on discrete devices: diodes, transistors, capacitors, etc. The advent of the photoemission microscope in mid-1980s made the imaging of the same devices possible, but now within the complex VLSI integrated circuits. If a failure site happened to be at the periphery of the VLSI IC (e.g. close to bonding pads) there were efficient techniques for fast and inexpensive fault analysis. For example, these included liquid-crystal hot-spot detection, electron and optical beam induced current (EBIC and OBIC) imaging. In the case of an "inside-the-chip" functional failure, complex and expensive equipment had to be used to identify the location of the failure site and the corresponding failure mechanism.

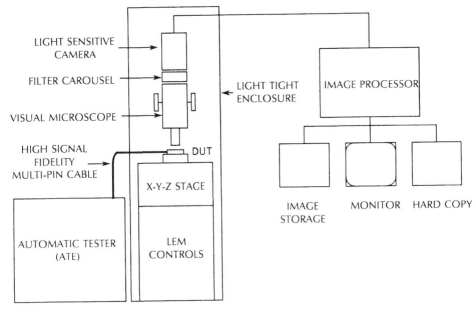

Fig. 1. Typical light-emission microscope (LEM) coupled to an automatic tester (ATE). Failed IC device under test (DUT) is connected to the ATE via a high-signal fidelity multi-pin cable. The faint light emitted from the chip is captured by a light-sensitive camera, and the image is processed by an image processor.

In order to locate the failure site it is necessary to electrically activate the structure responsible for the failure. The applied voltage has to provide an adequate current flow, which will create enough photons to be detected by the photoemission microscope (at least 10^{-8} A/μm^2). The most complicated functional failures are usually caused by a malfunctioning circuitry in the interior of an integrated circuit. This type of failure manifests itself as a logic discrepancy at the IC output pins. Thus, it is necessary to use a complex test generator which will generate a combination of input signals which reproduce the failed output combination. The interface of an automatic testing equipment (ATE) to the light-emission microscope (LEM) is shown in Figure 1. The device under test (DUT) is mounted in an electrical socket, and its surface exposed to the microscope and the camera. All test signals (test vectors) are supplied through a high-signal quality cable.

The main disadvantage of the system described above is the need to use lengthy high-pin count cable for electrical coupling of the failed IC to the test generator. High signal quality has to be preserved, especially at high frequency testing. Additionally, it may be necessary to use different cables for different device packages, or design complex cable-to-socket adapters. To overcome these difficulties, a direct coupling of the microscope and the camera to the DUT was implemented. The direct *in situ* interface eliminates the need for the cable, and makes it easier to use existing ATE programs.

Identification of the failure mechanism is easily addressed by photoemission microscopy because of the characteristic wavelength distribution of the light-emitting structure. It was postulated by Wallinger (1991) that all light-emitting VLSI failures can be described by one of the four basic models: forward-biased junction, reverse-biased junction, saturated transistor and oxide capacitor. Another classification by Wills (1990) of light-emitting events in silicon VLSI ICs, was done according to the device structure: transistor-related failures, substrate-related failures and oxide leakage current. Photoemission events due to transistor failures and operation are: hot electron injection, saturation and MOSFET breakdown. Crystal defects (e.g. stacking faults), CMOS latch-up and the electrical over-stress EOS/ESD damage are typical examples of substrate-related failures. The dielectric leakage current path in an oxide capacitor, which creates photoemission, can be caused by oxide pinholes, cracks, ionic contaminants, particles, pores and EOS/ESD damage.

In addition to the photoemitting IC failures, various non-emitting failures occur: ohmic shorts (e.g. shorted metal interconnections), surface inversion, timing discrepancies, etc. These failures serve as a proof that although a very capable technique, photoemission microscopy is not a universal panacea for all VLSI IC failure analysis problems.

4. NOTE

The work described in this paper was performed at the Electronics Division of the Xerox Corporation in El Segundo, California.

5. REFERENCES

Chynoweth A G and McKay K G 1956 Physical Review 102 pp 369-76
Khurana N and Chiang C L 1986 Proc. IEEE 24th IRPS pp 189-94
Khurana N and Chiang C L1987 Proc. IEEE 25th IRPS, 72-6
Newman R 1955 Physical Review 100 pp 700-3
Ong T C, Terrill K W, Tam S and Hu C 1983 IEEE Electron Dev. Lett. EDL-4 pp 460-2
Wallinger T 1991 Proc. ASM 17th ISTFA pp 325-34
Wills K S 1990 Proc. ASM 16th ISTFA-Workshop pp 73-7

Inst. Phys. Conf. Ser. No 130: Chapter 6
Paper presented at Int. Congr. X-ray Optics and Microanalysis, Manchester, 1992

469

Lithography and polymers: An XPS study

K S Robinson, F J Street, I W Drummond

Kratos Analytical Ltd. Barton Dock Road, Urmston, Manchester M31 2LD, UK

ABSTRACT: Damage to polymers by ionising radiations including X-rays and argon ions has been studied using X-ray photoelectron spectroscopy (XPS). Patterns have been imprinted on the surface of fluoropolymers using a masking technique to prevent radiation damage to certain areas. The resulting patterns have then been analysed using imaging XPS and small area XPS spectral analysis.

1. INTRODUCTION

The interaction of ionising radiation (eg ions, electrons and X-rays) with polymers is of great scientific and technological interest. The surface properties of the polymer may be modified by such radiation. Patterns may be registered on the surface to selectively modify only a part of the surface, effectively allowing microchemical engineering. XPS enables chemical state information to be obtained from the top atomic layers of the polymers so that the nature of the degraded products may be determined.

The analysis area of commercial XPS spectrometers has decreased during the last few years and the new use of magnetic lens technology (Kratos AXIS HS) enables routine analysis from areas less than 30μm..

Fluoropolymers were chosen because the C1s spectra show relatively large chemical shifts with minimal overlap, making mapping of there different chemical states straightforward. This study investigates the damage caused to polytetrafluoroethylene (PTFE) by magnesium Kα non-monochromatic radiation and to polyvinyldifluoride (PVDF) by 1keV argon ions.

2. EXPERIMENTAL

Polymer sheets were used without further cleaning prior to analysis since the initial XPS spectra showed minimal contamination. Analysis was performed using a Kratos AXIS HS spectrometer. A nickel grid was mounted in contact with the surface of the sample. The sample was then irradiated with the incident radiation normal to the surface. The PTFE was exposed to magnesium Kα X-rays at a distance of 15mm and operated at 12kV 15mA emission for 12 hours. The PVDF was exposed to argon ions with a current density of $0.38\mu Acm^{-2}$ for 300 seconds. After irradiation, the grids were removed from the samples and the samples were analysed by imaging XPS, with an analysis spot size of 60μm.

3. RESULTS AND DISCUSSION

Ion beam degradation of PVDF reduces the intensity of the F1s peak and reduces the relative intensity of the C-F_2 peak and introduces a new peak at a binding energy of 284.8eV, which may be attributed to graphitic carbon species. Figure 1 shows C1s maps obtained from the PVDF using the C-F_2 peak and the graphitic peak. These maps were obtained using magnesium X-radiation, but the analysis time was kept short to minimise X-ray degradation. Spectra were also acquired from points in the degraded area and masked areas (Figure 2). Note that the spectrum from the masked area does show some degradation indicating that the ion beam had not been entirely restricted by the mask.

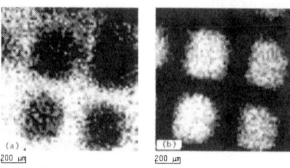

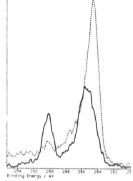

Fig 1. C1s XPS Maps (Mg source) of ion beam degraded PVDF (a) C-F$_2$ peak BE=290.2eV (b) graphitic peak BE=284.8eV

Fig 2. C1s spectra from masked area (solid line) and irradiated area (dotted line)

Figure 3 shows C1s maps obtained at binding energies of 293eV (C-F$_2$) and 288eV (degradation product) obtained from the X-ray degraded PTFE. These maps were obtained using a monochromatic X-ray source. Again, the degraded area is clearly visible. However, the C1s spectrum (Figure 4) from within the masked area shows minimal degradation with largely a single C-F$_2$ peak.

In both cases the resolution at the edges of the masked areas is relatively poor indicating that the masking was not very efficient.

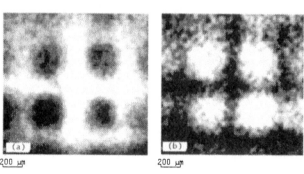

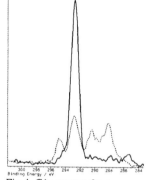

Fig 3. C1s XPS Maps of X-Ray Degraded PTFE. (a) C-F$_2$ peak BE=293eV (b) C-C-F peak BE=288eV

Fig 4. C1s spectra from masked area (solid line) and irradiated area (dotted line)

4. CONCLUSION

An imaging XPS spectrometer has successfully been used to analyse lithographic patterns formed on fluoropolymers by selective ion beam and X-ray degradation. Analysis is possible on a timescale during which X-ray degradation is minimised.

Inst. Phys. Conf. Ser. No 130: Chapter 7
Paper presented at Int. Congr. X-ray Optics and Microanalysis, Manchester, 1992

471

Laser-plasma XUV sources, advances in performance

F. Bijkerk

FOM-Institute for Plasma Physics Rijnhuizen, Edisonbaan 14, 3439 MN Nieuwegein,
The Netherlands,

ABSTRACT: Radiative characteristics of laser-plasma x-ray sources, and the dependence of plasma parameters on the heating conditions are reviewed. A comparison is given with other x-ray sources like electron storage rings. Scaling of laser plasma x-ray sources to high average x-ray powers and methods to eliminate the contamination by target debris are discussed.

1. INTRODUCTION

For more than two decades laser plasmas generated by diverse laser systems have been employed with considerable success as sources of XUV radiation. Extended fields of research have been investigated, i.e. diagnostics of inertial-confinement fusion, XUV spectroscopy, XUV laser research, x-ray microscopy and x-ray lithography. The hundreds of publications resulting witness the continuous improvement of this XUV source and the laser to generate it. High power lasers have become available with well focussable beams, and comprehensive knowledge on the physics of laser plasmas resulted in prescriptions on optimum irradiance conditions of this source. Have laser plasma XUV sources nowadays reached their ultimate level of performance or can one expect further progress?

Two factors are likely to decide this question. First the industrial/commercial potential of some of the applications (mainly lithography and microscopy), and second the ongoing rapid development in high power laser engineering. The use of laser plasma XUV sources has traditionally been stimulated by progress in laser technology. High power and compact lasers, enabling construction of small-scale XUV sources, have brought new applications within the reach of many researchers. Application-specific neodimium lasers in slab geometry and excimer lasers are likely to further accelerate this trend due to their intrinsic capability of operating at high repetition rates, and therefore enhanced average yield of XUV radiation.

Excimer lasers have the additional advantage of operating at short wavelengths and are therefore able to couple laser energy more favourably into a plasma than is the case with solid state lasers. The increased absorption observed with short-wavelength laser radiation leads to a high conversion efficiency of up to 50 % of laser light into nanometer (sub-keV) XUV radiation. High XUV fluxes make the laser plasma an alternative to electron storage rings.

The utilization of laser-plasma XUV sources is also stimulated by the progress in XUV optical elements such as multilayer mirrors. These mirrors are capable of collimating the isotropic plasma radiation using large acceptance angles. In this way XUV radiation can be focussed to unequalled power densities. Again excimer lasers have advantages because of their improved conversion of laser light particularly in the spectral range where normal incidence multilayer mirrors have high reflectivities.

At present, there are several laboratories carrying out optimization studies on topics such as laser repetition rate, pulse modulation, beam quality and XUV source characteristics. In addition, there is a significant effort towards multikilowatt excimer lasers. These programmes are supported by various international projects, e.g. AMMTRA (Advanced Material Processing and Machining Technology Research Association) in Japan and EUREKA in Europe. At present 750 W has been realized with a XeCl laser operated at 500 Hz [1]. Spin-off technology developed in these programmes benefits the state of the art commercial laser systems suitable to generate XUV-plasma radiation.

2. RADIATIVE CHARACTERISTICS OF LASER PLASMAS

If laser radiation is focussed onto the surface of a solid target with powers $q = 10^{11}$ - 10^{15} W/cm^2, a hot plasma is created with an electron temperature $T_e = 50$ - 1000 eV. The electron density n_e in such plasmas may reach levels up to the density of solids. Close to the level of critical density n_e^c ($n_e < n_e^c$), most laser radiation is absorbed and the electron temperature T_e is maximum. The critical density n_e^c is related to the laser wavelength λ_l according to

$$n_e^c = 10^{21} \cdot (\lambda_l)^{-2} \quad [\text{cm}^{-3}] \tag{1}$$

where λ_l is expressed in µm. In a plasma at this density and sufficiently high temperature, ions are formed which have a charge up to 40 depending upon the atomic number Z_A of the target material. Emission spectra of such ions are in the extreme ultra-violet (EUV) to soft x-ray spectral band (together known as the XUV) and range from several tens to tenths of nanometers. For heavy target elements ($Z_A = 50$ - 80) the conversion of laser energy into XUV radiation can be as high as 40 - 70 %.

XUV-radiation from laser plasmas has been studied in detail both experimentally and theoretically. See for example the review in ref. 2. In the experimental work emphasis is on $i)$ the relative and absolute spectral yield versus the atomic number of the target Z_A [3-5], $ii)$ the effective radiative temperature of laser plasmas [6], and $iii)$ the ratio of laser energy to XUV energy, i.e. the conversion efficiency η, as a function of the target atomic number Z_A [7,8], the laser wavelength λ_l [6], the laser pulse length [9], and the target irradiance q [10].

The conversion efficiency η exhibits a strong dependence on the atomic number Z_A of the laser target [7]. This is caused by significant spectral contributions to the total radiated intensity from line radiation from K-, M-, or L-shell spectra. This phenomenon enables one to optimize the XUV emission in a particular wavelength band by selecting the target material and the proper state of plasma ionization, determined by the irradiance.

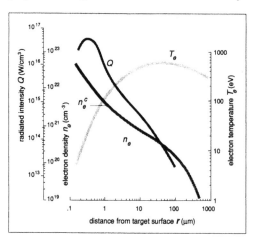

Fig. 1. Profiles of electron temperature T_e, electron density n_e, and intensity Q radiated per unit of volume, close to the surface of a Cu target heated by laser radiation at $q = 10^{13}$ W/cm^2 [11].

Comprehensive theoretical laser-plasma models allow one to evaluate the radiative plasma characteristics over a wide range of laser parameters and target materials. Calculations show that the total XUV power Q radiated per unit of volume for all types of emission (line radiation, photorecombination, dielectronic recombination and bremsstrahlung) is restricted to a narrow plasma layer of only several microns thick with an electron density n_e of 3 - 10 × the critical density n_e^c (Fig. 1). This shallow over-dense layer is primarily heated by electron thermal conductivity; the laser light itself is reflected at a plasma layer with the critical density n_e^c. In the so-called plasma corona region ($n_e < n_e^c$) the electron temperature T_e and the degree of ionization are maximum. Nevertheless, the radiated intensity Q per unit of volume decreases in this region due to the decrease of the electron density n_e (Q is proportional to $(n_e)^2$). Closer to the target, i.e. where the density $n_e \gg n_e^c$, the total radiated intensity is also reduced due to a decrease in the electron temperature T_e and an increase of the optical thickness of the plasma.

Can alterations to this general picture be expected if we consider plasmas generated using excimer lasers? For plasmas produced by short wavelength lasers the value for the critical density n_e^c is higher, according to (1). Shorter wavelength lasers will deposit their energy in a higher density region of the plasma. The radiated intensity Q will be higher in this case because it scales with the square of the density n_e. However, closer to the plasma core the electron temperature T_e will be lower. The peak temperature scales as $T_e \sim \lambda^{0.3}$, while at the critical density the temperature scales more strongly: $T_e \sim \lambda^{0.9}$. This shift in temperature will move the maximum of the XUV emission to longer wavelengths, generally above 1.5 to 3 nm. An example of the increased conversion efficiency η observed for short wavelength laser radiation is given in ref. 12. Here it was found that the conversion efficiency of incident laser light into the wavelength band from 12 nm to 0.8 nm increases by a factor of three when going from a laser wavelength of 1.05 µm to 0.26 µm. Most of this increase is due to increased emission above 1.4 nm. Emission at shorter wavelengths (between 0.8 and 1.4 nm) remains almost constant with laser wavelength.

The enhanced conversion efficiency at longer XUV wavelengths observed with short wavelength lasers is important when considering XUV optics. Normal-incidence multilayer XUV optics have been demonstrated with considerable success particularly for wavelengths above the carbon-K edge (4.36 nm).

3. APPLICATION-SPECIFIC EXCIMER LASERS

Excimer lasers suitable to generate laser-plasma radiation usually consist of two or three laser systems to achieve high pulse energy and good focussability. Due to a continuous development in laser engineering these lasers can now be made compact and reliable even at high power levels. An example of such a modern laser is a two module unit presently installed at FOM-Rijnhuizen. It supplies 1.5 J or 70 W in power oscillator-power amplifier mode, and 1 J or 43 W in injection-locked mode [13]. See Fig. 2. Although the focussability is such that power densities up to 10^{13} W/cm² can be achieved (sufficient to generate sub-nm x-rays) this laser is primarily intended as a driver for a source of soft x-rays for the application of XUV optics. Another example is a three-stage laser situated in the Sandia National Laboratories [14]. The laser system consists of an oscillator and two single-pass amplifiers and produces 1.5 J pulses of 25-30 ns at 100 Hz repetition rate. The laser energy is focussed on a rotating cylindrical target, which is Au-coated. The power density in the ~250 µm focus amounts 10^{11} W/cm². At this irradiance a relatively cool plasma is created, which shows a high conversion efficiency in the EUV range of η = 35-40 %. The XUV source has a dedicated high-throughput monochromator, which delivers an irradiance to the sample comparable to that of monochromators of e-storage rings (Tab. I).

Shorter wavelength radiation ($\lambda \leq 2$ nm), is usually generated with lower conversion efficiency. The typical 20-30 ns pulse length from excimer lasers leads to x-ray emission which falls off after ~10 ns with the result that only the initial part of the laser pulse is efficiently converted to x-rays [15]. Several pulse modification techniques have been successfully investigated: injection mode-locking with trains of ~100 ps pulses [9], and passive pulse shortening by a H_2-Raman cell and a saturable dye jet, resulting in 3.5 ns pulses of 400 mJ [16]. The wavelength range from 2 to 0.8 nm (0.6 to 1.4 keV) is of interest for proximity print x-ray lithography, a candidate technique for mass production of sub-0.5 micron semiconductor devices. X-ray lithography systems driven by application-specific excimer lasers have considerable potentials for industrial utilization [17].

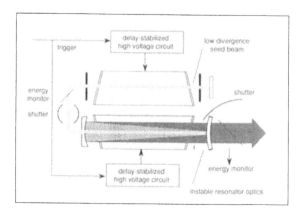

Fig. 2. Layout of a state-of-the-art two-stage KrF laser designed to construct a high average power plasma XUV source. The laser has a pulse energy up to 1.5 J and 70 W output power and a beam divergence better than 0.12 mrad.

4. COMPARISON WITH OTHER XUV SOURCES

The interest in laser plasmas has been stimulated greatly by the enhanced availability of synchrotron radiation facilities. In the past decade these facilities have led to a significant increase in research activity in many differing areas such as biology, chemistry and medicine. At present the laser plasma serves as an interesting alternative facility with specific capabilities.

The laser plasma XUV source is most suitable for applications requiring very high instantaneous fluxes for single, short pulse experiments. The peak brightness of laser plasma sources can be three orders of magnitude higher than for electron storage rings (Tab. I). In experiments requiring the accumulation of high doses of particularly short wavelength radiation, synchrotron facilities, being quasi-continuous, are at an advantage. However, if one takes into account typical geometrical factors arising from the source-to-sample distances, and the possibility to use excimer lasers with high repetition rates (~100 Hz), a comparable average sample irradiance is obtained in both cases. The actual number of photons (per unit of time, bandwith, and sample area) available from two examples of high throughput monochromators is of similar order (Tab. I).

The use of synchrotron radiation is still expanding considerably. New electron storage rings are being built, some dedicated to specific applications (such as x-ray lithography [18]), others for general research purposes offering beamlines for a large variety of experiments. Intensities of synchrotron radiation are being increased by the use of insertion devices such as undulators and wigglers. For short wavelength radiation down to 0.01 nm (~100 keV) synchrotron facilities have definite advantages over any laboratory XUV source. The final choice between large-scale user facilities such as synchrotron radiation sources and laboratory-scale devices such as laser-plasma XUV sources depends on the specific physical task as well as economical aspects.

In general terms, the advantages of laser-plasma XUV sources can be described as:

i) High brightness in the XUV spectral band, due to efficient conversion of laser light into XUV radiation and high pulse energy available from laser drivers,

ii) High average brightness with duty-cycles up to several Hz (solid state lasers) or up to several hundreds of Hz (excimer lasers),

iii) Short pulse duration, equal to the laser pulse duration (down to ~ps [19]), enabling study of time-resolved processes and 'pump and probe' experiments,

iv) Reduced capital cost and floor-space requirements compared with an electron storage ring, which makes commercial applications and small-scale laboratory set-ups possible,

v) Small source size of 10 to 100 μm and high pulse-to-pulse reproducibility both in source position and in XUV yield, *vi)* Wide spectral band ranging from several tenths to tens of nanometers, tunable according to the needs of specific applications by choosing the target material and irradiance conditions,

vii) Simple wavelength calibration of spectra using target-specific spectral lines,

viii) Reduced production of target debris compared with electrical discharge sources, possibilities to suppress debris by applying thin foil laser targets or generating the plasma in an environment of buffer gas,

ix) No UHV requirements and the possibility of operating in sub-atmospheric pressure of inert gasses.

In particular excimer lasers add to the advantages *i)* and *ii)* due to their high conversion efficiency into XUV radiation and their high repetition rate. General disadvantages of laser-plasma XUV sources are the absence of polarization of the plasma radiation and the lower limit in the spectral range at several tenths of nanometers. Compared with sources of synchrotron radiation, laser plasmas have a sample irradiance which is, in general, an order of magnitude lower. High repetition rate excimer lasers can exhibit a sample irradiance in the XUV range which is comparable to that of electron storage rings (Tab. 1).

Tab. I. Comparison of different XUV sources: electron storage rings and laser plasmas. The data (sets of two numbers) for brightness and irradiance refer to the wavelength range indicated.

	BESSY *)	DORIS **)	MPI #)	FOM ##)	SNL ###)
source size (mm^2)	0.8×1	1×10	$\pi(0.05)^2$	$\pi(0.05)^2$	0.25×0.25
pulse duration (ns)	0.1	0.15	3	15	28
repetition rate (Hz)	5×10^8	10^8	1/60	1/30	10^2
source-to-sample distance (m)	15	10	0.1	0.1	0.2-1.5
wavelength range (nm)	1-10	1-10	1-10	0.9-1.7	8-35
peak brightness (phot/s.nm.sr.mm^2)	2×10^{23}-2×10^{22}	5×10^{23}-5×10^{21}	10^{25}-5×10^{25}	5×10^{24}	4×10^{22}-10^{23}
average brightness (phot/s.nm.sr.mm^2)	9×10^{21}-10^{21}	7×10^{21}-7×10^{19}	6×10^{14}-3×10^{15}	3×10^{15}	10^{17}-3×10^{17}
peak sample irradiance (phot/s.nm.cm^2)	6×10^{16}-8×10^{15}	5×10^{18}-5×10^{16}	2×10^{21}-7×10^{21}	4×10^{20}	3×10^{19}-6×10^{19}
average sample irradiance (phot/s.nm.cm^2)	3×10^{15}-4×10^{14}	7×10^{16}-7×10^{14}	5×10^{10}-3×10^{11}	2×10^{11}	7×10^{13}-2×10^{14}
monochromator	SX-700				dedicated
peak sample irradiance (phot/s.nm.cm^2)	10^{17}-4×10^{16}				3×10^{20}-3×10^{19}
average sample irradiance (phot/s.nm.cm^2)	6×10^{15}-2×10^{15}				10^{15}-10^{14}

*) BESSY - Electron storage ring energy 0.8 GeV, ring current 300 mA [5 ,20].
**) DORIS - Electron storage ring energy 3 GeV, ring current 500 mA [21].
#) MPI - Nd:Glass laser (Max Planck Institute), λ_l = 0.53 μm, E = 7 J, q = 3 × 10^{13} W/cm^2 [8].
##) FOM - Nd:Glass laser (FOM-Institute Rijnhuizen), λ_l = 0.53 μm, E = 2.2 J, q = 2.5 × 10^{12} W/cm^2, Mg target [22].
###) SNL - KrF laser (Sandia National Laboratories), λ_l = 0.248 μm, E = 1.5 J, q = 1 × 10^{11} W/cm^2, Au target [14].

5. CONCLUSIONS

Laser plasmas generated with compact and powerful laser systems are versatile laboratory-scale sources of XUV radiation. The outstanding properties of these sources are the very high instantaneous brightness in the XUV range, the possibility of picosecond pulse duration, and the small and reproducible source size. Laser-plasma sources impose low requirements on capital investment and floor space. In particular, excimer lasers significantly increase the duty-cycle of these sources and generate time-averaged sample irradiances which are as high as radiation from bending magnet beamlines of electron storage rings.

The high conversion efficiency of laser light to XUV radiation, which is observed with short-wavelength laser radiation, is particularly large for XUV radiation above 1.5 nm. It is therefore possible to fully exploit the use of normal-incidence multilayer XUV optics. With the combination of an excimer-laser driven XUV source and multilayer XUV optics challenging applications can be realized, for example in soft x-ray projection lithography, x-ray microscopy, and spectroscopy. Future work is also directed to optimization of the XUV yield and further reduction of target debris. Returning to the earlier question about the likeliness of further advances in the performance of laser plasma XUV sources, the answer is clearly positive.

6. ACKNOWLEDGMENTS

This work is part of the programme of FOM (the Foundation for Fundamental Research on Matter) and STW (the Netherlands Technology Foundation). The work is made possible by financial support from NWO (the Netherlands Organization for Scientific Research) and the Netherlands Government in the framework of EUREKA.

Dr. G.D. Kubiak (Sandia Livermore Laboratories, Livermore, CA, USA), Dr. F. Schaefers (BESSY, Berlin, FRG), Prof. M.J. van der Wiel and Dr. E. Louis (FOM-Institute Rijnhuizen) are acknowledged for their kind cooperation in the realization of this paper.

7. REFERENCES

[1]   E. Müller-Horsche, P. Oesterlin, D. Basting, ECO4 Conf. Proc. (1991)
[2]   F. O'Neill, in Laser-plasma interactions 4, Ed. M.B. Hooper, Proc. 35th Scottish Universities Summer School in Physics, (1988) pp.285-315
[3]   P.K. Carroll, E.T. Kennedy, G. O'Sullivan, Appl. Opt, 19 9 (1980) pp.1454-1462
[4]   T. Mochizuki, T. Yabe, K. Okada, M. Hamada, N. Ikeda, S. Kiyokawa, and C. Yamanaka, Phys. Rev. A33 1 (1986) pp.525-539
[5]   J. Fischer, M. Kühne, B. Wende, Appl. Opt. 23 23 (1984) 4252-4260
[6]   H. Nishimura, F. Matsuoka, M. Yagi, K. Yamada, S. Nakai, G.H. McCall, and C. Yamanaka, Phys. Fluids. 26 6 (1983) pp.1688-1692
[7]   H. C. Gerritsen, H. van Brug, F, Bijkerk, M.J. van der Wiel, J. Appl. Phys. 59 7 (1986) pp.2337-2344
[8]   K. Eidmann, T. Kishimoto, Appl. Phys. Lett. 49 7 (1986) pp.377-378
[9]   F. O'Neill, I.C.E. Turcu, D. Xenakis, M.H.R. Hutchinson, Appl. Phys. Lett. 55 25 (1989) pp.2603
[10] W.C. Mead, E.M. Campbell, K.G. Estabrook, R.E. Turner, W.L. Kruer, P.H.Y. Lee, B. Pruet, V.C. Rupert, K.G. Tirsell, G.L. Stradling, F. Ze, C.E. Max, M.D. Rosen, and B.L. Lasinski, Phys. Fluids 26 8 (1983) pp.2316-2331
[11] A. V. Vinogradov, V.N. Shlyaptsev, Sov. J. Quant. Electr. 17 1 (1987) pp. 5-26
[12] R. Kodama, K. Okada, N. Ikeda, M. Mineo, K.A. Tanaka, T. Mochizuki, and C. Yamanaka, J. Appl. Phys. 59 9 (1986) pp.3050
[13] F. Bijkerk, L. Shmaenok, E. Louis, B. Nikolaus, F. Voß and R. Desor, Proc. SPIE 1810 (1992) submitted
[14] G.D. Kubiak, Proc. SPIE 1343 (1990) pp.283-291
[15] G.M. Davis. M.C. Gower, F. O'Neill, and I.C.E. Turcu, Appl. Phys. Lett. 53 17 (1988) pp. 1583-1585
[16] A. Tünnermann, K. Wrede, and B. Wellegehausen, Appl. Phys. B 50 (1990) pp. 361-364

[17] F. Bijkerk, G.E. van Dorssen and M.J. van der Wiel, Microelectr. Eng. **9** (1989) pp.121-126

[18] D.E. Andrews, M.N. Wilson, A.I. Smith, V.C. Kempson, A.L. Purvis, R.J. Andersen, A.S. Bhutta, A.R. Jorden, SPIE Conf Proc. **1263** (1990)

[19] R.R. Freeman, L.D. van Woerkom, T.J. McIlrath, W.E. Cooke, to be published

[20] J. Feldhaus, SPIE Conf. Proc. **984** (1988) 11-22

[21] Synchrotron Radiation Techniques and Applications, Ed. C. Kunz, Springer Verlag, (1979)

[22] F. Bijkerk, E. Louis, A.P. Shevelko, A.A. Vasilyev, submitted to J. Appl. Phys.

Inst. Phys. Conf. Ser. No 130: Chapter 7
Paper presented at Int. Congr. X-ray Optics and Microanalysis, Manchester, 1992

479

Pinch plasmas as intense X-ray sources for laboratory applications

D. Rothweiler[1], W. Neff[2], R. Lebert[1], F. Richter[3], M. Diehl[4]

[1]RWTH Aachen, Lehrstuhl für Lasertechnik, [2]Fraunhofer Institut für Lasertechnik, Steinbachstraße 15, 5100 Aachen, Germany, [3] Karl Süss KG, Garching, [4]Universität Göttingen, Göttingen, Germany

ABSTRACT: Pinch plasma x-ray sources have been developed for x-ray lithography and x-ray microscopy. The lithography source is compatible with the electron storage ring printing process with respect to spectrum and enables full depth exposures in 1 μm thick 60 mJ/cm^2 sensitivity resist at sub 0.2 μm resolution within 10 minutes. The source for microscopy applications provides flash imaging of biological specimen with sub-optical resolution (0.1 μm - 0.2 μm) at nanosecond exposure time.

1. INTRODUCTION

Pulsed pinch plasmas are intense sources of x-radiation with output energies up to several joule per pulse in a single line. Their spectrum peaks at soft x-ray wavelengths where conventional x-ray tubes provide poor intensity. With correct optimization, both continuous radiadion or line radiation with $\lambda/\Delta\lambda > 1000$ can be produced for broad and narrowband applications, respectively. Due to their low cost and their compact size pinch plasmas seem well suited to supplement research activities based on synchrotron radiation.

2. REPETITIVE PINCH PLASMA SOURCE FOR X-RAY LITHOGRAPHY

High throughput production of sub half micron structures by x-ray lithography (XRL) is generally planned around electron storage rings, whose synchrotron radiation exhibits near ideal properties for its use in lithography. Besides, the need for research to accompany manufacturing activities requires the availability of compact low cost x-ray point sources.

With this background, a repetitive pinch plasma x-ray source was developed in cooperation with Karl Suss, Munich. The major guideline was to achieve full compatibility with the manufacturing process at electron storage rings. Spectrum, beam line design, and source size are of particular interest in this concern.

The spectrum is essentially determined by the x-ray mask technology, typically 1 μm gold patterns on thin silicon foils (2 μm) (Heuberger 1988). Sufficient contrast is correspondingly achieved at wavelengths in the range 0.67 nm - 1.20 nm, the lower limit given by the silicon K-edge and the upper limit caused by absorption in the mask material. Further constraints that favour this spectral range refer to the XRL process latitude, heat load on the mask and wafer/mask handling as discussed by Cullmann (1989).

The beam line represents the interface between the vacuum system of the source and the mask wafer alignment placed under atmospheric conditions. If related to point sources the beam line concept must provide a minimum distance of at least 400 mm between mask/wafer and x-ray source to reach the structural resolution achievable by proximity printing (Richter 1989). Distortion of off-axis structures caused by proximity gap variations in conjunction with the central projection by point sources can thus be reduced to less than 0.1 μm, assuming a typical proximity gap of 40 μm between mask and wafer, gap control tolerances (2 μm) of commercial x-ray steppers and an exposure field size of one square inch. Penumbral blurring due to the finite source size should not exceed 0.1 μm, consequently. This restricts the source diameter to less than 1 mm at the above mentioned values for gap and mask source distance.

The pinch plasma device (size: 1 m * 2 m) consists of a plasma focus type electrode system powered by a fast 5 kJ capacitor bank (Richter 1989) capable of operation at 3 Hz. A three window beam line compatible with a commercial x-ray stepper provides both protection from erosion products and plasma debris and a plane exit window. The latter is required to prevent differences in absorption by air for homogenous illumination of the mask. Neon is used as gaseous discharge load. Its emission spectrum consists of K-shell lines and recombination continua of both heliumlike and hydrogenlike ions bearing a strong resemblance to the continuum offered by synchrotron radiation (Figure 1). The source diameter is less than 0.9 mm, as reported by Richter (1989).

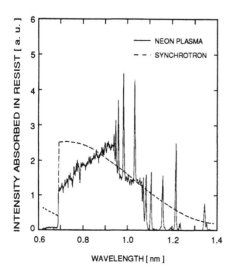

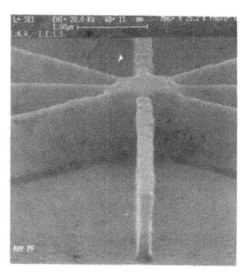

Fig. 1. The neon spectrum as transmitted by the beam line and absorbed within the x-ray resist. For comparison, the corresponding synchrotron continuum.

Fig. 2. Scanning electron micrograph (SEM) of 0.2 μm structures in 60 mJ/cm^2 sensitivity x-ray resist. Mask: CNR, Rome

The source allows for exposures in 1 μm thick 60 mJ/cm^2 sensitivity resist (RAY-PF) within 10 minutes at a repetition rate of 2 Hz, appropriate for test purposes in mask and resist development as well as circuit design (Cullmann 1987). Figure 2 shows an SEM of 0.2 μm structures. The resolution is better than 0.2 μm, without changes in the full exposure field of one square inch. Edge form and aspect ratio reveal that the process performance achieved is reasonably close to conditions at electron storage rings.

3. PINCH PLASMA SOURCE FOR A LABORATORY X-RAY MICROSCOPE

State of the art experiments in the field of x-ray microscopy take place at synchrotron radiation sources where imaging of biological specimen in their natural environment is demonstrated with sub 0.1 μm resolution. However, it will be difficult for x-ray microscopes to become widely available, if they are based on synchrotron sources only.

As a complement to research activities at synchrotron facilities a laboratory x-ray micro-scope with a pinch plasma x-ray source was developed as a joint effort of the Forschung-seinrichtung Röntgenphysik, Universität Göttingen, Carl Zeiss, Oberkochen, the Fraunho-fer Institut für Lasertechnik and the Lehrstuhl für Lasertechnik, Aachen.

The set up of this system was presented by Schmahl et al (1990). A high resolution mi-cro zone plate forms the image of the specimen and a mirror condenser provides the illumi-nation. Details on both design and manufacture of the x-ray optical components are dis-cussed by Thieme (1992).

The requirements to the x-ray source are strongly related with the properties of the mi-croscopes x-ray optical arrangement ; most important are spectral distribution of the emis-sion and x-ray yield.

The use of zone plate optics implies that radiation must be narrowband with a reciprocal bandwidth of at least $\lambda/\Delta\lambda = 200$, ideally, line radiation. A more restrictive requirement is introduced by the use of the mirror condenser that provides no wavelength selection. Addi-tional x-ray lines or continuous radiation within the close spectral neighbourhood of the de-sired x-ray line must hence be avoided. Contrast is achieved if radiation within the water window range, i.e. 2.3 nm $< \lambda <$ 4.4 nm, is applied. Especially the shorter wavelength re-gion close to the oxygen K-edge at $\lambda = 2.3$ nm is best suited to study thick specimen in their wet natural state, according to absorption by water.

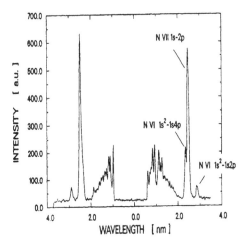

Fig. 3. Nitrogen spectrum as transmitted through the beam line. The spectrum was obtained by means of transmission grating (10000 lines/mm) and CCD.

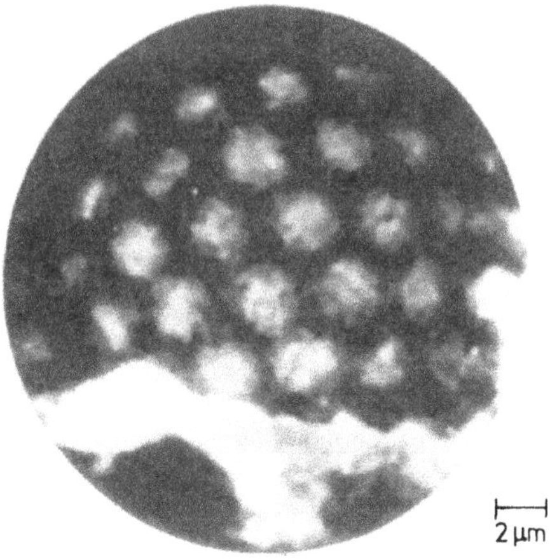

Fig. 4. X-ray image of a diatome structure. wavelength: 2.5 nm, x-ray magnification: 200, exposure time: 5 ns

The source, i.e. a 2.5 kJ plasma focus device (size: 1.5 m * 1 m) (Holz 1989), is opti-mized to produce Lyman-α line emission of hydrogenic nitrogen at 2.48 nm wavelength. A second transition, the N VI: $1s^2$-$1s3p$ transition of heliumlike nitrogen at 2.49 nm wave-length, coincides within a reciprocal bandwidth of $\lambda/\Delta\lambda = 250$ and can thus be used for experiments simultaneously. Undesired radiation at wavelengths less than 2.3 nm is filtered by an oxygen counter stream in the beam line that also protects the x-ray optical arrange-ment from plasma debris and erosion products. The corresponding spectrum (Figure 3) mainly consists of the x-ray line of interest as a result of beam line transmission.

The x-ray yield to be achieved is the main challenge. A sufficient number of photons has to be produced in one single pulse, such that the image is formed before motion or radiation damage can interfere. However, only photons originating from a finite spot of the source can effectively contribute to image formation; the corresponding spot diameter is related to the image field size. Therefore, the number of photons emitted per pulse, per unit source surface and per steradian into the x-ray line of interest is the figure of merit. This number was optimized with respect to filling pressure of the nitrogen discharge load and storage energy, based on both theoretical and experimental investigations (Lebert et al 1990). In addition, further improvements were recently made by studies of the electrical power feed by the capacitor bank. The x-ray line yield into 4π was measured by means of transmission grating and calibrated film. It amounts to 3 J ($4*10^{16}$ photons) for the full source diameter of 0.7 mm, and to 30 mJ ($4*10^{14}$ photons) for a small spot of 0.1 mm diameter. The latter effectively contributes to image formation.

Figure 4 shows the x-ray image of a diatome structure taken with the above mentioned set up of the laboratory x-ray microscope at a beam line transmission of 40 %. A CCD was used for high sensitivity detection. The resolution achieved is 0.1 μm - 0.2 μm and not limited by the optical system. One single x-ray pulse was applied for exposure at an x-ray magnification of 200. The pulse duration of 5 ns (Holz et al. 1989) is considered short compared to the time constants related to motion and radiation damage.

4. SUMMARY

Pinch plasma x-ray sources were developed and optimized with respect to two different requirements, namely repetitive broadband emission for x-ray lithography and single pulse line emission for x-ray microscopy. Both systems could successfully be applied for typical synchrotron radiation techniques.

5. ACKNOWLEDGEMENTS

The cooperation with the Karl Suss KG, the Forschungseinrichtung Röntgenphysik, Universität Göttingen and the Carl Zeiss company is gratefully acknowledged. We wish to express our thanks to C. David for providing the 10000 lines/mm transmission grating. This work was supported by the German Ministry of Research and Technology (BMFT) under contract numbers 13N53290, 13N5680 and 13N5838.

6. REFERENCES

Cullmann E, Kunneth T, Neff W and Stephan K H J. Vac. Sci. Technol. B5 (3) 638-640 (1987)
Cullmann E 1989 Suss Report Vol.3 (4) 1989
Heuberger A 1988 J. Vac. Sci. Technol. B6(1), 107-121 (1988)
Lebert R, Holz R, Rothweiler D, Richter F and Neff W 1990 X-Ray Microscopy III (Belin: Springer) pp 62-5
Richter F, Eberle J, Holz R, Neff W and Lebert R AIP Conf. 195 Proc. 2nd int. Conf. on High Density Pinches ed N R Pereira, J Davis, N Rostocker pp 515-9
Schmahl G, Niemann B, Rudolph D, Diehl M, Thieme J, Neff W, Holz R, Lebert R, Richter F and Herziger G 1990 X-Ray Microscopy III (Berlin: Springer) pp 66-9
Thieme J this Volume

Inst. Phys. Conf. Ser. No 130: Chapter 7
Paper presented at Int. Congr. X-ray Optics and Microanalysis, Manchester, 1992

483

X-ray microscopy using a laser-plasma source

A G Michette, I C E Turcu,* M S Schulz† and P Fluck

Physics Department, King's College, Strand, London WC2R 2LS UK
*Central Laser Facility, Rutherford Appleton Laboratory, Oxon. OX11 0QX UK

ABSTRACT: High-resolution soft x-ray microscopy has previously only been possible using synchrotrons. The first use of a scanning x-ray microscope with a laser-plasma source is reported here. Spatial resolutions were limited to $\approx 650\,nm$ by electrical noise in the detector, but single shot per pixel images were obtained of test and real specimens. With an improved system routine single shot per pixel imaging at $< 50\,nm$ will be possible.

1. INTRODUCTION

Several high-resolution ($< 0.1\,\mu m$) x-ray microscopes are now operating, one of which is the scanning transmission x-ray microscope (STXM) built at King's College London (KCL) and used on the undulator beamline at the Synchrotron Radiation Source (SRS), Daresbury Laboratory (Morris *et al* 1991). In this, a zone plate with outermost zone width $d_N = 30 - 100\,nm$ is used to focus monochromatic radiation with $\lambda = 2.3 - 4.4\,nm$ to a spot of size given by the convolution of the demagnified monochromator exit slit and d_N. A specimen is mechanically scanned across this spot and the transmitted X rays are detected by a flow proportional counter. Images with resolutions of around $50\,nm$ have been obtained.

Synchrotron x-ray sources are expensive, remote, multi-user facilities, and are not suited for laboratory-scale x-ray microscopy; alternative sources such as high repetition rate laser-generated plasmas (LGPs) must therefore be considered. LGP sources are ideal for laboratory-based STXMs, since (i) the source size can be $\sim 10\,\mu m$, allowing optimum geometric coupling to the zone plate without the need for a pinhole, which would reject photons; (ii) the source need be pulsed only when pixel information is being obtained, and not when stepping from pixel to pixel; (iii) a suitable choice of target material (e.g. carbon) gives a quasi-monochromatic line spectrum and, using filters, a single line can be selected to illuminate the zone plate, as required to prevent chromatic aberration; (iv) it is straightforward to obtain a vertical beam for horizontal specimen containment.

The KCL/Daresbury microscope has been used for preliminary experiments over two periods of two weeks with the LGP source at the Rutherford Appleton Laboratory which uses two Questek 2440 KrF excimer lasers. A fuller discussion of the work presented here has been submitted for publication (Michette *et al* 1992).

---

† Present address: Institute of Laser Engineering, Osaka University, Japan

2. THE SOURCE

For this experiment one laser was used as an oscillator, with integral unstable cavity, and the other as an amplifier, as described by Turcu *et al* (1991). The laser beam, with $\approx$ 500mJ in each 25 ns pulse, was focused by a $\approx$ 80% transmissive fused silica aspheric singlet $f = 7$ cm lens, to a spot on target to give a source size of $\approx$ 40 μm. The renewable target was a 20 μm thick, 4 mm wide mylar tape moving at up to 4 cm s^{-1}, driven by a DC motor between two spools. The tape was supported by a 16 mm diameter cylinder which kept it to within $\approx \pm 2\,\mu$m of focus. A 3 mm slot in the cylinder allowed free expansion of debris ejected through the tape. The tape target was used to allow a fresh target surface to be used for every laser pulse, for a large number of pulses, and to reduce the debris associated with the plasma. The effects of the debris were further reduced by 20 torr of helium with a flow of 5 litres per minute.

The laser beam was incident at 45° to the normal to the cylinder in the horizontal plane with the STXM at 90° to the beam. A 100 nm thick silicon nitride window separated the target chamber from the atmospheric environment of the microscope. A schematic diagram of the arrangement is shown in figure 1.

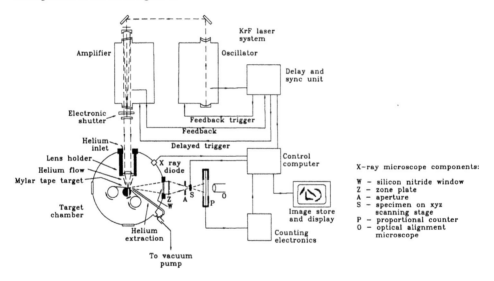

Fig. 1. A schematic diagram of the source/microscope combination.

The x-ray emission was measured with a calibrated 0.25 μm thick silicon window PIN x-ray diode (XRD). The conversion efficiency from laser energy to water-window X rays was $\approx$ 0.2% sr^{-1} for the mylar tape target. The x-ray spectrum was measured with a 44 $\times$ 44 μm^2 transmission grating with period 114 nm made by contamination lithography (Buckley *et al* 1985) on a 100 nm thick Si$_3$N$_4$ substrate. The lines were shadowed with 30 nm of platinum. Spectra were recorded on calibrated Kodak S101 x-ray film; $\approx$ 3 $\times$ 10^{12} photons sr^{-1} per shot were emitted in each of the He$_\alpha$ and L$_\alpha$ lines, with a total in the water window of $\approx$ 9 $\times$ 10^{12} photons sr^{-1} per shot. The water-window x-ray flux measured from these spectra is in good agreement with the XRD measurements.

Because zone plates require monochromatic radiation, nitrogen gas was used as a filter in the atmospheric environment of the microscope. The absorption of the nitrogen, with an edge at

3.1 nm, effectively removed all the spectral lines except the L_α line at 3.37 nm. The flux in this line, which was used for the results described below, was reduced by a factor of two.

3. THE DETECTOR AND IMAGE ACQUISITION

When a LGP source is used the X rays arrive in pulses lasting a few nanoseconds and individual photons cannot be counted. Thus a pulse height to digital converter was used to allow the proportional counter to work in proportional mode.

The image acquisition system used with a synchrotron, utilising an IBM PC and a Microlink crate, allows images to be obtained with a fixed dwell time per pixel. For the LGP source the control computer was used to fire the laser at a repetition rate of, typically 20 – 40 Hz, and the data for each pixel were acquired for a set number of laser pulses, usually one.

4. PERFORMANCE OF THE MICROSCOPE

The zone plate was a gold replica of a master pattern made in carbon by contamination lithography (Charalambous and Morris 1992), with a diameter of 70 μm and $d_N = 65$ nm, giving a focal length of 1.35 mm at $\lambda = 3.37$ nm. To determine the focal position a 3 μm pinhole was scanned across the x-ray beam; far from focus the characteristic doughnut pattern, caused because the zone plate was centrally obstructed, was observed, whilst near focus there was a central bright region due to the focused X rays. The position of best focus was determined by minimising the size of this bright region.

The linearity of the detecting system was determined to be better than 2% by scanning a 25 μm aperture across the edge of the target chamber exit window. The major problem with the detector was electronic noise from firing the lasers. This was reduced but not eliminated by screening and had the effect of limiting the apparent spatial resolution. Signal noise arose due to source-emission variation of ≈ 15% from shot to shot and because of shot noise due to the number of photons collected during each pixel. The monitored source emission per pixel (I_0) was removed from the images by dividing the data for each pixel by the relevant value of I_0.

A 64 × 64 × 160 nm image of a 6 μm pinhole (figure 2), taken with one shot per pixel at a repetition rate of 20 Hz, was used to determine the spatial resolution of the system. The averaged profile indicates an edge resolution of ≈ 650 nm, limited, as described above, by electrical noise. A 128 × 128 × 1.25 μm image of silica gel taken under the same conditions is shown in figure 3. This image shows similar features to higher resolution images of similar specimens obtained at the SRS.

5. CONCLUSIONS

It has been demonstrated that a LGP source is suitable for scanning x-ray microscopy. Images can be obtained with one shot per pixel and thus imaging times can be shortened by using higher laser repetition rates; using the system described here the source performance suffers little degradation up to 100 Hz. At $\lambda = 3.37$ nm the source is comparable in brilliance with synchrotron sources such as the SRS, where spatial resolutions of better than 50 nm are routinely obtained.

ACKNOWLEDGEMENTS

The support of M T Browne, G R Morrison, C J Buckley and G F Foster in modifying and

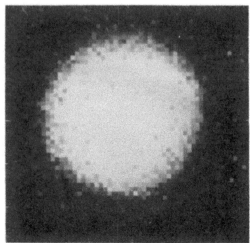

Fig. 2. A 64 × 64 × 160 nm image of a 6 μm diameter hole, recorded with 1 shot per pixel at $\lambda = 3.37$ nm.

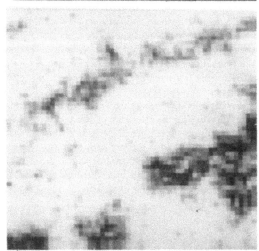

Fig. 3. A 64 × 64 × 1.25 μm image of silica gel, recorded with 1 shot per pixel at $\lambda = 3.37$ nm.

running the microscope was vital for these experiments. The attachment of the STXM to the LGP source was designed by A Damarell of RAL, and made in the workshops there.

REFERENCES

Michette A G, Turcu I C E, Schulz M S, Browne M T, Morrison G R, Fluck P, Buckley C J and Foster G F 1992 submitted to *J. Appl. Phys*

Morris D, Buckley C J, Morrison G R, Michette A G, Anastasi P A F, Browne M T, Burge R E, Charalambous P S, Foster G F, Palmer J R and Duke P J 1991 *Scanning* **13** 7-10

Turcu I C E, Gower M C, Reason C J, Huntingdon P, Schulz M, Michette A G, Bijkerk F, Louis E , Tallents G J and Al-Hadithi Y 1991 *Proc. SPIE* **1503** 391-404

Buckley C J, Browne M T and Charalambous P 1985 *Proc. SPIE* **537** 213-7

Charalambous P S and Morris D 1992 in *X-Ray Microscopy III: Springer Series in Optical Sciences* **67** eds. A G Michette, G R Morrison and C J Buckley (Berlin: Springer) pp79-82

Imaging with the Vulcan X-ray laser

R.E. Burge*, M.T. Browne*, P. Charalambous*, G. Slark*, P. Smith*, C.L.S. Lewis[+] and D. Neely[+].

* Wheatstone Physics Laboratory, King's College London, Strand, London WC2R 2LS, U.K.

[+] School of Mathematics and Physics, The Queen's University of Belfast, Belfast BT7 1NN, Northern Ireland.

ABSTRACT: An imaging microscope, comprising a Schwarzchild condenser and zone plate optical arrangement, has been established on the Vulcan Nd-glass laser system at the Rutherford Appleton Laboratory (RAL). Images of simple test structures have been taken in X-ray transmission using doublet X-ray laser radiation at 23.2 nm and 23.6 nm from collisionally pumped Ne-like germanium. Image resolution of about 0.15 μm has been measured.

1. INTRODUCTION

The first, preliminary, full field X-ray images with X-ray laser radiation were taken by us in April 1991 (Lewis et al, 1991) at RAL using germanium radiation (23.2 nm and 23.6 nm) and a double target configuration. Improved images, illustrated here, were gained in a second experiment in September 1991.

The work is directed towards the evaluation of the potential benefits of imaging and holography using X-ray lasers especially for specimens subject to radiation damage. There are two particular benefits that are anticipated, first the provision of a bright field image at sub-optical resolution in a sub-nanosecond duration pulse short in duration as compared with damage processes in, e.g. biological tissue, and second the generation of three-dimensional image information by X-ray laser holography.

Collisionally excited Ne-like germanium radiation is produced in the form of sub-nanosecond pulses with a doublet structure where the components have approximately the same intensity. Such radiation is far from ideal for imaging because of its long wavelength and correspondingly low penetration, but does provide a means to investigate the physics of imaging and holography with X-ray lasers, in advance, in the U.K., of further development towards harder X-radiation.

Due to the higher laser pump power available at the Nova X-ray laser
(McGowan et al, 1990), Ta radiation at 4.5 nm wavelength is available, with
considerable practical gains for imaging as against the softer radiation.
This group, in January 1992, gained the first preliminary images of intact
rat spermatozoa.

2. X-RAY MICROSCOPE

The optical layout of the imaging microscope is shown in Figure 1.

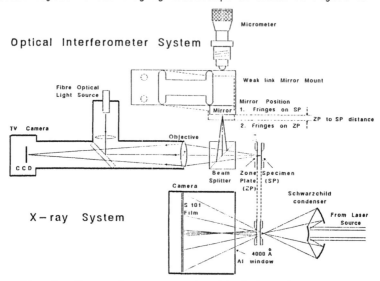

Fig. 1. X-ray microscope and optical interferometer

To accomplish the accurate placement of zone plate and specimen a white
light interferometer of the Michelson type (Figure 1) was constructed and
fringes were viewed through an optical microscope connected to a CCD camera.
Both the specimen and the zone plate were constructed on very flat 50 nm
membranes of silicon nitride which acted (separately), without the need
for reflective coatings, as mirrors in one arm of the interferometer.
The second arm of the interferometer carried a mirror driven under computer
control by a differential micrometer and stepper motor. By establishing
fringes for the two positions of the movable mirror, the zone plate to
test object distance was set to about one-third of a fringe separation
for a double-pass optical system, i.e. about 0.1 μm. The position of the
Schwarzchild condenser focus was set to within 2 μm by projection images
of the specimen recorded on film and taken in the absence of the zone plate
objective.   The Schwarzchild condenser is a X15 reflecting objective

commercially coated with 50 equal thickness layer pairs of Si/Mo. Measurement of the reflectivity and bandpass were made using BESSY synchrotron at Berlin at the PTB laboratory (Michette et al, 1992), giving a reflectivity, accounting for the two reflections of the Schwarzchild system, of 3 ± 1% at a wavelength of 23.3 nm, with a bandpass of 2.4 ± 0.1%.

The zone plate was fabricated by the King's College contamination technique (Charalambous & Morris, 1991), to have an outer zone width of 41 nm as required to match the numerical aperture (N.A. = 0.28) of the condenser at the mean germanium wavelength. Correction to the zone positions was made for spherical aberration. The focal lengths of the zone plate for the two wavelengths are 158.8 μm and 156.2 nm respectively for zone plate of 89.9 μm diameter. Images were recorded on Kodak S101 photographic film located 8 mm downstream from the zone plate at a magnification of about 60 times.

3. IMAGE RECORDING ANALYSIS

X-ray laser images of regions of a test pattern of gold islands, 800 nm square, on a 50 nm thick silicon nitride film were taken in single shots of the Vulcan laser with doublet Ge radiation. The best images were produced when the condenser was defocused, Figure 1, by some 16 to 20 μm, and the specimen was illuminated by a hollow cone. This procedure provided a condition where the zone plate was not fully filled by radiation, thus increasing its depth of focus.

The objects were essentially binary patterns. The contribution to the image from incoherent plasma radiation transmitted via the band-pass of the condenser was small compared with the energy of the coherent radiation (Lewis et al, 1991a).

The defocusing of the condenser had the further advantage that the two images corresponding to the two wavelengths were laterally displaced, thus allowing an unambiguous determination of the image resolution from single-

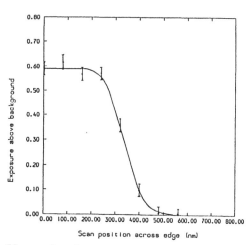

Figure 2. Average of 13 edge scenes obtained by densitometry of gold islands on S101 film.

wavelength images of the edges of the gold islands.

An average scan across such an edge is shown in Figure 2. Assuming the point spread function for the zone plate is gaussian in form, then the 12% to 88% levels correspond to the FWHM, corresponding to an image resolution of 0.15 ± 0.02 μm.

4. CONCLUSION

The illuminated region (irradiances from 10^{11} to 2 x 10^{12} W cm^{-2}) was destroyed by each pulse of radiation, but the object structure was successfully recorded by the transmitted X-ray wavefront; this may have useful consequences for the recording in a single X-ray laser shot of images of biological specimens without beam damage.

5. ACKNOWLEDGEMENTS

The work was carried out with support from the Science and Engineering Research Council. We thank Dr J. Bönisch for E-beam fabrication of the gold test specimens. The work was supported by NATO Collaborative Research Grant CRG 910493.

6. REFERENCES

Charalambous, P & Morris D, in "X-ray Microscopy III", Eds. A G Michette, G R Morrison and C J Buckley, 1992. Springer series in Optical Sciences 67, 79-82

Lewis, C L S, O'Neil, D M, Neely, D, Uhomoibhi, J O, Burge, R, Slark G, Browne, M, Michette, A, Jaegle, P, Klisnick, A, Carillon, A, Dhez, P, Jamelon, G, Kodama, R, Norreys, P, Rose, S J, Zhang, J, Pert, G J, Ramsden, S A, "Collisionally excited X-ray laser schemes: Progress at Rutherford Appleton Laboratory", SPIE Meeting, San Diego, July 1991.

McGowan, B, Maxon, S, Da Silva, L B, Field, D J, Kean C J, Matthers, D L, Osterheld, A L, Scofield, J H, Shimkaveg G, Stone, G F, 1990. Phys. Rev. Lett. 65, 420.

Michette, A G, Buckley, C J, Kuhne, M, Muller, P, in "X-ray Microscopy III", Eds. A G Michette, G R Morrison & C J Buckley 1992. Springer series in Optical Sciences, 67, 125-127.

Inst. Phys. Conf. Ser. No 130: Chapter 7
Paper presented at Int. Congr. X-ray Optics and Microanalysis, Manchester, 1992

491

Using soft X-rays from laser generated plasmas for the contact imaging of living biological specimens

R.A.Cotton[1], M.D.Dooley[2], J.H.Fletcher[3], C.E.Webb[3], A.D.Stead[1] & T.W.Ford[1].

[1]Royal Holloway and Bedford New College, University of London, Egham, Surrey, TW20 0EX, UK.
[2]Rutherford Appleton Laboratory, Chilton, Didcot, Oxon, OX11 0QX, UK.
[3]Dept. of Atomic and Laser Physics, Clarendon Laboratory, University of Oxford, Parks Road, Oxford OX1 3PU, UK.

ABSTRACT: Soft X-ray contact microscopy (SXCM) is a powerful imaging technique which enables the internal structure of living biological material to be recorded in an X-ray sensitive photoresist at a resolution beyond that of light microscopy. This paper describes advances in two areas of the technique. Firstly, the setting up of two separate small-scale laser facilities which have been optimised for the generation of soft X-rays for microscopy and secondly, the use of an atomic force microscope (AFM) to view the resist surface at high magnification.

1. INTRODUCTION

The technique of SXCM has been developed since the late 1970's with the aim of imaging hydrated biological specimens at sub-optical resolution. Initial work employed synchrotron radiation as the X-ray source (Feder 1977), although the low intensities available led to exposure times in excess of several minutes, and meant that problems associated with radiation damage and natural movement of the specimen necessitated the fixation and/or dehydration of the biological material. With the advent of pulsed plasma sources (Z-pinch and laser generated plasmas), images can be recorded on a time scale of nanoseconds, before changes in the specimen can occur. While the first lasers to be employed to generate soft X-rays for contact microscopy were large and usually part of a multi-user national facility, it has become increasingly clear that for the technique of SXCM to progress to the stage where it could become as widely accessible to laboratory users as electron microscopes are today, a small-scale dedicated laser system was needed.

2. THE TECHNIQUE OF SOFT X-RAY CONTACT MICROSCOPY

SXCM uses soft X-rays in the water window region (0.28-0.53keV), lying between the oxygen and carbon K-absorption edges, where there is a large difference between the absorption of carbon (in the form of proteins, carbohydrates etc) and that of oxygen as water. This provides natural contrast and means that, unlike electron microscopy, no staining is required. Living biological specimens can be imaged in a natural aqueous environment without any prior pretreatment.

The technique is conceptually very simple. The specimen is placed in contact with a silicon wafer coated with an X-ray sensitive photoresist material, for example the polymer polymethyl-methacrylate (PMMA), which acts as the recording medium. Exposure of the resist to soft X-rays results in the breakage of chemical bonds in the polymer chains, reducing the localised molecular weight, thereby increasing its solubility in the developer. The result is a highly detailed X-ray absorption profile of the specimen; i.e. higher elevations in the resist correspond to a higher absorption by the specimen. This is shown schematically in Fig 1. A more detailed review of SXCM is given by Ford (1991). Once the photoresist has undergone sufficient chemical development, the unity magnification image in the resist then needs to be magnified. Until recently a scanning electron microscope (SEM) has been employed to view the surface, although more recently an atomic force microscope (AFM) has been used which offers a potentially higher resolution (Tomie 1991, Cotton 1992).

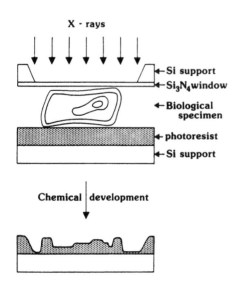

X - rays

←Si support
←Si₃N₄window
←Biological specimen
←photoresist
←Si support

Chemical development

Figure 1. The contact imaging method. X-rays are differentially absorbed by cells which is reflected in the pattern of bond breakage in the underlying photoresist. Development of the photoresist by a solvent produces a highly detailed relief profile of the cell as the exposed photoresist is dissolved.

3. LASER GENERATED PLASMA X-RAY SOURCE

By focusing a high powered, pulsed laser onto a solid target, a high temperature plasma can be created which is an intense X-ray source typically lasting between 1 to 25ns (depending on the laser system). By generating the correct intensity of laser radiation and by choosing a suitable target material, the X-rays emitted can be made to fall largely within the water window. In the UK a consortium to exploit SXCM, employing two separate laser systems optimised for the generation of soft X-rays. Both facilities are designed to be readily accessible for the rapid imaging of biological specimens. The two facilities are described below.

The first laser, based at the Clarendon Laboratory, is a specially designed compact, UV pre-ionised, discharge pumped krypton fluoride laser, operating in the ultra-violet at 249nm. It employs an injection controlled amplifier arrangement, with coupled confocal cavities which results in an output beam of 2.3J in a 20ns pulse (FWHM) with a divergence of only 2.5 times that of the diffraction limit. By focusing this beam with a 9cm focal length aspheric lens, intensities in excess of $10^{13}Wcm^{-2}$ can be achieved. The laser is relatively simple and inexpensive, small enough to pass though a standard laboratory door (Fletcher 1992).

The second laser uses a small beam line ("Target Area 4") from the large infra-red Nd:GLASS laser "Vulcan" at the Rutherford Appleton Laboratory. The laser delivers over 5J of energy on target in a pulse duration of between 0.6 and 3ns. Although the system is part of a multi-user facility and so is not permanently dedicated to SXCM, the system can be set up very rapidly and its availability is now much improved. Many images, from a wide variety of specimens, have been successfully recorded (Cotton 1992a).

Both systems use either gold, tungsten or titanium target material, with conversion efficiencies of up to 10% from laser radiation to water window X-rays being measured. The harder X-ray conversion efficiency (>530ev) is low (>1% for the KrF system and ~few % for TA4) and the XUV\UV emission is filtered at the specimen using aluminium coated silicon nitride windows. A paper comparing the suitability of each system to SXCM is currently in preparation.

4. RESIST EXAMINATION

An additional problem which has retarded progress has been the difficulty associated with obtaining a high magnification image of the resist surface. Until recently, the use of a scanning electron microscope (SEM) has been the only viable method for resist examination, although it has several disadvantages for this application. Firstly, a metallic coating must be applied to the resist to make it electrically conducting, and once this coating has been applied, the resist cannot be further developed. Due to low resolution of conventional light microscopy (LM) used to monitor resist development, it is difficult to estimate when the "optimal" level has been achieved. The poor resolution and relatively low magnification from LM are indeed the very limitations which it is hoped SXCM will be able to overcome. This meant that resist development was rather a "black art" with much of the information in the resist frequently being lost because of insufficient or over-development. In addition to this, severe damage to the resist by the electron beam of the SEM meant that it was limited to low accelerating voltages, so in many instances, imaging these relatively low contrast features, the performance of our SEM (Cambridge S100) was quite probably a limiting factor in the observation of fine detail.

Two possible alternatives to conventional SEM for resist examination have been investigated, namely the field emission scanning electron microscope (FESEM) and the AFM. Both are far superior to conventional SEM, but it is the AFM which is seen as a very major advance for SXCM.

4.1 The atomic force microscope:

The AFM operates by scanning a probe tip attached to a flexible cantilever across the resist surface. Deflection of the cantilever is caused by the interatomic repulsion between atoms on the cantilever tip and atoms on the resist surface. By monitoring the deflection of the cantilever, a high resolution three dimensional map of the surface is obtained. The AFM was first applied to the examination of the photoresist for SXCM by Tomie 1991. The AFM is now becoming widely used in the material science field, although is still largely unfamiliar to the life-sciences community. An AFM allows the surface of non-conducting materials to be imaged at a resolution approaching atomic dimensions. Experiments have revealed that there is no damage to the resist surface and an accurate height profile of the resist can be obtained from the computer software, which can also be used to enhance low contrast images, enabling the AFM to deal better with smaller height differences within the resist than SEM. The major advantage of using an AFM is that uncoated resists can be viewed, allowing a series of images at various stages of development to be taken.

4.2 Field Emission scanning electron microscopes:

The FESEM, by virtue of the high brightness of the electron source, can also view uncoated resists. However the operating voltage was found to be limited to a few kilovolts due to charging of and damage to the resist by the electron beam. This sets the resolution of the instrument with this type of specimen to approximately 7nm. The electron probe also causes resist mass loss. A further disadvantage of the FESEM for this purpose arises from the electron beam itself exposing the photoresist to the same effect as the soft X-ray flux, and therefore if subsequent development of the resist is necessary the image may be degraded.

5. BIOLOGICAL IMAGING

5.1 Studies of bacterial spores

As an example of the use of the technique, bacterial spores have been imaged. They are characterised by extreme dormancy and by their resistance to chemical and physical agents. They normally resist temperatures about 40°C higher than the corresponding vegetative cell, although the spores contents (enzymes etc.), once extracted from the spore, are no more resistant than those in the vegetative cell; ie the spore somehow imposes this enormous resistance on its contents. The exact dormancy and heat resistance mechanisms are not known, but there is evidence that the internal structure of spores is important (Gould 1977).
From studies of thin sections, using transmission electron microscopy, the spore is known to consist of a dense central core region surrounded by less dense "cortex", which is encased in a thin protein-rich, enzyme-resistant "coat". The SXCM image (Fig 2) shows the spore to be oval in shape and approximately 2-3μm in length. The central carbon-dense region and cortex are clearly discernable.

These spores contain about 10% of their dry weight as calcium dipicolinate, and this is known to be important in the maintenance of dormancy, but it is not known how. There is good evidence that it resides in the central core region, but there is very little indication of its state (ie. dispersed; interacting with other molecules; isolated as crystals etc.). The SXCM image (Fig 2) shows that within the core region, four X-ray dense structures, in a square-like arrangement can be seen. This structure has been seen in several other SXCM images of bacterial spores. The interpretation of these initial images is as yet uncertain, but it has been suggested that these structures are possibly crystals of calcium dipicolinate existing within the spore. If this is the correct interpretation this will be of considerable biological interest. Experiments are currently under way to investigate this hypothesis.

5.2 Yeast cells

An SXCM image of *Schizosaccharromyces pombe* (commonly known as fission yeast) as recorded in the resist is shown in Fig 3. A central granular structure with the outer cell wall is visible and septum formation can be seen as the cell undergoes division.

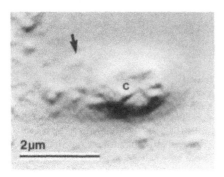

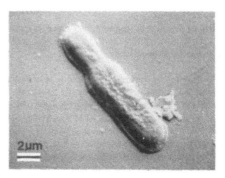

Figure 2. An SXCM image of a bacterial spore *Bacillus cereus*, taken using the RAL X-ray source, with an X-ray fluence of approximately 50mJcm^{-2} within the water window. The central carbon-dense core (c) is clearly discernable with the outer edge of the spore indicated by an arrow.

Figure 3. An SXCM image of a fission yeast cell taken with the Clarendon X-ray source with an X-ray fluence of approximately 10mJcm^{-2} in the water window.

6. ACKNOWLEDGEMENTS

The authors would like to express their gratitude to the SERC (grant GR/F 81514) and to the Royal Society's Paul Instrument fund which provided the grant for building the KrF laser and SXCM system at Oxford. We wish also wish to thank the support staff on the "Vulcan" laser at RAL and Pete Anastasi at Kings College for their help with this work. We would like to thank Brendon Mahoney at the microbiology unit, University of Oxford and Dr. G.W. Gould of Unilever (UK) for providing biological specimens and thank Dr Roger Emch of Park Scientific Instruments (Geneva), and Joel (UK), for access to instruments.

7. REFERENCES

Cotton R.A., Dooley M.D., Page A.M., Stead A.D. & Ford T.W., 1992a *RAL report*, pp 50-53

Cotton R.A., Dooley M.D., Fletcher J.H., Stead A.D. and Ford T.W., 1992b To be published by *SPIE, Soft X-ray Microscopy* **1741**

Feder R., Spiller E., Topalian J., Broers A.N., Gudat W., Panessa B.J., Zadunaisky Z.A., & Sedat J., *Science* **197**, pp 259-260.

Fletcher J.H., Cotton R.A. & Webb C.E., 1992 To be published by *SPIE, Soft X-ray Microscopy* **1741**.

Ford T.W., Stead A.D. & Cotton R.A., 1991, *Electron Microsc. Rev.*, **4**, pp269-292.

Gould G.W., 1977 ,*J. Appl. Bacteriology*, **42**, pp297-309.

Tomie T., Shimizu H., Majima T., Yamada M., Kanayama K., Kondo H., Yano M. & Ono M., 1991 *Science* **252**, pp 691-693.

X-ray optics by the use of 2-D bent crystals

Eckhart Förster, Ingo Uschmann

Max-Planck-Arbeitsgruppe "Röntgenoptik" an der
Friedrich-Schiller-Universität Jena
Max-Wien-Platz 1
D-O-6900 Jena, F.R.Germany

ABSTRACT: An X-ray microscope using bent crystals has been developed to investigate continuum and line radiation of laser produced
plasmas in the range 0.3...1.0 nm. By optical contact of crystal
wafers with spherical or toroidal glass surfaces two dimensionally
bent crystals have been obtained. To fulfil the Bragg condition for
selected X-ray lines, aberrations of spherically or toroidally bent
crystals were calculated for Bragg angles of $45°...89°$. X-ray images
of a counterstreaming Al-plasma for the lines w 1-n (n=2,3,4),
which have been obtained with a three-channel microscope, indicate
population inversion between levels n=4 and 3.

1. INTRODUCTION

X-ray microscopy is routinely applied to applied to laser-generated
plasmas as one of the important diagnostic tools. By analyzing the self-
emission of high-temperature plasmas, spatial distribution of plasma
emission is obtained. Usually, techniques with large spectral window -
like pinholes, multi-channel plates, and grazing-incidence optics - are
used, which yield only spectrally integrated images. The use of Bragg
reflections on two-dimensionally bent crystals enables the x-ray
microscopical imaging in narrow spectral ranges comparable with the
width of spectral lines. By matching the x-ray microscope to selected
lines one obtaines two-dimensionally spatially resolved images and after
evaluation 2D plots of plasma parameters.

2. X-RAY MICROSCOPY IN SELECTED SPECTRAL LINES

The crystal net planes must approximate confocal ellipsoids (Berreman
et al. 1977) with the centre of source and image in their focal points.
Actually the crystal shape is a toroidal one for medium Bragg angles
($\theta > 45°$) and a spherical one for nearly normal incidence ($\theta > 85°$). Fig.1
shows the schematic of an x-ray crystal microscope with object plane,
bent crystal, light protection, and image plane. The ratio of vertical
(sagittal) and horizontal (meridional) radii of curvature R_v / R_h depends
on the Bragg angle θ as following

$$R_v / R_h = \sin^2\theta.$$

The spectral window of imaging depends on horizontal and vertical divergences α, Φ and on the magnication factor k as well (Förster et al.
1991). It is very small if the source is near to the Rowland circle
($k \approx 1$).

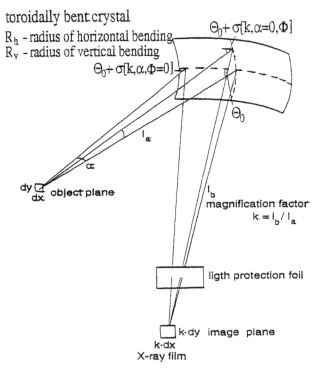

Figure 1: Schematic of an x-ray crystal microscope (focal length
is $f = (R \cdot \sin \theta)/2$)

3. MANUFACTURING AND TESTING OF BENT CRYSTALS

To achieve a spatial resolution of a few micrometer the fluctuation of net plane orientation from the required one should not exceed 5 seconds of arc. Some details of bending a ca. 100 μm thin crystalline wafer by a replica technique were described elsewhere (Förster et al. 1991). In Fig. 2 both the the scheme of testing (a), an x-ray image of an test object (b) and the corresponding photometer scan (c) are contained. Spatial resolution was obtained better than 10 μm, whereby Bragg reflection took place in a thin surface layer, normally disturbed by surface finishing.

(a) <u>Scheme</u>

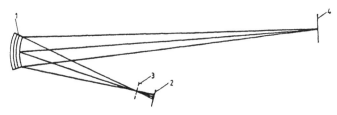

1- *bent crystal,*   2- *X-ray source,*
3- *test object,*   4- *X-ray film*

<u>X-ray image with a toroidally bent quartz (10.0)</u>

$R_v = 126.7\ mm,\ R_h = 152.8\ mm,\ \lambda = 0.7757\ nm$

(b) <u>test lattice</u>   (c) <u>photometer scan</u>

lattice constant 100 μm
(lines 40 μm, holes 60 μm)

   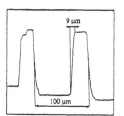

Figure 2: Scheme of crystal testing (a), x-ray image of a test object (b), and corresponding photometer scan (c)

4. THREE-CHANNEL X-RAY MICROSCOPY

For experiments at the laser facility NIXE of the Central Institute of
Optics and Spectroscopy Berlin a three-channel microscop was fabricated
by us. A description of the instrument was published recently (Glas et
al.1991, Förster et al. 1992). By focussing of intense laser radiation
(duration 6 ns, intensity 10^{13} W/cm^2) two separated aluminium plasmas
were produced (energy 26 J and 36 J, resp.). The x-ray microscopic images shown in Fig, 3 have been performed in the wavelength range from
0.4 nm to 0.85 nm, e.g., by using transitions of the type 1s^2 - 1s np
(n= 2,3,4) in helium-like aluminium ions. The two plasmas, 250 μm
distant from each other, and a counter-streaming region can be seen. Results indicate that under the special conditions of this experiment both
the size of the emitting region and the brightness of the transitions
show specific features which are very interesting for the study of the
level populations in these ions.

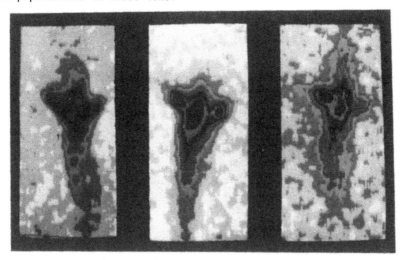

Figure 3: X-ray images of a double plasma obtained by using the
1s^2 - 1s np (n=2,3,4) transitions in helium-like aluminium ions (from
left to right).

REFERENCES

Berreman D W et al. 1977 Appl.Opt. 16 2081
Förster E et al. 1991 Laser Part. Beams 9 135
Förster E et al. 1992 X-ray Microscopy III (Berlin: Springer)
  pp 202-5
Glas P et al.1991 Opt.Commun. 86 271

Inst. Phys. Conf. Ser. No 130: Chapter 7
Paper presented at Int. Congr. X-ray Optics and Microanalysis, Manchester, 1992

499

Compact laboratory X-ray beam line with focusing microoptics

R. Hudec[1], U. Arndt[2], A. Inneman[3], L. Pína[4]

[1] Astronomical Institute, Czech Academy
    of Sciences, 251 65 Ondrejov, Czechoslovakia

[2] MRC Laboratory of Molecular Biology,
    Hills Road, Cambridge, CB2 2QH,
    England

[3] Koma Co., K lesu 964, 140 00 Praha 4 - Kamyk,
    Czechoslovakia

[4] Faculty of Nuclear Engineering, Czech Technical
    University, Prague, Czechoslovakia

ABSTRACT: A compact laboratory X-ray beam line with
focusing X-ray microoptics is described. The X-ray
intensity at the sample is expected to be many times
greater than obtainable by conventional methods. The focusing microcollimators are produced by electroformed
replication.

1. INTRODUCTION

The wide-spread use of synchrotron radiation for X-ray
crystallographic studies has demonstrated the benefits of
small brilliant X-ray sources used with focusing collimators.
Similar advantages are obtainable with laboratory sources,
when tuneability is not a prime consideration and when 8keV
X-rays from a copper target are satisfactory. These techniques
require novel design features for the X-ray tube and for the
focusing X-ray mirrors. We present here an account of the
present state of a research project in which groups at Prague,
Czechoslovakia, Cambridge, UK and Warwick University, UK are
collaborating; its aims are to produce a compact laboratory
X-ray beam line which should deliver an X-ray intensity at the
sample many times greater than that obtainable by conventional
methods.

2. X-RAY TUBES FOR USE WITH FOCUSING COLLIMATORS

In X-ray structure determinations using single crystals a typical requirement is for a beam diameter, a, of about 400 µm and a crossfire β of less than 10^{-3} rad. at the sample. The most efficient use is made of the X-ray source if the X-ray collimator subtends the largest possible solid angle at the source. Circularly symmetrical toroidal collimators in which the X-rays are totally reflected can subtend planar angles, α, at the source up to 0.75 of the theoretical maximum which is four times the critical angle θ_c. For 8keV X-rays and a gold or platinum surface $\theta_c \approx 10^{-2}$ rad. A divergence at the source which is twenty times greater than the convergence at the sample implied that the toroidal mirror must produce a magnified image of the source at the sample: for the required beam size a source diameter, f, of 20 µm is indicated, since $f\alpha = a\beta$. It is further necessary that the mirror-collimator be mounted as close as possible to the X-ray tube focus so as to avoid very long distances between mirror and sample: the X-ray tube windows must thus be very close to the electron focus; we are making this distance 10 mm.

Preliminary computer ray-tracing has shown that an ellipsoid of revolution with the dimensions shown in Table 1 will produce an adequate image (Hudec et al., 1993).

Table 1. **Ellipsoidal Mirror For 8keV X-rays**

Length	26 mm
Smallest Dia	0.394 mm
Largest Dia	0.730 mm
Length of Major Axis	600 mm
Magnification	x30
Solid Angle of Collection	8.9×10^{-4} sterad
Surface	Au ($\theta_c = 10^{-2}$ rad.)
RMS Surface Roughness	< 1.5 nm
Source to Mirror	10 mm
Source Dimensions	20 µm x 200 µm

The equality $f\alpha = a\beta$ is a statement of Liouville's Theorem; a and β are fixed by the requirements of the experiment and thus so is the product $f\alpha$. The intensity collected by a collimator, and thus the intensity at the sample, is proportional to $W\alpha^2$, where W is the maximum permissible power dissipation in the target, that is, the maximum power which can be conducted away from the focal spot without melting the target. This power is known to be approximately inversely proportional to the linear dimensions of the focal spot, (that is, the power loading per unit area is proportional to f). Consequently, the intensity at the sample is proportional to f^{-1} and the total power of the X-ray tube to f. A 20 µm square fore-shortened X-ray tube focus is about twenty times smaller than used in conventional crystallographic X-ray tubes used

with non-focusing collimators, where f ≈ a and α ≈ β. We thus expect a twenty-fold increase of X-ray intensity and a twenty-fold reduction in tube power.
We have designed, in conjunction with Drs P. Duncumb and J.V.P. Long, and are in the process of constructing, a magnetically focused microfocus X-ray tube with a rotating anode and a re-entrant X-ray window.

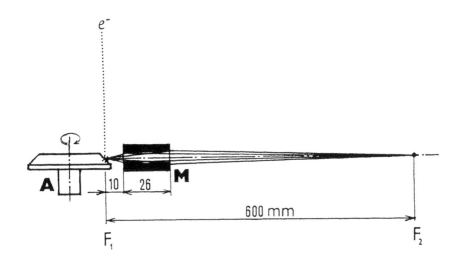

Fig. 1. The X-ray source with microfocusing optics. A – rotating Cu anode, M-X-ray mirror, e-electron beam, F_1-focus of the objective coinciding with the X-ray source,F_2-the focal plane of the mirror.

3. FOCUSING X-RAY MICROMIRRORS-COLLIMATORS

3.1. The replication technology

It is rather difficult and expensive to produce high-quality polished internal cavity (needed for X-ray micro-optics geo-metry) with diameters below 10 mm. The problem can be effici-ently solved thanks to the method already applied with success in astronomical X-ray optics (Hudec and Valnicek, 1986) in which electroformed replicas are made of a master mandrel which has an external shape which is that of the cavity of the final mirror. This solution has two main advantages: (i) it is not necessary to grind / polish the cavity (ii) many copies from only one master can be produced.Altogether 50 grazing in-cidence replica mirrors were developed in Czechoslovakia since 1970, with apertures between 1.7 and 24 cm and grazing angles between 5 arcmin. and 5 degrees (Hudec and Valnicek, 1986, Hudec et al. 1988) . Nickel or gold metal layers with good reflectivity (e.g. 60% at 0.83 nm and 1° incidence) served as reflecting surfaces.Unlike all the thin evaporated layers, the electroformed nickel layer is highly resistant to all environ-mental and laboratory influences.

3.2. The production of the micromirrors

(1) **The master**.The needle-like metal master with diameter below 1 mm and with an ellipsoidal geometry is produced on a computer-controlled machine by diamond turning. The surface roughness of the model obtained this way is better than 10 nm.The turned master is treated by a lacquer coating technique in order to achieve a better roughness of the surface (below 1.5 nm).

(2) **The shell.** The production of the shell consists of following steps:

(A) gold-coating of the master
(B) reinforcing of gold layer by electroforming
(C) electroforming of a nickel shell
(D) electroplating of external plastic surface
(E) external surface treatment by electroforming
(F) separation of the mirror from the master

3.3. The computer modelling and test facility

A ray-tracing program has been written. It calculates the distribution of the irradiance in the chosen focal plane. The calculation of the dependence of the maximal power density in the focus F2 on the surface roughness for the given geometrical configuration is also possible. The simulations shows that the final image of the X-ray source should have a reasonably uniform intensity distribution (Hudec et al., 1993).
The test facility for measurements on masters and shells is based on a microfocus X-ray tube with the X-ray source diameter of order of 10 μm with the possibility of scanning the e-beam. An 1D CCD array serves as a detecting system, the measured signal being evaluated by a computer and displayed on a PC monitor.

4. CONCLUSIONS

There are few remaining technological problems left to be overcome and we are confident that within the next year we shall a chieve our object of producing a laboratory X-ray beam line exceeding conventional sources in intensity by more than an order of magnitude with a large reduction in the power consumption and complexicity of the X-ray source.

Acknowledgements: We acknowledge support of Drs P. Duncumb and J.V.P. Long in design and construction of the magnetically focused X-ray tube.

6. REFERENCES

Hudec R. and Valnicek B. 1986 SPIE Proceedings No.597, 111.
Hudec R et al. 1988 Applied Optics 27, No.8, 1453.
Hudec R. et al. 1993 in preparation (the extended version of this paper)

Inst. Phys. Conf. Ser. No 130: Chapter 7
Paper presented at Int. Congr. X-ray Optics and Microanalysis, Manchester, 1992

503

Preparation of soft X-ray monochromators by laser pulse vapour deposition (LPVD)

H. Mai, R. Dietsch and W. Pompe

Fraunhofer-Einrichtung für Werkstoffphysik und Schichttechnologie (IWS)

O-8027 Dresden, Helmholtzstr. 20, Germany

ABSTRACT: The paper describes the design of our latest equipment for LPVD of multilayer structures. In detail an UHV-device (base pressure typically 10^{-10} mbar) has been equipped with a new target/substrate manipulator system that allows the deposition of individual layers of good thickness homogeneity. The ablation of the targets is induced by a Nd-YAG-laser (wavelength 1.064 μm, pulse energy 0.9 J, pulse length 7 ns, repetition rate 30 Hz). The deposition process is controlled by an iPC-486 computer i. e. target and substrate are stepper motor driven with reference to quartz oscillator monitoring. Model specimens of Ni/C- and W/C-layer stacks have been prepared and characterized by various techniques.

1. INTRODUCTION

Layered synthetic microstructures (LSM) are used in soft X-ray optics for various applications, e.g. monochromators, mirrors, focusing elements. In contrast to conventional Langmuir - Blodgett - films these layer stacks show - even for near normal incidence of X-rays - an outstanding reflectivity of typically 10 through 50 %. High quality X-ray monochromators are nowadays prepared by electron beam evaporation, sputter deposition and LPVD. In particular the latter technique is applied merely in a few laboratories throughout of the world (Gaponov 1992, Grudski 1989, Mai 1992). It shows a number of advantages (very smooth interfaces, high purity layers, precise stoichiometry) compared to the other methods (see e.g. Stock 1992, Boher 1989). A serious drawback of LPVD was - up to the present - the rather limited thin film area involving a good thickness uniformity. Even for very extended target/substrate distances the X-ray mirror size has been restricted to an area of typically 50 mm in diameter. The aim of the present paper is the description of LPVD applications that will provide multilayer preparation of decimetre- dimension involving appropriate lateral uniformity and low interface roughness.

2. EXPERIMENTAL

In LPVD the target material to be used for thin film coating is irradiated by a focused laser beam consisting of coherent light pulses of typically 1...5 J pulse energy, 10 ns pulse length and 1...30 Hz repetition rate. The power density created at the target surface approaches a level of typically 10^9 Wcm^{-2}. The interaction of such a laser pulse with the target causes a steep rise of temperature and pressure in the near surface region so that critical levels are approached. As a consequence the target material explodes and forms a plasma plume leaving the target surface in near normal direction. After a time of flight τ_p of some μs the particles condense at the substrate surface which is usually held close to room temperature. A laterally rather inhomogeneous layer growth is induced by that procedure since the particle flux is concentrated within a solid angle of typically 1/3 sr. Due to high fluencies of particles (10^{18}...19^{19} cm^{-2}) and particle energies (10^{19}...10^{20} eVcm^{-2}) within τ_p the condensation process can be thought to be of a rapid quenching type. Thus, this intrinsically non-equilibrium regime should additionally result in particular properties of layer growth different from those observed for continuous deposition techniques proceeding close to thermal equilibrium. Actually the rapid quenching character should induce an atomic random order within the layers and the initial energy of the plasma particles should cause an appropriate atomic mixing at the layer interfaces.

2.1 Deposition regime

For LSM synthesis the conventional principle of LPVD (Fig.1) - where plain target and substrate are arranged at a certain distance from each other and where planar motion of both is applied - has been modified for achieving a satisfactory thickness uniformity even for larger substrates. The schematic diagram in figure 2 shows the modified version where a curved surface of the target causes a plasma plume wagging across a large-scale substrate as a relative motion of laser spot and target happens. The inset demonstrates the thickness uniformity proved by ellipsometric measurement.

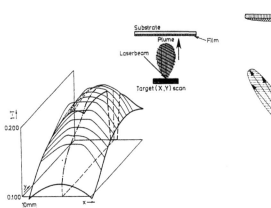

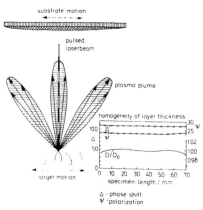

Fig.1 Typical thickness distribution observed for conventional LPVD arrangement
Inset: Schematic diagram of conventional LPVD configuration

Fig.2 Modified LPVD-principle where spatial characteristics of plasma plume follows surface normal orientation
Inset: Achieved thickness uniformity

Various multilayered specimens (area: 70 x 20 mm², Ni/C and W/C) have been prepared by this technique. The alternate deposition of metal (absorber) and carbon (spacer) layers (up to 80 individual layers) resulted in d-spacings

from 30 through 100 Å. Particular d_c/d_{Ni} -ratios have been provided to
achieve suppression of higher Bragg-orders. The characterization of the
LSM's has been carried out by XRD, HRTEM , AES and SNMS.

2.2 Typical results
There are three major properties that have to be provided by the LSM's if
their X-ray optical use is intended
- high reflectivity
- moderate spectral resolution and
- satisfactory long term stability.
Thus morphological perfectness and chemical adequacy of the layer stacks are
the prerequisites to meet these demands. The layer growth process that is
realized by the various techniques is forced to guarantee a specimen morphology (i.e. regularity of location, vanishing roughness and infinite slope
of chemical gradients of the interfaces) that will approach the ideally
shaped multilayer. The selected results to be presented here should demonstrate the efficiency of LPVD in this respect.

* XRD
The diffraction curve of a 9-period model - specimen under gracing incidence
(Cu-K$_\beta$) is given in figure 3a. It is easily to be seen that major and satellite peaks show a rather high regularity so that a moderate stack uniformity
can be suggested (satellite peaks above 2nd order are prevented from detection by background noise). This has been proven by computer simulation where
merely a standard deviation of period thickness $\sigma_d \approx 1.6$ % could be deduced
from (Fig. 3b).

* TEM
TEM cross section micrographs obtained in spatial high resolution
mode were used to gather information about interface morphology. In
particular boundary conditions characterizing roughness and atomic
number gradients should be attainable. Figure 4 represents a typical
picture of the layer arrangement in
an appropriate 20-period stack.
Again a good uniformity is observable. By image processing interfacial roughness and layer thickness
changes have been determined in
dependence on interface location
within the stack order.

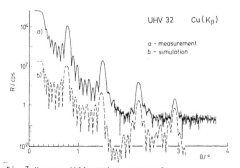

Fig.3 X- ray diffraction curve of
9 period LSM (Ni/C)
a - measured for Cu-Kβ,grazing incidence
b - computer simulation

Figure 5 shows no monotonic increase of σ with increasing number of Ni-
layers i.e. decreasing specimen thickness.

Thus the actual interface roughness of both C/Ni - and Ni/C - interfaces do
not exceed 0.1...0.15 nm. A similar range can be suggested for the Z-gradients since adjacent layers are distinguished by a very sharp contrast.
Less pronounced interlayer transitions are often described in the literature
(see e.g.   Petford-Long 1987, Ruterana 1989, Houdy 1989) for other preparation techniques.

Fig.4 TEM- micrograph of 20- period LSM
(Ni/C), d= 5.7 nm
Inset: HRTEM of interface region

Fig.5 Interfacial roughness and layer thickness
deviations in Ni/C- LSM (40 periods)

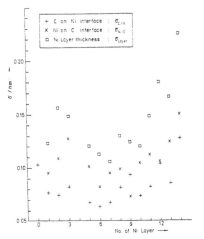

3. CONCLUSIONS

A modified LPVD-device for LSM preparation has been developed. Its efficiency is demonstrated by results of morphology characterization of selected
multilayer specimens. The basic concept of this preparation technique is a
prerequisite for the realization of large-scale X-ray mirrors even by LPVD.

ACKNOWLEDGEMENTS
The authors are indebted to Dr. B. Wehner (TU, Dresden), Mrs. S. Hopfe,
Dr. R. Scholz (MPI, Halle) and Dr. S. Völlmar (IWS, Dresden) for XRD -
and TEM - applications and simulation programming, resp.
Financial support has been provided by the Minister of Research and Technology of the FRG under contract-no.: 13 N 5945
REFERENCES:
Boher P, Houdy P, Barchewitz R and Joud J C 1989 Proc.
    12th ICXOM Cracow (Cracow: Acad. of Mining) pp. 379-385
Gaponov S V, Kluenkov E B, Platonov Y Y and Salashchenko N N
    1992 Physics of X-Ray Multilayer Structures, Proc. Meeting
    Jackson Hole, Wyoming (Washington DC: Optical Society of America)
    pp.66-69
Grudsky A Y, Brytov I A and Lazovski A G 1989 Proc.
    12th ICXOM Cracow (Cracow: Acad. of Mining) pp. 399-402
Houdy P, Boher P and Schiller C 1989
    Thin Solid Films 175
Mai H and Pompe W 1992 Applied Surface Sci. 54 215-226
Petford-Long A K, Stearns M B, Chang C H and Nutt S R 1987
    J. Appl. Phys. 61 1422
Ruterana P, Chevalier J P and Houdy P 1989 J. Appl. Phys. 65 3907
Stock H J, Kleineberg U, Kloidt A, Schmiedeskamp B and
    Heinzmann U 1992 J. Appl. Phys. in press

Coherent X-ray laser and its applications

Dao I Chiu
Ganzhou Research Institute of Non-Ferrous Metallurgy, CNNC Ganzhou, Jiangxi 341000
CHINA

ABSTRACT: The theoretical analysis of coherent X-ray laser will be presented in this paper. Furthermore, some important applications of coherent X-ray laser will also be discussed.

1.   INTRODUCTION

For a source to be coherent it must have time and space coherence. Time and space coherence can be summarized as follows: when we speak of time coherence we are saying that the relative phases between two points in time must remain constant over some long time interval; space coherence, on the other hand, involves the relative phases between two points in space remaining constant again over some long time interval. The term 'constant' implies a time long enough to perform some operation on the wavefront such as observation with the eye or photography. The beam of light emitted by a laser can have the property of being almost completely coherent. We must now discuss the subject of coherent X-ray lasers.

2.   THEORETICAL ANALYSIS

We use an electromagnetic treatment of the laser radiation to calculate its coherence, which is characterized by the mutual coherence function (MCF)

$$\Gamma_{12}(\tau) \equiv \ < E(r_1,t)E^*(r_2,t + \tau) > \tag{1}$$

where E is the complex electric field and $< >$ denotes an ensemble or time average. The intensity is a special case of the mutual coherence function: $I(r,t) = \Gamma_{11}(0)$. Longitudinal coherence is described by $\Gamma_{11}(\tau)$, while transverse coherence is described by $\Gamma_{12}(0)$. As an example, consider a two-slit interferometer placed some distance from a partially coherent source. For very small slit separations, the visibility of the fringes will be unity. However, as one increases the slit separation, the visibility drops. The radiation is thus coherent over small transverse distances but incoherent over large distances.

It is easy to show that the visibility (V) is equal to the absolute value of a normalized form of the MCF, called the complex coherence factor (μ):

$$V = |\mu|, \text{ where } \mu \equiv \frac{\Gamma_{12}(0)}{(\Gamma_{11}(0)\Gamma_{22}(0))^{1/2}} \tag{2}$$

A coherence length (L_c) is defined as the distance $|r_2 - r_1|$ at which $|\mu|$ has dropped to a characteristic value, typically taken to be ≈ 0.85. A coherence area is defined as $A_c \equiv \int \mu dx dy (\sim L_c^2)$, where x and y are the transverse coordinates.

The coherent power is identified as that passing through an area A_c:

$$P_c = I\Omega_s A_c = S\Delta VegL\lambda^2(gL)^{-1/2} \tag{3}$$

The main determinant of P_c is the gain length (gL); where Ω_s is the solid angle of the source subtended at the end of the laser ($\Omega_s = A_s/L^2$). The number of modes in the system is identified as A_s/A_c, where S is the source function describing the spontaneous emission, ΔV is the line profile width and g is the gain coefficient, the $(gL)^{-\frac{1}{2}}$ factor accounts for gain narrowing.

We discuss the modal method and present some of the numerical method, we begin with the electromagnetic wave equation

$$\nabla^2 E - \frac{1}{c^2}\frac{\partial^2 E}{\partial t^2} = \frac{4\pi}{c^2}\frac{\partial}{\partial t}\left[\frac{\partial P}{\partial t} + J\right] \tag{4}$$

where P is the atomic polarization and J is the current density.

We consider a 2-D laser medium, elongated in the Z direction and with transverse direction x. We assume a steady state and make the paraxial approximation ($|\partial^2 E/\partial z^2| \ll |2K\partial E/\partial z|$ or $|\partial^2 E/\partial x^2|$, where $K = 2\pi/\lambda$). We factor E and P each into a slowly varying envelope times a travelling wave:

$$E(z,x,t) = \text{Real} \left[E(z,x)e^{i(wt - Kz)} \right] \tag{5}$$

In the modal analysis the mode expansion

$$E(z,x) = \sum_n c_n(z)u_n(x) \tag{6}$$

is assumed and substituted into the paraxial wave equation. Assuming certain boundary conditions, we find a natural separation into two equations, a transverse eigenvalue-eigenfunction equation:

$$\left[\frac{1}{K}\frac{d^2}{dx^2} - k(x) + ig(x)\right]u_n(x) = 2iq_n u_n(x) \tag{7}$$

and a longitudinal (transfer-like) equation:

$$\frac{dc_n(z)}{dz} = q_n c_n(z) - iP_n(z) \tag{8}$$

Here h is a scaled electron density ($\equiv wp^2/KC^2$), g is the gain, q_n is an eigenvalue and P_n is the projection of the spontaneous emission source polarization P_{sp} onto the basis set u_n.

$$P_n(z) = \int P_{sp}(x,z)u_n(x)dx \tag{9}$$

The transvers eigenvalue equation is the same as the time-independent Schrödinger equation, with a complex potential. The analogy to quantum mechanics has been found helpful. Using Eq. (6), we express the MCF in terms of the eigenfunctions:

$$\Gamma_{12}(z,x_1,x_2) = \sum_{n,m} \langle c_n(z)c_m^*(z)\rangle u_n(x_1)u_m^*(x_2) \tag{10}$$

To evaluate the MCF we assume a δ function correlation of the spontaneous emission polarization and solve the longitudinal Eq. (8) to get

$$\langle c_n(z)c_m^*(z)\rangle = \text{const } B_{nm}[\exp(Q_{nm}z) - 1] \tag{11}$$

where

$$B_{nm} \equiv \int u_n u_m^* dx, \quad \text{and} \quad Q_{nm} \equiv q_n + q_m^*. \tag{12}$$

3. APPLICATIONS

(1) Microscopy, a 10^{-15} second pulse could 'freeze' any type of motion, such as phase transitions.

(2) Photography, for flash radiography and rapid crystallography.

(3) Remote analysis, for detecting various elements through X-ray fluorescence or diffraction.

(4) Holography, particularly Fraunhofer holography where a reference beam is unnecessary.

(5) Nuclear studies, an X-ray laser could coherently excite atoms in a crystal so they might in turn accelerate atoms.

(6)   Radiation-damage studies, by making dislocations in crystals.

(7)   Medical studies, especially the application to eye surgery and tomography of secondary X-rays that are intense and focusable.

(8)   Weapons, taking advantage of X-ray penetration permitting relatively low powers to damage or disrupt solid state electronics.

(9)   Optical technology, for making microscopic interference patterns.

(10)   Nonlinear optical phenomena studies, a single mode X-ray laser could be focused entirely onto one atom.

4.   CONCLUSIONS

The longitudinal coherence length of typical X-ray lines is of the order of 1 μm. The transverse coherence length must be larger than the extent of the sample. X-ray holography will be able to capture the invasion of a virus into a living cell, or the division of DNA in a cell.

REFERENCES

Bloembergen N, Science, 216 (1982), No. 4550, 1057.
Matthews D, UCRL-JC-104404, Jan. 14, 1991.
Chiu D.I., Proceedings of 12 ICXOM, CRACOW, 1989, 440.

Advances in X-ray lasers

Dao I Chiu
Ganzhou Research Institute of Non-Ferrous Metallurgy, CNNC Ganzhou, Jiangxi 341000
CHINA

ABSTRACT: A brief introduction is given on the research of collisional excitation and recombination inversion schemes as well as ultra-short pulse sources.

The main thrust of current efforts is in generating shorter-wavelength lasing. Other research is directed toward trying to obtain similar gains with smaller driving systems, i.e., increasing the efficiency of the X-ray laser.

Laser-produced plasmas have been extensively used as suitable media for soft X-ray lasers. Gain has been reported in experiments using electron excitation schemes as well as recombination schemes. The highest measured total gain was achieved at the Lawrence Livermore National Laboratory (LLNL) using the Novette laser system. The gain coefficient depends mainly on the plasma temperature and density and on the nuclear charge Z of the target material. In these X-ray laser experiments, thin exploding targets were used in order to avoid the steep density gradient characteristic of plasma expansion from thick targets. The thin target explodes and generates a roughly cylindrical plasma having a nearly flat electron density profile of about 3-5 x 10^{20} cm^{-3} and a flat temperature profile. This is a steady-state inversion mechanism and the lasing duration is limited by the hydrodynamic expansion. Since the lateral scale-length is of the order of 100 to 150 μm, rays can propagate several centimetres in such a profile before being deflected out of the lasing region. There is, however, a price to be paid for reducing refraction in this geometry: in order to achieve a symmetrically expanding plasma, the laser must burn through the foil early. To overcome these limitations, new geometries were studied, which, in addition to increasing the density, also result in a wider lasing channel or a concave lateral density profile that reduces refraction losses even further.

The recombination scheme has been extensively studied and modest gains have been observed in several laboratories. Experiments at the Rutherford//Appleton Laboratory have used thin fibres as targets in order to enhance the cooling rate due to the cylindrical expansion. In experiments at Princeton a plasma is created by CO_2 laser irradiation of a target, which then expands in a confining magnetic field. The confinement keeps the density from falling too fast, while the cooling is effected by radiation losses.

Scaling to shorter-wavelength lasing requires higher laser powers because the temperature generally has to be higher. However, since the optimal density usually rises also, a higher gain coefficient can be achieved if such higher densities can be realized. This permits a shorter gain length, which reduces power requirements. Some reduction in lasing wavelength without an increase in power requirements can be achieved by going to higher shell transitions in higher-Z ions. e.g., in nickel-like rare earths. For recombination schemes the laser pulse also has to be shorter, because the recombination of higher-charge ions is faster, dictating a faster cooling rate.

The last few years have seen an extensive effort in developing high-power lasers of short pulse width as potential drivers for X-ray laser systems. These efforts take one of two general approaches. In one, the goal is to focus the laser to a sufficiently small spot so as to achieve a very high irradiance. In the other approach, the short pulse is used to either create a plasma which serves as a lasing medium or as an X-ray source to pump the lasing medium. In this case, the optimal irradiance is usually not very high and the merit of the short pulse lies chiefly in its ability to pump a short-lived level in self-terminating lasing schemes.   The photoionization schemes typically entail pumping a neutral gas rather than a plasma; shorter lasing wavelengths would then be achieved by going to deeper-shell electrons rather than to higher ionization states.

The high power is achieved by amplifying a very short dye laser pulse with excimer amplifiers.   Alternatively, a Nd:glass laser pulse can be compressed in time to produce the high power.   Under this arrangement, a low-power pulse propagates down a very long single-mode optical fibre so that self-phase modulation plus group velocity dispersion cause chirping.   The resulting long, low-power pulse can be amplified, then compressed by diffraction from a pair of gratings.   It also should be noted that transmission phase gratings instead of reflective gratings can be employed.   At an irradiance of 10^{17} W/cm^2, the laser electric field is of the order of the atomic electric field that binds the outer electrons. Indeed, experiments show efficient multielectron ionization under high irradiance conditions (up to about 100 photons absorbed per one ionized atom).   The so-called "above threshold ionization" indicates that the interaction cannot be described by perturbation approximations.   The most likely ionization process is not that which requires the minimum number of absorbed photons, but, rather, a higher number.   For sufficiently high irradiances, the whole outer electronic shell can be thought of as classically oscillating in the laser field and interacting with inner-shell electrons, possibly leading to inversion. Direct inner-shell excitation by multiphoton absorption can also lead to inversion, especially if a suitable resonant excitation transition is found, so that the outer shell is not likely to be completely ionized at the same time.   In a situation where the outer electrons were accelerated classically at irradiances higher than 10^{17} W/cm^2, they would acquire energies of the order of 1 Kev over atomic-size distances.   Clearly, they can act as projectiles in ejecting inner-shell electrons.   Furthermore, the oscillatory laser field coupled nonlinearly to an ensemble of atomic electrons gives rise to the creation of intense high harmonic components that can couple to the inner electrons.

In other investigations, short pulse lasers have been used to generate a short X-ray pulse which in turn is used to photopump an inner-shell electron in the lasing medium. The difficulty in materializing these schemes stems primarily from the requirement for a very short X-ray pulse. This is because the inversion is transient (the lower laser level is longer-lived than the upper level), and the lifetime decreases for shorter lasing wavelengths. In addition, collisions by the ejected electrons are capable of destroying the inversion. An ultrashort laser pulse can mitigate both of these deleterious effects and previous unsuccessful attempts can then be revisited. A short, high-power pulse can be used not only for the purpose of scaling to shorter wavelengths, but also to higher efficiencies, by using lower laser energies for the same transition. For shorter pulses, the pressure of the lasing medium can be increased before collisions become detrimental again, thus increasing the gain per unit length.

Therefore, the total required pump energy, which scales as the length, will be lower (for a given total gain). The X-ray pulsewidth will be longer than the pulsewidth of the primary laser (by a factor that increase upon moving to softer radiation). Also, travelling-wave excitation is necessary for short pulses because the laser transit time is likely to be longer than the inversion time.

REFERENCES

Matthews D, et al., (1991), Collisional Excitation X-ray Lasers, UCRL-JC-104404.

Inst. Phys. Conf. Ser. No 130: Chapter 7
Paper presented at Int. Congr. X-ray Optics and Microanalysis, Manchester, 1992

515

Laser plasma X-ray source with a gas puff target

H Fiedorowicz, A Bartnik, P Parys and Z Patron

Institute of Plasma Physics and Laser Microfusion,
00-908 Warsaw 49, P.O.box 49, Poland

ABSTRACT: The contamination free (without the target debris production)
laser plasma X-ray source is proposed. The X-ray emission from plasmas
produced by laser irradiation of the gas puff targets has been
experimentaly studied for the first time. Spatial, spectral and temporal
measurements show that intense soft X-ray radiation is produced from the
laser-irradiated gas puff targets.

1. INTRODUCTION

Large interest in the use of high-intensity soft X-rays for microtechnology,
biology, material studies, optics and atomic physics has stimulated development and use of laser-produced plasmas as an extremely bright source of soft
X-ray emission.

In recent years the soft X-ray emission from laser plasma has been extensively investigated in a wide range of parameters such as laser wavelength,
laser intensity, pulse duration, target material, and geometry of target
irradiation. Most works on developing laser plasma X-ray sources have been
carried out to characterize the soft X-ray emission from plasma and to
optimize the conversion efficiency of laser light into X-rays (Gerritsen et
al. 1986, Kodama et al. 1986, Nakano and Kuroda 1987, Popil et al. 1987,
O'Neill et al. 1988, Toubhans et al. 1989, Juraszek et al. 1991, Eidmann and
Schwanda 1991, Mehlman et al. 1991, Morsell et al. 1992, van Dorssen et al.
1992).

A great disadvantage of the laser plasma X-ray sources is the production of
target debris by the laser plasma that may damage the laser optics and the
X-ray irradiated specimens placed in the plasma chamber. Two methods to
reduce the debris production were applied, namely the use of the He buffer
gas in the plasma chamber and the replacement of the bulk solid target by
a thin foil (O'Neill et al. 1989, Louis et al. 1991).

In the present paper we describe the first use of a gas puf type target to produce the laser plasma. Radiation emitted from the plasma has been measured in the X-ray region. Spatial, spectral and temporal measurements show that intense soft X-ray radiation is produced. This type of target is of interest because it is potentially a very promising method to develop a contamination free laser plasma X-ray source.

2. EXPERIMENTAL SETUP

The experimental setup is shown in Figure 1. A Nd-glass laser system produced laser pulses of 1ns duration and an energy of nominally 15J. The pulses were focused on the gas puff type target which was made by expanding gas (SF_6 or Kr) from reservoir through an orifice of 0.8mm in diameter into a vacuum chamber. The flow was initiated by a solenoid valve synchronized with a laser pulse. The diameter of the laser focus in the orifice plane was about 100µm.

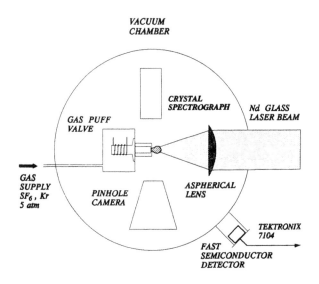

Fig. 1. Outline of experimental setup.

The spatial distribution of the plasma X-ray emission was measured with a pinhole camera. The pinhole of 50µm in diameter was covered with 15µm thick Be filter. X-ray spectra were analyzed by a spectrograph with flat ADP or CsAP crystal. The spectrograph was equipped with a 0.5mm slit oriented along the dispersion direction. The temporal measurements of X-ray emission were performed with the use of the subnanosecond semiconductor detector equipped with Be filter of 10µm thickness.

3. RESULTS AND DISCUSSION

Pinhole camera X-ray images of the plasma produced in result of laser irradiation of the gas puff targets are presented in Figure 2. Figure 2(a) presents the image for SF_6 gas target and Figure 2(b) - for Kr gas target. For comparison X-ray image of the plasma produced after

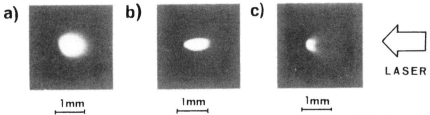

Fig. 2. Pinhole camera X-ray images of the plasma produced by laser
irradiation of (a) SF$_6$ and (b) Kr gas puff target and (c)
bulk teflon target.

irradiation of the bulk teflon target is shown in Figure 2(c). The most
prominent feature of the X-ray images [Figures 2(a) and 2(b)] is the
formation of a highly uniform hot plasma region, which emits strong X-ray
radiation. As can be cleary seen, the laser light is absorbed in larger
volume comparing to the ordinary laser plasma [Figure 2(c)].

Figure 3 shows the X-ray
spectra from (a) SF$_6$ and (b)
Kr gas puff target irradiated
with a laser pulse. Spectrum
for the SF$_6$ gas target
presents the characteristic K-
line emission of H- and He-
like fluorine ions. Spectrum
for the Kr gas target shows a
tangle of overlapping lines,
largely from the L-shell of F-
and Ne-like krypton ions.

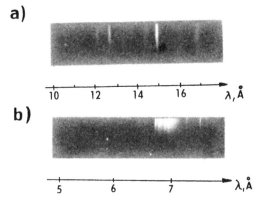

Fig. 3. X-ray spectra from (a) SF$_6$ and (b)
Kr gas puff targets irradiated
with a laser pulse.

The temporal evolution of the
X-ray radiation emitted from
the SF$_6$ gas target is shown in
Figure 4(a). Figure 4(b) shows
for comparison the X-ray signal from the laser irradiated
teflon target. It is clearly
seen that while the X-ray
pulse from the teflon target
follows the laser pulse and
rapidly falls off after about
1ns [Figure 4(b)], the X-ray
signal from the SF$_6$ gas target
[Figure 4(a)] lasts for about
2ns and than decreases very
slowly.

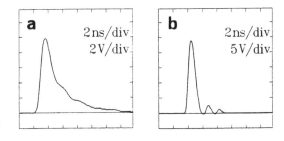

Fig. 4. Temporal evolution of the X-ray
radiation from (a) SF$_6$ gas and (b)
teflon targets. Sweep rate for
both oscillograms 2ns/div.

Rough estimation of the X-ray fluxes from the SF_6 gas and teflon targets, made by integration of the measured X-ray pulses [Figures 4(a) and (b)], shows that plasma produced by the laser irradiation of the gas puff type target emits approximately the same amount of soft X-ray radiation as the ordinary laser plasma.

ACKNOWLEDGMENTS

The authors appreciate the assistance of J Galik, A Klimek and A Ulinowicz in operating the Nd-glass laser.

This work was supported by the State Committee for Scientific Research of Polish Government under grant 2 0141 91/p01.

REFERENCES

Eidmann K and Schwanda W 1991 Laser and Particle Beams, vol. 9, pp.551
Gerritsen H C, van Brug H, Bijkerk F and der Wiel M J 1986, J. Appl.
    Phys.vol. 59, pp.2337
Juraszek D, Bayer C, Bernard M, Bocher J L, Garaude F and Thiell G 1991,
    J. Appl. Phys. vol. 70, pp.1980
Komada R, Okada K, Ikeda N, Mineo M, Tanaka K A, Mochizuki T and
    Yamanaka C 1986, J. Appl. Phys. vol. 59, pp.3050
Louis E, Bijkerk F, van Dorssen G E, Turcu I C E 1991 Proc. 21st European
    Conference on Laser Interaction with Matter ECLIM'91 (Warsaw: IFPILM)
Mehlman G, Burkhalter P G, Newman D A and Ripin B H 1991 J. Quant.
    Spectrosc. Radiat. Transfer vol. 45, pp.225
Morsell A L, Powers M and Shields H 1992 Appl. Phys. Lett. vol. 60,pp.425
Nakano N and Kuroda H 1987 Phys. Rev. A, Vol. 35, 4712
O'Neill F, Davis G M, Gover M C, Turcu I C E, Lawless M and Williams M
    1987, SPIE vol.831 X Rays from Laser Plasmas
O'Neill F, Turcu I C E, Tallents G J, Dickerson J and Lindsay T 1989,
    SPIE vol. 1140 X-ray Instrumentation in Medicine and Biology, Plasma
    Physics,Astophysics and Synchrotron Radiation
Popil R, Gupta P D, Fedosejevs R and Offenberger A A 1987 Phys. Rev. A,
    vol. 35, pp. 3874
Toubhans I, Fabbro R, Gauthier J C, Chaker M and Pepin H 1989 SPIE vol.
    1140 X-ray Instrumentation in Medicine and Biology, Plasma Physics,
    Astrophysics and Synchrotron Radiation
van Dorssen G E, Louis E and Bijkerk F 1992 Laser and Particles Beams
    (accepted for publication)

Inst. Phys. Conf. Ser. No 130: Chapter 7
Paper presented at Int. Congr. X-ray Optics and Microanalysis, Manchester, 1992

519

X-ray crystal spectrometer with CCD linear array

R Arendzikowski, H Fiedorowicz, P Parys and Z Patron

Institute of Plasma Physics and Laser Microfusion
00-908 Warsaw 49, P.O.box 49, Poland

ABSTRACT: An active-recording crystal spectrometer with a charge-coupled device (CCD) has been built to measure the X-ray spectra from laser-produced plasmas. The spectrometer uses the flat ADP crystal and is developed to obtain the X-ray spectra in the 6-8 Å wavelength range. The spectra are recorded with the use of the CCD linear array consisting of 1728 elements. The X-ray radiation can be detected both in the photosensors and in the transport registers, that extends the dynamic range of the spectrometer.

1. INTRODUCTION

Crystal spectrometers are commonly used to measure the spectral distributions of soft X-rays emitted by laser-produced plasmas. In these instruments photographic films are typically used to register the intensity of analyzed radiation. Film as an X-ray detector is cheap, sensitive, and gives high resolution but in some applications it is not enough precise in quantitative measurements of the X-ray intensities. Moreover, the processing of the film to recover X-ray data is very time-consuming. It is sensible disadvantage in monitoring the laser plasma, particularly when it is used as a laser plasma X-ray source.

One approach to solve this problem is the use of a charge-coupled device (CCD) as an X-ray detector. CCDs are photosensitive linear or area arrays widely used in the visible light region. Recently, some groups have reported validity of CCDs for the soft X-ray energy region (about 1 keV). CCD arrays have been used in imaging and spectrometric X-ray instruments applied in plasma diagnostics, astrophysics, topography (Peckerar et al. 1977, Marsh et al. 1984, Matsushima et al. 1986, Tsunemi et al. 1988, Sterzik et al. 1989, Twichell et al. 1990, Pina et al. 1991, Tuomi et al. 1992, Schirmann et al. 1992).

In this paper, we present an active-recording crystal spectrometer with a CCD linear array. The construction of the spectrometer, data-processing system and application results for laser-produced plasma are briefly described.

2. SPECTROMETER DESIGN

The spectrometer was originally designed for spectral measurements of the soft X-ray emission from laser-produced plasmas. The main components of the instrument are a miniature crystal spectrograph with a flat ADP analyzing crystal and a CCD camera for X-ray registration. A 38 mm long crystal, inclined at an angle of 40° with respect to the line of sight to the laser plasma, is placed in the spectrograph at the distance of 50 mm from the plasma. The entrance window is covered with a 10 μm thick Be foil, which protects the X-ray detector against visible light.

To the spectrograph the CCD camera of our design and production is attached. Figure 1 shows the general view of the spactrometer. In the camera a standard commercial CCD linear array (type FCCD123DC Fairchild Westons Systems, Inc.) is used. The cover glass is removed from the chip to allow X-rays to reach the array. The length of the array is 17 mm. It consists of 1728 photosensing elements (pixels). The pixel size is 10 μm by 13 μm on 10 μm pitch. The transport registers are covered with aluminium to shield them from visible light. Because this aluminium is transmitting for X-rays, the transport registers may also be used to detect X-rays. As the area of the transport register elements is much larger than the pixels', it is possible to exnhance the sensitivity and to extend the dynamic range of the spectrometer. Our CCD camera allows to deliver information from the pixels and the transport registers independently.

Fig. 1. X-ray crystal spectrometer with CCD linear array.

Output signals from the camera are transmitted to the data-acquisition system, which mainly consists of the A/D converter, the buffer memory and the D/A converter for displaying the stored signal on the oscilloscope. As the laser-produced plasma is a single pulse event, the data-acquisition system is synchronized with the laser. Data from the memory are sent to the IBM PC computer by the data transmission card. Schematic of the spectrometer and the data-acquisition system is shown in Figure 2.

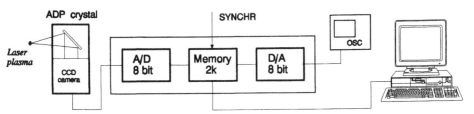

Fig. 2. Schematic of the spectrometer and the data-acquisition system.

3. RESULTS OF MEASUREMENTS

The spectrometer has been tested using the laser-produced plasma as an X-ray source. A Nd-glass laser produced 5-10 J, 1 ns pulses, which were focused onto targets made from metal foils. Figure 3 shows the examples of the spectra of radiation in 6.8-8.3 Å wavelength range obtained for the aluminium (a,b) and the iron (c,d) targets. Spectra presented in Figures 3a and 3c are recorded by the pixel arrays and in Figures 3b and 3d - by the transport registers arrays. The spectral lines for the aluminium target correspond to the K-shell emission of the H-and He-like aluminium ions and the lines for the iron targets correspond to the L-shell emission of the Li-like iron ions.

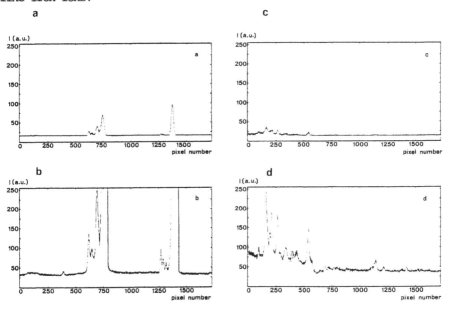

Fig. 3. Typical X-ray spectra for laser-heated aluminium (a,b) and iron (c,d) targets. Spectra (a,c) were recorded by the pixel array and spectra (b,d) - by the transport register array.

From comparison of the intensities of the appropriate lines one can see that the sensitivity of the transport registers array for the wavelength below the K-edge of aluminium (pixel numbers from 585 to 1728) is about 10 times greater than the sensitivity of the pixel array. For the wavelength higher than the K-edge of aluminium (pixel numbers from 0 to 585) the sensitivity of the transport registers array is about 12 times greater than of the pixel array. The difference of this values corresponds to the 1 μm thick aluminium layer shielding the transport registers.

ACKNOWLEDGEMENTS

The authors appreciate the assistance of J. Galik, A. Klimek and A. Ulinowicz in operating the Nd-glass laser.

This work was supported by the State Committee for Scientific Research of Polish Government under grant 2 0141 91/p01.

REFERENCES

Marsh K, Joshi C, Janesick J and Collins S 1984 Rev.Sci.Instrum.
   vol. 56, p. 837
Matsushima I, Koyama K, Tanimoto M and Yano M 1986 Rev.Sci.Instrum.
   vol. 58, p. 600
Peckerar M C, Baker W D and Nagel D 1977 J.Appl.Phys. vol. 48, p.2565
Pina L, Svoboda V, Rus B, Koshevoy M O, Rupasov A A, Shikanov A S and
   Fiedorowicz H 1991 Laser and Particle Beams vol. 6, p. 579
Salieres P, Mens A, Mazataud D, Schirmann D and Benattar R 1992 Proc. 3rd
   Int. Colloq. on X-ray Lasers, Schliersee
Schirmann D, Bocher J L, Le Breton J P, Mens A and Sauneuf R 1992 Laser
   and Particle Beams (accepted to publication)
Sterzik M, Brauninger H, Predehl P, Reppin C, Schuster K and Krumrey M
   1989 SPIE vol. 1159 EUV, X-ray and Gamma-ray Instrumentation for
   Astronomy and Atomic Physics
Tsunemi H, Mizukata K and Hiramatsu M 1988 Jap.J.Appl.Phys. vol. 27,p.670
Tuomi T, Partanen J and Simomaa K 1992 Rev.Sci.Instrum. vol.63,p.682
Twichell J C, Burke B E, Reich R K, McGonagle W H, Huang C M, Bautz M W,
   Doty J P, Ricker G R, Mountain R W and Dolat V S 1990 Rev.Sci. Instrum.
   vol.61. p.2744

Inst. Phys. Conf. Ser. No 130: Chapter 7
Paper presented at Int. Congr. X-ray Optics and Microanalysis, Manchester, 1992

523

X-ray microprobes based on Bragg–Fresnel crystal optics for high energy X-rays

V.V.Aristov[1], Yu.A.Basov[1], Ya.M.Hartman[1], C.Riekel[2], A.A.Snigirev[1]

[1] Institute of Microelectronics Technology and High Purity Materials, Russian Academy of Sciences, 142432 Chernogolovka, Moscow District, Russia.
[2] ESRF, BR 220, F-38043, Grenoble Cedex, France

ABSTRACT: One- and two-dimensional microprobes based on Bragg-Fresnel lenses (BFL) was suggested and realized for high energy X-rays. A 1D microprobe was tested using synchrotron radiation (SR) and conventional X-ray sources. Two schemes of 2D microprobes based on linear BFLs in the Kirkparick-Baez geometry were realized using seal-off X-ray tube and microfocus X-ray generator. The possibility of applying this kind of microprobe at a SR source is discussed.

1. INTRODUCTION

Traditional high energy X-ray microprobe techniques using cross grazing incidence mirrors, bent crystals and curved multilayers does not reach a sub-μ resolution limit due to a strong aberrations. Considering a grazing incidence it should be noted that a capillary optics developed by Engström et al (1990) demonstrated a sub-μ resolution, but an exact coincidence of the focus spot with an exit window of the optics element makes it difficult for practical applications. The use of a phase diffraction optics such as amplitude-phase sputtered-sliced zone plates (Bionta et al 1990) and Bragg-Fresnel optics (Aristov et al 1986,1989 and Erko et al 1990) is expected to be a perspective approach in modern microprobe techniques. Both the sputtered-sliced zone plates and the linear BFLs in a meridional setup are designed only for a fixed wavelength and should demonstrate an efficiency loss due to a strong amplitude modulation. In addition, the reach of the sub-μ resolution using BFLs in the meridional configuration is complicated due to the high aberrations. The use of linear pure phase aberration-free energy tunable crystal BFLs (Fig.1) in a sagittal setup permits to reach both the sub-μ resolution and a high efficiency (up to 40%).

2. 1D MICROPROBE

A number of diagnostics problems requires an X-ray beam collimation in an one direction and the focusing in an other one. The one linear phase BFL satisfies these requirements for both SR and conventional sources. The collimation with angle selectivity $\Delta\theta \sim 10^{-4}$ and monochromatization with $\Delta E/E = \Delta\theta ctg\ \theta_B$ (θ_B is the Bragg angle) is realized in the meridional direction and

Fig.1. Typical linear phase BFL structure

the source imaging with a demagnification coefficient corresponded to an optical scheme is realized in the sagittal direction. A minimum resolution is defined by the outermost zone width and now is as good as 0.5μm. The typical BFL parameters are presented in Table 1. Fig.2a,b shows a SR and microfocus X-ray generator sources imaged by the linear BFL. It should be mentioned here that the problem of the zero order does not exist for the pure phase Bragg-Fresnel optics. So a simplicity of the optical scheme is the main advantage of the 1D microprobe.

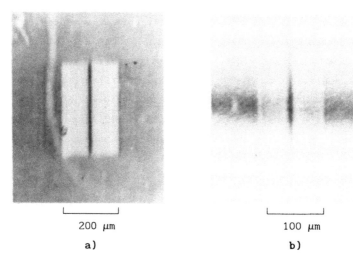

200 μm

a)

100 μm

b)

Fig 1. Linear focusing of (a) SR with E = 10 keV (b) CuK$_\alpha$ radiation.

Table 1. Typical BFL parameters.

Innermost zone radius (μm)	Outermost zone size (μm)	Aperture (μm)	Focal length λ = 1.54 Å (cm)
2.77		24	5
3.92		50	10
4.81	0.3	76	15
5.55		101	20
6.20		127	25

3. 2D MICROPROBE BASED ON TWO CROSS LINEAR BFLs.

The first experiment of the X-ray focusing by two cross linear BFLs was realized by Snigirev et al (1992) using a conventional seal-off X-ray tube. The minimum resolution of this microprobe is nearly 10 μm. The main disadvantage of this scheme is a loss of an incident beam direction. In order to keep a vertical X-ray incident beam direction a triple crystal setup was proposed. Fig. 3 shows an experimental setup of the triple crystal hard X-ray microprobe.

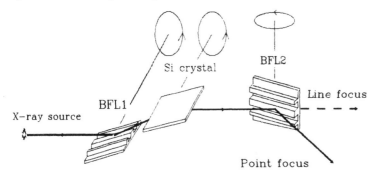

Layout parameters

BFL1 focal length – 15 cm
BFL2 focal length – 10 cm
Si(111) flat crystal
Microfocus X-ray generator with
    Cu anode

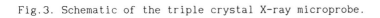

Fig.3. Schematic of the triple crystal X-ray microprobe.

The focus spot size is nearly 3×5 μm^2. As the 1D microprobe the 2D one may be applied for both SR and conventional X-ray sources. Estimated parameters for ESRF microfocus beamline are presented in Table 2. The triple crystal X-ray microprobe may by designed as a one computer-drive module used a set of the linear BFLs.

Table 2. Estimated microprobe parameters for ESRF microfocus beamline

Source	low β - undulator and Si monochromator
Flux density in the focus spot	$3.4*10^6$ phot*$(sec*\mu m^2*BW*mA)^{-1}$
Number of photons in the focus spot	$1.6*10^6$ phot*$(sec*mA)^{-1}$
Focus size for E = 8 keV	0.77×0.52 μm^2
Energy region	6-60 keV

4. SUMMARY

The scheme of 1D hard X-ray focusing was developed and successfully tested using 10 keV SR and CuK$_\alpha$ radiation of seal-off tube. The triple crystal 2D hard X-ray microprobe was realized using microfocus X-ray generator. It is evident that this optical system can be used for a high energy microprobe at SR sources. Advantages of such microprobes are: high efficiency, energy tunability, simplicity of operation, reality of sub-μ resolution. Besides the use of a crystal BFLs is expected to provide a high radiation and thermal stability of an optical system especially using a thin crystal membranes as a BFL substrate. In addition, a Bragg-Fresnel optics fabrication using usual microelectronics technology processes is rather cheaper than other types of high energy X-rays optics.

REFERENCES

Aristov V.V. et al 1986 AIP Conf. Proceed. 147 pp.253-259
Aristov V.V. et al 1989 NIM A274 pp.390-393
Bionta R.M. et al 1990 Opt. Ing. 29 6 pp.576-580
Engström P., Buttkewich A. et al 1990 NIM A302 pp.547-552
Erko A. et al 1990 X-Ray Microscopy III Third Int. Conf. Proceed.
    pp.217-219
Snigirev A.A., Bonse U., Riekel C. 1992 Rev. Sci. Inst. 63 1 pp.622-624

Inst. Phys. Conf. Ser. No 130: Chapter 7
Paper presented at Int. Congr. X-ray Optics and Microanalysis, Manchester, 1992

527

High resolution zone plates with high efficiencies for the Göttingen X-ray microscopes

J.Thieme, C. David, B. Kaulich, P. Guttmann, D. Rudolph, G.Schmahl

Forschungseinrichtung Röntgenphysik, University of Göttingen,

Geiststraße 11, 3400 Göttingen, Germany

ABSTRACT: Zone plates with high resolution and with high efficiencies at the same time are needed in x-ray microscopy for imaging with high contrast and low radiation dose applied to the object. Therefore, phase zone plates with zone widths down to 30 nm and with measured groove efficiencies up to 13% have been fabricated for the Göttingen X-ray microscopes. Structures of 30 nm size can be seen clearly in X-ray images.

1. INTRODUCTION

For imaging wet specimens in the wavelength region of $2.4 \, \text{nm} \leq \lambda \leq 4.5 \, \text{nm}$, the water window, X-radiation of $\lambda = 2.4 \, \text{nm}$ is well suited, as the transmission of water is the highest at this wavelength so that reasonable thick specimens can be investigated. Due to the same reason the wavelength of $\lambda = 0.3 \, \text{nm}$ is well suited for investigations with phase contrast X-ray microscopy.

2. FABRICATION OF ZONE PLATES

Two methods of zone plate fabrication have up to now provided zone plates for X-ray microscopy experiments, namely a holographic lithography method with visible and UV laser light and electron beam lithography. With holographic lithography, two wavefronts of laser light are superimposed to generate the zone plate pattern. The zone plate pattern is recorded into an about 45 nm thick layer of photoresist. This method allows the generation of micro zone plates and condenser zone plates with outermost zone widths down to about 55 nm (Schmahl et al. 1984, 1992). With electron beam lithography it is possible to construct zone plates with smaller outermost zone widths. Up to now, micro zone plates with outermost zone widths down to about 30 nm and zone numbers of some hundreds have been built with this technique (David et al. 1992, Anderson and Kern 1992, Charalambous and Morris 1992).

The zone plate pattern is recorded into a very thin layer of electron sensitive resist such as poly-methyl-methacralate (PMMA).

An important parameter characterizing a zone plate is the efficiency. The groove efficiency E_G of a zone plate is the ratio I/I_0 between the radiation diffracted in a specific order and the incident radiation I_0. A high efficiency can be achieved if a zone plate material with low absorption introduces a phase shift to the radiation penetrating the zones (Kirz 1974). A germanium phase zone plate has a maximum groove efficiency of $E_G = 18.7$ % at a thickness of t = 380 nm at the wavelength $\lambda = 2.4$ nm. Tantalum of 870 nm thickness gives rise to 24 % groove efficiency at $\lambda = 0.3$ nm. The absolute efficiency E_A of a zone plate is given by $E_A = E_G \cdot T$ with T the transmission of the support foil of the zone plate.

The size of a resolution element is approximately equal to the outermost zone width of a zone plate and the aquired phase shift is a function of the thickness of the material. This leads to high aspect ratios, i.e. to high ratios of the height to the width of the zone.

For optically thick zone plates, i.e., zone plates with structures of very high aspect ratios, the efficiency can be evaluated by application of coupled wave theory (Maser 1992). Such zone plates - which will have very high resolution and high efficiencies - are under construction using a sputtering slicing technique (Rudolph et al. 1980, Hilkenbach 1992).

3. PATTERN TRANSFER

Germanium has been chosen as a material for phase zone plates, as it can be structured with reactive ion etching (RIE). With the optimum thickness t = 380 nm of a zone plate made of germanium and with an outermost zone width of $dr_n = 30$ nm this leads to an aspect ratio

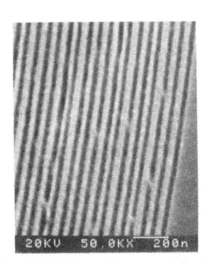

$t/dr_n = 12.7$. As the pattern transfer of structures with this aspect ratio is extremely demanding, zone plates with $dr_n = 30$ nm have been made up to now with germanium layers of up to 200-280 nm (David et al. 1992). For the fabrication of condenser zone plates the optimum height of 380 nm is used. In addition, first experiments have been performed to transfer a zone plate pattern with outermost zone widths down to 50 nm into 300 nm thick tantalum, which is a suitable material for phase zone plates at $\lambda = 0.3$ nm. To meet the requirements of transferring the pattern into the zone plate material with rectangular profiles highly selective tri-level processes have been used (Tennant et al. 1990, Thieme et al. 1992).

Fig. 1. SEM image of 30 nm zones of a micro zone plate in 200 nm thick germanium

Fig. 2. SEM picture of tantalum zone structures with zone widths of 70-90 nm.

First, the pattern is etched into a some nm thin intermediate layer, e.g. titanium or germanium, using BCl_3 or $CBrF_3$. This layer serves as a very stable mask for the etching of about 110 nm polyimide or hard baked resist with oxygen. This polymer layer then forms the mask for the subsequent structuring of the phase shifting zone plate material, e.g. germanium or tantalum, using $CBrF_3$ as etch gas. Figure 1 is an SEM image of 30 nm zones of a micro zone plate in germanium, Figure 2 shows a SEM picture of tantalum zone structures with zone widths of 70-90 nm.

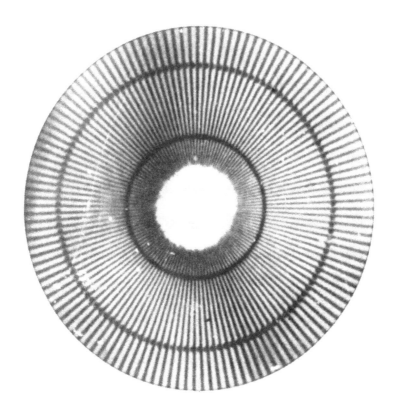

Fig. 3. X-Ray image of gold test structures, $\lambda = 2.4$ nm. Features 30 nm in size can clearly be resolved.

5. MEASUREMENTS OF EFFICIENCY AND RESOLUTION

The efficiency of the germanium phase zone plates has been measured at the electron storage ring BESSY in Berlin. A groove efficiency of up to 13 % has been achieved for zone plates with outermost zone widths of about 55 nm. For smaller zone widths the efficiency decreases as the task of structuring grows harder. The electron beam written micro zone plates for the Göttingen x-ray microscopes show groove efficiencies of up to 10 %.

The imaging performance of the micro zone plates has been investigated with the Göttingen X-ray microscope at the BESSY storage ring. Figure 2 shows an x-ray image of an electron beam written Siemens star consisting of 128 gold spokes on a thin silicon nitride membrane. The central part with features 30 nm in size is clearly resolved.

ACKNOWLEDGEMENTS

The current research is being funded by the German Federal Minister of Research and Technology (BMFT) under contract numbers 055MGDXB6 and 13N5328A0.

REFERENCES

Anderson E H and Kern D 1992 X-Ray Microscopy III ed A G Michette, G R Morrison and C J Buckley (Berlin: Springer) pp 75-8
Charalambous P S and Morris D 1992 X-Ray Microscopy III ed A G Michette, G R Morrison and C J Buckley (Berlin: Springer) pp 79-82
David C, Thieme J, Guttmann P, Schneider G, Rudolph D, Schmahl G 1992 Optik in press
Hilkenbach R 1992 X-Ray Microscopy III ed A G Michette, G R Morrison and C J Buckley (Berlin: Springer) pp 94-7
Kirz J 1974 J.Opt.Soc.Am. Vol.64 pp 301-9
Maser J 1992 X-Ray Microscopy III ed A G Michette, G R Morrison and C J Buckley (Berlin: Springer) pp 104-6
Rudolph D and Schmahl G 1980 Ann. New York Acad. of Sciences Vol 342 pp 94-104
Schmahl G, Rudolph D, Guttmann P, Christ O 1984 X-Ray Microscopy ed G Schmahl and D Rudolph (Berlin: Springer) pp 63-74
Schmahl G, Rudolph D, Niemann B, Guttmann P, Thieme J, Schneider G, David C, Diehl M, Wilhein T 1992 Optik in press
Tennant D M, Raab E L, Becker M M, O'Malley M L, Bjorkholm J E and Epworth R W 1990 J.Vac.Sci.Technol.B6(8) pp 1970-4
Thieme J, Rudolph D, Schmahl G, Guttmann P, Greinke B and Diehl M 1992 X-Ray Microscopy III ed A G Michette, G R Morrison and C J Buckley (Berlin: Springer) pp 83-6

Inst. Phys. Conf. Ser. No 130: Chapter 7
Paper presented at Int. Congr. X-ray Optics and Microanalysis, Manchester, 1992

Investigation of short period multilayers

L.L.Balakireva, A.I.Fedorenko[†], V.V.Kondratenko[†], I.A.Kopealets[†],
I.V.Kozhevnikov, A.Yu.Sipatov[†], S.A.Yulin[†], E.N.Zubarev[†], and A.V.Vinogradov

P.N.Lebedev Physics Institute, Leninsky pr. 53, Moscow, Russia
[†]*Kharkov Polytechnical Institute, Frunze street 21, Kharkov, Ukraine*

ABSTRACT: Problems of short period multilayer mirrors reflectivity and fabrication are discussed. Results of synthesis of multilayer structures with nanometer and subnanometer period are presented. The shortest period observed is 13Å for $W - Si$ sputtered multilayer and 6Å for $EuS - PbS$ epitaxial structure.

1 INTRODUCTION

One of the main goals of multilayer X-ray optics now is to fill in the gap which still exists between natural crystals having usually the period less than 5Å and good quality multilayer structures the period of which is not less than 25Å. Present state-of-art multilayer quality can be characterized by normal incidence reflectivity $R \sim 7\%$ at $\lambda \approx 44$Å which have been achieved with multilayer structure having $d \approx 22$Å (Underwood 1990; McGowan 1992). But this result is yet 6-9 times lower then theoretical predictions. The difference can be attributed to nonperfectness of substrate, layers and interfaces.

The pair of materials which can be used for multilayer fabrication should fit each other from two points of view: (a) they should have appropriate optical constants to provide high enough reflectivity in the chosen wavelength range, (b) they should be able to form very smooth, stable and reproducible interfaces in thin layers down to the thicknesses of 15Å-3Å we are speaking about.

Fig.1 shows upper theoretical limit for multilayer reflectivity at 6Å$< \lambda <$44Å for several materials.

Mirrors based on well-established spacers Si and C have rather low reflectivity due to absorption of these materials below 44Å. The use of Be as spacer increases reflectivity by a factor of 2. This material have been used for multilayer fabrication by Vinogradov et al (1987) and Utsumi et al (1988).

The elements Ca, Sc, V, Ti can also be good spacers but in certain rather limited wavelength ranges. The reflectivity in this case reaches 40-60%. The multilayers with these materials have been fabricated and tested by Akhsakhalyan et al (1987) and Nagata (1990).

The highest reflectivity in principle can be obtained with LiH and LiF spacers, which have the lowest absorption in the wavelength range considered. However, we are unaware of any attempts to use these materials for multilayer fabrication.

So one can see from Fig.1 that the reflectivity about 50% in the range 44Å- 6Å can be achieved by various materials. This is true strictly speaking only for ideal multilayers.

However, in practice for fabrication of short wavelength mirrors some other materials can be used, which have lower values of theoretical reflectivity, but which provide more smooth and stable layers and interfaces.

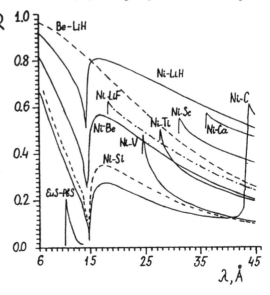

Fig.1. Calculated normal incidence multilayer reflectivity. The optimization of period and layers thickness is made according to Kozhevnikov and Vinogradov (1987).

2   SHORT PERIOD MULTILAYER SYNTHESIS

The problems of production of short period multilayers with period $d \leq 25$Å are the problems of: 1) supersmooth substrate and interface production with surface roughness of the order of atomic dimensions; 2) the layer intermixing due to interdiffusion and implantation of atoms and molecules during deposition and subsequent treatments of mirrors. Both factors, roghness and intermixing spoil the mirror.

Several $W - Si$ and $W - B_4C$ multilayers were produced by magnetron sputtering at Ar pressure $\sim 3 \cdot 10^{-3}$ Torr on commercial Si wafers and float glass plates with RMS surface roughness about 7Å.

Let us discuss the second of the above mentioned problems in connection with magnetron sputtering method.

It is known (Holloway et al 1989) that under Ar magnetron sputtering of high Z materials (W) the kinetic energy of incoming W atoms is enough for their implantation into Si layer with formation of ~ 10Å thick alloyed $W - Si$ transition zone between W and Si layers. In this case the deposition of high reflectivity X-ray mirrors with period less than ~ 10Å is impossible.

Fig. 2 illustrates this conclusion. In this figure the first order reflectivity dependence on multilayer period is given for $\lambda = 1.54$Å. At $d \simeq 26$Å the density modulation is still quite high as it was observed on multilayer cross-section electron microscopy image and electron diffraction pattern. In accordance with this fact the multilayer reflectivity keeps to be high (see Fig.2). With decreasing d the amplitude of the density modulation of $W - Si$ multilayer is going down and at $d \simeq 10$Å it disappears due to almost complete alloying of W and Si layers during magnetron sputtering deposition of mirrors. At the same time the multilayer reflectivity as it is seen from Fig.2 drops below 1%.

There are several possibilities to overcome this problem: (a) to reduce the kinetic energy of deposited particles (by using more heavy sputtering gas, for example Xe, or by using thermal and e-beam evaporation technique instead of magnetron sputtering), (b) to use as layer materials the compounds which can be evaporated without dissociation of molecules.

In Fig.2 one can see that reflectivity of $W - B_4C$ multilayers is higher in comparison with $W - Si$ ones at $\lambda = 1.54$Å as a result of replacement of Si by compound B_4C, which evaporates partially by molecular complexes.

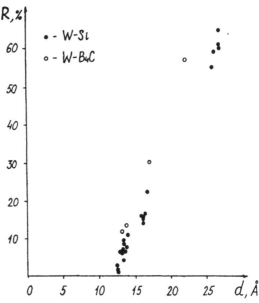

To study the problem of short-period multilayers we used thermal and e-beam evaporation. This technique gives possibility to produce epitaxial superlattices of compounds of lead and rare earth metals chalcogenides. Applying the most appropriate conditions of material evaporation and monolayer by monolayer mode of growth we tried to achieve as small superlattices periods d_{min} as possible.

It was found that lattice misfit $f = 2(a - b)/(a + b)$ (where a and b are native periods of lattice cells of two compounds epitaxially grown one on another in superlattice) is very important parameter of superlattice pair of materials for production of short period multilayers. In Fig.3 one can see

Fig.2. Measured reflectivity of sputtered $W - Si$ and $W - B_4C$ multilayers for $\lambda = 1.54$Å as a function of multilayer period. Number of periods is 200-300 for $W - Si$ multilayers and 100-180 for $W - B_4C$ multilayers. Thickness of W layers is about 0.3 of the period. Calculated reflectivity is equal 77-84% for $W - Si$ mirrors and 63-86% for $W - B_4C$ ones.

that d_{min} of superlattice is proportional to lattice misfit f of two compounds.

The reason for dependence of d_{min} on f is development of interface roughness due to inhomogeneous introduction of misfit dislocation in interface of two epitaxial layers with nonzero lattice misfit to relax elastic strain at some critical thickness of layers. Inhomogeneous introduction of misfit dislocation causes redistribution of depositing material on the surface of growing layer, which develops interface roughness and hinder the deposition of multilayer with small period (Borisova et al 1989). The more is f the more monolayers are needed to form a high quality layer which can be reproduced in a superlattice.

So to close up the gap between the periods of multilayers and natural crystals it is necessary to use two compounds with near zero lattice misfit. In this case the value of the shortest period d_{min} will be defined by surface roughness of substrate, as one can see in Fig.3 from interception of dependence of $d_{min}(f)$ with ordinate axis.

The short period superlattice to be a good mirror in addition to perfect layers and interfaces structure should be constructed of such materials the optical constants of which could provide high enough reflectivity. For example, $PbS - EuS$ superlattice gives calculated normal incidence reflectivity of 15-20% at $\lambda \simeq 11 - 12$Å. Probably other pairs of materials with better reflectivity and small misfit can be found.

3   SUMMARY

The analysis of materials for multilayer mirrors fabrication predicts the normal incidence reflectivity of 50-97% at 44Å-6Å wavelength range.

The results of testing with 1.54Å radiation show that the existing magnetron sputtering technology enables to fabricate multilayers with the periods down to 15Å.

The use of compounds as materials for multilayer fabrication results in 30-40% increase of reflectivity.

The application of epitaxial growth technology for production multilayers with shortest possible periods is studied. The linear dependence of minimum multilayer period on lattice misfit parameter is found. The shortest multilayer period achieved by epitaxial growth technology is ~6Å, and the calculated normal incidence reflectivity for $\lambda \simeq 11 - 12$Å is 15-20%.

The authors are indebted to Prof. R.V.Serov for fruitful discussions.

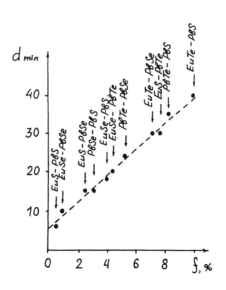

Fig.3. The minimum achieved period of epitaxial structure as a function of lattice misfit.

REFERENCES

Akhsakhalyan A D 1987 *Nucl.Instr.Meth.* **A261** 75

Borisova S S et al 1989 *Krystallography* **34** 716 (in Russian)

Holloway K et al 1989 *J.Appl.Phys.* **65** 474

Kozhevnikov I V and Vinogradov A V 1987 *Phys.Scripta* **T17** 137

McGowan B 1992 *3rd Int.Conf. on X-Ray Lasers* (Schliersee)

Nagata M 1990 *Jap.J.Appl.Phys.* **29** 1215

Utsumi Y et al 1988 *Appl.Opt.* **27** 3933

Vinogradov A V et al 1987 *Sov.Tech.Phys.Lett.* **13** 53

New doubly curved diffractor geometries and their use in microanalysis

D. B. Wittry

Department of Materials Science and Engineering
 and Department of Electrical Engineering
University of Southern California, Los Angeles, CA USA

ABSTRACT: X-ray diffractor geometries that have been used for parallel detection of x-ray spectra are reviewed, a basis for comparing spectrometers having parallel detection with scanning monochromators is suggested, and new spectrometers based on oriented polycrystalline films deposited on doubly curved substrates are discussed.

1. PARALLEL DETECTION BY FLAT OR SINGLY CURVED DIFFRACTORS

In some respects, parallel detection of x-ray spectra by a wavelength dispersive spectrometry is simpler than detection of x-ray spectra by a high efficiency scanning monochromator. For the latter case, Bragg's law and $\Theta_i = \Theta_r = \Theta_B$ must be satisfied over a significant portion of the diffractor's surface. This requires that the diffracting planes are not at constant angle relative to the surface (Johansson 1933, Wittry and Sun 1992), the d spacing vary over the surface of the diffractor (Carr and Romig 1990), or the diffractor have doubly curved surface steps (Wittry 1992).

In contrast, parallel detection of x-ray spectra can employ diffractors having the surface parallel to the diffracting planes because various parts of the diffractor diffract radiation of different wavelength. In this case, high collection efficiency for each wavelength is sacrificed in order to get a large wavelength coverage. The effect of this trade-off is discussed in a subsequent section of this paper.

The geometries that have been previously described are indicated in Fig. 1. In these figures S is the source, D is the diffractor, and R is the recording plane. Von Hamos (1934) used a cylindrically curved crystal whose axis passed through the source as shown in Fig. 1a. Schnopper and Kalata (1969) described the use of a cylindrically curved diffractor in a geometry similar to the Johann geometry but with the recording plane located between the diffractor and the usual image point as shown in Fig. 1b. Birks (1970) used a convex cylindrical diffractor as shown in Fig. 1c. Henke et al (1983) described spectrometers based on elliptically curved cylinders as shown in Fig. 1d. Zaluzec and Strauss (1989) described the possible use of flat crystals as shown in Fig. 1e. Fiori et al (1991) discussed the use of a Johansson crystal with the source and detecting plane close to the diffractor than the usual

*This work was supported by NIH under grant no. 1R01GM47303-1

Johansson geometry as shown in Fig. 1f.

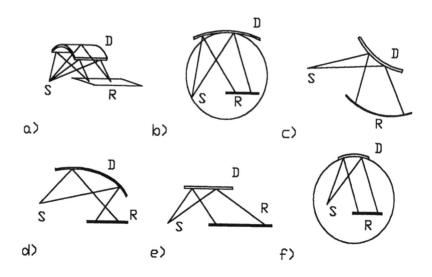

Fig. 1. Parallel detection spectrometer geometries

Except for Henke's work, all of the geometries shown in Fig. 1 suffer from the lack of a crossover of the x-rays between the diffractor and the recording plane. This means that excess background due to scattering and fluorescence of the diffractor cannot be eliminated. However, if doubly-curved diffractors are used, it is easy to overcome this limitation as will be shown. Moreover, the use of double curvature and suitable location of the recording plane makes it possible to control the size of the image in the direction perpendicular to the dispersion. This is desirable for use with some position-sensitive detector arrays that may saturate when the intensities are too high.

2. COMPARISON OF PARALLEL DETECTION SPECTROMETERS AND SCANNING MONOCHROMATORS

In addition to the need for a crossover of the x-rays, an x-ray spectrometer for parallel detection must have adequate wavelength resolution to provide for the following: a) unambiguous identification of characteristic x-ray lines for quantitative analysis, b) accurate determination of the intensity of the peak of the lines in order to perform quantitative analysis, and c) high signal/background ratio in order to obtain low detection limits. The wavelength resolution is determined by the diffracting material's rocking curve width, the distance from diffractor to the recording plane, and the pixel size of the position-sensitive detector. Suitable values of these contributions to the spectral line width can be obtained so that the resolution of the parallel detection spectrometer is adequate for microanalysis, although it may be less than the resolution typically obtained with scanning monochromators.

The resolution required may be evaluated by the following considerations:

For most elements, the natural width of the unresolved $K\alpha$ doublet corresponds to $\Delta\lambda/\lambda = 10^{-3}$. Assuming that 4 data points are required over each peak in order to determine its value accurately, the resolution required is about $(4000)^{-1}$ and the recording system must have about 4000 pixels in the direction of dispersion. However, for light elements, $\Delta\lambda/\lambda = 10^{-2}$ and about 400 pixels would be adequate.

Assuming adequate spectral resolution, a parallel detection spectrometer can be compared with a scanning monochromator by considering the relative collection solid angle and the efficiency of the diffractor. Let w_p and w_s be the solid angles for collection of a given spectral line for the parallel detection spectrometer and the scanning monochromator, respectively, and e_p and e_s be the corresponding diffractor efficiencies. The two systems will give comparable detection limits for a given measurement time if:

$$\frac{1}{e_p w_p} = \frac{2N(1+f)}{e_s w_s} \tag{1}$$

where N is the number of elements to be determined, the scanning monochromator is used for two measurements of equal time for signal and background determination, and the slewing time between characteristic lines is assumed to be equal to a fraction f of the measurement time. Thus, if $e_p = e_s$, the parallel detection system is superior only if $w_p > w_s/[2N(1 + f)]$.

In practical cases, w_p may be of the order of $10^{-2} w_s$ or less, except for long wavelengths and diffractor materials having a large rocking curve width. Thus, unless it is necessary to measure the concentration of a large number of elements or only light elements are to be studied, the parallel detection spectrometer will not be competitive in performance with a scanning monochromator driven by computer control. Its principle advantage, therefore, is the lack of moving parts and the possibility of smaller physical size. Both of these characteristics make it attractive for incorporation into transmission electron microscopes.

3. A SIMPLE PARALLEL DETECTION SPECTROMETER WITH DOUBLY CURVED DIFFRACTOR

Possible geometries for doubly curved diffractors include a sphere, toroid, ellipsoid of revolution, and other geometries obtained by rotating a conic section about a line passing though the source. Since the diffracting planes can be parallel to the surface, these doubly curved diffractors are particularly suited to fabrication by the use of oriented polycrystalline films (OPF) (Wittry et al 1992). These films can provide smaller d spacing than LSM and hence are useful for the x-ray wavelengths from 0.7-10 A as are commonly used for microanalysis with characteristic x-rays for most of the elements in the periodic table.

As an example of a simple parallel detection spectrometer, we may consider a spherical surface on which an oriented polycrystalline film has been deposited. This simple geometry has the advantage that a spherical surface is easily obtained by machining or by optical polishing. As shown in Fig. 2 the x-ray source is located on a line passing through the center of the sphere. By symmetry, the diffracted rays will have a crossover along this line and an exit slit placed at the crossover can reduce the background due to scattering and the fluorescence of the diffractor. The detector is located a distance d below the other side of the crossover where d is selected to obtain an appropriate length of the curved focal lines for the use of a CCD-based detector as proposed by Zaluzec and Strauss (1989).

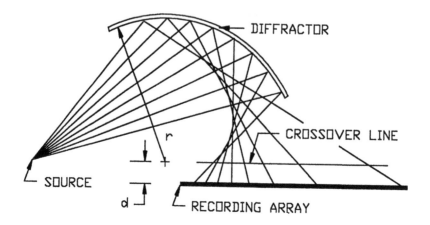

Fig. 2. Parallel detection spectrometer based on OPF

For the geometry described, the source may be located at various distances from the center curvature of the diffractor. In Fig. 2 the source is located at a distance r from the center of the sphere where r is the radius of the sphere. Larger values of this distance facilitate obtaining smaller Bragg angles and a larger wavelength range for a detector array of given size.

REFERENCES

Birks L S 1970 Rev. Sci. Instrum. 41 1129
Carr M J and Romig A D Jr 1990 US Patent No. 4,916,721
Fiori C E, Wright S A and Romig A D 1991 Microbeam Analysis - 1991 ed D G Howitt (San Francisco Press, San Francisco) pp 327-87
Hamos L von 1934 Ann. Phys. 5 252
Henke B L, Yamada H T and Tanaka T J 1983 Rev. Sci. Instrum. 54 1311
Johansson T 1933 Z. Physik 82 507
Schnopper H W and Kalata K 1969 US Patent No. 3,628,040
Wittry D B 1992 US Patent No. 5,127,028
Wittry D B and Sun S 1992 J. Appl. Phys. 71 1
Wittry D B, Chang W Z, and Barbee T W 1992 EMSA/MAS National Conference, Boston, MA
Zaluzec N J and Strauss M G 1989 Ultramicroscopy 28 131

Inst. Phys. Conf. Ser. No 130: Chapter 7
Paper presented at Int. Congr. X-ray Optics and Microanalysis, Manchester, 1992

539

High resolution zone plates for X-ray microscopy

D.Morris[1], M.Gentili[2], M.Baciocchi[2], S.Contarini[3], P.DeGasperis[2], C.Gariazzo[3], R.Maggiora[2], P.Melpignano[1], N.Minnaja[3], P.Nataletti[3], R.Rosei[1]

[1] Sincrotrone Trieste, Padriciano 99, 34012 Trieste, Italy
[2] Istituto Elettronica dello Stato Solido (IESS), Via Cineto Romano 42, 00156 Rome, Italy
[3] Eniricerche, Via Ramarini 32, 00016 Monterotondo, Italy

ABSTRACT: Gold zone plates have been fabricated by electron beam lithography at IESS in Rome. Finest line widths below 100 nm have been achieved, and tests of the zone plates on the x-ray microscopy beamline at the Brookhaven National Laboratory, New York, showed a first order resolution of $\sim$120 nm and a second order resolution below 70 nm. The measured second order resolution is smaller than the finest line width in the zone plate, illustrating that the resolution achievable with a zone plate is not necessarily limited by the finest line width that can be drawn.

1. INTRODUCTION

Zone plates are circular diffraction gratings with radially increasing line densities. They currently provide the highest spatial resolution of any optics for focusing soft x-rays, with measured resolutions below 50 nm (Anderson 1992). Electron beam lithography is now established as the best method for the fabrication of high resolution zone plates, and a number of groups worldwide have active research programs in this area (Charalambous 1992, Ingwerson 1992, Tennant 1992). The zone plates produced are used for a wide variety of x-ray microprobe experiments, mainly on synchrotron light sources (Ade 1990, Morris 1991a, Nataletti 1992, Niemann 1992).

Although they offer the advantage of high resolution, zone plates are achromatic and have relatively small focal lengths, depths of focus and numerical apertures. Typical zone plates in use at present have diameters of the order of 100 μm and outer zone widths of the order of 100 nm (N_{zones} = 250). For the first diffraction order, an illuminating wavelength of 4 nm yields focal lengths of the order of 2 mm, depths of focus of the order of 10 μm and numerical apertures of the order of 0.02.

Diffraction orders other than the first offer the advantage of higher spatial resolution (resolution $\propto 1/m$, where m is the diffraction order) but are seldom used, since in addition to the reduced focal length and depth of focus ($\propto 1/m$) there is also a reduced diffraction efficiency. The diffraction efficiency for an ideal zone plate is inversely proportional to the square of the diffraction order. For example, the maximum achievable diffraction efficiency for an amplitude zone plate in the first order is approximately 10%, but for the second order it is only 2.5%.

2. THE ZONE PLATE FABRICATION PROCESS

Zone plates are fabricated at IESS in Rome by electron beam lithography. The zone plate pattern is drawn in a thin layer of PMMA photoresist, then the developed pattern is electroplated. The machine used for the fabrication is a Leika Cambridge EBMF10 CS/120. The main machine parameters of the EBMF machine when it is set up for drawing zone plates are an accelerating voltage of 40 kV, a beam current of 0.25 nA and an electron beam spot size of 55 nm.

The substrates used for the zone plates are silicon nitride membranes, supplied by the physics department of King's College London. The membranes are 250 μm square and 100 nm thick. On top of the silicon nitride is deposited a 5 nm adhesion layer of chromium and a 5 nm plating base of gold. A 200 nm layer of PMMA is then spun on top, and the zone plate pattern is drawn by the EBMF machine. A zone plate with a diameter of 100 μm and an outermost zone width of 100 nm takes approximately 2.5 minutes to draw. Once the pattern has been drawn, the PMMA is developed for 2 minutes at room temperature in a mixture of 1 part methyl-isobutyl ketone to 3 parts isopropylic alcohol. The developed resist is then baked in an oven at 90°C for 30 minutes to ensure solvent evaporation and to stabilize the resist image. The specimen is then placed in an electroplating bath for 65 seconds, at an electroplating current of 2.5 mA.cm^{-2}, to electroplate the pattern in gold to a height of approximately 120 nm. The remaining PMMA is then removed with a suitable etch.

The finest line/space periods that can be drawn with this machine in this configuration are limited by forward-scattering of the electrons in the resist layer and by the size of the electron beam itself. With a 1:1 line-to-space ratio, the smallest pitch that has so far been recorded is approximately 160 nm, giving an outermost zone width of 80 nm. Figure 1 shows electron micrographs of one of the gold zone plates fabricated by IESS.

 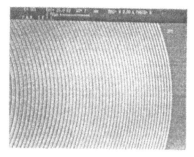

Figure 1. Electron micrographs showing details of an IESS gold zone plate.

3. TEST RESULTS

A number of zone plates fabricated by IESS were tested for 1st and 2nd order resolution and diffraction efficiency on the X1 x-ray microscopy beamline at the NSLS, Brookhaven National

Laboratory, New York. The testing of a typical zone plate is reported here. The parameters of the zone plate are shown in table 1. The wavelength of the illuminating radiation for these tests was 3.7 nm.

Diameter	104 μm
Outermost Zone Width	95 nm
Zone Material	Gold
Zone Height	~120 nm
Apodization Diameter	52 μm
Apodization Thickness	600 nm
Line-to-Space Ratio	Varying 1:1–3:1

Table 1. Parameters of one of the IESS zone plates.

The diffraction efficiencies that were measured were consistent with the theoretical values, taking into consideration the varying line-to-space ratio across the zone plate (Morris 1991b). The first order diffraction efficiency was in the range 3–4%, whilst the second order diffraction efficiency was approximately 1%. Previous efficiency tests with other IESS gold zone plates on the Synchrotron Radiation Source at the Daresbury Laboratory had shown first order diffraction efficiencies over 9%, and negligible second order diffraction efficiency. This implies a very good line-to-space ratio and well-defined zone profiles.

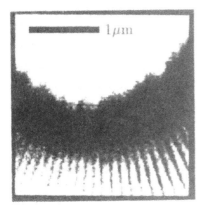
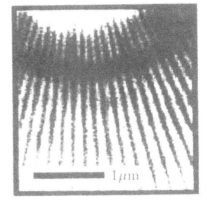

Figure 2. (a) An image of a gold test pattern using the first order focus of the zone plate and (b) an image of the same test pattern using the second order focus.

The first and second order resolutions of the zone plate were tested by using the zone plate to image a well-characterized test specimen. The resulting images (unprocessed) are shown in figure 2. Figure 2(a) shows the test pattern imaged with the first diffraction order of the zone plate. Line/space periods of approximately 120 nm are visible, which is consistent with the expected value. Figure 2(b) shows the same test pattern imaged with the same zone plate, but this time using the second diffraction order. In this image line/space periods below 70 nm are resolved.

From §1 it might seem that higher resolutions are automatically achieved when imaging with higher diffraction orders. This is not the case, however, since the resolution that is achieved is also dependent upon the accuracy to which the zones are positioned across the zone plate. Non-linearities, drifts or non-orthogonalities present during the zone plate fabrication can cause the experimentally measured resolutions to be far worse than those predicted by theory. Successful higher order imaging places proportionally stricter tolerances on the pattern placement accuracy required, and in some systems this higher accuracy cannot be achieved.

4. CONCLUSIONS

The results shown in section 3 illustrate the way in which higher order imaging with zone plates can be used to achieve spatial resolutions below the feature-size limitations imposed by the fabrication process. The main disadvantage of using higher-order imaging is the reduced diffraction efficiency, which leads to a reduced flux in the focal spot. This is not necessarily a limitation for some applications on second and third generation synchrotron sources, when data acquisition times can be limited by other experimental parameters, e.g. the speed of the electronics or the scanning stages.

5. ACKNOWLEDGEMENTS

The authors wish to thank Janos Kirz and Harald Ade for their help in performing the tests at Brookhaven, and Pambos Charalambous and Mike Browne for carrying out the efficiency tests at Daresbury.

REFERENCES

Ade H, Kirz J, Hulbert S, Johnson E, Anderson E, Kern D 1990 Nucl. Instrum. Meth. Phys. Res. A291 pp 126-131
Anderson E and Kern D 1992 X-ray Microscopy III (Springer-Verlag: Berlin) pp 75-78
Charalambous P and Morris D 1992 X-ray Microscopy III (Springer-Verlag: Berlin) pp 79-82
Ingwerson J 1992 X-ray Microscopy III (Springer-Verlag: Berlin) pp 98-100
Morris D, Buckley C J, Morrison G R, Michette A G, Anastasi P A F, Browne M T, Burge R E, Charalambous P S, Foster G F, Palmer J R and Duke P J 1991a Scanning 13 pp 7-10
Morris D 1991b PhD thesis (University of London)
Nataletti P, Contarini S, Gariazzo C, Minnaja N, Musicanti M, Jark W, Kiskinova M, Melpignano P, Morris D and Rosei R 1992 to be published in Surface and Interface Analysis
Niemann B 1992 X-ray Microscopy III (Springer-Verlag: Berlin) pp 143-146
Tennant D M, Raab E L, Becker M M, O'Malley M L, Bjorkholm J E and Epworth R W 1992 to be published in J. Vac. Sci. Tech. B

Inst. Phys. Conf. Ser. No 130: Chapter 7
Paper presented at Int. Congr. X-ray Optics and Microanalysis, Manchester, 1992

543

The computer-aided design of the hard X-ray diffraction lenses

V Aristov, S Kuznetsov, A Snigirev

Institute of Microelectronics Technology and High Purity Materials, Russian Academy of Sciences, Chernogolovka, Moscow District 142432, Russia

ABSTRACT: This paper is devoted to the computer-aided design of the hard X-ray diffraction lenses based on the crystals with modulated lattice deformation which are able to transform an incident radia- on into a predetermined wavefront. The main procedure is the solu- tion of the inverse problem of determination of the two-dimensional strain distribution in the crystal from the known dependence of the amplitudes of the incident and diffracted waves on the surface of the crystal upon the angular deviation from the Bragg angle. The examples of the computer-aided design of such lenses are presented.

1. INTRODUCTION

The development of the new types of the hard X-ray diffraction optical elements requires the numerical solution of a number of inverse problems of X-ray crystal optics. One of these problems is the computer-aided de- sign of the hard X-ray optical diffraction lenses based on the crystals with modulated lattice deformation which are able to transform an inci- dent radiation into a predetermined wavefront or to provide a required distribution of the intensity at a fixed point. The common way to create hard X-ray lenses like in the case of light optics is unattainable due to the fact that X-ray radiation ($\lambda < 1$ nm) is not in practice refracted by materials ($1-n \sim 10^{-5}-10^{-6}$). Otherwise a crystal lattice at the Bragg position can be used as an effective reflection element for X-ray beams. Recently the possibilities of the Bragg-Fresnel zone structures were de- monstrated by Aristov et al (1986) and X-ray diffraction elements on the base of heterostructures were proposed by Aristov and Shulakov (1988). The present work is devoted to the computer-aided design of the hard

X-ray diffraction lenses on the base of crystals with artificially modulated two-dimensional lattice deformation.

2. THEORY

The X-ray wave diffracted in the crystal in the case of Bragg reflection can be calculated on the base of the system of Takagi equations (Afanas'ev and Kohn 1971)

$$\partial E_0/\partial s_0 = iK\chi_0 E_0/2 + iK\chi_1/2\exp(iHu(r))E_1 \tag{1}$$

$$\partial E_1/\partial s_1 = iK\chi_1/2\exp(-iHu(r))E_0 + iK/2(\chi_0-\alpha)E_1 \tag{2}$$

with boundary conditions (Epelboyn and Riglet 1979), where E_0, E_1 – scalar complex amplitudes of the incident and diffracted waves, $K=2\pi/\lambda$ – absolute value of the wave vector, λ – wave length, χ_1, χ_0 – polarizability coefficients, H – reciprocal-lattice vector, $u(r)$ – deformation vector, α – deviation parameter, $\alpha=2\Delta\vartheta\sin2\vartheta_b$.

Let us divide the region in the crystal by lines parallel to the directions s_0 and s_1. We calculate the function $u(r)$ on the obtained grid in the kinematical approximation that neglects multiple scattering in the crystal. Taking into account the boundary conditions at $t=0$ and $t=L(s_0)$ one can obtain from (2) the next Fredholm integral equation of the 1-st kind for unknown function $f(t)$ (Aristov et al 1992):

$$(iK\chi_1/2)\int_0^L f(t)E_0(t)\exp[iK(\alpha-\chi_0)t/2]dt = E_1(L,\alpha)\exp[iK(\alpha-\chi_0)L/2] \tag{3}$$

where t is a variable along the direction s_1, $L(s_0)$ is the length of integration along the direction s_1, $f(t)=\exp(-iHu(t))$. By solving integral equation (3) on each line parallel to s_1 on the grid we obtain a two-dimensional strain profile in the crystal. Therefore the solution of this inverse problem allows unknown function $u(r)$ to be found from the dependence of the amplitudes $E(\alpha)$ of the incident and diffracted waves on the surface of the crystal upon the angular deviation α from the Bragg angle ϑ_b.

3. NUMERICAL EXAMPLES

The proposed method was used to calculate the displacements in a silicon crystal in the case of 111 CuK$_\alpha$ reflection and plane incident wave. All following figures correspond to a diffraction lens with 40 cm focusing distance and 250 μm aperture. Fig.1 is the contour map of the distribution of the lattice deformation Δd/d in the crystal, the corresponding displacement of the surface from the position in the perfect crystal is shown in fig.2 (dashed line). It is clear that such comparatively large displacements can't be obtained easily, to avoid these difficulties one can use piecewise approximation

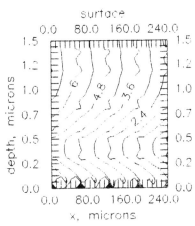

Fig.1   Contour map of the lattice deformations Δd/d×10^{-4}

(that resembles kinoform structure) as shown in fig.2 (solid line). In any case it is necessary to check the obtained strain profile by numerical simulation. A direct calculation of the intensity distribution in the focal plane has been carried out on the base of the system of equations (1) and (2) with the substitution of the obtained profile u(r). The calculated intensity in the case of completely irradiated aperture 250 μm is shown in fig.3 (dot line). The shape of the distribution of the intensity in the focus spot is due to the inadequate description of the X-ray diffraction in the vicinity of end points of the triangular mesh in numerical calculations. To overcome this shortcoming one can decrease the irradiated surface (or in other words to use the diaphragm). The smallest size of the focus spot 1.96 μm has been obtained in the case of 75 μm aperture and the corresponding intensity distribution is shown in fig.3 (solid line). For comparison the intensity distribution in the focal plane was calculated for the case of the ideal distribution of the spherical wave amplitude

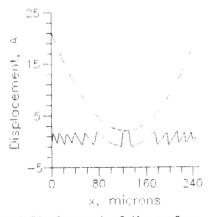

Fig.2 Displacement of the surface

on the exit surface of the lens and
is shown in fig.3 (dashed line).
Here the size of the focus spot is
about 1.87 μm at the 75 μm aperture.
A good correlation between calculated and ideal intensity distributions means that the proposed algorithm provides precise solution of
the inverse problem and rather high
accuracy of the deformation profile
calculation.

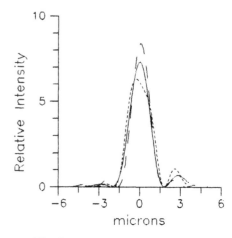

Fig.3   Intensity distribution
in the focal plane

3. CONCLUDING REMARKS

Numerical calculations have shown
that the maximum value of the lattice deformations $\Delta d/d$ is about 6×10^{-4}
for these lenses. Therefore the desired profile of deformations in the
crystal can be obtained by various methods such as ion implantation or
epitaxial growing. On the other hand it should be necessary to control
the obtained two-dimensional profile of deformations in the lenses after
fabrication. One of the ways to solve this problem is to use the triple-
crystal diffraction (TCD) technique and the numerical analysis of TCD
spectra as suggested by Aristov et al (1992). Thus the proposed method
can be implemented in the CAD systems for the hard X-ray diffraction
lenses fabrication.

REFERENCES

Afanas'ev A M and Kohn V G 1971 Acta Cryst. A27 pp 421-30
Aristov V V, Snigirev A A, Basov Yu A, Nikulin A Yu 1986 AIP Conf.
   Proc. 147 (New York: AIP) pp 253-59
Aristov V V and Shulakov E V 1988 Opt. Commun. 65 6
Aristov V V, Kuznetsov S M, Nikulin A Yu and Snigirev A A 1992
   Semicond. Sci. Technol. 7 pp A168-70
Epelboyn Y and Riglet P 1979 Phys. Stat. Sol. A54 pp 547-556

Inst. Phys. Conf. Ser. No 130: Chapter 7
Paper presented at Int. Congr. X-ray Optics and Microanalysis, Manchester, 1992

547

X-ray holography at the National Synchrotron Light Source

Chris Jacobsen, and Steve Lindaas

Physics Department, SUNY, Stony Brook, NY 11794, USA

Malcolm Howells

Advanced Light Source, Lawrence Berkeley Laboratory, Berkeley, CA 94720, USA

Ian McNulty

Advanced Photon Source, Argonne National Laboratory Argonne, IL 60439, USA

ABSTRACT: Using a soft x-ray undulator at the National Synchrotron Light Source at Brookhaven National Laboratory, we are able to record high resolution holograms in minutes at 1.5–4.5 nm wavelengths. Synchrotrons now under development will both reduce the required exposure time and expand the available wavelength range, and x-ray lasers and free-electron lasers are under development which will offer the capability to record x-ray holograms with sub-nsec exposure times, thus capturing images on a timescale shorter than that of radiation damage.

1. INTRODUCTION

The possibilty of using holography for x-ray imaging was first proposed by Baez (1952) only a few years after Gabor's original proposal (1948), and the first experiments were carried out soon afterwards [the early history of x-ray holography is described in Aoki (1974) and Jacobsen (1990a)]. Until about five years ago, a lack of available coherent flux limited experimental demonstrations of x-ray holography to resolutions greater than 1 μm. More recently, the availability of undulators at low emittance storage rings have led to experimental demonstrations at < 100 nm resolution (Jacobsen 1990b and McNulty 1992a) and efforts aimed at high resolution, high throughput optical reconstruction (Joyeux 1992). Several groups are developing different approaches to x-ray holography, both with synchrotron sources and with x-ray lasers (Trebes 1992).

These various efforts are aimed at developing methods for high resolution imaging of thick and, ultimately, wet and initially living biological specimens using x-rays within (Wolter 1952, Sayre 1977) or just outside (London 1989, Schmahl 1987) the 2.3–4.4 nm "water window" spectral region. While other types of x-ray microscopes are farther along the way towards that goal, holography offers unique capabilities which may be advantageous in certain cases: the ability to obtain amplitude *and* phase contrast information in a single exposure, reduced requirements on sample prealignment in many cases, and (perhaps most importantly) straightforward scaling to flash exposures such as may be obtained using x-ray lasers. The absence of three-dimensional imaging from this list should be made explicit; high quality 3D images require multiple viewing angles (Devaney 1989).

2. COHERENT X-RAYS AND EXPOSURE TIMES

Optical holography was greatly advanced by the development of the laser; similarly, x-ray holography is benefitting from the development of high-coherent-flux sources. While the coherence requirements for different holographic geometries vary considerably, most work thus far has required a radius of good mutual coherence of ~ 50 μm and a temporal coherence requirement of $\lambda/\Delta\lambda \sim 200\text{--}500$. These coherence requirements are in fact identical to those of Fresnel zone plate optics of similar resolution (indeed, zone plates can be thought of as holographic optical elements). If a quasimonochromatic but spatially incoherent x-ray source is used to form a hologram, the Van Cittert-Zernicke theorem (Born 1980) leads to the conclusion that the accepted illumination size-angle product (or phase space area) must be restricted to about λ in each dimension in order to have an adequate degree of mutual coherence at the hologram. For example, with an x-ray source diameter of 100 μm (such as at the exit slit of a synchrotron beamline monochromator), a $\lambda = 3$ nm hologram recording should be done at a distance of ~ 2 m from the exit slit if good mutual coherence is required over a transverse distance of 50 μm. We therefore conclude that for synchrotron sources, one must take the spectral brightness of the source (usually stated with 0.1% BW or $\lambda/\Delta\lambda = 1000$) and multiply it by λ^2 to obtain an estimate of the coherent flux. Such an estimate is provided in Fig. 1.

Fig. 1. Coherent flux (defined as flux into a 2D phase space area of λ^2 and $\lambda/\Delta\lambda = 1000$) from several synchrotron sources. "NSLS x-ray bend" refers to a bending magnet and "NSLS X-1 undulator" refers to a soft x-ray undulator on the X-ray Ring of the National Synchrotron Light Source at Brookhaven Laboratory, "ALS U3.9 undulator" refers to an undulator planned for the Advanced Light Source at Lawrence Berkeley Laboratory, and "APS undulator A" refers to an undulator planned for the Advanced Photon Source at Argonne National Laboratory. An undulator source similar to NSLS X-1 is available in Japan, and sources similar to the ALS and APS are under construction at several sites worldwide.

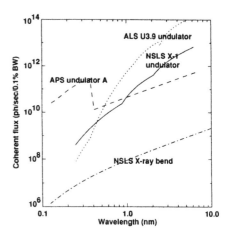

In recording a high resolution hologram, one must illuminate each resolution element with $\sim 10^3$ or more photons (depending, of course, on sample contrast, recording geometry, and so on). If one wishes to obtain 50 nm resolution in a 50 μm wide coherent field, an exposure of at least 10^9 coherent photons is required. In practice, this number must be increased to compensate for the quantum efficiency of the holographic recording medium ($\sim 10\%$ for photoresists, slightly better for CCDs), absorption in sample support windows and overlying water layers ($\sim 3\text{--}10\times$), and efficiencies of optics such as a grating monochromator (typically 10%). From this estimate of a source coherent flux of $> 10^{12}$ photons, one might expect to be able to record a hologram in a matter of seconds using the NSLS X-1 undulator; in practice, Gabor holograms have been recorded in ~ 1 minute, and a 5× speedup is expected with vertical refocussing beamline optics (Zhang 1992). Scaling to higher resolution δ increases the exposure time as δ^4 for thin samples and as δ^6 for thick samples (Howells 1988, Jacobsen 1990b).

From these considerations, one can contemplate high resolution hologram recording times in the 100 msec range using advanced synchrotron sources. For wet biological specimens, one will need to pay careful attention to make sure that heat conduction is sufficient to limit temperature increases to acceptable levels (Greinke 1992), but a more crucial question is raised by the problem of radiation damage. While studies of soft x-ray damage on sample morphology are only beginning, Williams *et al.* (1992) have found that fixed, hydrated chromosomes from the beam plant *V. faba* undergo mass loss and visible shrinkage following exposures of order 10^6 Gy such as might be expected in < 50 nm resolution holograms, and that the timescale for such damage may be in the 0.01–10 second range. Reducing the hologram exposure time to minimize the detrimental effects of radiolytical damage may thus become crucial. (Slower exposure times are compatible with the study of dry or frozen hydrated specimens). Another possibility for holographic microscopy of wet samples is to use < 1 nsec exposures for flash imaging on a timescale shorter than that of thermal blurring (Solem 1986, London 1989). X-ray sources for flash holography are under development, both based on plasma-based x-ray lasers (Tallents 1991) and free-electron lasers (Cornacchia and Winick 1992). The challenges involved in obtaining sufficient coherent flux for flash imaging are formidable, but the potential payoff is high.

3. GABOR HOLOGRAPHY

Gabor or in-line holography is one approach to holographic microscopy. In Gabor holography, a plane or weakly spherical coherent wave serves both as the reference wave and as the sample illuminating wave (see Fig. 2). The sample field must be largely empty or of low contrast. The hologram records the interference pattern between the reference wave and the complex wavefield emerging from the specimen. For high resolution features which diffract the wavefield to high angles, the increasing angular separation between the sample and reference wave leads to high fringe frequencies on the hologram. Recognizing that a zone plate is like a Gabor hologram of a point object, Baez (1952) pointed out that the resolution of the reconstructed image in Gabor holography can be no better than the resolution of the holographic recording. For this reason, all recent work in Gabor x-ray holography has made use of photoresist detectors. The resist poly(methylmethacrylate) or PMMA is thought to have an intrinsic resolution of 5–10 nm and a detective quantum efficiency of $\sim 10\%$ (Spiller 1977, Shinozaki 1988), and other resists offer higher sensitivity with perhaps lower intrinsic resolution. We are also exploring the use of alkali halide films as ultrahigh resolution resists.

Two approaches have been used in the reconstruction of Gabor x-ray holograms recorded on photoresist. Joyeux and Polack *et al.* (1989, 1992) have concentrated on developing optical reconstruction schemes for high throughput and ease of use, with a resolution goal of 0.05–0.1 μm. They are developing a system whereby the developed photoresist is aluminum-coated so as to become reflective, and a UV laser and specially designed UV optic is used to reconstruct one point of the image at a time in a scanning system. Our efforts have been centered on reaching the resolution limit of the photoresist, and so we have taken the approach of high resolution hologram digitization and numerical reconstruction. We have previously used a transmission electron microscope (TEM) to magnify holograms recorded on thin resist layers on 100 nm Si_3N_4 substrates (Jacobsen 1990b). In this way, we have obtained 60 nm resolution images of air-dried rat pancreatic zymogen granules (see Fig. 3), and power spectral analysis of the TEM-magnified holograms suggests that 16 nm information is present in the holograms. This approach suffers from the fact that the hologram must be metal coated prior to TEM examination, and the TEM exposure damages the photoresist. More importantly, the commercial TEMs we have used suffer from spiral distortions when used at the low (1000–2000×) magnifications we require for readout of a hologram coherence area, and these distortions may in fact be the limiting factor in the resolution obtained thus far (Jacobsen 1991).

Gabor   Fourier transform

Fig. 2. Two geometries used for x-ray holography. In Gabor holography (left), the object is placed in the illumination path, and the hologram records the interference between the wavefield emerging from the specimen and the plane or weakly spherical coherent reference wave. In Fourier transform holography (right), the hologram records the interference between the wavefield emerging from the specimen and a spherical reference wave generated nearby. The spherical reference wave can be generated by a focussing optic (a zone plate is shown here) or by a point scatterer.

More recently, we have begun construction of a scanning force microscope for hologram readout (Lindaas 1992). The system is being designed to produce linear field readouts at 5 nm resolution, and thus should provide a nearly nondestructive and easy-to-use method for digitizing holograms at the photoresist resolution limit. Because the hologram will no longer need to be largely transparent to 100 keV electrons, we can work with rigid substrates (such as glass) and the somewhat thicker resist layers (with increased exposure lattitude) we believe are needed to reliably record high resolution holograms of wet biological specimens (Jacobsen 1991). Once the hologram is digitized, one can also use a variety of iterative algorithms (Liu 1987, Jacobsen 1992a, Koren 1992) to reduce the effects of twin image noise in Gabor hologram reconstructions.

Both the optical and numerical reconstruction approaches to Gabor holography have in common the feature that the sample-to-hologram working distance can be varied over a large range with little consequence. Because the image is focussed in the reconstruction stage, Gabor holography offers easy sample prealignment. In addition, if a spatially incoherent source is used, the resolution of a Gabor hologram is limited by the radius of good mutual coherence (the foot print of one spatially coherent "mode" on the hologram), but the field of view can be as large as the detector area (1 mm is convenient).

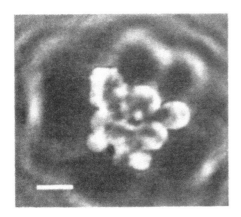 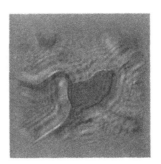

Fig. 3. Images from Gabor x-ray holograms recorded on PMMA, enlarged using a transmission electron microscope, and reconstructed numerically. (Left) Rat pancreatic zymogen granules recorded using 2.5 nm x-rays. The appearance of non-uniform structure in granules suspended in a sucrose-containing buffer (in this case, prior to fixation and air-drying) is being studied by S. Rothman of UC San Francisco as evidence towards a picture of protein transport across the granule membrane. (Right) Hydrated, fixed hippocampal cell provided by J. Pine of CalTech. The hologram was developed to show good fringes in the area outside the cell, so the cell boundaries and processes reconstructed well. However, because a thin resist layer is required for TEM readout, the additional development needed to bring out detail within the cell body would have developed away the more strongly exposed fringes contributing to this reconstruction.

4. FOURIER TRANSFORM HOLOGRAPHY

Fourier transform holography is another widely considered geometry for x-ray holographic microscopy (see references in Jacobsen 1990b and McNulty 1992a). In it, the sample is placed near a source of spherical reference waves (see Fig. 2). By placing the recording medium at a large distance from the sample, one can obtain fringes of low enough spatial frequency that an electronic hologram detector such as a CCD camera can be used. Therefore, whereas the detector resolution sets the image resolution limit in Gabor holography, the size of the reference source [or, in principle, the size scale on which the reference source is well characterized (Stroke 1965)] sets the resolution limit in Fourier transform holography. With a planar object and a point reference located in the same transverse plane, the image can then be reconstructed by simply taking an inverse Fourier transform of the hologram. The turnaround time from hologram recording to image reconstruction can be quick enough so that one image can be evaluated before another hologram is acquired.

Research in Fourier transform x-ray holography at the NSLS has centered on the use of a Fresnel zone plate for generation of the spherical reference wave (McNulty 1992a). The NSLS X-1A undulator beamline is used to coherently illuminate a Fresnel zone plate with $\lambda = 3.4$ nm x-rays. The specimen is placed near the zone plate focus, and is illuminated by the undiffracted fraction of the light transmitted by the zone plate. A phosphor-coated CCD chip is used to record the hologram with exposure times of 10–30 minutes; with such exposures, the combined thermal and readout noise is 45 electrons per pixel, the equivalent of 15 x-ray photons of noise per pixel. The outer zone width of the Fresnel zone plate used was 50 nm, and images of electron beam fabricated test patterns faithfully reproduce the 60–125 nm features present (see Fig. 4). With the 576 × 384 pixel CCD used at present, the field of view at 50 nm resolution is limited to 12 μm. The field of view for a given resolution can be increased by using the 2048 × 2048 pixel CCDs now available.

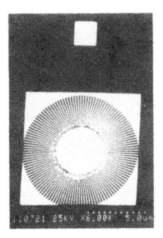 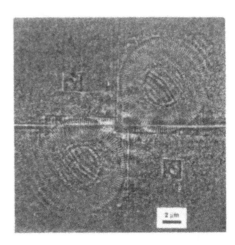

Fig. 4. Microfabricated test pattern (left) and image (right) reconstructed from a Fourier transform hologram of the test pattern (from McNulty 1992a). The focus of a Fresnel zone plate was placed in the corner of the large square holding the spoked pattern. The twin image in Fourier transform holography is a spatially centered conjugate image.

The implementation described above represents one of several ways to do Fourier transform holography. Haddad *et al.* (1988) have proposed to use grazing incidence reflection from microspheres to form a point source reference wave; only a small surface area will be oriented to reflect the reference wave towards a given part of the hologram. While in principle this may lead to very high resolution holograms, consideration must be given to surface roughness of the microsphere and to the possibility that weak bulk scattering from the sphere will contribute a signal comparable to or stronger than efficient reflection from the small reflecting surface area. Another approach is to use a resolution element sized scatterer to produce the reference wave; this approach has been adopted in preliminary experiments aimed at 3D imaging via multiple view holography (McNulty 1992b). Other possibilties include the use of more generalized reference objects, or ultrafine pinholes located at the focus of a condenser optic. In any case, attention must be paid to obtaining an appropriate balance of reference and specimen waves so that the holographic intermodulation term remains weak unless it can be spatially separated (Collier 1971). In all these approaches, Fourier transform holography combines the ability to use lower resolution electronic detectors with a requirement that the sample be closely aligned (especially in the longitudinal direction as limited by the longitudinal coherence length $\lambda^2/\Delta\lambda$) with the source of spherical reference waves. Finally, because the reference and signal waves are spread out at the detector plane, Fourier transform holography may offer more robustness against detector damage in flash imaging experiments than may be the case for Gabor holography.

5. OTHER APPLICATIONS OF X-RAY HOLOGRAPHY

Research in x-ray holography is not limited to applications in biological imaging. Holography with harder x-rays may prove valuable for materials science applications and, for example, for imaging defects in microcircuits. Polack and Joyeux have drawn from their experience in holographic microscopy to develop a Fresnel mirror method for measuring the real part of the soft x-ray refractive index (Polack 1992). Howells and Jacobsen have developed an approach to projection x-ray lithography which could replace the shadow mask used in proximity lithography with a diffractive mask designed using holographic principles; the fact that x-rays are both absorbed and phase-shifted in matter would be exploited to reduce twin-image noise in an in-line holographic system to levels low enough to produce no error bits in the lithographic image (Howells 1991, Jacobsen 1992b). The main feature of this approach is that sub-0.1 μm linewidth features can in principle be printed with conveniently large (> 50 μm) mask-to-wafer separation distances.

6. ACKNOWLEDGEMENTS

We thank Denis Joyeux, Janos Kirz, Francois Polack, and Jim Trebes for many helpful discussions. We thank Erik Anderson and Dieter Kern for fabrication of the zone plates used for Fourier transform holography, and David Attwood for supportting this effort. Supported by the National Science Foundation under grants DIR-9005893 and BIR-9112062 (C.J., S.L.), and by the Department of Energy, Office of Health and Environmental Research contract DE-FG02-89ER60858 (M.H.), Office of Basic Energy Sciences contract W-31-109-ENG-38 (I.M.), and Advanced Energy Projects Division contract W-7405-ENG-48 (I.M.).

8. REFERENCES

Aoki S and Kikuta S 1974 Japan J. Appl. Phys. 13, 1385.
Baez A 1952 J. Opt. Soc. Am. 42, 756.
Born M and Wolf E 1980 Principles of Optics (Pergamon, Oxford).
Collier RJ, Burckhardt CB, and Lin LH 1971 Optical Holography (Academic, New York).
Cornacchia M and Winick H, eds. Stanford Synchrotron Radiation Laboratory report SSRL 92/02, Stanford, CA.
Devaney AJ 1989 Phys. Rev. Lett. 62, 2385.
Gabor D 1948 Nature 161, 777.
Greinke B and Gölz G 1992 in Michette *op. cit.*, pp 316–318.
Haddad WS Cullen D Solem JC Boyer K and Rhodes CK 1988 OSA Proceedings on Short Wavelength Coherent Radiation: Generation and Applications eds. Falcone RW and Kirz J (Optical Society of American, Washington), pp 284–289.
Howells MR 1988 X-ray Microscopy II eds. Sayre D, Kirz J, Howells M, and Rarback H (Springer-Verlag, Berlin, 1988), pp 263–271.
Howells M and Jacobsen C 1991 OSA Proceedings on Soft X-ray Projection Lithography, ed. Bokor J (Optical Society of American, Washington), pp 108–112.

Jacobsen C 1990a X-ray Microscopy in Biology and Medicine ed. Shinohara K, Yada K, Kihara H, Saito T (Springer, Berlin) pp 167–177.

Jacobsen C, Howells M, Kirz J, Rothman S 1990b J. Opt. Soc. Am. 7, 1847.

Jacobsen C, Lindaas S, and Howells M 1991 OSA Proceedings on Short Wavelength Coherent Radiation eds. Bucksbaum P. and Ceglio N (Optical Society of America, Washington, DC, 1991), pp 130–134.

Jacobsen C, Lindaas S, and Howells M 1992a in Michette *op. cit.*, pp 244–250.

Jacobsen C and Howells MR 1992b J. Appl. Phys. 71, 2993.

Joyeux D and Polack F, 1988 OSA Proceedings on Short Wavelength Coherent Radiation: Generation and Applications ed. Falcone RW and Kirz J (Optical Society of American, Washington), pp 295–302.

Joyeux D 1992 in Proc. SPIE 1741 (in press).

Koren G, Polack F, and Joyeux D, 1992 in Proc. SPIE 1741 (in press).

Lindaas S, Jacobsen C, Franck K, and Howells MR, in Proc. SPIE 1741 (in press).

Liu G and Scott PD 1987 J. Opt. Soc. Am. A 4, 159.

London RA, Rosen MD, and Trebes J, 1988 Appl. Opt. 28, 3397.

McNulty I, Kirz J, Jacobsen C, Anderson EH, Howells MR, and Kern DP, 1992a Science 256, 1009.

McNulty I, Trebes J, Brase J, Yorkey T, Anderson EH, Attwood DT, Kern DP, and Jacobsen C, 1992b Proc. SPIE 1741 (in press).

Michette AG, Morrison GR, and Buckley CJ eds. 1992 X-ray Microscopy III (Springer Series in Optical Sciences 67, Springer-Verlag, Berlin).

Polack F and Joyeux D 1992 in Michette *op. cit.*, pp 301–305.

Schmahl G and Rudolph D 1987 in X-ray Microscopy: Instrumentation and Biological Applications, eds. Cheng PC and Jan GJ (Springer, Berlin) pp 231–238.

Shinozaki DM 1988 in X-ray Microscopy II eds. Sayre D, Howells M, Kirz J, and Rarback H (Springer, Berlin) pp 118–123.

Solem JC 1986 J. Opt. Soc. Am. B3, 1551.

Spiller E and Feder R 1977 in X-ray Optics ed. Queisser HJ (Springer, Berlin) pp 35–92.

Stroke GW 1965 Appl. Phys. Lett. 6, 201 and Stroke GW, Restrick R, Funkhouser A, Brumm D 1965 Phys. Lett. 18, 274.

Tallents GJ 1991 ed. X-ray Lasers 1990 (Institute of Physics Conference Series, London).

Trebes J, Annese C, Birdsall D, Brase J, Gray J, Lane S, London R, Matthews D, Peters D, Pinkel D, Stone G, Rapp D, Rosen M, Weier U, and Yorkey T 1992 in Michette op. cit. pp 255–258.

Williams S, Zhang X, Jacobsen C, Kirz J, Van't Hof J, Lamm SS 1992 submitted for publication.

Wolter H 1952 Ann. Phys. 10, 94.

Zhang X, Jacobsen C, and Williams S 1992 in Proc. SPIE 1741 (in press).

Inst. Phys. Conf. Ser. No 130: Chapter 7
Paper presented at Int. Congr. X-ray Optics and Microanalysis, Manchester, 1992

555

X-ray microscopy studies with the Göttingen X-ray microscopes

J.Thieme, G.Schmahl, D.Rudolph, B.Niemann, P.Guttmann,
G.Schneider, M.Diehl, T.Wilhein

Forschungseinrichtung Röntgenphysik, University of Göttingen,
Geiststraße 11, 3400 Göttingen, Germany

ABSTRACT: For the investigation of specimens in their natural wet state the wavelength range of the water window is well suited. The transmission of water is the best at $\lambda = 2.4$ nm, i.e. objects in a water layer of up to 10 μm thickness can be investigated. Studies with an X-ray microscope at the electron storage ring BESSY in Berlin were performed. An X-ray microscope with a pulsed plasma source is under development.

1. INTRODUCTION

X-ray photons in the wavelength range of about 0.3 nm - 5 nm are best suited for X-ray microscopy studies. X-ray microscopy provides higher resolution than optical microscopy and higher penetration ability than electron microscopy. Most importantly, X-ray microscopy has the potential for imaging hydrated specimens with high resolution (Rudolph et al. 1992). The two dominating processes causing the contrast are photoelectric absorption and phase shift. In the wavelength range $\lambda \geq 2.4$ nm, the absorption cross-sections are about one order of magnitude smaller than the cross-sections for electrons in the energy range in which electron microscopy is performed. These absorption cross-sections for soft X-rays are appropriate for high-resolution investigation of up to 10 μm thickness. The wavelength range between the K absorption edges of oxygen at $\lambda = 2.34$ nm and carbon at $\lambda = 4.38$ nm is especially interesting because in this wavelength range the radiation is weakly absorbed by water but strongly absorbed by e.g. organic matter or silicates resulting in a good amplitude contrast of wet specimens (Wolter 1952).

2. ZONE PLATE OPTICS

For X-ray microscopy two types of zone plates are needed. The object has to be illuminated by a condenser or a collector-condenser system which has to collect as much radiation as possible from the X-ray source. For X-ray microscopy with synchrotron radiation at electron storage rings condenser zone plates with diameters of 2 to 9 mm and rather large zone numbers of 10^3 to about $4 \cdot 10^4$ are required. The object is imaged by a high-resolution zone plate, i.e., a zone plate with outermost zone widths dr_n as small as possible. The size of the resolution element is nearly equal to the width of its outermost zone. The micro zone plates have only a few hundred zones in order not to restrict the usable X-ray bandwidth too much. Therefore, micro zone plates have very small diameters in the region of 10 to 100 µm (Schmahl et al. 1984). More details on the properties of zone plates are reported by Thieme et al. (1992).

3. THE X-RAY MICROSCOPE AT THE ELECTRON STORAGE RING BESSY

The Forschungseinrichtung Röntgenphysik has developed two X-ray microscopes with zone plate optics, an X-ray microcope and a scanning X-ray microscope. The two microscopes are installed at a bending magnet of the electron storage ring BESSY in Berlin. In addition, an X-ray microscope with a pulsed plasma X-ray source is under development.

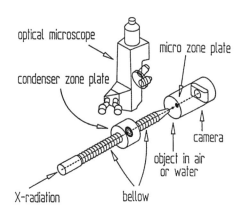

Fig. 1. Scheme of the X-ray microscope at BESSY

Figure 1 shows the scheme of the X-ray microscope at BESSY. The condenser zone plate is mounted in a bellow which ends in an Al_2O_3-foil transparent to X-radiation. This foil separates the vacuum of the bellow from air. The object field is limited by a pinhole with a diameter of about 20 µm. Due to the wavelength dependence of the focal length of a zone plate the condenser - in combination with the pinhole - acts as a linear monochromator. It reduces the bandwidth of the polychromatic synchrotron radiation. The object is mounted in an environmental chamber - not shown in Fig. 1 - where it can be prepared in its natural state. The next compartment contains the micro zone plate and the camera under prevacuum conditions. The micro zone plate as a high resolution X-ray objective generates a magnified image of the object in the image field. The central stop of the condenser zone plate prevents that the object is illuminated by broad band zero order radiation of the condenser. In addition, due to the stop the image field is free of zero order radiation of the micro zone plate. The enlarged X-ray image can be recorded either with a photographic plate or with a CCD-detector. In addition, a micro-channel-plate is installed so the image may be viewed directly. An optical microscope on a swivel-mount is installed to adjust and to prefocus the object. Therefore, the bellow containing the condenser zone plate can be bent downwards (Schmahl et al. 1992b).

4. THE X-RAY MICROSCOPE WITH A PULSED PLASMA X-RAY SOURCE

To perform X-ray microscopy experiments independent of electron storage rings as X-ray sources an X-ray microscope with a pulsed plasma source is under development (Schmahl et al. 1992a). Details on the pulsed plasma X-ray source are discussed by Rothweiler et al. (1992). The condenser is a 115 mm long part of an ellipsoid of revolution, chosen in a way that the opening angle of the focused radiation fits to the aperture of a micro zone plate whith an outermost zone width of 30 nm. This micro zone plate creates the magnified image of the object. As the plasma X-ray source emits mainly line radiation, a reduction of the bandwidth of the radiation is not necessary. Object and micro zone plate together can be swiveled into an inverse optical microscope. This allows adjustment and prefocusing. Figure 2 shows the image of a diatom, taken with a CCD-camera with a single X-ray pulse. The X-ray magnification was 380 x. The CCD was covered with an X-ray to visible converter Gd_2O_2S:Tb with 8 μm grains.

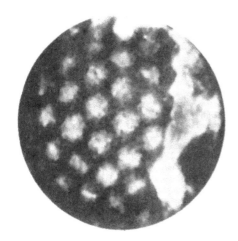

Fig. 2. X-ray image of a diatom taken with a single X-ray pulse with a CCD-camera. The image field has a diameter of 20 μm.

5. X-RAY MICROSCOPY STUDIES WITH THE MICROSCOPE AT BESSY

The ability of X-ray microscopy to contribute to research has been demonstrated in several examples: First, the structure of chromosomes in a range between the 10 nm scale and the μm scale is of great functional importance. However, there is little knowledge about the relation between form and function in this range. X-ray microscopy can contribute to close this gap (Guttmann et al. 1992). Secondly, resurrection plants or parts of them are an example for objects that can be investigated with X-rays in their inactive dry state and in their different states between dry and becoming active by soaking up water. The third example shows the examination of a three layer mineral as the montmorillonit. They are characterized by their large specific surface , their high adsorption power and their ability to form a gel, which is widely used to alter the viscosity of solutions. Figure 3 shows an X-ray image of the "book - house"- structure of a montmorillonit gel. Direct imaging of this confi-guration of clay platelets in aqueous solution could not be performed till now by other means (Thieme et al. 1992).

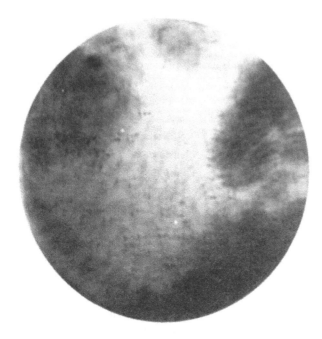

ACKNOWLEDGE-MENTS

The current research is being funded by the German Federal Minister of Research and Technology (BMFT) under contract numbers 055MGDXB6 and 13N5328A0.

Fig. 3. X-ray image of the "book-house"-structure of a montmorillonit gel, λ = 2.4 nm. The image field has a diameter of 7.6 µm

REFERENCES

Guttmann P, Schneider G, Robert-Nicoud M, Niemann B, Rudolph D, Thieme J, Jovin T M and Schmahl G 1992 X-Ray Microscopy III ed A G Michette, G R Morrison and C J Buckley (Berlin: Springer) pp 404-7
Rothweiler D, Neff W, Lebert R, Richter F and Diehl M 1992 this volume
Rudolph D, Schneider G, Guttmann P, Schmahl G, Niemann B and Thieme J 1992 X-Ray Microscopy III ed A G Michette, G R Morrison and C J Buckley (Berlin: Springer) pp 392-6
Schmahl G, Rudolph D, Guttmann P, Christ O 1984 X-Ray Microscopy ed G Schmahl and D Rudolph (Berlin: Springer) pp 63-74
Schmahl G, Niemann B, Rudolph D, Diehl D, Thieme J, Neff W, Holz R, Lebert R, Richter F and Herzige G, 1992a X-Ray Microscopy III ed A G Michette, G R Morrison and C J Buckley (Berlin: Springer) pp 66-9
Schmahl G, Rudolph D, Niemann B, Guttmann P, Thieme J, Schneider G, David C, Diehl M, Wllhein T 1992b Optik in press
Thieme J, Guttmann P, Niemeyer J, Schneider G, David C, Niemann B, Rudolph D and Schmahl G 1992 Nachr.Chem.Tech.Lab. 40 Nr.5 pp 562-3
Thieme J, David C, Kaulich B, Guttmann P, Rudolph D and Schmahl G 1992 this volume
Wolter H 1952 Ann.Phys.6.Folge Bd.10 pp 94-114

Inst. Phys. Conf. Ser. No 130: Chapter 7
Paper presented at Int. Congr. X-ray Optics and Microanalysis, Manchester, 1992

559

Review on the development of cone-beam X-ray microtomography

P. C. Cheng, T. H. Lin, G. Wang, D. M. Shinozaki*, H. G. Kim† and S. P. Newberry‡

Advanced Microscopy and Imaging Laboratory (AMIL) and Advanced Real Time System Laboratory (ARTS), Department of Electrical and Computer Engineering, State University of New York, Buffalo, NY 14260, USA

*Department of Materials Engineering, Faculty of Engineering Science, University of Western Ontario, London, Ontario, Canada N6A 5B7

†Laboratory for Laser Energetics, University of Rochester, 250 East River Road, Rochester, NY 14623-1299, USA

‡CBI Labs, S. Westcott Road, Schenectady, NY 12306 USA

ABSTRACT: Considering the characteristics of the microscope system of SUNY at Buffalo (NY, USA) and the limitations of current cone-beam reconstruction algorithms, a general cone-beam reconstruction algorithm is developed at SUNY at Buffalo, which can be used to reconstruct spheric, rod-shaped and plate-like specimens. The spatially variant PSF of the general cone-beam reconstruction algorithm is derived, and several properties related to the exactness of reconstruction are proved. Furthermore, An error analysis is performed via theoretical analysis and computer simulation.

1   INTRODUCTION

An X-ray shadow projection microscope using a scannable point source of X-rays is under development at AMIL-ARTS, SUNY at Buffalo [Cheng et al., 1989, Cheng et al., 1990, Schmahl and Cheng, 1991, Cheng and Jan, 1987, Cheng et al., 1991b, Cheng et al., 1991a] [Wang et al., 1991d] (Figure 1). The point source is generated by a focussed electron beam, which can be steered electromagnetically in a plane perpendicular to the optical axis of the microscope. A specimen is mounted on a rotatable and shiftable mechanical stage for microtomography. An elaborate feed-back system is being implemented to measure and correct the motion error of the mechanical stage dynamically.

The cone-beam reconstruction has been studied for more than thirty years. Two recent review papers were written by Smith [Smith, 1990a] and Gullberg et al. [Gullberg et al., 1992]. Existing cone-beam algorithms can be categorized into two broad classes: analytic and algebraic. Algebraic algorithms demand more computing resources than analytic ones. Because a high reconstruction speed is very important in practice, analytic algorithms are always preferred. Feldkamp et al. [Feldkamp et al., 1984, Kak and Slaney, 1987] proposed a cone-beam reconstruction algorithm, which is an extension of the equispatial fan-beam reconstruction algorithm, where 2D projection data from different angles are filtered and back-projected to voxels along projection beams. The reconstructed value of a voxel is the sum of contributions from all tilted fan-beams passing through the voxel. For cone-beam X-ray microtomography, the main drawbacks of Feldkamp's algorithm, shortcomings of most of other algorithms as well, are that the specimen must be contained in a sphere-like reconstruction region and that the X-ray source must be moved along a circle in the specimen coordinate system. In addition, Feldkamp's algorithm suffers from a larger inaccuracy in reconstructing off-midplane structures.

2   GENERAL CONE-BEAM ALGORITHM

Considering the characteristics of our X-ray microscope system and the limitations of available cone-beam algorithms, a general cone-beam image reconstruction algorithm is developed at SUNY at Buffalo. In general cone-beam reconstruction, there are mainly two scanning modes, planar and helix-like. A typical case of planar scanning loci is a circle, which is used in the conventional cone-beam reconstruction. A helix-like scanning locus is used to deal with rod-like specimens. Without lose of generality, a locus turn of the X-ray source can be defined in cylindrical coordinates by the following equation:

$$\rho = \rho(\beta), \ h = h(\beta), \ \beta \in [0, 2\pi]. \tag{1}$$

In the first scanning mode, $h(\beta) = 0$, $\beta \in [0, 2\pi]$. In the second mode, $h(\beta) \in [z - \frac{h_p}{2}, z + \frac{h_p}{2}]$, $\beta \in [0, 2\pi]$, where z is the vertical axis coordinate of a horizontal slice in an interesting specimen and h_p is the pitch of the helix-like scanning locus. We define the mid-plane is the x-y plane in the first scanning mode, the plane with a vertical axis coordinate z in the second mode.

The general cone-beam formula can be derived by modifying the convolution and back-projection formula for equispatial fan-beam data. In a 3D coordinate system (x, y, z), the cone-beam reconstruction is performed by tilting the fan out of the mid-plane, thus changing the coordinate system (t, s), where (t, s) denotes the rotated coordinate system described by:

$$t = x \cos \beta + y \sin \beta, \ s = -x \sin \beta + y \cos \beta. \tag{2}$$

A new coordinate system $(\tilde{t}, \tilde{s})$ is defined to represent the location of the reconstructed point with respect to the tilted fan. Because of the geometry of the tilted fan, both the source-to-origin distance and the angular differential must be modified. After the modification, we obtain

$$g(x, y, z) = \frac{1}{2} \int_0^{2\pi} \frac{\rho(\beta)^2}{(\rho(\beta) - s)^2} \int_{-\infty}^{\infty} R_\beta(p, \zeta) f(\frac{\rho(\beta)t}{\rho(\beta) - s} - p) \frac{\rho(\beta)}{\sqrt{\rho(\beta)^2 + p^2 + \zeta^2}} dp d\beta, \tag{3}$$

where $R_\beta(p, \zeta)$ is the 2D projection data, $f(\cdot)$ is the reconstruction filter,

$$\zeta = \frac{\rho(\beta)\tilde{z}(\beta)}{\rho(\beta) - s},$$

and the local coordinate system $(x, y, \tilde{z}(\beta))$ associated with β is defined by:

$$\tilde{z}(\beta) = z - h(\beta).$$

The projection data collected from the turn described by equation (1) are used to reconstruct the image in the mid-plane. The equation (3) is referred to as the general cone-beam reconstruction formula [Wang et al., 1991b, Wang et al., 1991c, Wang et al., 1992d].

3 SPECIAL VERSIONS

Several special versions of the general cone-beam reconstruction algorithm can be used for dealing with spherical, rod-shaped and plate-like specimens [Wang et al., 1991b, Wang et al., 1991c, Wang et al., 1992d].

3.1 Polygonal Algorithm: Spherical Specimens

A polygonal cone-beam reconstruction algorithm is obtained by setting

$$
\begin{cases}
\rho = \dfrac{\rho_0}{\cos(\beta - \frac{2\pi}{N_S}\lfloor\frac{N_S\beta}{2\pi}+0.5\rfloor)}; \\
h = 0,
\end{cases}
\tag{4}
$$

where ρ_0 is the minimum distance between the source and the origin, N_S is the number of sides of the scanning polygon, and $\lfloor\cdot\rfloor$ represents a floor function [Wang et al., 1991a]. The polygonal cone-beam reconstruction algorithm is used to reduce the number of recalibrations of the mechanical axis position.

For tomographic imaging, a large number of 2D projection images must be recorded at various orientations. The specimen therefore must be rotated, and a mechanical stage is installed for this purpose. Because the variation of rotation axis position of the mechanical stage is one of the major error sources in microtomography, an elaborate feed-back system is required to minimize the error. However, experiments showed that the calibration of the rotation axis position is both time-consuming and complicated. On the other hand, the X-ray source of our microscope system can be precisely moved by steering the e-beam electromagnetically. With the polygonal cone-beam algorithm, the number of the stage rotation can be greatly reduced. In the reconstruction, the straight line portions of the polygon are achieved by the X-ray source movement, and the number of the polygon sides determines the number of the stage rotations.

3.2 Dashed-line Algorithm: Rod-shaped Specimens

In the dashed-line cone-beam reconstruction, the scanning turn is described as:

$$
\begin{cases}
\rho = \dfrac{\rho_0}{\cos(\beta - \frac{2\pi}{N_S}\lfloor\frac{N_S\beta}{2\pi}+0.5\rfloor)}; \\
h = \dfrac{h_p}{N_S}\lfloor\frac{N_S\beta}{2\pi}\rfloor,
\end{cases}
\tag{5}
$$

where h_p is the pitch of the scanning locus. The scanning locus required by the dashed-line algorithm is similar to "winding stairs". When dealing with rod-shaped specimens, the dashed-line algorithm requires a much less number of vertical translations of the mechanical stage, compared with the helical algorithm [Lin et al., 1991].

3.3 Placing Detector Array Horizontally: Plate-like Specimens

In practical cases of microscopy, many specimens are rather plate-shaped, as in typical thick film sections. The reconstruction of a plate-like sample appears to be a limited angle problem which can be addressed using standard methods described in the literature. Since the methods based on a point source and projection microscopy have great advantages experimentally, a novel scheme was developed to reconstruct plate-like specimens by placing the detector array horizontally and using the general cone-beam algorithm [Wang et al., 1991c, Wang et al., 1992c].

The method is explained as follows (Figure 2). Assume that the locus of the X-ray source is a 2D curve (for example, it may be a circle, polygon, or a rectangle). The plate-like specimen (or thick film) is placed parallel to the plane of the source locus. Arrange the relative position so that the perpendicular passing through the center of the scanning locus passes through the center of the specimen area of interest. This perpendicular is labeled the principal axis. The specific locus used is made to be similar to the shape of the area of interest, and substantially larger than that area. The detector plane is placed behind the specimen, parallel to the specimen, and the 2D projection data recorded for each X-ray source position. After each frame of these projection data is mapped onto the plane facing the X-ray source and passing through the principal axis, the general cone-beam algorithm is applied to the modified projection data. In this way the problems associated with the standard limited angle approaches are avoided. An example of cone-beam reconstruction of plate-like specimens is shown in Figure 3. For more details, please see [Wang et al., 1991c, Wang et al., 1992c].

4   ANALYTIC STUDIES

In this section, the PSF of our general cone-beam reconstruction algorithm is given. Although our general algorithm is not exact in general, under some conditions three properties of the general cone-beam algorithm are proved, which are related to the exactness of reconstruction.

4.1   Point Spread Function

In order to study the PSF of the general cone-beam reconstruction algorithm, we take advantage of the 2D projection formula of a 3D point source [Yan and Leahy, 1991]. We require that the point source be in the interior of a cylinder of radius R. Let $i(\mathbf{x})$ represent a specimen to be reconstructed where $\mathbf{x} = (x, y, z)^t$ and the operation $(\cdot)^t$ represents the transpose of a vector. Performing a band-limiting operation on the projection data of a 3D δ function and substituting the band-limited projection data into the general cone-beam formula, we obtain the band-limited PSF as follows:

$$\bar{s}(x, y, z; x_0, y_0, z_0) = \frac{1}{2} \int_0^{2\pi} \frac{\rho^4(\beta)}{(\rho(\beta) - s)^2 \hat{\alpha}_{02}^2} \frac{\sin \frac{\pi(\zeta - \frac{\hat{\alpha}_{03}}{\hat{\alpha}_{02}}\rho(\beta))}{\tau}}{\pi(\zeta - \frac{\hat{\alpha}_{03}}{\hat{\alpha}_{02}}\rho(\beta))} \bar{f} \left[\rho(\beta)(\frac{t}{\rho(\beta) - s} - \frac{\hat{\alpha}_{01}}{\hat{\alpha}_{02}}) \right] d\beta, \quad (6)$$

where

$$\bar{f}(x) = \frac{1}{2\tau^2} \frac{\sin \frac{\pi x}{\tau}}{\frac{\pi x}{\tau}} - \frac{1}{4\tau^2} \left(\frac{\sin \frac{\pi x}{2\tau}}{\frac{\pi x}{2\tau}} \right)^2,$$

and τ is the sampling interval. The above band-limited PSF formula was used in our simulation for computing the PSFs of the polygonal cone-beam algorithm and the dashed-line cone-beam algorithm. For more details, please see [Wang et al., 1992e]. Note that in derivation of the PSF [Wang et al., 1992e], some intermediate integrals should be interpreted in the generalized sense when necessary [Yan and Leahy, 1991, Kak and Slaney, 1987, Smith, 1985, Smith, 1990b].

4.2 Exactness in Reconstruction

Although the general cone-beam algorithm is not exact, it has several nice analytic properties related to the exactness of reconstruction, if the reconstruction is performed using projection data collected from a turn of a scanning locus whose horizontal projection meets the conditions required by our derivative-free noncircular fan-beam reconstruction formula [Wang et al., 1992a]. First reconstruction is exact on the midplane in a planar scanning mode [Wang et al., 1992b, Wang et al., 1992a]. Second, the vertical integral of a reconstructed image is equal to that of the actual image [Wang et al., 1992e]. Third, reconstruction is exact if an actual image is independent of the rotation-axis coordinate [Wang et al., 1992f].

5 ERROR ANALYSIS

The primary error source in cone-beam reconstruction is the theoretical precision limitation. A reconstruction error formula has been derived which expresses the error in terms of the specimen structure and various imaging parameters [Wang et al., 1992f, Lin et al., 1992]. Based on the reconstruction error formula, some qualitative observations can be made. Roughly speaking, in the planar scanning mode, $e(x, y, z)$ is proportional to the height of a voxel away from the midplane; in the helix-like scanning mode, $e(x, y, z)$ is proportional to the pitch of a helix-like scanning locus; in both cases, $e(x, y, z)$ is inversely proportional to the horizontal size of the scanning locus and is proportional to the vertical variation of a specimen.

In order to evaluate the reconstruction error quantitatively, a series of simulation experiments were performed using computer-generated stochastic images [Wang et al., 1992f, Lin et al., 1992]. The influence of various practical factors including the system noise and inaccuracy of the scanning locus measurement were incorporated into the reconstruction error formula. Simulated reconstruction error distributions indicated some interesting features. First, the reconstruction error is reasonably small. On average, it is just several percentage in our experiments. Second, the dashed-line cone-beam algorithm significantly reduces the maximum of averaged reconstruction errors of horizontal slices. Although the exact reconstruction can be achieved on the midplane using the polygonal cone-beam algorithm, the reconstruction error away from the midplane can be several times larger than that of the dash-line cone-beam algorithm. Third, the general cone-beam reconstruction algorithm is not sensitive to various random interferences.

6 CONCLUSIONS

The general cone-beam algorithm described in this paper takes full advantage of the hardware characteristics of the X-ray microscope system at SUNY at Buffalo (NY, USA), overcomes the limitations of existing cone-beam reconstruction algorithms, allows various scanning loci, reconstructs spheric, rod-shaped and plate-like specimens, and facilitates near real-time reconstruction by providing the same efficiency and parallelism as Feldkamp's algorithm. This algorithm not only gives satisfactory reconstruction images but also has the some nice properties related to the exactness of reconstruction. A precision accessment for our general cone-beam algorithm has been performed through theoretical analysis and computer simulation. It has been found that the reconstruction error is both reasonably small and quite robust.

References

[Cheng and Jan, 1987] Cheng, P. C. and Jan, G. J. (1987). Springer-Verlag.

[Cheng et al., 1991a] Cheng, P. C., Lin, T. H., Shinozaki, D. M., and Newberry, S. P. (1991a). *J. Scanning Microscopy*, 13 (I):10-11.

[Cheng et al., 1989] Cheng, P. C., Newberry, S. P., Kim, H. G., and Wittman, M. D. (1989). *Europ. J. Cell Biology*, 48(25):169-172.

[Cheng et al., 1990] Cheng, P. C., Newberry, S. P., Kim, H. G., Wittman, M. D., and Hwang, I.-S. (1990). In Duke, P. and Michette, A., editors, *Modern microscopy*, pages 87-117. Plenum Press.

[Cheng et al., 1991b] Cheng, P. C., Shinozaki, D. M., Lin, T. H., Newberry, S. P., Sridhar, R., Tarng, W., Chen, M. T., and Chen, L. H. (1991b). In Michette, A., Morries, G. R., and Buckley, C. J., editors, *X-ray microscopy - III*. Springer-Verlag, Berlin, in press.

[Feldkamp et al., 1984] Feldkamp, L. A., Davis, L. C., and Kress, J. W. (1984). *J. Opt. Soc. Am.*, 1 (A):612-619.

[Gullberg et al., 1992] Gullberg, G. T., Zeng, G. L., Datz, F. L., Christian, P. E., Tung, C. H., and Morgan, H. T. (1992). *Phys. Med. Biol.*, 37(3):507-534.

[Kak and Slaney, 1987] Kak, A. C. and Slaney, M. (1987). IEEE Press, USA.

[Lin et al., 1992] Lin, T. H., Wang, G., and Cheng, P. C. (1992). *J. Scanning Microscopy, Suppl. II (Addendum)*, pages 20-23.

[Lin et al., 1991] Lin, T. H., Wang, G., Cheng, P. C., and Shinozaki, D. M. (1991). *J. Scanning Microscopy*, 13 (I):11-13.

[Schmahl and Cheng, 1991] Schmahl, G. and Cheng, P. C. (1991). In E. E. Koch, T. S. and Winick, H., editors, *Handbook on Synchrotron Radiation*, volume 4, pages 481-536. North-Holland.

[Smith, 1985] Smith, B. D. (1985). *IEEE Trans. Med. Imag.*, MI-4:14-28.

[Smith, 1990a] Smith, B. D. (1990a). In *Proceedings of the symposium on Advanced Tomographic Imaging Methods for the Analysis of Materials*.

[Smith, 1990b] Smith, B. D. (1990b). *Opt. Engineering*, 29(5):524-534.

[Wang et al., 1992a] Wang, G., Lin, T. H., and Cheng, P. C. (1992a). *IEEE Trans. on Image Processing*, to appear.

[Wang et al., 1992b] Wang, G., Lin, T. H., and Cheng, P. C. (1992b). In *Proc. of SPIE, vol. 1660*, pages 262-273.

[Wang et al., 1991a] Wang, G., Lin, T. H., Cheng, P. C., and Shinozaki, D. M. (1991a). *J. Scanning Microscopy*, 13 (I):126-128.

[Wang et al., 1991b] Wang, G., Lin, T. H., Cheng, P. C., and Shinozaki, D. M. (1991b). In *Proc. of SPIE, vol. 1556*, pages 99-113.

[Wang et al.. 1991c] Wang. G.. Lin. T. H.. Cheng. P. C.. and Shinozaki. D. M. (1991c). In *Proc. of SimTec '91*, pages 177–182.

[Wang et al.. 1992c] Wang. G.. Lin. T. H.. Cheng. P. C.. and Shinozaki. D. M. (1992c). *J. Scanning Microscopy*. 14(5). to appear.

[Wang et al.. 1992d] Wang. G.. Lin. T. H.. Cheng. P. C.. and Shinozaki. D. M. (1992d). *IEEE Trans. on Med. Imag.* to appear.

[Wang et al.. 1992e] Wang. G.. Lin. T. H.. Cheng. P. C.. and Shinozaki. D. M. (1992e). *J. Scanning Microscopy*. 14(4):187–193.

[Wang et al.. 1992f] Wang. G.. Lin. T. H.. Cheng. P. C.. and Shinozaki. D. M. (1992f). In *Proc. of SPIE. vol. 1660.* pages 274–285.

[Wang et al.. 1991d] Wang. G.. Lin. T. H.. Cheng. P. C.. Shinozaki. D. M.. and Newberry. S. P. (1991d). In *Proc of SPIE. vol. 1398.* pages 180–190.

[Yan and Leahy. 1991] Yan. X. H. and Leahy. R. M. (1991). *IEEE Trans. Med. Imag.* 10(3):462–472.

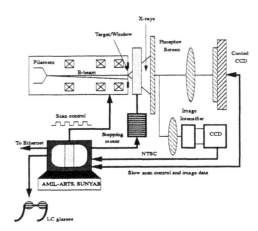

Figure 1: The schematic diagram showing the X-ray shadow projection microscope at SUNY at Buffalo (NY. USA).

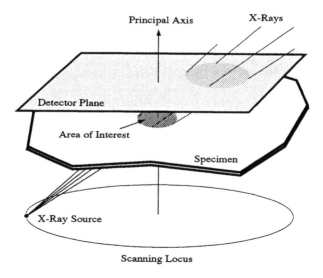

Figure 2: Schematic diagram illustrating the arrangement for cone-beam reconstruction of a plate-like specimen.

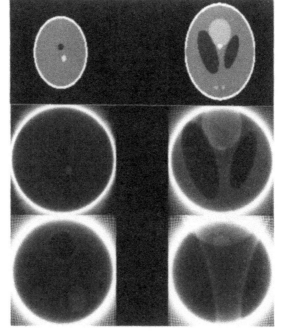

Figure 3: Two sets of the phantom and reconstructed images of plate-like specimens. Shepp and Logan's 3D phantom was used after compressed along z axis by a factor $k = 10$ and shifted by $d = 0.5$, and scaled along x and y axes by $s = 2.5$ respectively. The first row shows two slices of the original phantom, and the rest two rows the corresponding reconstructed images. The left column corresponds to the an upper horizontal section of the phantom, and the right column a lower section. The scanning locus is a circle of diameter 10.

Inst. Phys. Conf. Ser. No 130: Chapter 7
Paper presented at Int. Congr. X-ray Optics and Microanalysis, Manchester, 1992

Progress in digital X-ray projection microscopy

D. ERRE, D. MOUZE, X. THOMAS, J. CAZAUX

LASSI/GRSM BP 347 Faculté des Sciences 51062 REIMS FRANCE

ABSTRACT : Relative to the shadow X-ray microscopes built in the
sixties, the main advantages of our equipement are to obtain rapidly
digital images with a large dynamic (using a CCD camera) and to change
rapidly the incident photon energy without moving the specimen These
advantages allow the optimization of the sensitivity of the method and
the quantification of the X-ray images (added to chemical mapping).
They will lead, in the near future, to X-ray microcinematography (for
dynamic studies) and to X-ray microtomography.

1. INTRODUCTION

The rapid expansion of the field of X-ray microscopy is associated to the
intrinsic advantage of the use of incident X-rays (instead of electrons
for instance) to see inside thick specimens (surrounded by controlled
environments if necessary) and to the technological development of new
sources, new optics and new detectors (Kirz, 1992). Unfortunately, the
simplest way to produce X-rays remains the electron bombardment of a
target and the efficient focusing of X-ray is often restricted to soft
X-rays. To take benefit of the intrinsic advantage of X-rays without the
complications associated to the use of new sources (such as synchrotron
radiation) or to the development of new focusing elements, we are
exploring the possibilities offered by the "old" X-ray shadow microscopy
but using modern detectors such as CCD cameras. For this goal, we have set
such a camera inside a conventional scanning electron microscope which
kept its ability to obtain secondary electron images and X-ray emission
spectras (using a Si,Li detector). The possibilities of changing rapidly
the target (for changing the energy of the incoming photons) without
moving the specimen (in order to obtain various images -and next
to combine them-) and to insert it in an environmental chamber are the two
main advantages of our apparatus. The final goals are the developments of
X-ray microtomography and of X-ray microcinematography (X. Thomas et al,
1992 ; D. Erre et al, 1992) but here we restrict our purpose to the
results obtained in digital X-ray microscopy combined to the change of the

incident photons energy.

2 EXPERIMENTAL ASPECTS

The experimental arrangement being used has been shown elsewhere (X.Thomas et al ; D.Erre et al, 1992). Relative to the use of the old photographic plates the advantage of the use of a CCD camera is related to its speed, its excellent linearity and its wide dynamic (up to 3.10^4). It allows one to obtain images on a TV screen at the video rate (to set the specimen in the optimum conditions) and it provides digital images that can be next processed. For example for quantitative imaging, 3 successives images are obtained ; the first, B, without neither the specimen nor the illumination (to obtain the background value due to the use of the CCD alone); the second one E, with the X-ray beam but not the specimen (to obtain, for each pixel, the incident photon beam intensity) and the third one, I, with the incident X-rays and the specimen. Taking next, pixel by pixel, the log of the ratio (I-B)/(E-B) a μt map of the specimen is then obtained (μ : linear absorption coefficient and t thickness of the specimen). The acquisition speed of the CCD camera (an image of good quality can be acquired in ten seconds) leads to further developments : X-ray microtomography and X-ray microcinematography.The advantages of changing the incident photons' energy are : the possibility of selecting the best radiation for optimizing the sensitivity to the thickness and the nature of the investigated specimen and to obtain several X-ray images of the same specimen using radiations situated of the two sides of the absorption edges of the elements to be imaged. Concerning the optimization of the sensitivity, it can be easily established that the minimum detectable μt change is given by $\Delta\mu t/\mu t \approx 3(N_o)^{-1/2}(e^{\mu t/2}/\mu t)$ and it corresponds to minimum of $e^{\mu t/2}/\mu t$ which is obtained (Fig.1) when t $\simeq 2/\mu$ at a minimum value close to the unity. When N_o (Number of incident photons per pixel) is of around 10^5, $\Delta\mu t/\mu t$ may reach 10^{-2} and the sensitivity in the concentration detection limit of medium or heavy elements (Cu, Fe,Ni) embedded in very light matrices (water, carbon) may reach 10 ppm because $\Delta C/C = (\Delta\mu t/\mu t).(\sigma_{mat}/\sigma_{imp})$; $(\sigma_{mat};/\sigma_{imp}$ being the photoabsorption coefficients of the matrix and the impurity respect).

The main drawback of our arrangement is a lateral resolution limited to the micron range mainly due to the corresponding dimension of the diffusion of incoming electrons in the target material.

3. RESULTS

Taken in material sciences, Figs. 2 to 6 give examples of X-ray projection images obtained with this instrument. Fig.2 obtained in 10 sec shown a graphite matrix in which SO_4H_2 has been partly intercalated (the central part A, is composed of pure graphite ; the intermediate region, B, contains 16 % in molecular concentration of $SO_4 H_2$ while the external part C contains 36 % of SO_4H_2. Fig. 3 (obtained in 60 sec) is that of composite material composed of a polypropylen matrix embedding silica fibers. Image 4 obtained in 15 sec., is that of a Al matrix (100 μm thick) containing copper precipitates (A : pure Al ; B : 10 % Cu + 90° Al ; C : 2 % Cu and 98 % Al). The series from Fig. 5a to 5e correspond to contact images (specimen set on the CCD camera) obtained using respectively TiKα (a) ; V Kα (b) ; Cr Kα (c) ; CoKα (d) and Ni Kα (e) radiations of a submarine polymetallic nodule containing Fe and Mn. The final goal is to extract characteristic maps of these two elements from the acquired serie (work in progress).

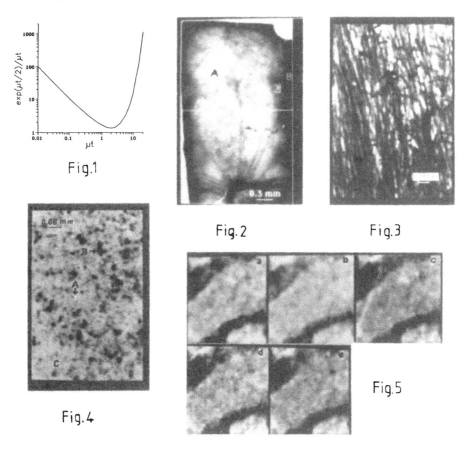

Fig.1

Fig.2

Fig.3

Fig.4

Fig.5

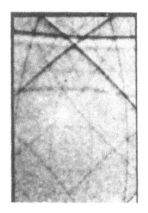

Fig. 6 is in fact the difference between two microradiographies (obtained, in 20 sec) of the same integrated circuit, but the X-ray source (via the e⁻ beam) has been shifted slightly. The numerical substraction exalts the contrast gradient (first difference image). At last Fig.7 shows a X-ray diffraction pattern (Kossel diagram obtained in 5 sec) of a LiF single crystal (0.45 mm thick) using Cu Kα radiation.

4 CONCLUSION

The use of modern CCD cameras permits the acquisition of digital images in X-ray shadow microscopy. Combined to the rapid change in energy of the incoming X-ray radiations, this progress opens the way of quantitative X-ray microscopy. The acquisition speed of digital images also leads to X-ray microtomography and to X-ray microcinematography. In this last field a series of images showing the dynamics of the intercalation process of SO_4H_2 into graphite (under an applied voltage and inside a wet cell) has been obtained very recently in our laboratory illustrating the wide field of possible applications of this new technique to diffusion processes (liquid/solid or solid/solid diffusion studies).

Erre D. et al (1992) Surface Interf. Anal. 19, 89.
Kirz J. (1992) in X-ray Microscopy III, Springer Series in Optical Sciences 67, 11.
Thomas X. et al (1992) Springer Series in Optical Sciences 67, 190.

Scanning soft X-ray microscopy

Chris Jacobsen, Harald Ade, Janos Kirz, Cheng-Hao Ko, Shawn Williams, and Xiaodong
  Zhang
Physics Department, SUNY, Stony Brook, NY 11794, USA
Erik Anderson
Center for X-ray Optics, Lawrence Berkeley Laboratory, Berkeley, CA 94720, USA
Dieter Kern
IBM Research Center, Yorktown Heights, NY 10598, USA

ABSTRACT: Scanning x-ray microscopes using Fresnel zone plate objective lenses
can serve as the basis of systems for minimum dose transmission microscopy of
biological specimens. Other signals can be used as well; for example, photoemission
can be used for imaging surface chemical states, and luminescence can be used to
map out dye binding sites.

1. INTRODUCTION

The first scheme to build a scanning x-ray microscope was proposed 40 years ago by
Pattee (1953), and the first scanning x-ray microscope to use synchrotron radiation was
reported 20 years ago by Horowitz and Howell (1972). It is, however, the spectacular
developments in bright x-ray sources and high resolution x-ray optics over the past
decade that have led to major advances in x-ray microscopy. (See for example Michette
et al. (1992)).

The ability to focus the radiation into a tiny spot is at the heart of scanning microscopy.
Since our interest in the energy range between the carbon K edge at 283 eV and 1 KeV,
we have chosen zone plates to form the x-ray microprobe. Similar instruments have been
built at Daresbury by the King's College group (Morrison *et al.* 1992), and at BESSY
by the Göttingen group (B. Niemann 1992).

For work at lower energies, where multilayer coated normal incidence mirrors are available, scanning microscopes using Schwarzschild objectives have been built by Trail and
Byer (1989), and by Ng *et al.* (1992). A given objective reflects only a narrow energy
band for which the coating was designed. Grazing incidence mirrors, by contrast, operate over a wide energy range, and the Hamburg scanning microscope makes full use of
this broad tuneability (Moewes 1992).

In scanning microscopy, the image is built pixel by pixel in a serial fashion. The speed at
which the information for each pixel is collected depends on the intensity of the probe,
and this in turn is directly proportional to the brightness of the source. It is for this

reason that scanning microscopes are prime beneficiaries of the development of bright undulator sources at electron storage rings. The image is stored in digital form, making subsequent quantitative analysis convenient.

Our work takes place at the Soft X-ray Undulator beamline X1A of the National Synchrotron Light Source at Brookhaven. The characteristics of the undulator and the beamline have been described by Rarback *et al.* (1990).

2. X-RAY FOCUSSING

The short wavelength of x-rays provides a path to obtaining small focal spots; the smallest focal spots have been made using soft x-rays because of their good match of interaction length to the size scale of microstructures in diffractive and multilayer optics. (Fine microprobes can also be created in the extreme near field of drawn optical fibers). Although smaller probes and higher resolution can be achieved with electron beams, x-rays offer a variety of advantages including higher penetrating power, good contrast mechanisms, less damage, and the possibility to image specimens at atmospheric pressure.

Because the refractive index for x rays is so close to unity, no refractive lenses can be made. Highest resolution is achieved by zone plates, which are diffractive optical elements. The focusing properties of zone plates are very much like those of a lens of the same numerical aperture. They have been known for over a century, and used to focus beams of photons, neutrons, atoms, and other types of waves. Their limitations are:

- they are highly chromatic, with the focal length proportional to the photon energy.
- they only focus a portion of the incident wave, with the rest either absorbed by the zones or thrown into other diffraction orders which could create a background if not intercepted.
- the zones must be formed and placed with high precision. For a perfect zone plate the size of the diffraction limited focal spot is approximately the width of the finest zone-ring.

Our zone plates are fabricated onto 100 nm thick silicon nitride membranes by electron beam writing into polymer resist, followed by electroplating. The fabrication process is described by Anderson and Kern (1992b), and by Anderson (1992a), while the performance is discussed by Jacobsen *et al.* (1991) and Zhang *et al.* (1992). The zones are made of about 120 nm thick nickel which, due to the slight deviation of the refractive index from unity, provides near optimal phase shift for good focusing efficiency. A thick central stop covers 20–25% of the area. It is in the shadow cast by this stop that the focus is formed. Unwanted orders are removed by placing an aperture between the plate and the focus (see Fig. 1). Depending on the application, two different geometries have been used. For experiments requiring the finest probe, the zone plate diameter is 90 μm, with a finest zone width of 45 nm. The Rayleigh resolution for these focusing elements is 50 nm, making it possible to observe structures with 35 nm lines and spaces. For

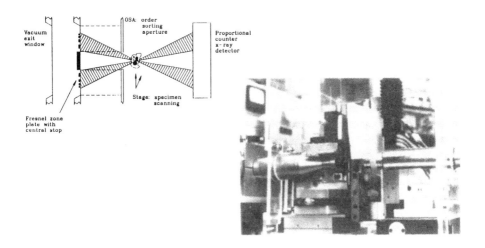

Fig. 1. Schematic diagram (left) and photograph (right) of the X-1A scanning transmission x-ray microscope.

experiments that require more working distance, and therefore a longer focal length, the zone plate diameter is 140 μm, with finest zone width of 60 nm.

Although zone plates with 25–30 nm finest zones have been fabricated in gold (Anderson and Kern 1992b), in germanium (Schmahl 1991), and in carbon (Charalambous and Morris 1992), the processing of these plates becomes harder, for the zone thickness must be kept fixed for good efficiency. This leads to very high and difficult-to-achieve aspect ratios. With the availability of ever more bright sources of x-rays, it becomes more attractive to consider using zone plates operating in the third diffraction order (see Fig. 2). The focal length is only 1/3 that of the first order, and the efficiency is typically 1/9, but the size of the focused probe is also reduced to 1/3 in the diffraction limit. To achieve this improvement particular attention must be paid to the placement of the zones, as the spherical aberration correction in third order is nine times larger than in first.

3. IMAGING AND MICROANALYSIS

The X1A beamline has two branches, and two scanning instruments (Rarback 1990). The two branches are served simultaneously by the same monochromator, with the scanning transmission microscope, STXM, occupying the long wavelength branch, while the scanning photoemission microscope, SPEM, using the second harmonic peak from the undulator is on the short wavelength branch.

The photoemission microscope is designed as a surface analysis tool with high spatial resolution, and analytical capabilities, including electron energy analysis and total yield measurement. The instrument is in a UHV chamber. The zone plate focuses the x-rays onto the surface to be investigated. Electron spectra can be obtained from areas as small as 0.15 μm in diameter (Ade *et al.* 1992a) by sweeping the kinetic energy window of the electron energy analyzer. Alternatively the kinetic energy window is fixed to a

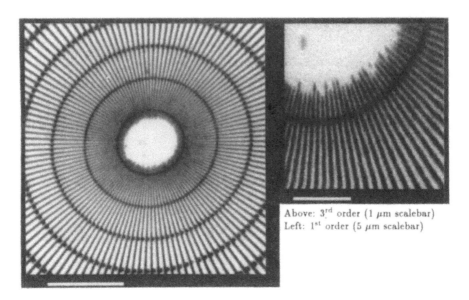

Above: 3rd order (1 μm scalebar)
Left: 1st order (5 μm scalebar)

Fig. 2. Scanning transmission x-ray micrograph of a microfabricated test pattern taken using the first order focus of a zone plate (left) and using the third order focus of the same 45 nm outer zone width zone plate (right). The third order image shows improved resolution, although to obtain optimal results a specially designed zone plate with proper spherical aberration correction would be required.

feature of interest and the specimen is scanned using stepper motors which act through bellows and a demagnifying flexure stage. By selecting a particular photoelectron or Auger peak, the distribution of a particular chemical species can be mapped on the surface. (Ade *et al.* 1990). The instrument is being upgraded, with the single channel cylindrical mirror electron analyzer being replaced by a hemispherical analyzer with multichannel detection. This will improve considerably both the energy resolution and the rate of data acquisition (Ko *et al.* 1992).

The scanning transmission microscope operates at atmospheric pressure in a helium-flushed environment (see Fig. 1). With the source and zone plate optics at the X1A beamline, we get up to 5×10^6 photons/sec in a 50 nm probe. In most applications, the sample is scanned through the zone plate focus with piezoelectric positioners, with the transmitted photons detected by a high rate gas flow counter. This provides a map of specimen absorptivity. This map is a strong function not only of the thickness, but also of the elemental and chemical constituents. With water relatively transparent in comparison with organic material between the carbon and oxygen absorption edges ("water window"), much of the interest is in the imaging of wet biological specimens (see Fig. 3) (Williams *et al.* 1992, Gilbert *et al.* 1992, Goncz and Rothman 1992). Images with x-ray energies at two or more of the sharp spectral structures in the neighborhood of x-ray absorption edges (XANES) allows the distribution of elements and chemical compounds of interest to be determined.

Elemental contrast in STXM has been used by Buckley *et al.* (1992) to map the distribution of calcified deposits in catrilage, while chemical contrast has been used by Ade *et al.* (1992b) to distinguish the constituents in high resolution images of a polymer

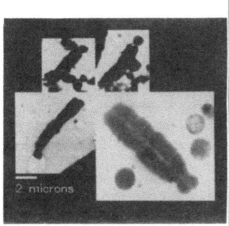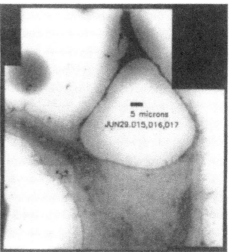

Fig. 3. Images obtained using the X-1A scanning transmission x-ray microscope. (Left) Bean plant *V. faba* chromosomes which have been prepared using three different freeze drying procedures (lower left, upper left, and upper right) compared to a fixed, wet, unsectioned chromosome (lower right). The structural artifacts caused by all three freeze drying procedures are considerable. (Right) HeLa cell culture imaged in the STXM following treatment according to a staining protocol for fluorescence microscopy which included an acetone rinse.

blend, and to form DNA maps of a chromosome. With the source and zone plate optics at the X1A beamline, we get up to 5×10^6 photons/sec in a 50 nm probe.

Other contrast mechanisms available to scanning x-ray microscopy include differential phase contrast, either using quadrant detectors as proposed by Palmer and Morrison (1991) or using optical lever schemes. X-ray luminescence microscopy has also been explored for its potential to to selectively highlight sites of biochemical activity (Jacobsen 1992); an example of this work is shown in Fig. 4.

4. CONCLUSIONS

Scanning soft x-ray microscopy is characterized by its simplicity and versatility. Once the microprobe is formed, it can be used as a static spectroscopic tool of the minute specimen area illuminated, or as an imaging instrument with a variety of detectible signals if the specimen is actually scanned. The information is acquired in digital form for convenient quantitative analysis.

It is important to keep in mind, however, that soft x-rays will damage radiation sensitive specimens. This is particularly true for unprotected biological samples. The damage depends on the radiation dose, and takes place over a time scale that may change the morphology of the specimen before a second picture could be taken. Furthermore, as the focusing optics improves, and higher resolution becomes technically possible, the dose to the specimen goes up roughly as the inverse cube of the spot size. Model calculations by Sayre *et al.* (1977) and others, as well as experimental studies by Williams *et al.*

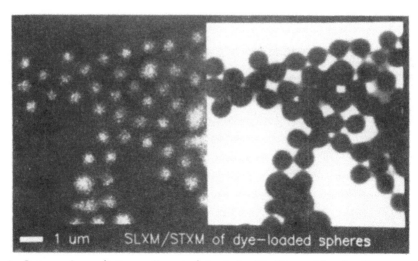

Fig. 4. Latex spheres (1.0 μm diameter) loaded with 50 μmol/g of a fluorescent dye with an excitation maximum at 490 nm and an emission maximum at 515 nm (Molecular Probes L-5181). The scanning transmission x-ray micrograph at right was taken immediately after the scanning luminescence x-ray micrograph at left. Images of P31 phosphor grains have shown that the spatial resolution of this technique is at the 50–75 nm level as determined by the zone plate.

(1992) and by Ade *et al.* (1992b) indicate that, except for the thinnest specimens, scanning soft x-ray microscopy involves less damage than electron microscopy and, with appropriate care, even wet biological samples can be imaged with good fidelity at 50 nm resolution. Gilbert *et al.* (1992) even succeeded in imaging initially live cells in culture at 90 nm resolution, although the radiation damage did cause death less than 3 hours after the exposure.

Because of the lack of post-specimen optics, and the near 100% efficiency of the transmitted x-ray detector, scanning transmission microscopy involves minimal dose to the specimen.

5. ACKNOWLEDGEMENTS

The development and operation of the X-1A scanning microscopes has benefitted from the efforts of Harvey Rarback, Christopher Buckley, Mark Rivers, Steve Lindaas, and Sue Wirick. We thank David Attwood for his support in connection with zone plate fabrication, and the staff of the National Synchrotron Light Source. Development of the STXM is supported by the National Science Foundation under grant DIR-9005893, and the Office of Health and Environmental Research, Department of Energy under contract DE-FG02-89ER60858.

6. REFERENCES

Ade H, Kirz J, Hulbert S L, Johnson E D, Anderson E, and Kern D 1990 Appl. Phys. Lett. 56, pp 1841–1843.

Ade H, Ko C-H, and Anderson E 1992a Appl. Phys. Lett. 60, pp 1040–1042.

Ade H, Zhang X, Cameron S, Costello C, Kirz J, and Williams S 1992b (submitted for publication).

Anderson E H 1992a SPIE Proc. 1741 (in press).

Anderson E H and Kern D 1992b, in Michette op. cit. pp 75–78.

Buckley C J, Ali S Y, Scotchford C A, and Rivers M L 1992 SPIE Proc. 1741 (in press).

Charalambous P S and Morris D 1992 in Michette op. cit. pp 79–82.

Gilbert J R, Pine J, Kirz J, Jacobsen C, Williams S, Buckley C J, and Rarback H 1992 in Michette op. cit. pp 388–391.

Goncz K K and Rothman S S 1992 Biochim. Biophys. Acta (in press)

Horowitz P and Howell J A 1972, Science 178, pp 608–611.

Jacobsen C, Williams S, Anderson E, Browne M T, Buckley C J, Kern D, Kirz J, Rivers M, and Zhang X 1991, Opt. Comm. 86, pp 351–364.

Jacobsen C, Lindaas S, Oehler V, Williams S, Wirick S, Zhang X, Guo S, and Spector I 1992 SPIE Proc. 1741 (in press).

Ko C-H, Ade H, Kirz J, Hulbert S L, Johnson E D, Anderson E H, and Kern D 1992 SPIE Proc. 1741 (in press).

Michette A G, Morrison G R and Buckley C J, eds 1992 X-Ray Microscopy III (Springer, Berlin).

Moewes A, Dadras H, Kunz C, Roy G, Sievers H, Storjohann I, Voss J, and Wongel H 1992 in Michette op. cit. pp 231–240.

Morrison G R, Anastasi P A F, Browne M T, Buckley C J, Burge R E, Charalambous P S, Foster G F, Michette A G, Morris D, Palmer J R, Slark G E, Bennett P B, and Duke P J 1992, in Michette op. cit. pp 139–142.

Ng W, Ray-Choudhuri A K, Liang S H, Welnak J, Cerrina F, Capasso C, Underwood J H, Kortright J B, and Perera R C 1992, SPIE Proc. 1741 (in press).

Niemann B 1992 in Michette 1992 op. cit. pp 143–146.

Palmer J, and Morrison G 1992 in Michette 1992 op. cit. pp 278–280.

Pattee, H H Jr. 1953, Phys. Rev. 43, pp 61–62.

Rarback H, Buckley C, Ade H, Camillo F, DiGennaro R, Hellman S, Howells M, Iskander N, Jacobsen C, Kirz J, Krinsky S, Lindaas S, McNulty I, Oversluizen M, Rothman S, Sayre D, Sharnoff M, and Shu D 1990 J. X-Ray Sci. Technol. 2, pp 274–296.

Sayre D, Kirz J, Feder R, Kim D M, and Spiller E 1977, Ultramicroscopy 2, pp 337–349.

Schmahl G 1991 private communication.

Trail J A and Byer R L 1989 Opt. Lett. 14, p 539.

Williams S, Zhang X, Jacobsen C, Kirz J, Lindaas S, Van't Hof J, and Lamm S S Ultramicroscopy (in press).

Zhang X, Jacobsen C, Williams S 1992 SPIE Proc. 1741 (in press).

Inst. Phys. Conf. Ser. No 130: Chapter 7
Paper presented at Int. Congr. X-ray Optics and Microanalysis, Manchester, 1992

579

X-ray diffraction contrast of polycrystalline material, visualized by scanning X-ray analytical microscope

S.Shimomura and H.Nakazawa

National Institute for Research in Inorganic Materials,1-1 Namiki,Tsukuba,Ibaraki,305,Japan

ABSTRACT: An X-ray diffraction image of polycrystalline material was obtained by using a scanning X-ray analytical microscope (SXAM): An X-ray microbeam illuminates a small area ($10\,\mu$ m ϕ) of a sample. A solid state detector (SSD) detects the secondary X-rays, and their intensities are stored in a micro computer. These processes are repeated for all points of a desired area. X-ray images can be constructed by using the intensity data of all the points as picture elements. The image of the distribution of crystallites of hydroxyapatite of a fish tooth demonstrates successfully SXAM's capabilities.

1.INTRODUCTION

To analyze chemically or crystallographically small areas of different types of samples, several microbeam techniques have been developed such as electron probe micro analysis (EPMA), scanning electron microscopy (SEM) and transmission electron microscopy (TEM). Use of electrons as a probe has the advantage of easy focusing by magnetic or electric fields. The minimum practical beam size used is about 10 nm in diameter. The use of an electron beam requires, however, a vaccum around the sample and it causes contamination of the sample. These two disadvantages are especially serious in the analysis of biological specimen or high purity samples such as electronic device. An X-ray microbeam is, therefore, more adequate for such samples and construction of an X-ray microscope and an X-ray micro area analyzer is required, although there are difficulties in focusing X-rays.

An X-ray guide tube (XGT) was used to guide X-rays from the source to the sample, using total external reflection at the inner wall of a glass tube (Nakazawa 1983; Nozaki and Nakazawa 1986). A conical-type XGT was applied to focus X-rays in the construction of a prototype of SXAM (Nakazawa et al. 1990). Along these lines, a new model of SXAM has been constructed by the present authors, with which an X-ray image showing a distribution of elements with spatial resolution of $10\,\mu$ m has been reported (Shimomura and Nakazawa 1991). The distribution of elements does not, however, give any structural information on the sample. To see and to visualize the distribution of a particular crystalline phase in a sample which contains different crystalline phases or that of an amorphous phase, a system which detects diffracted X-rays and makes an image using their intensities is needed as a compliment method of imaging the element distribution. Our SXAM can be used for such a purpose.

In the present paper, a thin section of a fish tooth was used as a sample to demonstrate the capability of the system in imaging both the element and the crystalline phase distributions. The distribution of hydroxyapatite crystals in the enameloid and dentine of the tooth is successfully imaged using the intensity data of diffracted X-rays.

2.EXPERIMENT

The SXAM presently used consists of the following components (fig.1): the X-ray generator is a micro focusing type (Rigaku microflex CN4180E2) and its focus size is 0.1mm × 0.1mm at a take off angle of 6° (Mo target, 30kV accelerating voltage and 2.5mA current). A conical-type XGT (0.25mm inner diameter at the entrance, 0.45mm at the exit and 220mm in length) is used for forming an intense microbeam (Nozaki and Nakazawa 1990). A pinhole (10 μ m× 10 μ m) is placed just before the sample. Assuming an ideal point source of X-rays and a total external reflection at the inner wall of the XGT, the divergence of X-rays through the pinhole is ± 0.86 × 10^{-3}rad. A small area (10 μ m× 10 μ m) of the sample is irradiated by the X-ray microbeam through the XGT and pinhole. The sample is placed on an x-z stage and is traversed two-dimensionally in the plane normal to the X-ray beam by computer controlled stepping motors. SSD (Li doped Si, 80mm^2 of working area and 3mm in thickness) is used for detecting the secondary X-rays from the sample. The x-z stage and SSD were set on the goniometer to arrange the SSD and a sample in a certain Bragg condition. The intensity of diffracted and fluorescent X-rays from a small area is measured by SSD and stored in a computer. Then the sample was moved by a small step (10 μ m), and the measurement repeated. After the measurements for all positions in a certain area of the sample, a mapping image could be obtained using the intensity data of the diffracted and fluorescent X-rays as picture elements.

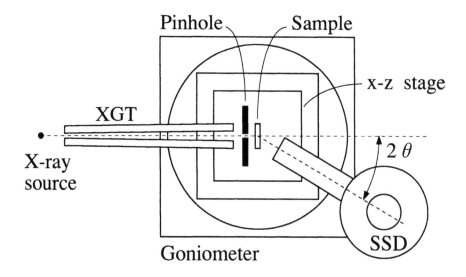

Fig. 1.   Scanning X-ray analytical microscope.

A thin section (60 μ m in thickness) of a tooth of a fish (surgeonfish) was prepared by polishing. The sample was mounted on the x-z stage perpendicular to the X-ray microbeam.

3.RESULTS

Before imaging of X-ray diffraction contrast, a Debye photograph of the tooth sample was taken to determine the crystalline phases in a small area and to observe their preferred orientation. An imaging plate (IP; FUJI imaging plate type BA) was used as a film which was placed behind the sample at a distance of 130mm. Because the diffraction intensity is always low when using X-ray microbeam, the high sensitivity and latitude of IP is covenient. As shown in fig.2, the Debye rings observed correspond well to those of hydroxyapatite crystals (Mo-K α, 30kV,1mA,108hrs of exposure) and indicate that the crystals were not oriented preferentially with in the beam size of about 10μ m $\times 10 \mu$ m.

Based on the observation of the Debye rings, the SSD was set at $2 \theta =14.5°$ which corresponded to the diffraction angle of 211 of hydroxyapatite for Mo-K α radiation.

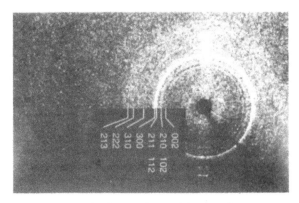

The intensity measurements of diffracted X-rays were synchronized with the sample traverse of 10μ m step in the area of 1.28mm $\times$ 1.28mm, thus, the number of measured points were 128 $\times$ 128.

Fig. 2. Debye rings of the sample.

An X-ray diffraction image of the sample was constructed using all those intensity data (fig.3(a)). Right bottom of the image is the tip of the fish tooth. An optical micrograph of the sample, at the same scale, is also shown in fig.3(b) for reference. The contrast in diffraction intensities from the surface and inner of the tooth is clearly imaged as well as that of the tip and the bottom.

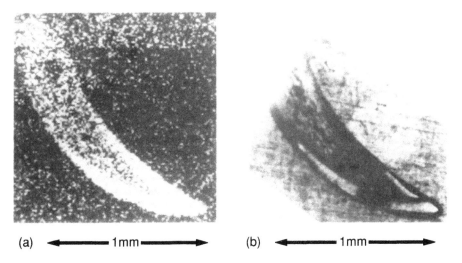

(a) ◄━━━━ 1mm ━━━━►   (b) ◄━━━━ 1mm ━━━━►

Fig.3. X-ray diffraction image of the sample (a). Optical micrograph of the sample (b).

582 X-Ray Optics and Microanalysis 1992

4.DISCUSSION

In the present geometry of SXAM, the intensity of diffracted X-rays from a samll area of the sample is in proportion to the amount of crystallites in that area. The contrast seen in fig.3(a) is, thus, that of the crystallite distribution. The contrast is well explained by the known fact that, in general, over 95% of enameloid (the outer part of the tooth) and about 65% of dentine (the inner part of the tooth) consists of hydroxyapatite crystallites (A.Makishima and H.Aoki 1984).

The present experiment successfully demonstrates that the distribution of a crystalline phase in a polycystalline sample can be visualized in an image using SXAM, of which spatial resolution is about 10 μ m. If a sample consists of different kinds of polycrystals, the distribution of each crystalline phase could be visualized by repeating similar intensity collections at other diffraction angles of other crystalline phases. This SXAM system will also be applied for imaging the crystal orientation distribution when the polycrystalline sample has a preferred orientation.

5.CONCLUSION

Imaging of X-ray diffraction contrast of polycrystalline material was successfully performed using SXAM. The image obtained in the present demonstration experiment represents well the distribution of hydroxyapatite crystallites in enameliod and dentine of a fish tooth as X-ray diffraction contrast.

6.REFERENCE

A.Makishima and H.Aoki 1984 Bioceramics Vol.7 Gihoudou Publishing House (in Japanese)
H.Nakazawa 1983 J.Appl.Cryst.16,pp239-241
H.Nozaki and H.Nakazawa 1986 J.Appl.Cryst.19,pp453-455
H.Nakazawa, Y.Kanazawa, H.Nozaki and S.Komatani 1990 X-ray Microscopy in Biology
    and Medicine. ed. by K.Shinohara, Japan Sci. Soc. Press, Tokyo/Springer-Berlin,pp 81-86
S.Shimomura and H.Nakazawa 1991 To be published in proceedings from PICXAM(1991)

Inst. Phys. Conf. Ser. No 130: Chapter 7
Paper presented at Int. Congr. X-ray Optics and Microanalysis, Manchester, 1992

583

Small d-spacing multilayer structures for the photon energy range E > 0.3 kev

Yu.Platonov, S.Andreev, N.Salashchenko, S.Shinkarev

Institute of Applied Physics, Uljanova 46, 603600, Nizhny Novgorod

ABSTRACT: Deposition possibility of small d-spacing (d~1-2 nm) multilayers on the basis of the material combinations W/Sb, W/B$_4$C, Cr/Sb, Cr/Sc, Fe/Sc and their utilization as dispersive and focusing elements for the photon energy range E>0.3 kev have been investigated.

1. INTRODUCTION

The interest in the develoment of small d-spacing (d<2nm) layered synthetic nanostructures (LSN's) arises from the possibility of their employment as focusing and imaging normal incidence optical devices as well as effective dispersive elements for monochromators for the photon energy range higher than the carbon K-edge (E > 0.3 kev). But there are two reasons which limit the utilization of LSN's. First is a relatively small spectral selectivity of LSN's, Which usally does not exceed E/ΔE=100 because of period fluctuation. The second restriction is caused by the influence of the interface roughness on the reflectivity:

$$R = R_o \exp(-16\,\pi^2\sigma^2\sin^2\vartheta_B\,/\,\lambda^2), \quad \sin\vartheta_B = m\lambda/2nd \qquad (1)$$

where R$_o$ is the peak reflectivity coefficient of an ideal LSN, σ is r.m.s.roughness, ϑ_B is the Bragg angle, λ is the wavelength of the radiation, nd is the optical period and m is the diffraction order. It follows that to preserve a large value of ϑ_B and to construct normal incidence LSN's for to shorter wavelengths it is necessary to have smaller d-spacing LSN's, but in this case the reflectivity of the LSN's is decreased dramatically by the influence of interface roughness. Using higher diffraction orders the reflectivity does not increase as follows from (1), but it may be done to improve the spectral selectivity and increase the layer thickness (G.van der Laan et al 1987 and F.Schafers et al 1988). The LSN working at the second diffraction order has twice higher spectral selectivity and is thus preferable if a high spectral resolution is needed. On the contrary, if a high integral coefficient of the reflectivity is needed (microscopy and lithography, for instance) utilization of the multilayers working at the first diffraction order is better. To obtain a high flux in the soft x-ray range (E>0.3 kev) we have investigated the possibility of a small d-spacing LSN deposition and the creation of effective focusing elements.

2. MATERIALS CHOICE AND LSN DEPOSITION

The knowledge of the optical constants (Henke et al 1982) allows to systematically carry out a reseach of the effective pair combinations of the materials for LSN construction. The main criterion is the possibility to produce high quality LSN's using available technology. In this work we used W/Sb LSN's despite the fact that the calculated reflectivity of these multilayers is lower than for Ni/Sb LSN's. The W/Sb pair can be used in a wider wavelength range and, as it turned out, it may be done with a superthin period (d=1.0nm). W/Si, W/B$_4$C, Cr/Sc and Fe/Sc multilayers have been also produced. Plane wafers of Si, plane glass and spherical surfaces of fused quartz were used as substrates. The magnetron sputtering technique in the high frequency and constant current mode have been employed. The diameter and the thickness of the targets were 46 mm and 2-2.5mm, respectively. The targets were sputtered in the atmosphere of pure Ar at a pressure of ~0.5 pa. The sputtering power was during B$_4$C deposition between 400W and for W 60W. The deposition rate for all materials was 0.2- 1.0 nm/sec. The substrates (S=60*30 mm-plane and D=30 mm-spherical) were mounted on a special table rotating under the targets at the height of 60 mm.

3.MEASUREMENT EQUIPMENT

The standard diffractometer "Dron 3M" with Cu x-ray tube was used to study of the LSN's by hard x-ray radiation. From the angular dependence of the reflectivity (ϑ-2ϑ mode) the following parameters of LSN's were determined-the layer thickness, the density of layer materials and the value of the interface roughness (A.D.Akhsakhalyan et al 1991). Two kinds of equipment were used for LSN's characterization in a soft x-ray range: a)spectrometer-monochromator RSM-500 with a grazing incidence grating and x-ray tube and b) Synchrotron Radiation from the storage ring in combination with a BESSY double cristal monochromator and reflectometer. The RSM-500 covers the spectral range from 0.8 to 50 nm with two spherical gratings of 2m and 6m radius and with 300 l/mm and 600 l/mm, respectively. The RSM-500 was equipped with a special vacuum chamber containing a goniometer which allows to measure the angular dependence of the reflectivity in the grazing incidence range of ϑ=0-85°. The optical scheme and description of the BESSY double crystal monochromator and the PTB reflectometer were presented by J.Fedhaus et al 1986 and M.Kuhne, P.Muller 1989. We used beryl crystals for the photon energy range 1.2<E<1.8 kev and crystals of InSb (111) for 1.8<E<4 kev.

4.RESULTS

The performance of the multilayers W/Si,W/Sb and W/B$_4$C at λ=0.154nm is presented in Table 1. The calculated reflectivity for an ideally sharp layer boundary and for a density of layers which is the same as in bulk material are also shown in Table 1.

Table 1.Experimental and calculated parameters of the LSN at λ=0.154nm. N is the number of periods; ϑ is the Bragg angle; $\Delta\vartheta$ is the Brag peak width at half maximum; "th." is for the calculated data and "ex." - for the experimental data.

Pair	W/Si	W/Sb			W/B$_4$C	
NN	M269	M302	M305	M307	M36	M31
N	100	100	100	110	100	60
ϑ, deg.	3.4	3.6	3.0	2.8	2.2	1.88
d, nm	1.31	1.25	1.48	1.59	2.04	2.39
β	0.5	0.56	0.44	0.41	0.42	0.35
$\Delta\vartheta$th., deg.	0.04	0.07	0.07	0.03	0.04	0.05
Rth., %	55.2	30.1	42.3	48.2	80.9	78.7
$\Delta\vartheta$ex., deg	0.04	0.05	0.05	0.03	0.03	0.04
ϑc, deg	0.42	0.47	0.45	0.44	0.39	0.37
Rex., %	1.3	4.32	6.9	17.2	25.6	41.6

From the reflectivity value for only the first Bragg diffraction order at λ=0.154nm it is impossible to unequivocally determine real LSN's parameters. So, we used the measured performance data of the multilayers in the soft x-ray range. In Fig.1 a),b) reflectivity and resolution data of the interested multilayers are shown (Platonov Yu. et al 1991). The data were obtained on the BESSY double crystal monochromator · using the reflectometer of the Physikalisch Technische Bundesanstalt (PTB). In the analysis we supposed that the transition layers are only limited by roughness, and interdiffusion was not taken into account. Of course, in this simplified model we can not expect a good correspondence between the experimental and the theoretical data. So we limited our accuracy to 10 %. We excluded from our consideration the range around $M_{4,5}$ absorption edges of W where the atomic factors of the W are not determined.

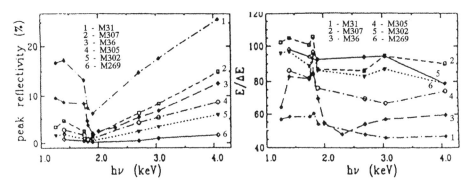

Fig.1. Reflectivity (a) and resolution (b) data for multilayers of different material combination and d-spacing.

The calculated results of the LSN's parameters are shown in Table 2. The interface roughness σ_{ideal} was calculated using data presented in the Table 1 supposing that the layer densities are the same as in bulk samples. As you may see from Table 2, the density in strongly absorbed layers (W) is 10-25 % smaller, but in the low absorbing ones, based on light elements (Si and $B_4 C$) it is 20 % more than in bulk samples. This is in agreement with the results of metal/carbon multilayers (A.D.Akhsakhalyan et al 1991). However, the density in low absorbing layers based on a heavy element (Sb) is 30% smaller than in bulk samples. The highest values of the interface roughness was achieved for W/$B_4 C$ (0.47nm) and W/Si (0.41nm).

Table 2. Layer densities and interface roughness of the LSN's W/Si, W/Sb and W/$B_4 C$ calculated from the data of Fig.1a,b and Table 1.

Pair	h₁,nm	h₂,nm	ρ₁,g/cm³	ρ₂,g/cm³	σ,nm	σideal,nm
W/Si	0.65	0.66	16.0∓1.5	3.0∓0.3	0.38∓0.03	0.43
W/Si	0.67∓0.3	0.77∓0.22	17.3∓1.0	4.5∓0.5	0.28∓0.02	0.32∓0.02
W/B_4C	0.85∓0.01	1.36∓0.19	16.0∓1.0	3.0∓0.3	0.43∓0.04	0.44∓0.02

h and ρ are the layer thickness and the density of the layer materials, respectively; signs "1" and "2" belong to the W layers and to low absorbing layers, respectively.
The roughness of all W/Sb structures does not exceed 0.3nm. From the data of Table 2 it follows that the multilayers W/Sb have the highest quality in comparison with the othe LSN's that were investigated. The calculated results of the performance of the LSN's containing Sb as the low absorbing layer show that such structures are promising at wavelengths just after the K-edge of oxygen-in a short wavelength range of the "water window". To study the possibility of creation of optical elements for this extremely important spectral range we have investigated LSN's which were deposited both on plane and spherical substrates. The experimental results of x-ray optical LSN's characterization by the RSM-500

are presented in Table 3.

Table 3. The characteristics of the multilayers W/Sb in the soft x-ray range.
S=tgϑ/Δϑ - LSN selectivity

LSN	d,nm	N	r,mm	E,ev	ϑ,deg.	R,%	S
				1254	33.2	0.8	55
M279	1.2	60		525	72.4	0.8	48
M290	1.3	80	320	525	60.7	0.4	46
M288	1.4	80	154	525	55.0	1.5	46
M285	1.8	80	154	525	40.0	1.8	40

In Table 4 the x-ray optical charachteristics of multilayers Cr(Fe)/Sc which are promising in the long wavelength range of the "water window" (0.3<E<0.4) are shown. The obtained W/Sb (d~3nm) and Cr/Sc (d~4.5nm) LSN's on flat substrates have a large reflectivity ,which permits their application as dispersion elements for x-ray fluorescent element analisis of O and N elements, as well as for contact micriscopy of biological objects. In the last case the whole "water window" range is recovered. Mirrors of periods 1.2...2.0 nm do not have a high reflectivity, especially in case of their deposition on spherical substrates. However, these mirrors can be used for imaging in the energy range 280...540 ev.

Table 4. The characteristics of the multilayers Cr(Fe)/Sc in the soft x-ray range. (conventional signs are as in the previous Tables)

Pair	NN	d,nm	N	r,mm	E,ev	ϑ,deg.	R,%	S
	M196	1.8	70		395	60.6	1.8	60
	M198	1.8	70	154	395	60.5	0.8	45
Cr/Sc	M199	1.9	70	154	395	54.2	1.0	45
					277	33.9	5.3	42
	M459	4.0	50		395	22.9	27.6	39
	M172	2.0	80		395	50.2	1.3	52
Fe/Sc					277	20.8	4.8	31
	M170	4.7	35		395	19.8	20.5	25

5.CONCLUSION

In the paper the performance of short period nanostructures (d<2nm) was studied. They are supposed to be used in the spectral range of the "water window". It has been found that W/Sb LSN's have the best quality for supershort periods of 1 nm. These LSN's can be used in the whole range 0.3<E<4kev, that has been studied in our experiments. For short period W/Sb and W/Si LSN's the spectral resolution λ/Δλ is close to 100 in the range 1<E<4kev. In the region around the W[4,5] absorption edges E(M)=1.88 and 1.81kev the reflectivity worsens for all W-containing LSN's. It is better to use LSN without W layers in this region. But we could not obtain high-quality structures with periods less than 2.5nm for other materials X/Sb (X=Cr,Ni,Fe). Besides the Sb-containing structures Cr(Fe)/Sc LSN's have been found to be the most effective. LSN's of the type W/Sb and Cr(Fe)/Sc recovered the whole "water window" range.

6.REFERENCES

Akhsakhaljan A,D. et al 1991 Thin Solid Films,203,pp 317-326
Fedhaus J.,Schafers F. and Peatman 1986 SPIE,v.733,pp 242-247
Henke B.L. et al 1982 Atomic data and nuclear data tables,27,pp 1-144
Kuhne M. and Muller P. 1989 SPIE,v.1140,pp 220-225
van der Laan G. et al 1987 Nucl.Instr.and Meth. in Phys.Reseach, A255,pp 592-597
Platonov Yu.,Salashchenko N.,Muller B.,Schaefers F. 1991,BESSY annual report
Schaefers F. et al 1988 SPIE, v.984,pp 23-30

Microspectroscopy and spectromicroscopy at the Hamburg focusing mirror scanning microscope

I Storjohann, C Kunz, A Moewes, J Voß

II. Institut für Experimentalphysik, University of Hamburg, Luruper Chaussee 149, D-2000 Hamburg 50, FRG

ABSTRACT: In the soft X-ray microscope with elliptical mirror optics at HASYLAB a hemisspherical electron analyser is applied for photoelectron spectroscopy. Over the photon energy range from 20 to 1300 eV spectroscopy on small areas ($2\mu m$) as well as microscopy based on chemical shift specific contrast are used. New results which demonstrate the advantages and the limits of the instrument are presented.

1. INTRODUCTION

The Hamburg focusing mirror scanning microscope (Voß et al 1992) demagnifies an entrance aperture by the factor of 35 by means of an ellipsoidal mirror. The whole range of the FLIPPER monochromator at the wiggler/undulator beamline (Senf et al 1986) of 20 - 1300 eV can be used to obtain various signals. A total/partial electron yield detector and a spherical electron analyser (SDA) can be used to collect photoelectrons. The SDA with a mean radius of 50 mm and a section angle of $150°$ has an energy resolution of $E/\Delta E = 75$. For microscopic image formation the sample can be scanned through the small light spot. With the light spot at a fixed position spectroscopic information can be collected by varying the photon energy, the kinetic energy of the detected electrons or both.

2. RESULTS

We investigated microstructures of an X-ray lithography mask, which consists of 50 nm thick Si_3N_4 structures on a Si single crystal support. The sample was sputtered and slightly annealed. The pressure in the preparation chamber was $7 \cdot 10^{-8}$ Pa and in the microscope chamber $1 \cdot 10^{-6}$ Pa.

Fig. 1 shows electron energy distribution curves (EDC's) obtained at different positions on the sample. The important aspect of these spectra is the energy shift of the $Si2p$ electrons due to their neighbour atoms (chemical shift). The difference in binding energy between the pure silicon 2p peak and the shifted peak in Si_3N_4 amounts to 2.3 eV. This shift in connection with the high count rate can be used to obtain spatially resolved

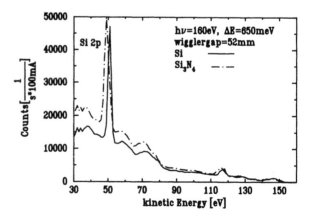

Figure 1: EDC's of different areas of the sample consisting of silicon and Si_3N_4.

pictures in reasonable time with high contrast. Fig. 2a shows an image of the sample taken with the anaylser tuned to the $Si2p$ electrons of elementary silicon and fig. 2b shows the corresponding image using the $Si2p$ electrons from Si_3N_4.

These pictures are put together from 9 single pictures, because the piezo driven scanning length is restricted to $135\,\mu m$. But due to the fact that the sample position can be measured with an accuracy of 40 nm over the whole range of manual prealignment (3 mm) the pictures could easily be added in one. Some difficulties occur in normalising intensity in the pictures with respect to each other and to the incoming photons (displacements of the beam in the storagering).

a)

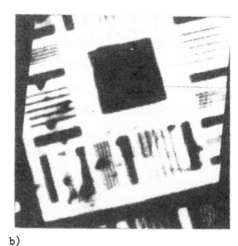

b)

Figure 2: Pictures of a Si/Si_3N_4 sample, photon energy = 160 eV, time per pixel = 0.03 s, $270 \cdot 270$ pixel, frame = $400 \cdot 410\,\mu m^2$.
a) $Si2p$ electrons of elementary silicon.
b) $Si2p$ electrons of Si_3N_4.

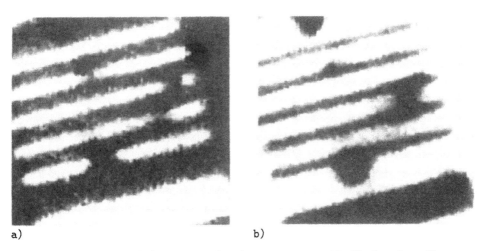

a)   b)

Figure 3: Pictures of a Si/Si_3N_4 sample, photon energy $= 160$ eV, time for taking up $\approx$ $2\,min$, $70 \cdot 65$ pixel, frame $= 107 \cdot 98\,\mu m^2$.
a) $Si2p$ electrons of elementary silicon.
b) $Si2p$ electrons of Si_3N_4.

Fig. 3 shows a detail of the sample. The smallest structures with a width of $2.0\,\mu m$ are given in fig. 4a.
The resolution with this geometry (microscope entrance aperture $= 100\mu m$) can be determined from a line scan across an edge of the central structure. The 25–75 % rise demonstrates a lateral resolution of $2.5\,\mu m$ (fig. 4b). A $55\,\mu m$ entrance aperture yields a value below $2.0\,\mu m$ but would give also a loss in intensity of 70%.

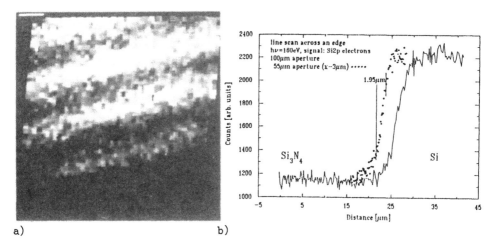

a)   b)

Figure 4: $h\nu = 160\,eV$ a) These $2\mu m$ wide Si structures are the smallest on the sample (frame $= 16 \cdot 17\mu m^2$). b) Line scan across an edge of the central structure (microscope entrance aperture $=100\mu m$).

In fig. 2 some extra "dirt" spots appear which are dark in both pictures and therefore are neither silicon nor siliconnitride. Fig. 5a shows an EDC obtained from such a spot. The spectrum displays a strong Na 2p peak, which is intense enough to obtain pictures and spectra typically in a few minutes. (fig. 5). Other elements (e.g. Ca and S could be fixed using different photon energies and the total yield detector.

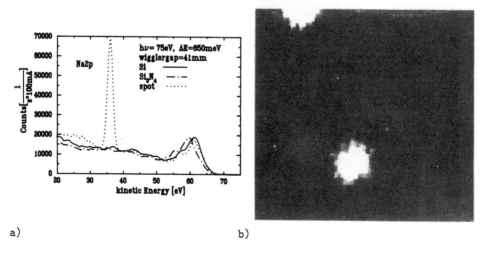

a)   b)

Figure 5: $h\nu = 75\,eV$. a) EDC light spot kept fixed on a dark spot (see fig. 2). b) The same sample detail as fig. 3, energy window $= 36\,eV$.

3. CONCLUSIONS

The results presented demonstrate a usable spatial resolution of 2.5 to $2.0\,\mu m$ in combination with sufficiently high count rates for photoemission experiments. Not only full spectroscopic information from small spots or samples is attainable, but also images can be taken in a reasonable time.

Acknowledgements
This work was supported by the German Federal Minister of Research and Technology (BMFT) under contract no. 05 5GUAX B 1, Teilprojekt 02.
The sample was provided by W.H.Brünger, Frauenhofer-Institut für Mikrostrukturtechnik, IMT, Berlin.

Voß J, Dadras H, Kunz C, Moewes A, Roy G, Sievers H, Storjohann I and Wongel H
    J. of X-ray Science and Technology 3, 85-108 (1992)
Senf F, Rautenfeld v. K B, Cramm S, Lamp J, Schmidt-May J, Voß J, Kunz C and
    Saile V Nucl. Instr. Meth. A246 (1986) 314

Aberration-corrected spherical and toroidal mirror systems for imaging and focusing of soft X-rays

J Voß and H Lübberstedt

II. Institut für Experimentalphysik, Universität Hamburg, HASYLAB

ABSTRACT: In the VUV and soft X-ray region glancing incidence mirrors are widely used for concentrating, focusing and imaging of radiation. Aspheric optical elements, paraboloids, ellipsoids or Wolter systems, which theoretically have the optimum imaging properties, are more difficult to manufacture compared to spherical, toroidal or cylindrical surfaces and are therefore more expensive and have worse surface accuracies.
The problem of the aberrations of the latter can be overcome by using corrected multiple mirror systems which are presented in a number of various combinations and geometries.

1. INTRODUCTION

Glancing incidence mirrors are an indispensable component in beamlines for synchrotron radiation. In order to monochromatize the polychromatic beam or to create a microfocus for X-ray microscopy curved elements are required. Aspheric surfaces of parabolic, elliptical or hyperbolic shape, spherical and toroidal mirrors can be used. Due to their strongly varying radii of curvature, aspheres, which in most cases show optimum theoretical imaging properties, usually can be figured only with less accuracy than spheres and toroids. On the other hand, reasonably-priced spherical and toroidal elements produce considerable aberrations.

To overcome this dilemma, Namioka et al (1970) proposed a single focusing, coma-corrected, double spherical mirror arrangement. The first mirror forms an intermediate focus which is then used as the object of the second one.

The concept presented here is based on the spherical approximation of an ellipse-hyperbola system. In this case, the second mirror forms a corrected focus by demagnifying the virtual image of the first mirror. In contrast to Namioka, the compensation equation (7) was derived starting from simple geometrical considerations. To exemplify the imaging properties of single and double focusing multiple mirror systems extracted from (7), the results of ray tracing calculations are discussed.

2. ABERRATION COMPENSATION OF DOUBLE TOROID SYSTEMS

The paraxial focusing properties of a toroidal mirror in meridian and sagittal direction (subscripts m and s) are described by the Coddington equations

$$\frac{1}{f_m} = \frac{1}{g} + \frac{1}{h_m} = \frac{2}{R_m \sin \alpha} \qquad \frac{1}{f_s} = \frac{1}{g} + \frac{1}{h_s} = \frac{2 \sin \alpha}{R_s} \qquad (1)$$

where α is the glancing angle of the principal ray, f is the focal length, g and h are the object and image distances and R is the radius of curvature in the appropriate direction.

The meridianal spherical aberration of a mirror of circular cross-section (see Figure 1) is ap-

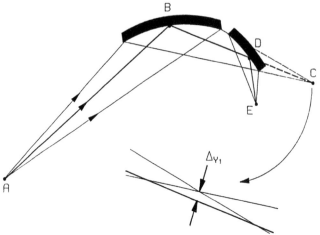

Figure 1: Meridian focusing, aberration corrected double mirror system.
The first mirror of circular cross section creates a virtual image C of the point object A. The second images C to the optimized real focus E.
Δy_1, the spherical aberration of the first mirror is shown in detail.
$ABDE$ is the principal ray, AB: g_1, BC: h_1, CD: g_2, DE: h_2.

proximately given by Kirkpatrick and Baez (1948) (subscript m suppressed)

$$\Delta y_1 = \frac{3}{8}\frac{L_1^2}{R_1^2}(M_1 - 1)\left(R_1 + \frac{M_1^2}{M_1^2 - 1}\frac{L}{\alpha_1}\right) \tag{2}$$

where L_1 is the illuminated length of the mirror, and $M_1 = h_1/g_1$ gives the magnification.

If the image of a point source created by the first mirror serves as the virtual object of the second one (see Figure 1), Δy_1 becomes magnified by $M_2 = h_2/g_2$. The image distance h_2 and the aberration Δy_2 of the second mirror can be calculated using the equations (1) and (2) if g_2 is treated as a negative value. The crucial parameter L_2 is determined approximately by

$$L_2 = (w_{21} + w_{22})\left(1 - \frac{\Delta y_1}{h_1\alpha_2}\right) \tag{3}$$

$$w_{21} = -R_2(a_{12} - \alpha_2) - \sqrt{R_2^2(a_{12} - \alpha_2)^2 - 2R_2 a_{12} g_2} \tag{4}$$

$$w_{22} = -R_2(a_{11} + \alpha_2) + \sqrt{R_2^2(a_{11} + \alpha_2)^2 + 2R_2 a_{11} g_2} \tag{5}$$

$$a_{11} = \frac{L_1}{2}\left(\alpha_1 + \frac{L_1}{4R_1}\right)\left(h_1 + \frac{L_1}{2}\right)^{-1} \qquad a_{12} = \frac{L_1}{2}\left(\alpha_1 - \frac{L_1}{4R_1}\right)\left(h_1 - \frac{L_1}{2}\right)^{-1} \tag{6}$$

The total aberration of the double mirror system reaches its minimum, if the compensation equation is fulfilled

$$M_2\Delta y_1 + \Delta y_2 = 0 \tag{7}$$

This equation can be solved for a required geometry. Equal radii of curvature, equal glancing angles, and minimum distances d between the mirrors, or any combination of these are possible.

In Figure 2 the results of ray tracing calculations of a single sphere (a) are compared with those of a corrected double mirror system consisting of identical spheres (b). Both systems have the same full vertical aperture of $\sigma_v = 230$ mrad and the same magnification $M = 0.033$. The residual aberrations $\Delta_r = 7.9$ μm of the corrected system are reduced by a factor of 33 proceeding from the single sphere with $\Delta = 262$ μm.

The simplest double focusing mirror is given by a "bicycle tyre" toroid. The strong aberrations of a single toroid are shown in fig. 2c. The focus has an absolute size of 350×380 μm^2 (vertical $\times$ horizontal).

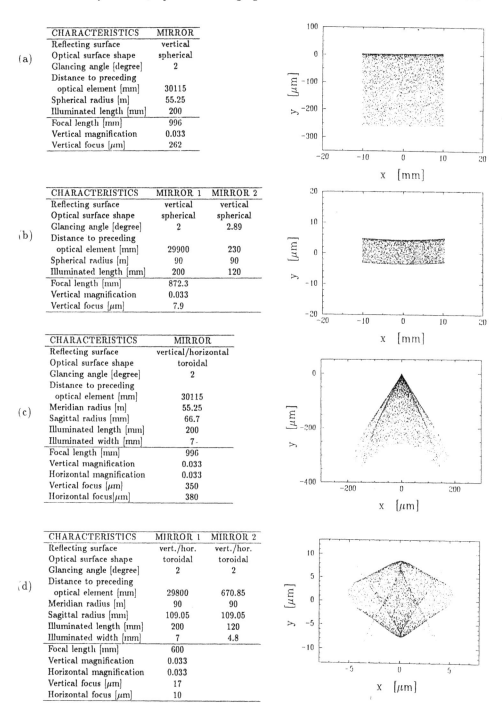

Figure 2: Results of raytracing. (a) single sphere, (b) single focusing double sphere system, (c) single "bicycle tyre" toroid, (d) double "bicycle tyre" toroid system. The "distance to preceding optical element" gives the distance to the source respectively the distance between mirror 1 and mirror 2.

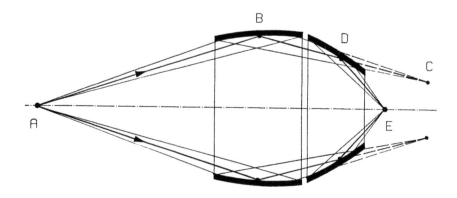

Figure 3: Wolter type double "apple core" toroid system.

A corrected double toroid system of the same aperture and magnification (see Figure 2d), consisting of two identical toroids, forms a focus of 17×10 μm^2.

A Kirkpatrick-Baez type system can be realized by a combination of two crossed pairs of spherical mirrors. A focus of 11×11 μm^2 can be obtained for an arrangement comparable with the double toroid.

Another promising mirror system arises if the spherical cross sections of two elements are rotated around the line AE (see Figure 3). It is similar to the Wolter objective (Wolter 1952), consisting of two confocal conicoidal surfaces, usually an ellipsoid and a hyperboloid. In general the object, the intermediate image and the focus of such a corrected double "apple core" toroid are not located on the line AE. Therefore, a meridian spherical approximation of a Wolter-system (e.g. Aoki et al 1980) does not satisfy the condition for the compensation of aberrations.

3. CONCLUSIONS

The imaging properties of spherical and toroidal mirrors can be improved drastically by using corrected multiple mirror arrangements. In the presented cases, the aberrations could be decreased by a factor of more than 30.

In particular the possibility of systems consisting of identical mirrors indicates a cheap way to avoid aspheric optics. The proposed combination of "apple core" toroids can be a real alternative to the Wolter objective.

ACKNOWLEDGEMENTS

This work was supported by the German Federal Minister of Research and Technology (BMFT) under contract no. 05 5GUAX B 1, Teilprojekt 02. The ray tracing calculations were performed with RAY, a program developed by Schaefers et al (1992) at BESSY.

References

Aoki S and Sakayanagi Y 1980 Annals of the New York Academy of Sciences **342** 158
Kirkpatrick P and Baez A V 1948 J. Opt. Soc. Am. **39** 766
Namioka T and Seya M 1970 Appl. Opt. **9** No.2 459
Schaefers F and Feldhaus J 1992 Technical Report BESSY
Wolter H 1952 Annalen der Physik **10** 97 and 107

Inst. Phys. Conf. Ser. No 130: Chapter 8
Paper presented at Int. Congr. X-ray Optics and Microanalysis, Manchester, 1992

595

The opportunities and challenges of using high brilliance X-ray synchrotron sources

P Pattison

Institut de Cristallographie, Université de Lausanne, BSP Dorigny, CH-1015 Lausanne, Switzerland

ABSTRACT: The trend towards the development of dedicated storage rings optimized as sources of synchrotron radiation is described. Various parameters are used to illustrate the dramatic and continuing progress being made in the performance of these facilities. This progress will shortly culminate in the operation of a new generation of storage rings, which will offer the combination of very low emittance and the availability of a wide variety of insertion devices. The design of X-ray optical components which can make optimum use of these enhanced source characteristics is presently the subject of intense scientific and technical activity.

1. INTRODUCTION

Although the potential of synchrotron radiation as a useful source of X-rays was already recognized by a few far-sighted individuals in the 1950's, it was not until the following decade that the first experimental work began in earnest. Much of the early work was in the soft X-ray/VUV region, since this type of radiation could be readily produced by many of the low energy machines in operation at that time. The use of harder X-rays depended upon the availability of much higher energy machines, which meant sharing (often a rather small share) of the time with the high energy physics community or else operating in a purely parasitic mode. Despite the difficulties of carrying out experiments in such an environment, some of the early work was very successful indeed. This led to the establishment of several permanent synchrotron laboratories within high energy physics establishments (e.g. at DESY, Hamburg and at CESR, Cornell), and eventually to the construction of high energy machines exclusively for synchrotron radiation work. The 2 GeV storage ring (SRS) at Daresbury Laboratory was one of the first large machines dedicated to the production of synchrotron radiation.

Many very productive laboratories do continue to operate on shared facilities, but it has become clear that most high energy physics machines are not particularly suitable for the production of synchrotron radiation. The size of the electron (or positron) beam may not be optimized at the source point, and the beam itself may not be sufficiently stable. The lifetime of the stored beam is often inconveniently short, resulting in a stop-go mode of data collection, and the beam characteristics may not be sufficiently reproducible between fills. All of these aspects have long been recognized as placing difficulties, sometime almost insurmountable difficulties, in the way of the experimentalist. Despite these problems, parasitic facilities have proved to be an extremely cost-effective method of producing a remarkable output of experimental data in a wide range of applications, ranging from biology to materials science and from geophysics to atomic physics. Nevertheless, the trend is either towards the design and construction of dedicated facilities or, in some cases, the parasite has actually been able to take over the host.

The aspects of synchrotron radiation which make it an attractive X-ray source are intensity, collimation, tunability, polarization and, for some experiments, time structure. These features have allowed dramatic improvements to be achieved in many well-established fields (e.g. single crystal X-ray diffraction and topography), while also enabling new techniques to be developed in such areas as absorption spectroscopy (EXAFS), surface diffraction and magnetic scattering. It is interesting to note that the success of these techniques using synchrotron radiation has not, as one might at first imagine, rendered conventional laboratory sources redundant. Rather the opposite, the excitement generated by the synchrotron work has often promoted efforts to repeat these experiments in the home laboratory. This has been the case for EXAFS, where many laboratory-based spectrometers are now operating, as well as for surface diffraction, X-ray interferometry, ultra-high resolution X-ray scattering (phonon scattering) and many others. This somewhat paradoxical development can be understood in the light of two important factors. On the one hand, technical improvements in conventional X-ray sources (e.g. rotating anodes and micro-focus tubes) and detectors (particularly one - and two-dimensional position sensitive detectors) have improved the competitiveness of the home laboratory. On the other hand, demand for synchrotron radiation has always far outstripped supply, while the sources themselves were often somewhat unreliable and usually situated at an inconvenient central facility. Hence, the importance of providing not only more synchrotron radiation sources, but also better quality sources with higher overall reliability. Fortunately, recent developments have led to dramatic improvements in the performance of synchrotron sources, which far outstrip the parallel development of laboratory sources. In addition, the experience built up during the operation of dedicated facilities over the last decade should ensure that the combined reliability of source, beamline and experimental equipment (all of which have to work at the same time) will also continue to improve.

Two factors have led to the improvements mentioned above in the performance of synchrotron sources. The design of the magnetic lattice (the array of magnets which constrain the electron beam) has been optimized for achieving low emittance, principally by reducing the cross section of the electron beam. The emittance of an electron beam is a parameter which combines the positional and angular distribution of electrons in the stored beam. A smaller source size produces a corresponding increase in the brilliance of the X-ray beam. The second factor was the development of insertion devices (wigglers and undulators), in which an array of dipole magnets is placed into long straight sections between the bending magnets. In contrast to earlier machines using only bending magnets as sources of radiation, the insertion devices allow synchrotron radiation to be produced with special characteristics e.g. increased total flux, wavelength shifted, quasi-monochromatic, reduced angular divergence, circular polarization etc. These factors have led to dramatic increases in flux, and particularly in flux per unit source area. Experiments can therefore be carried out on samples where the scattering is very weak indeed, either because of tiny sample volumes (e.g. scattering from surfaces or micro-crystals) or because the small scattering cross sections (e.g. magnetic scattering, plasmon scattering and nuclear resonance scattering). Hence the *second generation* of storage rings based on the use of bending magnet sources are now being replaced by *third generation* rings which are optimized for low emittance and the use of insertion devices. Since undulators are low field devices, the new storage rings for X-ray applications will tend to operate at much higher energies (6 - 8 GeV) than the earlier ones (1.5 - 2.5 GeV).

These developments in synchrotron sources have been accompanied by similar progress in the design of X-ray optical components for *beam conditioning* i.e. the selection of photons with the appropriate characteristics (energy, direction and polarization) and, equally important, the rejection of unwanted photons. These optical components are situated within the *beamlines*, which carry the X-ray beam from the source to the sample. The design and construction of the beamlines turns out to combine a whole array of technical problems related to positional stability (at the level of microns), angular precision (μrad), thermal stability under extreme heat loads and very high precision mechanics, all within a high vacuum or UHV environment and operating under remote control. Finally, the equipment used for collecting the experimental data must fulfill tougher specifications, if the advantages of synchrotron radiation are not to be wasted. For example, when it is possible to carry out a diffraction experiment on a sample whose dimensions are only be a few microns, then the tolerances on the stability, accuracy and reproducibility of the diffractometer goniometry must also be at the micron level. Similarly, the very high fluxes can generate very strong diffraction or fluorescence signals, often in the MHz range. Detectors must be able to extract relatively subtle variations in intensity from this strong signal. This is a typical situation, for example, in a fluorescence EXAFS experiment, and it places new and difficult demands on the performance of X-ray detectors and their associated counting chains. Indeed, the whole field of X-ray detector development, particularly those detectors with position sensitivity, is a crucial subject for many synchrotron applications (e.g. small-angle scattering, protein crystallography, powder diffraction, and topography). The demands on detectors will only increase, as more and more experiments require on-line information in "real-time".

In the following section, the basic characteristics of synchrotron sources are summarized. The requirements which these sources place upon the X-ray optics will then be discussed. As an example of the application of these techniques, a typical beamline design will be described and its expected performance assessed.

2. SYNCHROTRON SOURCES

This section will be used to define some parameters related to synchrotron sources, and to illustrate the scale of these parameters for typical facilities. The notation is the same as that used in the ESRF Foundation Phase Report (ESRF 1987), and examples will be used for the ESRF, Grenoble and the SRS at Daresbury Laboratory.

The spectral flux (expressed in units of photons/sec/mrad$_H$/0.1%$\partial E/E$) produced by an ESRF bending magnet is shown in Fig. 1. The bending magnets produce a broad spectrum of radiation peaking around the critical energy, Ec, expressed by,

$$E_c \text{ [keV]} = 0.665 \cdot E^2 \cdot B \tag{1}$$

where the electron beam energy, E, is expressed in units of GeV and the magnetic field, B, in Tesla. The bending magnet radiation is emitted in a broad horizontal fan, and the appropriate width of beam is selected by apertures and slits in the beamline. The vertical divergence is constrained by the natural opening angle of the radiation to be of the order of $1/\gamma$, where

$$\gamma = 1957 \cdot E \tag{2}$$

For the ESRF with 6 GeV energy, for example, γ has a value of 11742, resulting in a FWHM of the vertical divergence of the radiation fan of 0.13 mrad (at the critical energy). In fact another contribution to the vertical divergence comes from the angular distribution of the electrons, but this factor is not significant in low emittance storage rings. Eq (2) illustrates one of the advantages of using a higher energy machine, since there will be a reduction in the vertical divergence of the beam. The brightness of the photon beam on the orbital plane and at the critical energy is given by,

$$\dot{\Phi}_\Omega = 1.92 \times 10^{13} \cdot I \cdot E^2 \tag{3}$$

where I is the circulating electron current in Amps, and $\dot{\Phi}_\Omega$ is expressed in photons/sec/mrad^2/0.1%$\partial E/E$. Hence, for a 100mA current, the ESRF bending magnet with a field of 0.85T produces a photon beam with a brightness at E_c of about 6.9×10^{13}.

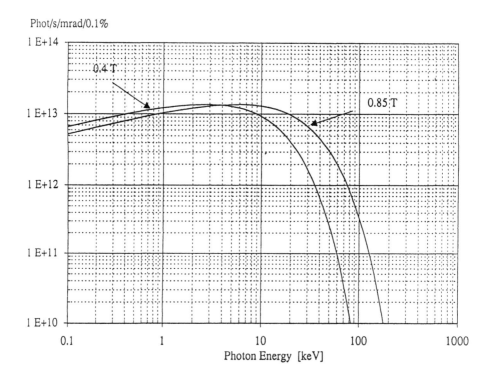

Phot/s/mrad/0.1%

Fig. 1. Spectral distribution of synchrotron radiation produced by a bending magnet at the ESRF

It is useful at this stage to compare this figure for brightness with the X-ray beam produced by a laboratory source. Following the procedure described by Honkimäki et al (1990), the flux of characteristic Cu radiation produced by a rotating anode X-ray generator operating at 300 mA and 40 KeV (12kW) is 4.5×10^{11} photons/sec into a solid angle of 10^{-3} sterad (i.e. a brightness of about 4.5×10^8 in the characteristic line). A typical solid angle that might be collected (i.e. focused) from a rotating anode generator is, say, 5 mrad$_H$ by 0.5 mrad$_V$, which therefore contains a flux of 1.1×10^9 photons/sec of characteristic radiation. If we take the example of the ESRF bending magnet referred to above, in a solid angle of 2.5 mrad$_H$ x 0.1 mrad$_V$, and take into account the bandpass, $\partial E/E$, of a Si(111) monochromator (about 1.3×10^{-4}), the resulting flux is 2.2×10^{12} photons/sec. Hence the bending magnet source is producing a flux of photons about 1000 times more intense than the rotating anode into a solid angle about a factor of 10 smaller. This calculation is not intended to be an accurate quantitative comparison between these two sources, since there are many more practical factors which would also have to be taken into account, but it serves to illustrate the potential advantages of a synchrotron experiment. However, it is interesting to note that, in those case where a large solid angle of radiation from the rotating anode generator can be used (say, 10^{-3} sterad), then the flux from the rotating anode X-ray source becomes much more competitive.

In fact, the X-ray beam from the synchrotron source carries so many photons in such a small solid angle that the thermal energy deposited into any solid object in its path is significant. The heat load, P_Θ, expressed in W/mrad$_H$, is given by,

$$P_\Theta = 4.224 \cdot B \cdot E^3 \cdot I \qquad (4)$$

Hence the photon beam from an ESRF bending magnet carries a thermal load of about 78 W/mrad$_H$, when the ring has a stored current of 100 mA. At a distance of about 30m, where optical components are installed in the beamline, this is equivalent to a power density of about 1 W/mm^2. Heat loads of this magnitude can already cause severe heating, deformation and even mechanical damage in uncooled components. A recent photograph of an uncooled metallic gate valve accidentally exposed to a bending magnet beam (Fauchet et al 1992), showing how the X-ray beam melted its way through the valve, provides a graphic illustration of the problems posed by these high power densities.

In a *multipole wiggler*, a series of dipole magnets are arranged in a straight section of the storage ring between the bending magnets. Although the relations governing the production of synchrotron radiation from wigglers are not the same as those for a bending magnet, the radiation produced is similar. Wigglers generally have higher fields than bending magnets, which also increases the critical energy of the synchrotron radiation. A typical wiggler in the ESRF, for example, will have a field of 1.2T and 24 poles. In a wiggler, radiation from the various dipole sources does not interfere, and the total effect is to produce a beam whose intensity is simply scaled up by the number of poles (compared to an equivalent bending magnet). Unfortunately, the heat loads are also scaled up accordingly, and hence the power density in the white beam of the ESRF wiggler is about 30 W/mm^2 at a distance of 30m from the source. The total power produced in a horizontal fan of 1.5 mrad is about 4.9 kW.

If the field is much lower and the period between magnetic poles is reduced, then interference effects between the dipole sources can result in a dramatic modification of the spectral distribution of the synchrotron beam. Such a device is called an *undulator*, and the effect on the spectral distribution is to concentrate the photons into a narrow fundamental energy band, and into its higher harmonics. Since the undulator is a low field device, the total power produced is generally much smaller than on a wiggler. However, the interference effects also concentrate this power into a narrow angular cone. This greatly increases the brightness of the beam (at the fundamental and its harmonics), but also produces power densities which can approach or even exceed those on a wiggler beamline.

An important parameter in any optical system is the brilliance of the source, which is the brightness per unit source area. Although the use of insertion devices has led to important improvements in the brightness of the beam, the low emittance of third generation sources provides an additional increase in the brilliance. A comparison of the SRS and the ESRF illustrates this point well. The original source size of the SRS (expressed in FWHM) was 14.4 mm (H) and 0.57 mm (V), which was improved by the installation of a new magnetic lattice in 1987 to be 2.6mm (H) and 0.24mm (V) as described by Suller et al (1988). The source size of the ESRF bending magnet is 0.38mm (H) and 0.31mm (V), and an ESRF undulator has a source of 0.13mm (H) and 0.09mm (V). The ESRF parameters illustrate the substantial improvement in brilliance resulting from the much smaller source sizes on the new machines. However, advantages accrue from the higher brilliance only if the X-ray focusing can be achieved without introducing significant optical aberrations, which would lead to a blurring of the focus. The design of the optical components which can fulfill these specifications, despite high thermal loading, is the subject of the following section.

3. BEAMLINE DESIGN

The interested reader is referred to a number of review articles on X-ray optics as well as the proceedings of recent international conferences on synchrotron radiation instrumentation, some of which are listed in the bibliography. In the present section, we will examine a particular type of beamline, and use this design to identify some of the specifications which the optical components must fulfill.

Some beam defining devices, such as slits and apertures, are used both for definition of the beam cross section and to reduce the amount of secondary scatter which propagates down the beamline. Other devices, called filters, attenuators or absorbers, serve the purpose of modifying the overall spectrum of radiation which reaches X-ray optical elements placed farther downstream (for example, to reduce the overall heat load in the beam by rejecting unwanted photons). Selection of the appropriate spectral range for a particular experiment is generally done with a crystal monochromator (in the X-ray range above about 5 keV). A multilayer (also called layered synthetic microstructure) can also be used for energy selection, although the energy bandpass is too wide for most spectroscopic and high resolution diffraction work. Although multilayers can also be used in conjunction with crystals, we shall restrict our discussion here to the use of perfect crystals (such as high purity silicon or germanium). Focusing of the X-ray beam onto the target is essential if proper advantage is to be taken of the high brilliance of the new sources. Focusing can be achieve either by bending the monochromator crystals or by using grazing incidence reflection from mirrors (or by a combination of both techniques).

We present a design for an ESRF bending magnet beamline using perfect crystals for X-ray energy selection and curved mirrors both to provide focusing and to improve the energy resolution. Similar designs are already in use at the Photon Factory in Japan and at the NSLS, Brookhaven, and several ESRF beamlines (e.g. the materials science beamline on a multipole wiggler) will use the same principles. This general-purpose beamline is being funded as a joint venture by national agencies and universities in Switzerland and Norway, and will be used for a wide variety of experiments ranging from materials science to macromolecular crystallography.

A double crystal, symmetrically cut, Si(111) monochromator will be used to select the appropriate energy in the range from about 5keV to 30keV. This is a very useful range for most X-ray diffraction, spectroscopy and topography applications, and also well suited to the spectral distribution produced by the ESRF bending magnet (as shown in Fig. 1). A double crystal arrangement is chosen because a single reflection from a monochromator would result in an inconvenient change in beam direction whenever a different energy was selected, whereas the direction of incident and reflected beams is conserved after two reflections. It is, however, necessary that the reflecting planes of the two crystals in this type of monochromator always remain parallel to within the rocking curve of the chosen reflection. In the case of Si(111) this amounts to a few arc seconds (10 - 30 μradians), which is a very stringent criterion indeed. A separated, double crystal arrangement is chosen, rather than the monolithic "channel-cut" design, because the gap between the two crystals can then be adjusted in such a way as to maintain a constant exit beam height. In addition, the second crystal can then be bent cylindrically (with the cylinder axis along the beam direction) in order to provide horizontal focusing of the beam. The aim is to collect several mrad of horizontal aperture at a distance of about 30m from the source point, and to focus the beam onto a sample situated about 15m away from the monochromator. The cylindrical bending radius depends upon the chosen photon energy, and is typically between 2m and 10m. Unfortunately, bending a thin plate (i.e. the crystal) into a cylinder induces an inverse bend in the other direction, so that the shape of the bent plate resembles a saddle. This is called the anticlastic curvature, and would result in an anticlastic radius of about 10m - 40m in our case. However, the requirement that the planes of the two diffracting crystals be parallel to within the rocking curve width implies that the anticlastic radius should be more than about 1km! Clearly some method of constraining the anticlastic bending is required. One method is to provide strengthening ribs along the rear face of the crystal. This is not an ideal solution, since the ribs distort the curvature away from the ideal cylinder, and several alternative schemes are presently under investigation.

The white beam incident upon the first crystal has a power density of about 1 W/mm^2, and a total thermal load over 2.5 mrad$_H$ of around 200W. Without an efficient method of cooling, the crystal would quickly reach a high temperature, and thermal distortion would lead to a drastic drop in the throughput of the monochromator (since the reflecting planes of the two crystals would no longer be parallel). Many different cooling methods are under development. With the heating loads described above, water cooling through a large-bore channel drilled through the silicon monochromator crystal will be sufficient. More efficient cooling can be provided when the water/silicon surface area is increased by micro-channeling or by cutting cooling fins in the silicon. The ESRF wiggler beamlines will be producing power densities of around 30 W/mm^2, and water cooling may not be adequate. Alternate schemes using various cryogenic fluids (including, for example, liquid N$_2$) are under investigation, as are methods of employing liquid metals (such as liquid gallium).

The vertical divergence of the synchrotron beam incident on the monochromator is about 0.13 mrad (FWHM) at energies around E_c. This divergence is almost an order of magnitude larger than the rocking curve width, which implies that the energy resolution of the X-ray beam selected by this monochromator will be dominated by the vertical opening angle of the radiation. Where photon flux rather than energy resolution is important, a choice of monochromator crystal with a larger rocking curve width would be advisable (e.g. the use of Ge rather than Si). In some cases, the use of silicon after heat treatment or surface abrasion has resulted in increased flux. In many experiments, such as X-ray spectroscopy, the energy resolution is crucial and the vertical divergence must be reduced. This can be achieved either by inserting vertical slits in front of the monochomator, which reduces the flux, or by collimating the incident beam. Beam collimation is clearly the better option, since good energy resolution can then be obtained without loss of flux. This collimation can be achieved by the use of total reflection in suitably curved mirrors set at grazing incidence of a few mrad. Some intensity is lost due to non-ideal reflection in the metal-coated mirrors, but reflectivities of 80% - 90% can readily be achieved. The technical problems are again quite challenging, since the mirrors must be long enough to intercept a substantial part of the beam (which implies lengths of a meter or more) and achieve a radius of curvature of the order of 15km - 20km. At these radii, the deflection at the centre of the mirror is only about 10 microns. The deflection of the mirror under its own weight (sagging) becomes significant at these very large radii, and

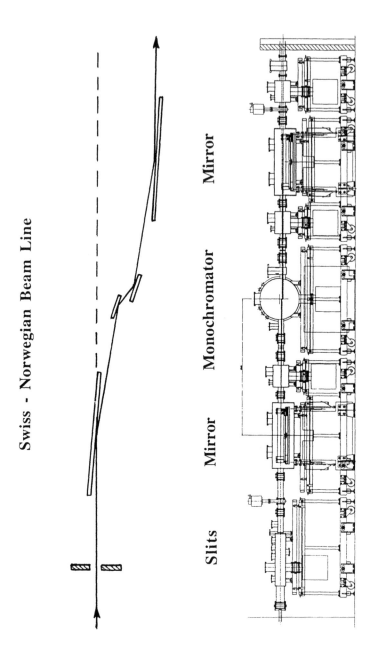

Fig. 2. Layout of the X-ray optics section of the Swiss-Norwegian Beam Line at the ESRF

the bending mechanism must therefore be able to cater for this self-weight effect. In addition, the mirror is exposed to the white synchrotron beam, and hence will distort under thermal load. For the relatively modest thermal loads present on the ESRF bending magnet beamlines, a suitable choice of low expansion coefficient material is sufficient to avoid serious thermal distortion. However, such materials would not survive in a wiggler beam, in which case appropriate cooling is required. As with monochromator cooling, various schemes are under evaluation, including cryogenic cooling. It should be noted that both the cooling and the bending must take place within a high vacuum environment, which complicates both the mechanical arrangement for the bending mechanism and the introduction of a suitable cooling medium into the body of the mirror or its surroundings. The mirror itself is manufactured as a flat, and bent to the appropriate radius. Because of the grazing incidence angles, the tolerances on microroughness (≤ 5Å RMS) and the tangential slope error (≤ 2.5 μrad) are extremely small. The X-ray beam emerging from the monochromator is converging horizontally, but should be as parallel as possible in the vertical plane. Clearly, the intensity arriving on a small target would be improved if it were also possible to focus vertically. This can be achieved with the use of a second mirror positioned after the monochromator. Since the parallel beam has a height of about 3mm, and a typical single crystal sample may be 200μm - 300μm in size, the gain in useful intensity could be up to an order of magnitude. Although the mirror reflectivity is less than 100%, and residual optical aberrations produce an imperfect focus, practical intensity gains of about a factor of 3 - 5 times are certainly possible. Moreover, the use of two mirrors rather than one allows the synchrotron beam to remain in the horizontal plane, and also enables the incident angle of both mirrors to be adjusted without causing a large displacement of the exit beam.

The final configuration of the beamline is shown in Fig. 2. After passing through a vessel containing various beam apertures, slits and a beam position monitor, the beam strikes first a mirror, then the double crystal monochromator, followed by the second mirror. The beam aperture is redefined after each optical element in order to eliminate unwanted scattering and hence to improve the signal-to-background in the beam arriving on the sample. The useful flux can be estimated in the following way. The brightness of the synchrotron beam from the ESRF bending magnet is about 6.9×10^{13} photons/sec/mrad^2/0.1%$\partial E/E$. It will be assumed that the radiation can be collected in a solid angle of about 2.5mrad$_H$ x 0.1mrad$_V$. Using a bandpass for Si(111) of 1.3×10^{-4}, and reflectivities of each mirror of 0.6 and of each monochromator of 0.9, then the flux arriving in the focal spot will be about 5×10^{11} photons/sec within an energy resolution $\partial E/E$ of about 2×10^{-4}. This flux is still orders of magnitude greater than the flux from a rotating anode X-ray generator, even without taking into account the losses which would occur in any focusing arrangement used on the conventional X-ray source. With the arrangement of X-ray optics shown in Fig. 2, the focal spot should be about 0.4mm x 0.4mm (FWHM). It is worth noting that the brightness produced by a multipole wiggler at the ESRF will be a factor of 10 - 30 times higher than the bending magnet (depending upon the number of poles) while the brightness of the undulator will be even higher. However, the problems presented by the correspondingly high heat loads produced by wigglers and undulators remain a technical challenge to the builders of synchrotron beamlines.

ACKNOWLEDGMENTS

The author is grateful for financial support from the Swiss National Science Foundation under Grant No. 21-28,948.90. Some of the work described in the publication forms part of the Swiss-Norwegian Beam Line Project at the ESRF, and the author would like to express his thanks to many colleagues in this project who are working on the design and construction of the beamline. This paper is No. 92-01 from the Swiss-Norwegian Beam Line.

BIBLIOGRAPHY

Proceedings of the 3rd International Conference on Synchrotron Radiation Instrumentation, 1989 Rev Sci Instrum 60 1373 - 2566

Proceedings of the 4th International Conference on Synchrotron Radiation Instrumentation, 1992 Rev Sci Instrum 63 283 - 1634

Handbook on Synchrotron Radiation Vol 1A 1983 (Amsterdam : North-Holland)

Handbook on Synchrotron Radiation Vol 3 1991 (Amsterdam : North-Holland)

REFERENCES

ESRF Foundation Phase Report 1987 (Grenoble: ESRF)

Fauchet A M, Biscardi R, Singh O, Yu L H, Stefan P 1992 Nucl. Instrum. Methods A319 8

Honkimäki V, Sleight J and Suortti P 1990 J. Appl. Cryst. 23 412

Suller V P, Corlett J N, Dykes D M, Hughes E A, Poole M W, Quinn P D, Mackay J S, Thompson S L, and Walker R P (1988) Daresbury Laboratory Report DL/SCI/P590A

Inst. Phys. Conf. Ser. No 130: Chapter 8
Paper presented at Int. Congr. X-ray Optics and Microanalysis, Manchester, 1992

605

Present status of and future prospects for synchrotron-based microtechniques that utilize X-ray fluorescence, absorption spectroscopy, diffraction and tomography

Joseph V. Smith

Center for Advanced Radiation Sources, University of Chicago, Chicago, Illinois, USA 60637

ABSTRACT: Current experimental stations at second-generation hard x-ray synchrotron sources permit the following: part-per-million sensitivity for energy-dispersive analysis of many elements using fluorescence from a 10-micrometer beam; Laue and Bragg diffraction from crystals down to 1-10 micrometer across; absorption spectroscopy for beam spots down to 0.1 mm for concentration down to 0.1 wt. %, with near edge spectroscopy much easier than for extended fine structure; absorption and fluorescence tomography. The design of experimental stations for the Chem/Mat-, Geo- & Soil/Enviro-CARS consortium at a third generation synchrotron source, the Advanced Photon Source at Argonne National Laboratory, is presented. Much higher sensitivity and/or faster data collection will be obtained by use of undulator and wiggler sources coupled with multi-element and area detectors.

1. INTRODUCTION

Hard x-ray synchrotron x-ray sources (SXSs) are transforming the long-standing techniques of *x-ray fluorescence analysis* (XRFA), especially for trace elements, *x-ray absorption spectroscopy* (XAS: includes both x-ray absorption near edge spectroscopy, XANES, for the electronic state, and extended x-ray absorption fine structure spectroscopy, EXAFS, for the near-neighbour coordination), *x-ray diffraction* (XRD: includes both powder diffraction, XRPD, and single crystal Bragg and Laue diffraction, XRSCB/LD), and *microtomography*. This paper briefly reviews the current status of these techniques at first- and second-generation SXSs, and concentrates on the design and operation of experimental stations at two sectors of a third-generation SXS - the Advanced Photon Source (APS) at Argonne National Laboratory, near Chicago, Illinois. (Note that the third-generation hard x-ray SXSs use 6 to 8 GeV to produce x-rays most useful from 3 to 200 keV, whereas the soft ones operating up to 1.5 GeV are most useful below 3 keV). The station designs are being developed by scientists and engineers belonging to three consortia representing the chemical and materials sciences (Chem/MatCARS), the geo- and cosmo-sciences (GeoCARS), and the agricultural and related environmental sciences (Soil/EnviroCARS). These consortia share technical expertise through the Consortium for Advanced Radiation Sources (CARS) which is managed by a department (UCCARS) at the University of Chicago. A fourth consortium, BioCARS, is developing a third sector for structural biology of viruses, and for time-resolved crystallography and small-angle scattering: its microcrystallography facilities will provide extra expertise via cross-sharing with the other two sectors. Smith et al (1986) reported at IXCOM-11 on XRFA with a microprobe at the National Synchrotron Light Source. For brevity, only references since 1986 will be cited here. Winick and Williams (1991) provide an overview of SXSs world-wide.

2. STATUS OF MICROTECHNIQUES AT FIRST- AND SECOND-GENERATION HARD X-RAY SYNCHROTRON SOURCES

First-generation hard x-ray sources were parasitic on accelerators used for high-energy physics. Several are still functioning well, and expanding operations (e. g. the Cornell High Energy Synchrotron Source on the 5.5 GeV CESR ring, and the Stanford Synchrotron Research Laboratory on the 3 GeV SPEAR ring now turned over for dedicated x-ray generation). Most of the radiation is of Bremsstrahlung type produced from *bending magnets* (BMs). The radiation is naturally condensed by a relativistic Doppler shift which increases with higher acceleration energy. In addition, the beam is fully polarized in the plane of the ring, except for minor depolarization from electrons not traveling exactly in the perfect circle. Some is produced by pioneering *insertion devices* (IDs) consisting of an array of N bending magnets several centimeters across of alternating polarity which cause the electrons to follow a snake-like track. One type of ID generates Bremsstrahlung beams which merely add up to give a combined beam with N times higher intensity (*wiggler*). Furthermore, the horizontal divergence is reduced over that for a BM with a large arc. The other type uses weaker magnetic fields. The Bremsstrahlung beams remain in the same phase space, and interfere with each other to produce a spectrum at generally lower energy than for a BM or a wiggler, but with harmonic peaks of greatly enhanced intensity (*undulator*). The x-ray beam is highly collimated in both the horizontal and vertical directions. Odd harmonics ($n = 1, 3$) are on-axis, and even ones are slightly off-axis. The band width of the first harmonic is ~1/N. The polarization is complex, and the detailed shape of the harmonic peaks is controlled by the profile of the electron bunches and the profile of the magnetic field. Hard magnets used as BMs operate near 1 Tesla. Superconducting magnets operating up to 6 Tesla shift the spectra to higher energy, but are expensive and hard to run.

Second-generation sources are dedicated to x-ray production, mostly from BMs. A few IDs have been retrofitted. UCCARS scientists collect data mainly at the National Synchrotron Light Source, Brookhaven National Laboratory, New York, using beamlines attached to BMs. The 2.5 GeV acceleration energy for the electrons and the 1.2 Tesla magnetic field produces Bremsstrahlung continuous spectra with *critical photon energy* (E_c) of 5.0 keV. Pioneering experiments are done on the X-17 beamline with a superconducting-magnet wiggler. Its continuous x-ray spectrum has $E_c = 21$ keV, and is very similar to that from the hard-magnet BM of the APS (19.5 keV).

A key feature for the usefulness of a SXS for microanalysis using a simple pinhole or collimator is the *brightness*, which is the number of photons per solid angle. The term *brilliance* , which is the brightness/source area, is the appropriate figure of merit for a beamline with properly designed focusing optics. The tighter the bunches of electrons, the smaller is the source area, and the focused spot on the sample.

For synchrotron-*XRFA*, the low background at 90° permits analysis of certain trace elements (mostly with $Z >20$) in certain thick samples by energy-dispersive techniques using the white beam from unmodified BM or wiggler radiation with a 10 μm spot at the part-per-million level using energy-dispersive detectors (EDSs). For thin specimens, the equivalent detection level is 1-10 femtogram. Selective filters are used to reduce the peaks from major elements. The K-spectrum has been used for rare earths and even heavier elements. Bragg diffraction peaks must be identified and subtracted. Careful selection of samples, proper mounting, and optical and other characterization are essential. The specimen thickness must be determined accurately. Standard-less analysis allows quantification for some elements in some samples to < 20%. Use of standards permits higher accuracy. Crystal monochromators allow lower detection levels in theory, but the loss of intensity must be compensated in practice by focussing the radiation or going from a BM to an ID source. Scanning images can be generated simultaneously for many elements using EDS. Published work includes: reviews (Giauque, Jaklevic and Thompson 1986; Jones and Gordon 1989; Rivers, Sutton and Jones 1991), analysis of fluid inclusions (Frantz et al 1988), trace elements in sulphides of East Midlands coal (White et al 1989), trace elements in feldspar standards (Lu et al 1989), diffusion profiles of Sr, Y, Zr and Nb between dacite and rhyolite rocks produced at 10 kbar and ~1700 K (Baker 1990), trace elements in carbonates from the Mississippi Valley ore zone (Kopp et al 1990), gallium in rat bone (Bockman et al 1990), lead in human bone (Jones et al 1990), Br, etc. in stratospheric dust particles (Flynn and Sutton 1991), extreme partitioning of Cu into a volcanic vapour (Lowenstern et al 1991), rare earths and other heavy elements in volcanic titanites (Candela, Piccoli and Rivers 1991), lower polarization from X-17 wiggler, and intensity increase from focusing mirrors (Rivers, Sutton and Jones 1992), the 15 keV microprobe at the Synchrotron Radiation Source, Daresbury (Przybylowicz et al 1992), trace element analysis of cancerous kidneys at Hasylab, Hamburg (Kwiatek et al 1992), quantitation of data collected at the Electron Stretch Accelerator, Bonn (Pantenburg et al 1992).

For synchrotron-XRD, it is sufficient to state briefly that there are only a few measurements of single crystals of micrometer size so far. Bachmann et al (1985) determined the structure of a 6 μm fluorite crystal, Harding (1990) listed structure determinations for 6 crystals with mean diameter of 20 to 80 μm, and Skelton et al (1991) obtained accurate cell dimensions for 0.4 cubic micrometer Bi wires 0.22 μm in diameter while under pressure. Lee et al (1991) determined the modulated structure of an 800 Å film of a $Bi_2Sr_2CaCu_2O_8$ superconductor. This points the way to micro-identification of small particles and surface films or deposits. Collection of diffraction intensities across absorption edges allows location of the crystallographic sites for species like Fe with different valence states (e. g. Warner et al 1992).

For synchrotron-XAS, there is very little work so far on small spots. Typically for transmission XAS, a sample is only a few μm thick, and the concentration of the chosen element is ~ 1%. Lower concentrations are tackled by detecting the fluorescence on the input side. A cluster of energy-dispersive detectors is used to get down to lower concentrations. Some recent reviews are: x-ray absorption fine structure of materials (Dobson and Greaves 1992), report on an instrumentation workshop (Greaves 1992). Some representative applications of broad-spot studies are: oxidation and deprotonation of biotite (Güttler et al 1989), Pb-complexes on wet γ-alumina powder (Chisholm-Brause et al 1990), Cr, Fe and Co in thiospinels (Charnock et al 1990), Mn and Sr in silicate glasses (Kohn et al 1990), Ni in dehydrated and reduced Y zeolite (Dooryhee et al 1991). Rotation of a crystal in a polarized x-ray beam yields information on the directional bonding (Waychunas and Brown 1990): two illustrations are on the Fe distribution among octahedral sites in mica (Manceau et al 1990) and on octahedral Fe-oxygen clusters in a centimeter crystal of diaspore (Hazemann et al 1992). Convergent-beam XAS (Allen et al 1992) permits monitoring of chemical changes on a millisecond scale: e. g. Couves et al (1990). A new technique to sharpen a XANES spectrum by using a detector with better resolution than the core hole lifetime (Hämäläinen et al 1991) unfortunately reduces the signal intensity severely, and will require third-generation SXSs for routine application. XAS study of micro-spots requires a focussed brilliant beam. Hayakawa et al (1991) obtained scanning images with a 19 x 12 μm focussed beam from a Photon Factory BM of XANES spectra for Cr, Fe, Ni and Zn from a peridotite rock, and Cr EXAFS from chromite. GeoCARS scientists are developing technology for small-spot/low-concentration XAS at BM beamline X26 and superconducting wiggler beamline X17 at NSLS. XANES of olivine crystals from lunar basalts (Sutton et al 1992) revealed that the Cr is divalent (0.1

mm spot, 0.1 wt. % Cr). Reconnaissance data for olivine inclusions in diamond indicate mostly trivalent Cr (unpublished). Micro-spot Fe XANES reveals major variations in the divalent/trivalent ratio for Fe in feldspars from the Earth, Moon and meteorites (Delaney et al 1992).

Synchrotron microtomography can, in principle, be based on any of the x-ray signals. The simplest type is based on the transmitted beam after absorption in the sample (Nußhardt et al 1991), and micrometer resolution can be obtained for certain samples with a light matrix and high contrast. Polyethylene with embedded SiO_2-supported Cr catalyst (Jones et al 1992) gave images with resolution of ~ 2 μm^3.

3. FUTURE PROSPECTS FOR MICROTECHNIQUES AT THIRD GENERATION SYNCHROTRON HARD X-RAY SOURCES

Three third-generation hard x-ray sources should be running this decade. The European Synchrotron Research Facility will operate at 6 GeV at Grenoble, France, beginning 1994. Spring-8 is designed for 8 GeV in Japan. The linac, booster synchrotron and part of the storage ring of the 7 GeV APS are well under way in August 1992, and Phase 1 commissioning is scheduled for 1995/6.

In the APS, bunches of positrons (note: lower interaction with residual molecules than for electrons) are accelerated to 7 GeV, and current R&D should permit replenishment of the positrons to maintain near-constant current and heat-load on the x-ray optical elements. The cycle time around the storage ring is 3.5 μs, and each bunch flashes for ~ 0.1 nanosecond. There are enough photons from one bunch to yield a usable diffraction pattern for a millimeter crystal. For samples only 10-100 μm across, various types of data should be obtained in much less than one second. Hence the prefix *micro* can be assigned for both space and time in some synchrotron experiments.

Each of the 34 sectors contains a *bending magnet* (BM) of ~1 Tesla which deflects the positrons around an arc of the ring thereby producing a narrow Doppler-shifted jet of Bremsstrahlung x-rays, and a 5-meter *straight section* containing one or more *insertion devices* (IDs), each containing a linear double-array of magnets with alternating polarity.

The BM x-ray beam has a *critical energy (E_c)* of 19.5 keV and *angular divergences* of 73 microradian (vertical) and 6 milliradian (horizontal). We will split each BM beam into two, each 2 mrad wide separated by 2 mrad. The effective source area is nearly square, 110 (vertical) x 150 (horizontal) μm. The peak on-axis brilliance at E_c is 10^{15} photon/(second 0.1% bandwidth mm^2 $mrad^2$). This corresponds to an integrated flux at the critical energy of 1.5 x 10^{13} photons/(s 0.1 % bw mrad horizontal). The flux is still high at 3 E_c. Because the magnetic deflection of the positrons is radial, the Bremsstrahlung x-rays are highly polarized. In the center of the beam, the polarization is nearly 100% horizontal. Off-center, the beam is less polarized because of slight non-circularity of the positron orbit.

There are two genera of IDs, each with a wide range of sub-types . We shall put a *wiggler* and an *undulator* in each straight section. A *wiggler* with N poles essentially deflects the positrons alternately left and right to produce Bremsstrahlung x-rays of N times the intensity as for a BM with the same magnetic field. Currently available are designs for types A and B of fixed-gap wigglers with critical energies of 33 and 10 keV, respectively. The two racks of magnets of an ID may be moved together to increase the field and increase the x-ray energy. We are planning R & D for a new variable-gap wiggler (type C), 2m long, 8 cm period, critical energy tunable from 10 to 50 keV, flux on axis (photons/(s 0.1% bw mradθ) = 3.1 x 10^{13} x G(E/E$_c$) x N, where G is ~ 0.6 near E$_c$. Addition of electromagnetic coils permits generation of elliptically polarized x-rays which are useful for surface scattering and chemical crystallography.

An *undulator* deliberately produces a magnetic field weak enough to allow the Bremstrahlung x-ray beams to interfere in the same phase space. The APS standard undulator A has 61 magnets with a 3.3 cm period. Its individual beams of Bremsstrahlung x-rays, which travel nearly in the same direction at almost the velocity of light, interfere at the Å range. This causes the resultant beam to be tightly defined both vertically and horizontally: thus the illuminated area at 50 meter is only 2 x 5 mm. The flux there is 10^{15} photons/mm^2/sec/0.1% bandpass on axis. The power of 195 Watt/mm^2 at 26 m from the source poses severe problems for any optical element placed about halfway between source and sample. Instead of the "whale's-back" continuous spectrum from the BM and wiggler sources, the spectrum from an undulator has harmonic (n) peaks with a band-width approximately 1/N. The 7 GeV accelerating potential of the APS was deliberately chosen so that the first harmonic of *undulator A* can range from 4.5 to 14 keV (= 2.8 to 0.9 Å) as the magnet gap is reduced to 11.5 mm, the limit of the vacuum wall of the storage ring. This allows full tuning up to ~40 keV with just the first and third harmonics. The actual shape of an undulator spectrum is complex because the magnets do not act as point scatterers, and the positron bunches have a finite width. The polarization is a complex and steep function of energy and deviation from the center line. We plan R&D to generate intense variable-polarization beams by modifying off-axis x-rays with special optical elements.

In the initial commissioning, the limiting gap is 15 mm, and the maximum E$_c$ for wiggler C is limited to 32 keV. The restricted gap reduces the tuning ranges for the first and third harmonics of undulator A (7.2-12.5 & 22-37), but the intervening energy range is covered by the second harmonic. Symmetrical tilting (usually called *tapering*) of the racks of an undulator broadens each harmonic from the ~ 1/N half-width for a parallel rack, and a width of about 10% is useful for many applications including XAS. In mature operation of the APS, tapering of an undulator should be allowed at any time, thus permitting scanning through an absorption edge for very sensitive XRF analysis.

A major concern with the inherently narrow beam from an undulator is its stability if the positrons shift position by more than a µm or so. Position monitoring, both global around the ring and local at a sector, will be extremely important. A wiggler should be more tolerant.

The figure shows the proposed layout of sectors 2 and 3 for Geo- and Chem/MatCARS respectively, with some shared use by Soil/EnviroCARS. Beamlines C and E have BM sources. Beamline D is served by an Undulator-A and a high-brightness tunable Wiggler C. Beamline F will be fitted with a tapered Undulator-A and a Wiggler-C modified to produce elliptically-polarized radiation.

Micro-XRFA at station D2 will profit greatly from the quasi-monochromatic harmonic peaks from Undulator-A. The first, second and third harmonics will excite the K or L lines of all elements. Detection levels for some elements in some samples will go down to 0.01-0.1 ppm. A key concern for R&D is the optics. A 1 µm pinhole delivers ~10^9 photons/s from an undulator harmonic; this suffices for energy-dispersive detectors. A 10 µm pinhole suffices for wavelength-dispersive detectors. Microfocussing optics can increase the flux, and decrease the spot size, or both. Installation of a high-resolution monochromator in the in the incident beam will permit one micrometer resolution for determination of the electronic state and chemical coordination in consort with chemical analysis. For the fluorescent detection mode, a solid-state detector with many crystals and independent electronics is essential. Candidate technologies for microfocussing optics are: *Fresnel zone plates* (now routine below 1 keV, and under development for 20 keV or even higher), *multi-layer coated mirrors in Kirkpatrick-Baez crossed geometry* (further progress in producing aspherical surfaces and graded multi-layer spacings should lower the current 5 µm spot size), and *tapered capillaries* (sub-micrometer beams have been produced, but alignment and short working distance pose severe challenges). The BM radiation at station C1 will prove valuable for many research problems.

Micro-XAS will be done at three stations. Micro-XAS at station D2 will be coupled with micro-XRF. The undulator will be tapered, and scannable monochromatic radiation will be obtained with a Laue-Bragg High Heat Load Monochromator. This optical device is particularly suitable because of its natural rejection of other harmonics. Microfocusing optics in the hutch will focus the monochromatized beam to 1 µm. Station C2 has a dual crystal monochromator with sagittal focusing at the second crystal. Energy-dispersive XAS is initially set up on station A1 (not shown in the figure). Eventually, a capability will be set up on sector 3.

Microtomography development to obtain a 1 μm^3 voxel will first be done at NSLS beamlines X-26A and X-17. Specimen stages precise to 0.1 µm translation and 0.001° rotation, and a 1 µm mechanical collimator will be used. The sample thickness will normally be less than a millimeter. Further software development will be needed.

Micro-XRSCD is planned with a kappa diffractometer mounted on a theta-omega base, normally in hutch C2 using BM radiation from 7 to 30 keV. Bragg data will be collected at fixed energy/ changing angle, or several energies for anomalous dispersion/ changing angle, or fixed angle/changing energy. Laue data will be collected for well-ordered crystals. Challenging problems involve beam stability, mounting and handling of microcrystals, orientation, background and software.

To conclude, there are many wonderful new research problems to be tackled with these stations on the third-generation sources. For micro-XRFA, we are particularly excited about brine and volcanic-melt inclusions in minerals, noble and other metals in ore minerals, atmospheric and extraterrestrial particles, trace-element partitioning and zoning, trace elements in plants, soils, and humans. For micro-XAS, CARS members are targeting speciation of metals in fluid inclusions, in mineral inclusions in diamonds, and in materials at high pressure-temperature in a diamond-anvil cell; study of toxic elements in atmospheric particles; and a host of geochemical and industrial-chemistry applications that relate to element partitioning between solids and liquids, surface reactions and catalytic phenomena.

4. ACKNOWLEDGEMENTS

This paper is based on many planning documents prepared by CARS members too numerous for full acknowledgement. Key staff work was done by P. James Viccaro, Wilfried Schildkamp, Mark L. Rivers, Stephen R. Sutton and Bing-xin Yang. Prose writers for the GeoCARS Planning Workshop of July 9-11, 1992, included William A. Bassett, John Parise, Charles T. Prewitt, Michael T. Vaughan and Donald J. Weidner. Current support from CARS is provided by The University of Chicago, Southern Illinois University at Carbondale, Northern Illinois University, Keck Foundation, NSF, DoE, NIH, and NASA. A proposal by Keith Jones for microtomography was also valuable.

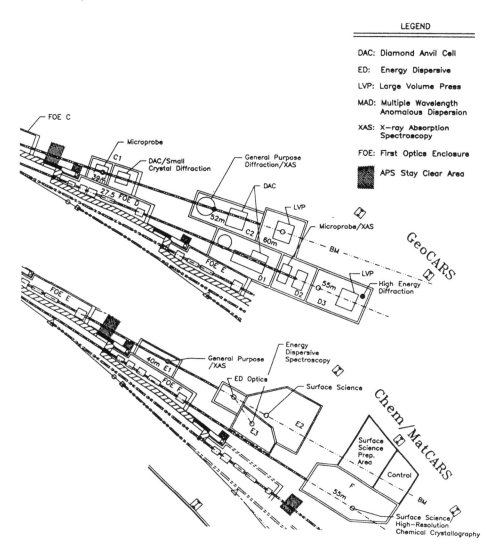

LEGEND

DAC: Diamond Anvil Cell

ED: Energy Dispersive

LVP: Large Volume Press

MAD: Multiple Wavelength
Anomalous Dispersion

XAS: X—ray Absorption
Spectroscopy

FOE: First Optics Enclosure

APS Stay Clear Area

4. REFERENCES

Allen P G, Conradsen S D and Penner-Hahn J E 1992 Synchr Rad News 5 16
Bachmann R, Kohler H, Schulz H and Weber H-P 1985 Acta Cryst A41 35
Baker D R 1990 Contrib Mineral Petrol 104 407-23
Bockman R S, Repo M A, Warrell R P Jr, Pounds J G, Schidlovsky G, Gordon B M and
  Jones K W 1990 Proc Nat Acad Sci 87 4149
Candela P A, Piccolo P M and Rivers M L 1991 Geol Soc Amer Abstr Progr 23 A465

Chisholm-Brause C J, Hayes K F, Roe A L, Brown G E Jr, Parks G A and Leckie J O 1990 Geochim Cosmochim Acta 54 1897

Couves J W, Thomas J M, Catlow C R A, Greaves G N, Baker G and Dent A J 1990 J Phys Chem 94 517

Delaney J D, Sutton S R, Bajt S and Smith J V 1992 Lunar Planet Sci XXIII 299

Dobson B R and Greaves G N 1992 Nucl Instr Meth Phys Res B68 111

Dooryhee E, Catlow C R A, Couves J W, Maddox P J, Thomas J M, Greaves G N, Steel A T and Townsend R P 1991 J Phys Chem 95 4514

Flynn G J Sutton S R 1991 Proc Lunar Planet Sci Conf 21 549

Frantz J D, Mao H K, Zhang Y-G, Wu Y Thompson A C, Underwood J H, Giauque R D, Jones J W and Rivers M L 1988 Chemical Geology 69 235

Giauque D, Jaklevic J M and Thompson A C 1986 Anal Chem 58 940

Greaves G N 1992 Rev Sci Instr 63 1625

Güttler B, Niemann W and Redfern S A T 1989 Mineral Mag 53 591

Hämäläinen K, Siddons D P, Hastings J B and Berman L E 1991 Phys Rev Lett 67 2850

Harding M M 1990 Chemistry in Britain October 956

Hayakawa S, Iida A, Aoki S and Sato H 1991 Rev Sci Instrum 62 2545

Hazemann J L, Manceau A, Sainctavit Ph and Malgrange C 1992 Phys Chem Minerals 19 25

Jones K W and Gordon B M 1989 Anal Chem 19 A341

Jones K W, Spanne P, Lindquist W B, Conner W C and Ferrero M 1992 Nucl Instr Meth Phys Res B68 105

Jones K W, Schidlovsky G, Burger D E, Milder F L and Hu H 1990 Advances in *in vivo* body composition studies, ed S Yasumura et al, (New York:Plenum)

Kohn S C, Charnock J M, Henderson C M B and Greaves G N 1990 Contr Mineral Petrol 105 359

Kopp O C, Reeves D K, Rivers M L and Smith J V 1990 Chemical Geology 81 337

Lee P, Graafsma H, Gao Y, Sheu H-S, Coppens P, Golden S J and Lange F F 1991 Acta Cryst A47 57

Lowenstern J B, Mahood G A, Rivers M L and Sutton S R 1991 Science 252 1405

Lu F-Q, Smith J V, Sutton S R, Rivers M L and Davis A M 1989 Chemical Geology 75 123

Manceau A, Bonnin D, Stone W E E and Sanz J 1990 Phys Chem Minerals 17 363

Nußhardt R, Bonse U, Busch F, Kinney J H, Saroyan R A and Nichols M C 1991 Synchr Rad News 4 21

Pantenburg FJ, Beier T, Hennrich F and Mommsen H 1992 Nucl Instr Meth Phys Res B68 125

Przybylowicz W, van Langevelde F, Kucha H, Lankosz M and Wyszomirski P 1992 Nucl Instr Meth Phys Res B68 115

Rivers M L, Sutton S R and Jones K W 1991 Synchrotron Radiation News 4 23

Rivers M L, Sutton S R and Jones K W 1992 X-ray Microscopy III 4 212

Skelton E F, Ayers J D, Qadri S B, Moulton N E, Cooper K P, Finger L W, Mao H K and Hu Z Science 253 1123

Smith J V , Rivers M L, Sutton S R, Jones K W, Hanson A L and Gordon B L 1986 IXCOM-11 163

Sutton S R, Jones K W, Gordon B M, Rivers M L and Smith J V 1992 Geochim Cosmochim Acta in press

Warner J K, Cheetham A K, Cox D E and von Dreele R B 1992 J Amer Chem Soc 114 6074

Waychunas G A and Brown G E Jr 1990 Phys Chem Mineral 17 420

White R N, Smith J V, Spears D A, Rivers M L and Sutton S R 1989 Fuel 68 1480

Winick H and Williams G P 1991 Synchrotron Radiation News 4 23.

X-ray optics for synchrotron radiation induced X-ray micro fluorescence at the European synchrotron radiation facility, Grenoble

L. Vincze, K. Janssens and F. Adams

Department of Chemistry, University of Antwerp, Universiteitsplein 1, B-2610 Wilrijk, Belgium.

Different optical designs for generating synchrotron X-ray micro beams suitable for use in an X-ray fluorescence microscope using an ESRF bending magnet X-ray source are compared. Attention is devoted to the spatial and energy distribution of the photons in the micro beam and to the minimum detection limits that are achievable with each alternative optical system.

1. INTRODUCTION

Micro-SRXRF (Synchroton Radiation X-ray Fluorescence) is a microanalytical technique which combines the sensitivity of more conventional microchemical methods such as Secondary Ion Microscopy (SIMS) and μ-PIXE (proton induced X-ray emission) with the non-destructive and quantitative character of X-ray fluorescence analysis. Via a suitable optical system, the white beam originating from the storage ring is conditioned into a micro beam which impinges on a sample, oriented at 45° to the exciting radiation. The sample is mounted on a motor-driven XYZ-stage permitting it to be scanned through the micro beam. Fluorescent radiation originating from the excited micro-volume on the sample surface is detected by an energy-dispersive Si(Li)-detector. The latter is oriented at 90° degrees to the incident beam in order to minimise the detection of scattered radiation. The detection limits attainable at current SRXRF-facilities are situated in the ppm (and in favourable cases the sub-ppm) range [1-5]. Possible applications of synchrotron X-ray microprobes are the mapping of essential and non-essential elements in biological tissues [1,2], investigation of element migration and partitioning in geological systems [1], the analysis of individual microscopic particles (aerosols [4], micro-meteorites [2]) and a variety of topics in applied research [1,2]. This paper compares the qualities of a number of possible optical designs for a μ-SRXRF instrument when such a device would be installed at a bending-magnet beamline of the European Synchrotron Radiation Facility (ESRF), Grenoble, France. The various optical systems considered are schematically represented in Fig. 1. These include focussing of the X-rays by means of double bent ellipsoidally bent Si(111) crystals (Figs. 1a and 1b) [2,3,5], simple collimation of the primary beam by means of a pin-hole (Fig. 1c) [1] and the use of capillary optics for beam demagnification (Fig. 1d) [6].

The ray-tracing code SHADOW [7] was used to estimate the spot-sizes and flux-densities which would be obtainable when the various above mentioned optical configurations would be installed at an ESRF 0.8 T bending magnet X-ray source. When appropriate, the effect of heat load, of the slope error and of the intrinsic mosaicity of the optical elements was included in the calculations and the results obtained in this way compared to those corresponding to geometrically perfect optics.

Since the interactions the photons undergo when they are inside the sample material (i.e., Compton and Rayleigh scattering, photo-electric effect) are very well known, by means of a detailed Monte Carlo simulation of these interactions [8], for each optical configuration, the corresponding fluorescent X-ray spectra were also simulated. As an example, XRF spectra from a biological reference material (National Institute of Science and Technology (NIST) Standard Reference Material (SRM) 1571 Orchard Leaves) were simulated. From the latter, the achievable lower limits of detection (LLD's) for each of the alternative optical designs were derived.

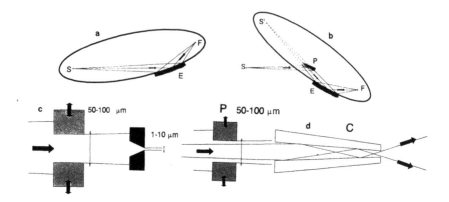

Fig. 1   X-ray optical configurations suitable for μ-SRXRF. (a) Ellipsoidal crystal, (b) planar + ellipsoidal crystal, (c) pin hole, (d) pin hole + conical capillary. A = aperture, C = glass capillary, E = ellipsoidal Si(111) crystal, F = focal point, S,S' = X-ray source, P = planar Si(111) crystal.

2. MICRO-FOCUS X-RAY OPTICS

Details on the optical configurations shown in Fig. 1a and b can be found elsewhere [3,5]. By means of a thin ellipsoidally bent Si(111) crystal an optical system can be created which Bragg-reflects the radiation of a particular energy coming from one focal point of the ellipsoid into the other focus. Demagnification factors of up to 200 can be obtained in this way. The energy of the focussed beam is related to the bending radii of the curved crystal [3,5] and is called the *design energy*. In the experimental setup at SRS [3], this type of crystal is employed to simultaneously monochromate and focus the primary beam. By using the crystal without pre-monochromator (Fig. 1a), Van Langevelde et al. obtained experimental spot sizes in the order of 10 x 20 μm, corresponding to a flux density of about 10^4 photons/s/mA/μm^2 for a monochromatic energy of 15 keV. In the configuration of Fig. 1b, a flat Si(111) pre-monochromator crystal is used upstream of the ellipsoidal crystal to protect the latter from the white synchrotron beam.

At the NSLS-ring [1] and at Hasylab [6], the setup shown in Fig. 1c is employed. The original beam is simply collimated by two sets of crossed slits. The first pair is motor driven and can be used to collimate the primary beam to dimensions of about 50 x 50 μm. The second collimator consists of two pairs of crossed Ta knife blades, fixed in a holder; the blades can typically be 2 to 20 μm apart. Using this setup, beam spots down to a few micrometers can be obtained. Of course, in this case, only a very small fraction of the total flux reaches the sample (< 0.01 %). As the sample is irradiated with the white synchrotron spectrum, a more uniform excitation of all elements in the sample is obtained. A disadvantage is the relatively high scatter background in the fluorescent spectra and the efficient excitation of the matrix lines, limiting the detectable count rate. The intensity of the matrix lines can be reduced by placing an absorber foil (e.g., Kapton, Al) or a second collimator between sample and detector.

The setup of Fig. 1d has also been employed in Hasylab [6] and uses the principle of total reflection. The primary beam is first collimated down to about 100x100 µm² and then enters a conically shaped capillary. If the X-rays of energy E strike the inner walls of the tube under an angle smaller than a critical value $\theta_c(E)$, total external reflection occurs; otherwise the photon is absorbed in the glass wall. Due to repeated reflection of the X-rays on the inner walls of the tube, the original beam diameter is reduced down to the inner dimensions of the far end of the capillary. Sub-micron beam diameters have been produced in this way [6]. The disadvantage of the capillary is that the divergency of the focussed beam is much larger than that of the primary beam which is in the order of 0.6 mrad in the vertical plane (in the ESRF case). Accordingly, if such a device is to be used for µ-SRXRF purposes, the end of the cappilary must be placed extremely close (ie. less than 1 mm) to the sample surface under investigation. Research in this field is now directed towards making cappilaries with paraboloidal and ellipsoidal inner surfaces [9]. An advantage of using this type of totally reflecting focussing device is that the capillary behaves as a high-energy filter since θ_c is proportional to $1/\sqrt{E}$. Accordingly, the higher energies in the primary spectrum which otherwise give rise to a significant Compton background in the low energy part of fluorescent spectra are removed. This is especially useful when high-energy sources such as the DORIS or ESRF rings are employed.

3. RESULTS and DISCUSSION

Table I summarises the results of the ray-tracing calculations and compares the obtained spot sizes and flux densities with those currently attainable at the NSLS and SRS SRXRF setups. The polychromatic setups (pin hole, capillary) were assumed to be at 30 m from the ring. The results shown in Table I refer to a capillary with start and end inner diameter of resp. 58 and 1 µm, having a length of 25 cm (see [6] for further details). Ellipsoidal crystals of different design energies (see [5] for a detailed description) were considered; these were assumed to be located at 50 m from the storage ring.

Table I. Characteristics of various micro-focus optical configurations.

Optical Configuration	Energy (keV)	Spot Size (µm²)	Flux Density (ph/s/µm²/100 mA)	Flux (ph/s/100 mA)
ellipsoidal crystal (Fig. 1a)	10	4 x 9	7.4 10⁹	2.3 10¹¹
	20	7 x 24	1.5 10⁹	2.0 10¹¹
	30	11 x 27	5.3 10⁸	1.2 10¹¹
planar + ellipsoidal crystal (Fig. 1b)	10	4 x 4	1.0 10⁸	1.6 10⁹
	20	8 x 6	6.2 10⁷	2.9 10⁹
	30	11 x 8	4.1 10⁷	2.7 10⁹
SRS microprobe	15	17 x 18	1.0 10⁶	6.2 10⁸
pin hole (Fig. 1c)	white	10 x 10*	1.3 10⁹	1.3 10¹¹
		8 x 8*	1.3 10⁹	8. 10¹⁰
		2 x 2*	1.3 10⁹	5. 10⁹
pin hole + capillary (Fig. 1d)	white	10 x 10 (at 1 mm)	7.4 10⁹ (at 1 mm)	2.2 10¹¹ (at 1 mm)
NSLS microprobe	white	8 x 8	3 10⁷	1.9 10⁹

*: user selectable spot size via collimator opening

It appears that ellipsoidal+planar and the pinhole configurations offer the most interesting balance between achievable spot size dimensions/lateral resultion and available flux/analytical sensitivity. The predicted LLD values achievable with these two configurations are compared in Fig. 2 with those currently available at the NSLS and SRS stations for the case of a biological matrix. Although the use of a single ellipsoidal crystal offers the highest monochromatic flux density, the achieved spot sizes are quite large, significantly limiting the scope and applicability of an X-ray microprobe equipped with such optics. In the case a capillary unit is used to concentrate the polychromatic radiation, a minimum spot size of 10 x 10 μm^2 is obtained when the sample is positioned at 1 mm from the end of the capillary. Taking into account the general character of the envisioned X-ray microprobe facility, requiring a uniform excitation of most constituents of sample materials, a collimated microprobe setup (pin hole configuration), optionally equipped with a conical capillary is considered to be more appropriate than the monochromatic configurations.

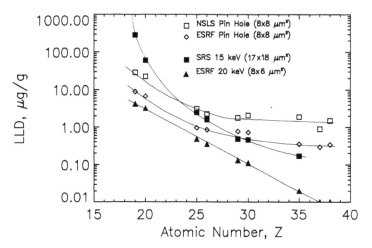

Fig. 2   Comparison of currently obtainable MDL's in a biological matrix (NIST SRM 1571 Orchard Leaves) with predicted values for the optical configurations shown in Fig. 1b and 1c. at an ESRF 0.8 T bending magnet beamline. Counting time is 600 sec.

References
[1]   K.W. Jones, D.M. Gordon, "Trace Element Determinations with Synchrotron-induced X-ray Emission", Anal. Chem., 61 (1989) 341A-358A.
[2]   R.D. Vis, in "Applications of Synchrotron Radiation", CRA Catlow, GN Greaves (eds.), Blackie & Son Ltd., 1990, Glasgow, UK, Chapter 13, "Trace Element Analysis".
[3]   F. van Langevelde and R.D. Vis, "Trace element determinations using a 15 keV synchrotron X-ray microprobe at the Synchrotron Radiation Source, Daresbury (UK)", Anal. Chem., 63 (1991) 2253.
[4]   K. Janssens, F. van Langevelde, F. Adams, R.D. Vis, K. Jones, M. Rivers, S. Sutton, D.K. Bowen, "Comparison of Synchrotron X-ray Microanalysis with Electron and Proton Microscopy for Individual Particle Analysis", Adv. X-ray Anal., 35 (1992) 1265-1273
[5]   F. van Langevelde, K. Janssens, F. Adams, R.D. Vis, "Prediction of the Optical and Analytical Qualities of an X-ray Fluorescence Microprobe at the European Synchrotron Radiation Facility (Grenoble).", Nucl. Instr. Meth. A317 (1992) 383-393.
[6]   P. Engstrom, S. Larsson, A. Ringby, A. Buttkewitz, S. Garbe, G. Gaul, A. Knochel and F. Lechtenberg, "A submicron synchrotron X-ray beam generated by capillary optics", Nucl. Instr. Meth. A310 (1991) 538-547.
[7]   B. Lai, K. Chapman, F. Cerrina,"SHADOW: new developments", Nucl. Instr. Meth., A266 (1988) 544-549.
[8]   L. Vincze, K. Janssens, P. Van Espen, F. Adams, "Monte Carlo simulation of conventional and synchrotron energy-dispersive X-ray spectrometers", X-ray Spectrometry, 1992, submitted.
[9]   F. Lechtenberg, A. Rindby, private communication, 1992.

Inst. Phys. Conf. Ser. No 130: Chapter 8
Paper presented at Int. Congr. X-ray Optics and Microanalysis, Manchester, 1992

617

Scanning X-ray microprobe with Bragg–Fresnel multilayer lens

P.Chevallier, P.Dhez, F.Legrand,
LURE, University Paris-Sud, Bat. 209D, F-91405, Orsay, France.
A.Erko,
Institute of Microelectronics Technology Russian Academy of Sciences (IMT),
Chernogolovka, Moscow dist., 142432, Mailbox 1957, Russia.
B.Vidal,
LOE, Faculty of Science St. Jerome (case 262), 13397, Marseille, France.

ABSTRACT: The preliminary design of a fluorescence X-Ray scanning microprobe with submicron resolution was tested at the LURE (France). A linear multilayer Bragg- Fresnel lens has been used for vertical focussing of the white X-ray synchrotron beam. In the energy range 6Kev to 14KeV, and with a bandwidth of $\delta\lambda/\lambda \approx 10^{-2}$, the vertical spot size checked with a photographic plate was 1.7µm. At 10KeV input photon energy the Ni(Ka) fluorescence signal registered during a knife-edge test with a test object shows a 3µm resolution.

1. INTRODUCTION

According to Sparks and Ice (1989) the real resolution limit of X-ray fluorescence microscope is about 50nm, due to the diffraction spread of a beam and the depth of its penetration. Taking into account the quantum yield of fluorescence and the signal-to-noise ratio they concluded that a microprobe with a spot diameter of 5nm to 1µm is preferable to other methods of microscopy for the majority of samples.

The first successful demonstration of a X-ray fluorescence microscope for specimen mapping was achieved by Horowitz and Howell (1972) by using a synchrotron radiation source. In this case a simple microprobe pinhole collimator with diameter 1-2µm was used. Fluorescence signals were recorded to map the distribution of various elements in their specimen.

Focusing X-ray optics have also been used in several operating fluorescence microprobes. Among them is the Kirkpatrick-Baez microprobe, built at the LBL Centre for X-ray Optics by Wu et al. (1990c), and based on a pair of small spherical multilayer mirrors. Compared to grazing incidence optics, a multilayer coating permits reduction of the mirror size because of its larger incidence angle. The calculated size of the focal spot was on the order of 3µm x 1µm, but due to spherical aberrations, the measured beam size was actually 6µm x 6µm. To obtain a resolution better than 6µm diffraction optics must be used. Bragg-Fresnel multilayer lenses (BFML) developed by Aristov and Erko (1990b, 1991) are a good candidate for such a focusing system. The first test of these lenses was done at LURE by Chevallier et al. (1990a) and exhibited a 2.3 x 3µm², focal spot size using two linear BFML in the Kirkpatrick-Baez focusing scheme.

2. LURE-IMT FLUORESCENCE MICROPROBE

Compared to the previously reported system of Chevallier et al. (1990a), a 100µm entrance pinhole was added 6.3m from the linear BMFL to create a smaller apparent source. The present microprobe was a single linear BMFL with focal length of 7cm at 8KeV photon energy and installed on the DCI storage ring (LURE, Orsay (France)). To adjust the linear BFML a goniometer with 5 degrees of freedom was located behind a collimating slit. To restrict the horizontal width of the vertically focused line, and to cut the undiffracted zero order beam, a 50μm diameter pinhole was positioned on half-focal length from the lens.

The test objects were positioned in the focal plane within an accuracy of 0.5μm and the focal distance was adjusted to within 1μm by using a remote-control stage having 3 degrees of freedom. A Si(Li) X-ray -detector was used in two positions: behind the object in transmission mode and in front of the object, at almost 90° to the incident beam, to monitor the fluorescence signal.

2.2 Lens fabrication.

As described in the book of Aristov and Erko (1991), BFML with straight structures which provide linear focusing were used. These were made by using electron beam lithography, optical lithography, and ion beam etching processes. The aperture of each lens was 1mm x 5.6mm and the external minimum zone width was 3.5μm. The BFML were prepared by ion etching a 141 layer 3.2nm period W/Si multilayer mirror coated by magnetron sputtering by one of us (B.V). The calculated average Bragg angle at 8Kev was 1.34° (23.4mrad).

Several specimens having different etching depths (0.05μm-0.15μm) were prepared for efficiency tests. The optimum profile depth corresponding to the maximum first order diffraction efficiency can be estimated by using the formula from Agafonov et al. (1992).

$$T_{opt} = \frac{\lambda^2}{8\sin(\theta_B)}\left[\delta_w T_w + \delta_{Si} T_{Si}\right]$$

where d_W and d_{Si} are the optical constants of the multilayer materials, T_i the thickness of the material (i), $\sin\theta_B$ the Bragg angle, and λ the wavelength.

For our 3.2nm multilayer period with $T_W = 1.0$nm and $T_{Si} = 2.2$nm, this equation predicts $T_{opt} = 41.6$nm, which corresponds to 14 bilayers.
In the case of an amplitude-phase grating the efficiency can be evaluated by a simple formula proposed by Kirz (1974) if one use the χ value:

$$\chi = \frac{\beta_w T_w + \beta_{Si} T_{Si}}{\delta_w T_w + \delta_{Si} T_{Si}}$$

Using the optical constants from the Henke tables (1982) ($\delta_W = 4.57\ 10^{-5}$, $\beta_W = 4\ 10^{-6}$ and $\delta_{Si} = 7.56\ 10^{-6}$, $\beta_{Si} = 1.7\ 10^{-7}$) our multilayer mirror has reflectivity ≈ 0.70 which would lead to a first order efficiency of ≈ 0.26 if our BFML was similar to a classical amplitude-phase grating.

2.3 Aberrations

The diffraction aberration characteristics of this type of optics can be calculated using aberration formulas given by Aristov and Erko (1991). Due to our 100μm pinhole, the view angle of the initial source is ≈ 1.5mrad. In such a case the diffraction aberration formula indicates that the maximum number of zones of the Bragg-Fresnel lens can be as high as 1500. Therefore, the maximum theoretical spatial resolution of our focusing system is less than 0.1μm. However the limit placed on the number of zones by spherical aberration is much higher than 1500. So in practice we are limited only by the technological problems and by the multilayer grating performance. For the focal distance of our lens (F = 7cm) the calculated demagnification of the entrance pinhole in the vertical direction is M = 1.3 10⁻². Consequently with our 100μm X-ray source the minimum spot size in the focal plane is ≈ 1.3μm.

The classical formula can be used to estimate the depth of focus. For the 10KeV energy and a 0.1μm pinhole image transfer, the depth of focus is in the range of +-1mm.

3. RESOLUTION TESTS.

3.1 Photographic resolution test.

Photographic resolution tests were performed in the range 8KeV-14KeV. High resolution (0.1μm) VRE holographic type photographic plates made in Russia have been used to test our linear X-ray microprobe spot configuration. Due to the short focal length of the BFML at 14KeV (λ = 0.084nm), the best value of the observed focusing was 1.7μm x 50μm. Figure 1 is the corresponding densitometry curve for the vertical focusing direction. The 50μm length observed in the horizontal direction corresponds to the last collimator pinhole size. Experimental efficiency in normalized units (the ratio of the focused energy to the total reflected energy) for one of the BFML was found to be 44% at 8 KeV (Chevallier, 1990a). When compared to the estimated value of 26% for a classical amplitude phase grating this indicates the phase character of the obtained diffraction. For our 60% reflectivity multilayer mirror this corresponds to a 25% absolute efficiency which coincides well with the theoretical prediction.

3.2 Transmission mode resolution test.

The resolution of the X-ray microprobe was first tested in transmission mode (absorption scanning microscopy). We used a specially prepared gold grating having a 1μm period supported by a 15μm period gold mask with 3μm bar width. This test object was placed in the focal plane of the optical system and vertically scanned with 0.5μm steps. The total flux in the focal spot was between 1 and 2 10⁻⁵ phot/sec. Figure 2 shows the variation of the absorbed X-ray flux versus the vertical position of the mesh. At an average beam energy of 8KeV the measured width of the vertical focus was 1.5μm.

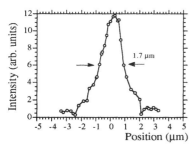

Fig.1 Intensity distribution in the BFML focus. λ = 0.084nm

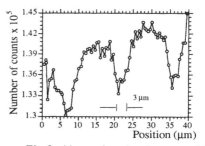

Fig.2 Absorption in the test-object when vertical position of the object is scanned.

3.3 Fluorescence mode test.

The resolution of the fluorescence X-ray microprobe was tested with our single linear BFML as a focusing element. For this purpose a knife edge test was achieved using the edge of a 20μm thick polished Ni foil. The fluorescence signal was detected by a Si(Li) detector linked to a computer working as a multichannel analyzer. The fluorescence photon counting rate was between 10 and 100 counts per second. The average count rate on the Ni spectral line versus the position of the knife edge is shown in figure 3. The 3μm half width peak appearing on the derivative curve demonstrates the achieved resolution.The high value of signal-to-noise ratio of the detected signal in the fig.4 is demonstrated. In this case steel wire (20μm in diameter) was placed on the focused beam axis.

4. CONCLUSION.

A linear BFML has been tested on the DCI synchrotron source, as an intermediate step toward using an ellipsoidal BFML with 2D focusing. The synchrotron source size has been collimated so as to check the spatial resolution limit. Imaging of specially prepared test objects has been achieved at around 10 KeV on photographic plates, and transmission and fluorescence tests have been successfully performed. These measurements confirm that the theoretical spatial resolution can be obtained with the present multilayer and BFML technologies. In addition, the efficiency measurements indicate the phase nature of the prepared gratings.

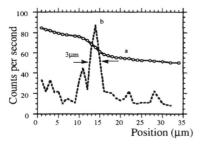
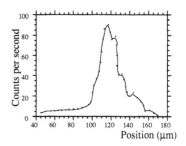

Fig.3a) Variation of the Ni(Kα) fluorescence signal vs the vertical knife edge position.b) Derivative of the curve in Figure 3a

Fig.4 Variation of the Cr(Kα) fluorescence signal vs the stainless steel wire position.

The authors gratefully acknowledge Dr. A. Freund (ESRF) for his assistance and Prof. V.V. Aristov (IMT) and Dr.M.Brunel (CNRS Crystallography laboratory) for useful discussion of the experimental results and help in organization of these experiments.

REFERENCES

Agafonov Yu.A., Brunel M., Erko A.I., Martynov V.V., Roshchupkin D.V., Vidal B., Vincent P., 1992 "Multilayer Gratings Efficiency: Numerical and Physical experiment", to be published.
Aristov V.V., Erko A.I., 1991, X-Ray Optics. "Nauka", Moscow, 150 P.
Chevallier P., Dhez P.,Erko A., Khzmalian E., Khan-Malek C., Freund A., Panchenko L., Redkin S., Zinenko V., Vidal B, 1990a, Proceed. of the 3-d Conference on X-Ray Microscopy. London.
Erko A.I. 1990b, Journal of X-Ray Science and Technology, 2, P.297-316
Henke B.L., Lee P., Tanaka I.J., et al., 1982, Atomic data and Nucl. data Tables. Vol.27, 1
Horowitz P., Howell J.A. 1972, Science Vol.178, p.608.
Kirz J., 1974, JOSA, Vol.64, P.301-309.
Sparks C.J., Ice G.E. 1989, Mat. Res. Soc. Symp.Proc. Vol.143, p.223-233.
Wu Y., Thompson A.C., Underwood J.H., Giauque R.D., Chapman K., Rivers M.L., Jones K.W.1990c, Nucl. Instrum. and Meth. Vol.291A, p.146.

Inst. Phys. Conf. Ser. No 130: Chapter 8
Paper presented at Int. Congr. X-ray Optics and Microanalysis, Manchester, 1992

621

X-ray probe mapping of calcium deposits in articular cartilage

C J Buckley, R E Burge, G F Foster, M. Rivers [†], S Y Ali[††] and C A Scotchford[††]

Physics Department, Kings College, Strand, London WC2R 2LS.
† Centre for Advanced Radiation Sources, University of Chicago, Chicago, Illinois 60637.
†† Institute of Orthopaedics, Brockley Hill, Stanmore, MIDDX, UK.

ABSTRACT: Elemental mapping with high signal to noise has been achieved by scanning x-ray microscopy. Absorption difference imaging and x-ray induced fluorescence was used to obtain calcium maps of mineral deposits in articular cartilage. The imaging times were a few minutes and a few hours respectively. Possiblities for mapping normal and arthritic cartilage at ionic concentrations are considered in respect of 3rd generation synchrotron sources.

1. INTRODUCTION

The marked variation in ionization cross section for the elements with x-ray energy allows for considerable contrast selection between the element of interest and the surrounding matrix. Two contrast modes for elemental mapping using x-ray probes are possible. These are: absorption difference imaging, and fluorescence imaging. With a sufficiently bright x-ray source, absorption difference imaging is rapid compared to EPMA, due to efficient ionisation and signal collection combined with detection rates in excess of 1MHz. The sensitivity of the technique is ultimately limited by photon induced mass loss, and has a limit of a few hundred parts per million (ppm), which is similar to the sensitivity limits of EPMA. Fluorescence imaging using monochromatic x-ray probes also benefits from efficient excitation, and offers a sensitivity below 1ppm due to the very low background asociated with the monochromatic probe. Like EPMA however, the mapping speed is restricted in part by the signal collection efficiency, and detection rate limits imposed by energy and wavelength dispursive detection schemes.

A key advantage in using x-ray probes over e-beam probes is the ability to map specimens with thicknesses up to a few tens of microns. Here we present an application of both absorption difference and fluorescent mode x-ray probe mapping. These techniques were used to study the distribution of mineral deposits in human articular cartilage via calcium mapping. We also discuss the extention of the study to mapping the free ionic calcium in cartilage via the possibilities opened up by the new third generation synchrotron x-ray sources.

2. CALCIUM DEPOSITS IN NORMAL AND ARTHRITIC ARTICULAR CARTILAGE

The articular cartilage from patients suffering from osteo-arthritis is found to contain sizeable deposits (0.1 to 1mm) of calcium pyrophosphate - which has a calcium to phosphorous ratio of 1:1. These pathogenic deposits are thought to be responsible for the break up and fragmentation of the cartilage. The origin of these deposits is not well understood, however, osteo-arthritic cartilage, is also found to contain another form of calcium phosphate deposit, namely calcium whitlockite (Scotchford *et al* 1992). The whitlockite deposits are considerabley smaller in size (ranging from 0.05 to 0.5μm) than the pyrophosphate deposits, and have a calcium to phosphorous ratio of 1.3:1. This ratio is in between calcium hydroxy apatite (1.6:1) and the pathogenic calcium pyrophosphate. These whitlockite microcrystals are also found in normal articular cartilage, and it has been suggested (Ali 1985) that under appropriate conditions these microcrystals may act as seed points for the development of the pathogenic pyrophosphate deposits. For this reason, it is of interest to map the whitlockite deposits in normal tissue, so that their distribution and possible origins can be linked to the morphology and related biochemistry of the cartilage tissue.

Articular cartilage is composed of a collagen/proteoglycan matrix which contains chondrocyte cells. The cartilage tissue is avascular, and relies on nutrients via diffusion from the blood barrier at the bone interface, and from the sinovial fluid at the articular surface. There are natural biochemical gradients in the bone/articular direction, and it is therefore of interest to map the distribution of the crystalites on the macro scale in this direction. Also, because the chondrocytes export much of the cartilage biochemistry, it is of interest to map the distribution of the any calcium deposits on the micros scale in the vicinity of the chondrocytes. The following section describes how x-ray probes have been used to map cartilage specimens in scanning microscopes, and the results are shown in section 4.

3. ELEMENTAL MAPPING USING X-RAY PROBES IN SCANNING MICROSCOPES

The previous section described the need to map the calcium deposits on the macro and micro scales. On the micro scale, it is necessary to create a probe with dimensions equal to the smallest microcrystal, ie about 50nm. This has been achieved by using modified Fresnel zone plates (made by e-beam lithography) to focus soft x-rays down to this size. Monochromatic x-rays from a synchrotron source provide the illumination, while imaging is performed by scanning the specimen in steps smaller than the probe size, and recording the transmitted flux at each pixel. Details of the scanning transmission x-ray microscope (STXM) used in this work are presented by (Kirz *et al* 1992) in this volume.

Calcium mapping using the STXM for absorption difference imaging is performed by taking two images of the same area either side of the calcium L absorption edge at 350eV. The calcium map is found by taking the log of the ratio of corresponding pixel counts. This results in an image, which when displayed via an appropriate frame store, indicates increasing calcium concentration by an increasing brightness level. Details of the technique are reported elsewhere (Buckley *et al* 1992). In the case of the STXM, specimens were fixed, embedded, and sectioned but not stained. The section thickness was 0.1μm which corresponds to the absorption thickness of the whitlockite microcrystals on the absorbing side of the Ca L edge.

On the macro scale, (mapping an area in excess of 1 mm^2), a larger probe of 10 by 10μm was used. Further, because an average distribution is required, it is convenient to use a greater sample thickness, which gives a larger signal. However, in order to penetrate greater thicknesses, higher energy x-rays are required. If a monochromatic beam which is tunable over the Ca K edge at 4KeV, then calcium mapping can be performed as described above by absorption difference imaging. Alternatively, the fluoresced calcium signal can be detected and used for mapping.

Here, a scanning fluorescence microscope (SFXM) was used with a "white" x-ray beam, from a synchrotron bending magnet source (Rivers *et al* 1992). was used to excite a 5μm thick section of articular cartilage, prepared as for the STXM above, and serial sectioned from the same sample block.

4. RESULTS OF MAPPING DEPOSITS IN CARTILAGE WITH X-RAY PROBES

The distribution of the crystalites in relation to chondrocytes is of considerable interest, as they may play a role in the crystalites formation. In other cartilage tissues, chondrocytes are thought to export vesicles from which the initial seed crystals of bone mineral are formed. If a vesicle system of seeding is responsible, then it might be expected that calcium rich vesicles would be found inside the chondrocytes or their lacunnae.

Several chondrocyte cells (prepared as described above) were imaged and mapped. In no case did we observe any calcific crystalites either inside the chondrocytes or their lacunaes. However, the crystalites were often found in clusters around chondrocytes, suggesting that the cells may not export the mineral directly, but are responsible for exporting the biochemistry involved in their creation. Fig. 1 shows a typical result.

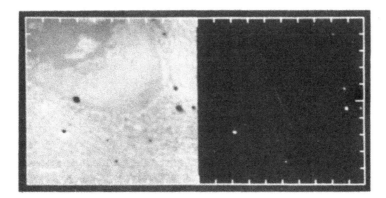

Figure 1. The image on the left is an x-ray transmission micrograph of a 0.1 micron thick unstained section of articular cartilage, as formed at an energy of 0.350 eV in an STXM. The scale bar is 1 micron. Several dense particles can be seen around the chondrocyte cell, and one inside the cell lacuna. The image on the right is a calcium map made by processing the image on the right, and shows only the location of calcium. Note that the deposit which appears in the cell lacuna (left image), is not a calcium deposit (right image).

Mapping by scanning x-ray fluorescence microscopy shows that the crystalites are distributed on the macro scale in a band within 20μm of the articulating surface. The surface zone of the articular cartilage matrix has been shown to differ from deeper matrix in both the collagen architecture and proteoglycan concentration. The collagen architecture changes from a radial to a tangential near the surface, and has a finer, denser structure in this region (Clark 1990). The proteoglycan concentration is reduced in this region, and it has been shown that proteoglycans play a role in inhibiting mineral formation. Thus it is possible that the cells produce less

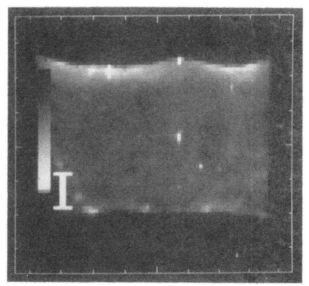

Figure 2. A calcium map of a 1.5 by 1 mm, 5 micron thick section of articular cartilage. The image was formed by detecting the characteristic fluorescent x-rays emitted by calcium when excited by a polychromatic x-ray probe in an SFXM. The image clearly shows that the highest concentration of calcium is located in a narrow band close to the articular surface. The scale bar is 220μm.

proteoglycan as a function of the mechanical stress near the surface, and this contributes to a condition which is opertune for the formation of the whitlockite crystalites.

5. PROGRESSION TO IONIC MAPPING WITH 3RD GENERATION SYNCHROTRON SOURCES

The data above has detailed the distribution of the calcium whitlockite crystals on the micro and macro scale. The data presented above, provides information on which to base theories of the origin of theses deposits. However, it is highly desirable to map cartilage tissue with a greater sensitvity than is presented here. Of great benefit would be the ability to map the ionic streangths of both calcium and phophorus in articular cartilage. Indeed, if an image could be made in which the brightness level was proportional to the *ratio* of calcium to phosphorous at ionic concentrations (a few milimolar), this would make possible the viewing of regions in which a particular form of mineralisation would be favoured.

Monochromatic x-ray probe fluorescence analysis has the potentail for sensitivities at concentrations below one part per million (ppm) (Sparks 1980). Clearly, such a technique could easily image at the 10ppm (sufficient for ionic calcium and phosphorous in articular cartilage). At present there is no operating x-ray source which can provide Ca/P ratio imaging at these concentrations in a realisticaly short imageing time. However, the advent of 3rd generation synchrotron x-ray sources will provide the opertunity for high sensitivity analysis, combined with short pixel times. Undulator beam lines for x-ray probe fluorescence imaging are proposed for the european synchrotron radiation facility (ESRF), and the advanced photon source (APS). These very bright tunable x-ray sources will make possible milimolar iamging with a spatial resolution of 0.1μm with pixel times of less than a second (see fig. 3). We belive that the application of SFXM (on these sources) to the study of firm tisuues will lead to a new and hitherto unobtainable perspective in the study of the causes of arthritic disease.

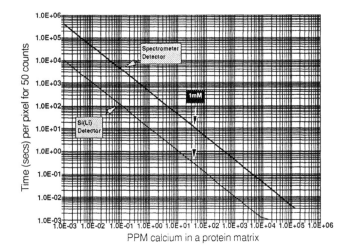

PPM calcium in a protein matrix

Figure 3. Pixel times against parts per million calcium in a protein matrix as calculated for a zone plate scanning fluorescence microscope (Howells *et al* 1992) which produces a 0.1μm diameter probe, operating on a 5 micron thick cartilage section.

6. CONCLUSIONS

Mapping by absorption difference imaging has proved both rapid and effective in mapping specimens with a resolution of 50nm. The application of this technique to the study calcium whitlockite mineral deposits in human articular cartilage has indicated that deposits are not found within chondrocyte cells. Also, x-ray probe fluorescence mapping has confirmed the macro scale distribution of the whitlockite crystals as being distributed in a band 20μm from the cartilage articulating surface. In the next few years, the new 3rd generation x-ray sources will provide sensitivities below 1ppm. This will facilitate Ca/P ratio mapping at the milimolar level with imaging times from minutes to a few hours. We expect that the use of these techniques will continue to provide new and relevant information in the study of the causes of arthritic disease.

ACKNOWLKEGEMENTS

The authors thank the management and technical staff of the national synchrotron light source (NSLS) for their efforts in running an excellent x-ray source. Also, we express our gratitute to the IDT members of the X1a and X26 beamlines who's efforts made our imaging possible.

The zone plates used in the creation of the STXM images here were fabricated by Eric Anderson of the centre for x-ray optics, LBL, Berkeley California.

Our collaboration with our friends at the NSLS was made possible by a NATO travel grant: number CRG965.

REFERENCES

Scotchford C. A., Greenwald S., and Ali S.Y. " Calcium phosphate crystal distribution in the superficial zone of human femoral head articular cartilage", Journal of Anatomy, currently in press.

Ali S.Y. "Apatite type crystal deposition in articular cartilage", Scanning Electron Microscopy IV, 1555 - 1566, 1985

Kirz J., and Jacobsen C.J. "Scanning X-ray Microscopy", this volume.

Buckley C.J., Fostre G.F., Burge R.E., Ali S.Y., and Scotchford C. A. " Elemental Mapping of Biological Tissue by Absorption Difference Imaging in the STXM", X-Ray Microscopy III, Springer Series in Optical Sciences, Springer 1992.

Rivers M.L., Sutton S.R., and Jones K., "X-Ray Fluorescence Microscopy", X-Ray Microscopy III, Springer Series in Optical Sciences, Springer 1992.

Clark J.M., "The organisation of Collagen fibrils in the Superficial Zones of Articular Cartilage", Journal of Anatomy, 171, 117-130, 1990.

Sparks C.J. "X-Ray Fluorescence Microprobe for Chemical Analysis", Synchrotron Radiation Research, Winnick H., and Donniach (eds), Plenum Press, New York 1980.

Howells M.R. "Conceptual Design for an X-Ray Microscope Facility at the ESRF", Internal report, ESRF 1992.

Properties and applications of soft X-ray undulators in structural and microstructural studies of the surfaces of materials

D.P.Woodruff

Physics Department, University of Warwick, Coventry CV4 7AL, UK

ABSTRACT: The special properties of undulators installed on third generation synchrotron radiation sources are described, and some illustrations are given of the potential applications of such sources in the soft x-ray energy range with particular reference to the structure and microstructure of surfaces of solid materials.

1. INTRODUCTION

One of the greatest strengths of conventional (bending magnet) synchrotron radiation is that it provides an extremely bright (high intensity per unit solid angle per unit source area) continuum source, so that combined with a suitable monochromator, it provides a means of delivering high fluxes of radiation to a sample with a tunable photon energy. This combination of high useable flux and tunability is particularly attractive in the soft x-ray range (say from photon energies of 300eV to 2keV) because of the total absence of laboratory sources with comparable properties. The fact that the radiation is a continuum is not without its problems, however, because it implies that the total radiation energy falling on the monochromator optics is very large compared with that associated with the narrow bandwidth of radiation which is actually being used. Indeed, most experiments use synchrotron radiation at wavelengths longer than the so-called critical wavelength for a given machine, because the flux output is rather weakly dependent on wavelength under these conditions, yet 50% of the total output power lies at wavelengths less than this critical value.

An alternative method of providing electromagnetic radiation from an electron storage ring which is now the basis of so-called third generation synchrotron radiation sources (first generation being parasitic use of synchrotron bending magnet radiation) is to design a ring with long straight sections between the bending magnet groups, and to insert magnetic structures into these straights which are each optimised for particular experiments using particular parts of the spectrum. This is a major change of philosophy in synchrotron radiation sources in that bending magnet sources are approximately identical for all users of a specific storage ring; by contrast, insertion devices are tailored to the needs of each user. This approach has some very important advantages, but also some limitations.

In this short review the basic mode of operation of these insertion devices, and particularly of undulators, is described in the next section. In section 3 some current and potential applications of these new sources are discussed, with a particular emphasis on techniques concerned with structural and microstructural characterisation of surfaces.

2. UNDULATORS AND WIGGLERS

Insertion devices generally comprise an array of dipole magnets, such as that illustrated in fig. 1, which causes the electrons passing through it to oscillate ('undulate' or 'wiggle') about the nominal straight path as they pass through. The characteristics of the radiation emitted along the axis of such a device depends on the angular amplitude of these oscillations. If this amplitude is large compared with the intrinsic opening angle of the synchrotron radiation produced by individual 'wiggles', then the output flux has the usual bending magnet continuum form but the intensity is increased by a factor equal to the number of wiggles (or strictly, the number of half-wiggles, although two separate sources will be seen, corresponding to the maximum positive and negative lateral displacements of the electron beam). This large amplitude perturbation corresponds to a large periodic magnetic field strength, and under these conditions it is common to use a field strength higher than that of the main bending magnets, so that not only is the intensity increased due to the increased number of bends, but also the critical wavelength is reduced and the whole spectrum is shifted to shorter wavelengths. A *multipole wiggler* of this type is therefore often also refered to as a *wavelength shifter*.

Fig. 1 Schematic magnet array for an undulator or multipole wiggler, also showing the electron trajectory.

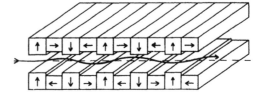

More interesting behaviour is found if the angular amplitude of the oscillations is small compared to the synchrotron radiation opening angle. In this case, one can obtain temporal coherence between the radiation emitted at equivalent points on the multiple undulations, and the constructive interference of these coherent components leads to a distinct harmonic structure. Such an *undulator* has much in common with the simple multiple slit interference problem in optics; the angular width is reduced by a factor $1/N$, and the axial brightness increases by a factor N^2, where N is the number of undulations. The wavelength of the radiation leading to constructive interference in these conditions is given by

$$L = (L_0/2n\Gamma^2)(1 + (K^2/2) + (\Gamma\Theta)^2) \qquad (1)$$

where L_0 is the periodicity of the magnetic structure, n is the harmonic number, Γ is the energy of the storage ring electrons in units of the electron rest mass, K is a parameter defining the strength of the oscillatory magnetic field, and Θ is the angle of view relative to the axis of the device. Notice that in the weak field ($K \to 0$) limit, the on-axis wavelengths are simply given by the first term, so for a machine energy of 1GeV ($\Gamma \approx 2000$), the first harmonic wavelength is approximately 10^{-7} times the wavelength of the magnetic structure. The exact spectrum obtained from such a device depends on the angular aperture (which constrains allowed values of Θ and allows even harmonics, forbidden on axis, to be seen), and on K. The spectral output from an undulator, integrated over Θ, and sketched on logarithmic axes,is shown schematically in fig. 2. Notice (eqn. 1) that the harmonic wavelengths *increase* as the field strength increases; by contrast the critical wavelength of bending magnet radiation decreases as the field strength increases. At low values of K (weak fields) most of the emitted intensity is in the first harmonic; increasing K leads to increased higher harmonic emission and to a rising continuum background which is the conventional bending magnet synchrotron radiation from the undulator. In order to compare the output with that of normal bending magnets, one can plot the locus of the harmonic peaks as the magnetic strength (K) is varied and the undulator is tuned. We can compare two quantities, the integrated output flux (in photons per second for a specified bandwidth and beam current), and the spectral brilliance or brightness (in photons per second per mrad² per mm² for the same current and bandwidth). The flux comparison requires certain assumptions about the acceptance angle from a bending magnet as this radiation is not intrinsically collimated in the horizontal plane, but the main results of such comparisons are that undulators offer flux gains over bending magnets of a factor of 10-100, but brilliance gains of 10^4 to 10^6.

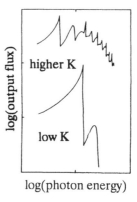

Fig. 2 Schematic graph showing the spectral output of an undulator, and the way this spectrum changes as the strength of the modulating magnetic field (and hence K) is varied. Low K means K<1; higher K is for K≥ 2-3.

log(photon energy)

This clear advantage of undulator radiation is offset by at least one restriction: their limited spectral range. The range of energies available from any specific undulator on a particular storage ring is determined by the K range over which it can be operated, and by the number of useable harmonics emitted. The extension in the range to be achieved by constructing different undulators on the same storage ring is also constrained by the range of useful magnetic periodicities (L_0). To understand this we should first note that the length of the straight section into which an undulator is installed is fixed by the machine, typically in the range 1-5 m. The upper limit of the period of any useful undulator is then set by the fact that one needs a minimum of say 10 periods if one is to make substantial use of the coherent interference characteristic. The lower limit is set by the fact that it is difficult to make a useful periodic magnetic structure with a period much less than the gap between the magnets, and the lower limit of this value is set by the minimum size of the storage ring vacuum vessel needed to store an electron beam; typically this value is 15-25 mm. The useful range of L_o is therefore no better than 2-50 cm even in the most favourable conditions. Short period undulators also offer rather narrow range tunability because the minimum gap also constrains the range of useable K (because K can only be increased by reducing the gap between the magnet arrays). There is also a general upper limit to K for practical operations if the user is to be able to adjust K, because high K devices have a significant disturbing effect on the storage ring electrons, making undulator tuning potentially disruptive. In addition, high K values lead to increasing continuum backgrounds and thus increasing unwanted power; ultimately, one produces a multipole wiggler, a high power, but nonselective, device. The overall impact of these constraints is to ensure that for a given machine energy it is difficult to operate true harmonic undulators over a total energy range of more than about a factor of 50. The only way to extend the range is by changing the machine energy (or the machine), in which case all the output photon energies change by a factor equal to the square of the machine energy ratio (see equation (1)). For undulator operation in the specific photon energy range which we earlier defined, 300-2000eV, the optimum machine energy is found to be about 2.5-3.0 GeV.

The fact that the greatest gains in performance offered by undulators occur in the spectral brilliance means that the full benefit of undulators is obtained in those experiments which depend on this parameter. In truth, experiments are normally limited by the useable flux which can be delivered to the sample, so the extent to which integrated flux or spectral brilliance is important depends on the characteristics of the optical transfer system which takes photons from the storage ring to the sample. Brilliance is a measure of the flux which can be focussed into a given spot size or into a narrowly convergent or divergent beam. This means that the kind of experiments which benefit most from high brilliance are those requiring very small focal spots on the sample, or very high spectral resolution which is ultimately dictated by the flux which can be transferred through defining apertures in the dispersion surface of a monochromator.

Two further aspects of undulator sources are relevant to their impact in scientific exploitation. One concerns the fact that, as we have already implied, undulator radiation has a high component of coherent radiation which can be advantageous for certain applications. The second is that they offer not only wavelength control, but also polarisation control. In normal bending magnet synchrotron radiation, the in-plane radiation from a perfect source is perfectly linearly polarised, the perpendicular polarisation component increasing (while the overall flux decreases) as one moves out-of-plane. A real (finite emittance) storage ring does have some degradation of this character, but it is basically a source of reasonably linearly polarised light, with some possibility of providing a weaker source of elliptically polarised light of lower intensity by deliberately taking light out-of-plane. A simple undulator of the type shown in fig. 1, which exploits coherence in dipole radiation, is an even better source of linearly polarised light.

In addition, however, it is possible to fabricate undulators with circular or controllable radiation. For example, by adjusting the separation of two, end-to-end in-line undulators, with their magnet structures rotated by 90° relative to the axis, one can change the phase and thus the form of output polarisation from such a device. Tunable polarisation at high spectral brilliance offers important advantages for some types of experiment as we shall see below.

3.   SCIENTIFIC EXPLOITATION POTENTIAL

In a very short review it is clearly not possible to do full justice to all the areas of science which are seen as likely to benefit from undulator sources. Brief examples are given here of a few cases, relevant to solid state materials and surface science which are expected, and in several cases are already being found, to benefit in this way. We should also stress that very few undulators are currently operating in the soft x-ray range, although many are planned; the examples are thus mainly based on extrapolation of current achievements.

3.1   Photoelectron and photoabsorption microscopy

A longstanding goal in surface analysis is that of chemical state (as well as element specific) imaging of surfaces using high spectral resolution core level photoemission. Synchrotron radiation, and particularly undulator radiation, offers some important potential advantages to achieving this goal relative to conventional laboratory soft x-ray sources (mainly Al and Mg K_α). Specifically, the need for high spectral resolution and high photon fluxes focussed into small spots on the sample are ideally matched by the high spectral brilliance of the undulator; additionally, sensitivity to specific elements can be optimised by tuning the photon energy to be close to the photoionisation threshold for the appropriate core level. One route to achieving this, by focussing undulator radiation with zone-plate optics (ideally suited to the source coherence) and mechanically scanning the sample, has already met with some success (Ade et al, 1990). The alternative route of imaging the emitted photoelectrons in an energy selected fashion to obtain the spatial resolution is also being investigated, but this aspect of the technique shares all the same problems as instruments based on laboratory x-ray sources.

By contrast, the tunability of synchrotron radiation can be exploited to obtain the same chemical state information via shifts in the x-ray absorption threshold energies rather than the photoelectron binding energies. This technique of μXANES (microscopic x-ray absorption near-edge structure) has been extensively demonstrated already by Tonner and coworkers (e.g. Tonner and Harp, 1989). In this method the chemical state imaging relies only on high spectral resolution in the photon flux and thus is ideally suited to undulator radiation; the electron imaging, using electrostatic optics, is simply of the low energy secondary electrons which can be imaged at high spatial resolution with no careful energy filtering. Indeed, a variant on the μXANES method, using the zone-plate focussed undulator radiation to obtain the spatial resolution, is also proving a powerful transmission microscopy method for biological and polymeric samples (Ade et al, 1992). The basic potential of the near-edge absorption spectra to provide the chemical state imaging is well-illustrated by results from a recent study of the thermal desorption of the native oxide from silicon surfaces by Harp et al (1990). Using μXANES from 10μm regions of the surface as the oxide is removed by a nucleation and growth process, intermediate oxidation states are clearly seen at the Si L-edge. Fig. 2 shows individual μXANES spectra for different oxidation states obtained in this way; although these spectra are enhanced by subtracting clean surface data, the source of image contrast and of spectral information is clear.

Fig.3. XANES spectra near the Si L-edge from different oxidation states of Si at the surface, taken in selected 10μm regions of the surface and subjected to background subtraction. The peaks are attributed to Si 2p core excitons. After Harp et al (1990)

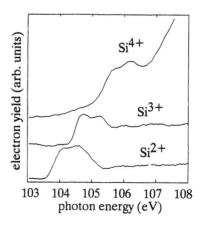

3.2   Magnetic spectroscopy and microscopy

One of the main uses of current undulators in the vacuum ultraviolet (typically 10-150eV) is as a source for spin-polarised angle resolved photoemission (SPARUPS) experiments. By specifying the energy, direction and spin of photoemitted electrons in this energy range from magnetised single crystal samples, one can map the spin-dependent electronic band structure. This experiment, which provides unique information on magnetic properties, is extremely photon-hungry because of the need to resolve the electrons in energy and direction, coupled with the low efficiency of spin detectors (Mott detectors or similar); the higher flux output of undulator radiation is therefore of huge value (e.g. Johnson et al, 1992). A rather different experiment which could prove of great interest for the exploitation of undulator radiation in the soft x-ray range is magnetic circular dichroism in both photoabsorption spectroscopy and microscopy. These experiments involve the measurement of the near-edge photoabsorption (XANES) in magetised samples using circularly polarised light. Substantial differences are found in these spectra depending on whether the photon polarisation vector lies parallel or anti-parallel to the magnetic (majority) electron spins of the sample, and recent observations of large effects in the L-edges of the first row transition elements (notably in Ni by Chen et al (1990)) have highlighted the value of measurements in this energy range for some of the key ferromagnetic metals. Bulk elements can be investigated quite successfully using bending magnet synchrotron radiation out of the plane of the storage ring, but variable polarisation undulators offer fascinating potential for the study of local magnetic effects in dilute alloys. Moreover, a recent experiment which couples this phenomenon with the μXANES technique of Tonner and coworkers, described above, promises to be of great value in practical studies of magnetic materials. Because of the pronounced difference in photoabsorption depending on the spin direction, imaging the secondary electrons from a magnetised sample produced by circularly polarised radiation at the adsorption edge provides a route to imaging magnetic domains on a sample. This idea has recently been brought to fruition in experiments at the Co and Ni L-edges ($\approx$780eV and 850eV) and magnetic domains have been imaged on a thin film magnetic recording disc with a spatial resolution of 1μm by Hermsmeier and Stöhr (1992) and coworkers.

3.3   Photoelectron diffraction and holography

A quite different aspect of surface science investigations is the determination of surface structure at an atomic level. In particular, if a surface has an adsorbed atom, molecule or molecular fragment on it, can we determine the atomic registry of this adsorbate relative to the substrate as well as finding if either the molecule or the substrate is modified by the interaction? One route to achieving this which appears to be particularly effective for adsorbed molecular layers which commonly lack long-range order is by photoelectron diffraction. In this technique the angle and energy dependence of photoemission from core levels in the adsorbate are measured; the coherent interference of the directly emitted electron wavefield with components elastically scattered by surrounding atoms leads to modulations in the measured intensity which can be related to the local geometry. Recently attention has been drawn to the fact that the angular distribution of electrons in this experiment are actually a photoelectron hologram (Barton, 1988), although it remains to be seen whether a practical route to direct imaging of atomic locations is offered by this technique (Dippel et al, 1992). Whether or not this is the case, the basic technique is now a proven route to structure analysis (Woodruff, 1992). One particular development of this method which will greatly enhance its range of applicability is to work at sufficient spectral resolution that emission from atoms of the same element but different structural environment can be distinguished by their photoelectron binding energy chemical shifts. This offers the possibility of determining atomic adsorption sites for different components of complex coadsorption systems, such as hydrocarbon fragments on model catalysts. Recently, the basic demonstration of this experiment has been achieved by Weiss et al (1992) for the two chemically distinct C atoms within adsorbed acetate (CH_3COO-) and fluoroacetate (CF_3COO-). Fig 4 shows a representative photoelectron energy spectrum from this system, in the vicinity of the two chemically shifted C 1s peaks, and the resulting photoelectron diffraction spectra obtained by measuring the intensity of each of these as a function of photoelectron kinetic energy in a specific direction. Also shown are theoretical simulations of these spectra based on the model structure shown in the inset. This model system, however, has rather large chemical shifts, and far more general application of the method will necessitate a combination of high flux and spectral resolution which is only likely to be achieved using undulator radiation. This example also highlights the importance of the soft x-ray energy range for surface chemistry, as the K-edges (or 1s binding energies) of the chemically important C, N and O species all fall in the 280-540eV energy range, and their further study is therefore a prime motivation for the development of soft x-ray undulators.

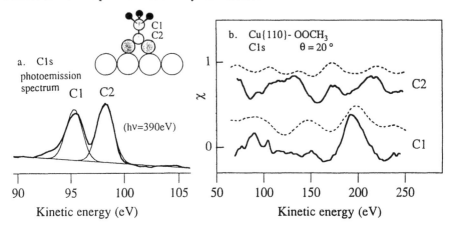

Fig. 4 Chemical shift photoelectron diffraction data from acetate on Cu(110). In (a) is shown the two chemically shifted C 1s peaks of a photoelectron energy spectrum, corresponding to the methyl and carboxyl carbons (see inset). In (b) is shown a comparison of the two resulting photoelectron diffraction spectra recorded from these peaks with the results of theory assuming the acetate bridges two nearest neighbour Cu atoms, as shown in the inset.

REFERENCES

Ade H, Kirz J, Hulbert S, Johnson E D, Anderson E and Kern D 1990
   Appl.Phys.Lett. 56 1841
Ade H, Zhang X, Cameron S, Costello C, Kirz J and Williams S 1992 Science
   submitted
Barton J J 1988 Phys.Rev.Lett. 61 1356
Chen C T, Sette F, Ma Y and Modesti S 1990 Phys.Rev.B 42 7262
Dippel R, Woodruff D P, Hu X-M, Asensio M C, Robinson A W, Schindler K-M,
   Weiss K-U, Gardner P and Bradshaw A M 1992 Phys.Rev.Lett. 68 1543
Harp G R, Han Z L and Tonner B P 1990 J.Vac.Sci.Technol. A8 2566
Hermsmeier B and Stöhr J 1992 private communication
Johnson P D, Brookes N B and Chang Y 1992 Mat.Res.Soc.Symp.Proc. 231 49
Tonner B P and Harp G R 1989 J.Vac.Sci.Technol.A 7 1
Weiss K-U, Dippel R, Schindler K-M, Gardner P, Fritzsche V, Bradshaw A M,
   Kilcoyne A L D and Woodruff D P 1992 submitted for publication
Woodruff D P 1992 Angle Resolved Photoemission ed S D Kevan
   (Amsterdam, Elsevier) p243-290

Inst. Phys. Conf. Ser. No 130: Chapter 8
Paper presented at Int. Congr. X-ray Optics and Microanalysis, Manchester, 1992

635

Combined time resolved SAXS/WAXS/DSC experiments

Wim Bras*, Gareth Derbyshire, Anthony Ryan[+], Geoff Mant, Rob Lewis, Andrew Felton, Neville Greaves.

SERC Daresbury Laboratory

*Daresbury Laboratory/Netherlands Organisation for Scientific Research (NWO)

+University of Manchester Institute of Science and Technology

Abstract: An instrument is developed which offers the possibility to perform simultaneous Differential Scanning Calorimetry and time resolved X-ray scattering experiments in the range from approximately 0.007 - 0.21 Å^{-1} and 0.31 - 4.2 Å^{-1}. A synchrotron Small Angle X-ray Scattering beamline is therefore equipped with two proportional position sensitive detectors. Ultimate system time resolution will be of the order of 0.01 s. Successful test experiments are performed on polymer samples.

1. INTRODUCTION

There are several fields in which it would be advantageous to collect wide and small angle X-ray scattering patterns simultaneously, thus gathering structural information from roughly atomic resolution to structures of about 1000 Å. One of the main applications is found in the study of phase transitions. The melting and deformation behaviour of polymers are a good example described by Ryan, as well as the temperature induced rearranging of lipids in membranes reviewed by Lis. If one manages to combine this structural information with simultaneously obtained thermodynamical information from Differential Scanning Calorimetry one has potentially a very powerful instrument.

One of the main problems if one wants to perform real time X-ray diffraction experiments is the intensity of the X-ray source. Practically this means that one has to use a synchrotron beamline. Several user groups have managed to combine SAXS with DSC experiments. Examples are given by Ungar and Russel. Attempts to combine SAXS and WAXS have been reported from the Brookhaven(Tashiro), Hamburg(Bark) and Tsukuba(Nojima) synchrotron radiation facilities. All the systems mentioned here, can be improved upon, either in time resolution, in the q-range to be studied or in the capacity to handle the X-ray fluxes available with synchrotron radiation.

2. Experimental equipment

We have assembled equipment that potentially can combine all the three techniques mentioned before and will provide a (system) time resolution that is more than adequate to follow fast phase transitions. The ultimate X-ray flux that this system is capable of handling as well as the sensitivity are at least one order of magnitude better than previously built systems and form a welcome addition to the facilities at Daresbury Laboratory.

The basis of the system is the Small Angle Diffraction beamline 8.2 at the Synchrotron Radiation Source (Daresbury UK). The beamline can generate a flux of 4×10^{10} photons/sec at the sample position with a wavelength of 1.52 Å. The bandwidth is $\Delta\lambda / \lambda \leq 4 \times 10^{-3}$. In the focal plane the beam size is 3×0.3 mm^2 (H x V). The horizontal and vertical divergence are respectively 5 and 0.25 mrad.

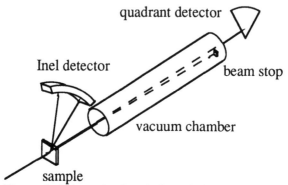

Figure 1. Schematic description of the experimental equipment. Optical elements not shown.

Detection of the small angle pattern can be performed with a proportional position sensitive delay line detector, either a quadrant or an area detector as described by Lewis. The sample to detector distance is set to 3.5 meter. The wide angle diffraction pattern is more difficult to detect properly. Linear delay line detectors in general have parallax problems, when photons are incident at large angle due to the finite depth inside the detector. Beside that, a long linear detector will not satisfy the requirement to cut through the theoretical Ewald sphere over its whole length. The solution that we have found for this problem is to use a commercially available curved Inel delay line detector. The detector was positioned such that its centre of curvature coincided with the sample. It was slightly offset so that the small angle pattern could pass by it unhindered. Obviously this leaves a part of the spectrum undetected but with careful alignment this can be reduced to a minimum.

The Inel detector has a dynamic range of approximately 1:10^4. The countrate limit is approximately 100 kHz over the whole detector, and the system spatial resolution is 0.1°. The opening angle is 120°, of which only 75°were used in these experiments.

The Daresbury quadrant has a dynamic range of 1:10^6, a system spatial resolution of 250 µ, and a countrate limit of 5×10^5 counts/sec. The X-ray beam was focussed at a position half way between the two detectors. To reduce the smearing due to under or over focus in the detector planes the distance between the two detectors was kept as short as possible. In order still to be able to detect the SAXS pattern the opening angle for parasitic scatter had to be reduced to a minimum by increasing the distance between the last two guard slit sets of the camera.

The standard Daresbury VME system was used to perform the time framing, essential for the following of real time processes. TTL pulses generated by the VME system were used to drive the PC, that was used to control the INEL detector, in a master - slave mode.

Differential Scanning Calorimetry equipment that is suitable for use in these experiments has been developped in collaboration with Linkam Scientific Instruments Ltd. The initial off-line experiments show that the sensitivity is comparable with several off-line instruments. The temperature range over which experiments can be performed is from - 196 to +600° C.

4   Results

In order to test the performance of the set up, experiments were performed on suitable polymer samples. Spatial calibration of the SAXS data was done with the help of a wet collagen sample. Calibration in the WAXS region was performed with the well known peaks of polyethylene. Scattering patterns were corrected as described by Mant.

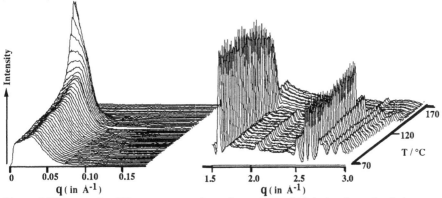

Figure 2 Time resolved X-ray patterns from the melting of high density polyethylene.

In figure 2 three dimensional plots of scattering vector versus intensity, obtained from an experiment on high density polyethylene, are shown. The material was annealed and melted by raising the temperature with 2.8°C/min from 70° to 150° C. The time resolution in this experiment was dictated by the experiment rather than the experimental equipment.

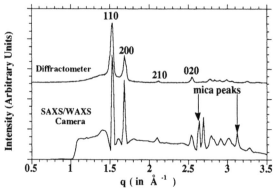

Figure 3 Comparison between the SAXS/WAXS camera and data obtained from a Philips θ/2θ diffractometer. The extra peaks (at q= 2.384 and 1.999 Å^{-1}) have to be attributed to hexagonal muscovite; the material used as cell windows.

The spatial resolution in the WAXS region compares favourably with data obtained from an X-ray diffractometer using CuK_α radiation. This can be explained by the lower divergence of the synchrotron radiation (0.25 versus 14 mrad). Unfortunately it can be seen that there are problems due to the exceeding of the local countrate in the strong diffraction peaks. This does not have to be a problem since the signal to noise ratio is very high, which means that the data rate can be lowered considerably by attenuating the detector. The performance of synchrotron beamlines used for SAXS experiments is in general superior to conventional sources.

We gratefully acknowledge the support from our colleagues at Daresbury Laboratory, especially Colin Morrel for his help in producing the drawings.

LITERATURE
Bark, M. et al, Polym.Prep.Am.Chem.Soc.Div.Polym.Chem.,1990,31(2),163
Bouwstra J.A. et al. Int.J.of Pharmaceutics 79(1992)141-148
Lewis R.A.et al , (1988)Nucl.Instrum.Methods Phys. Res.A,273,773-777
Lis L.J., et al., J.Appl.Cryst.24,48(1991)
Mant G.R., private communication
Nojima, S. et al.,Macromolecules 1992,25,2237-2242
Russel, T.P. et al.,J.Polymer Sci. Vol23,1109-1115(1985)
Ryan, A.J. et al. Macromolecules,1991,24,2883
Tashiro, K. et al., Macromolecules 1992,25,1809-1815
Ungar, G and Feijoo,J., Mol.Cryst.Liq.Cryst,1990,180B,281

Inst. Phys. Conf. Ser. No 130: Chapter 9
Paper presented at Int. Congr. X-ray Optics and Microanalysis, Manchester, 1992

639

Summary and highlights of XIII ICXOM92—What next?

Kurt F. J. Heinrich, 804 Blossom Drive, Rockville MD USA

ABSTRACT: The present state of electron probe microanalysis is reviewed, with special attention to the material presented in this congress. It will be attempted to predict future developments, and requirements for further progress.

INTRODUCTION

Recent years have seen a remarkable revival in the development of electron probe microanalysis (EPMA) techniques, which is well reflected in the communications presented at the XIIIth ICXOM congress[1] as well as in other recent conferences in Europe and in the USA. This may be a good time to point out the new, and sometimes diverging, paths this branch of analysis has taken. To predict future events is somewhat riskier; what one can do in this regard is to present an opinion as to which lines of development seem worthwile to follow.

WHAT WE NEED

To perform analyses, we need the <u>instruments</u> and devices that perform measurements, <u>standard materials</u> of known composition, <u>data reduction procedures</u> ('correction procedures') and the <u>expressions</u>, <u>parameters and constants</u> which aim to define the physical processes that form the basis for the corrections.

INSTRUMENTS

The main new instrumental characteristics are greater stability, new efficient spectrometers, particularly for long-wavelength (E<1keV) x-rays, both energy- (ED) and wavelength-dispersive (WD), and the availability of fast and powerful computers for on-line treatment of the experimental data. Wavelength-dispersive analysis of soft x-rays can now take advantage of more efficient element multilayer diffractors [2.4.9]. The resolution of solid-state energy dispersive detectors is still limited to 130-140 eV [1.3.3], but impressive advances are achieved in the identification and separation of lines and bands and in background reduction (Fiori et al 1992), and the energy range of solid state detectors has been extended to boron Ka.

The introduction of solid-state detectors in scanning and transmission electron microscopes (SEM, TEM) has widened the range of application of x-ray analysis, and special techniques and data reduction schemes were developed for them. In the scanning instruments (including the scanning electron probe microanalyser), quantitative area scanning techniques have become possible due to the speed and the storage capacities of on-line computers. On the other hand, the need for rapid answers renders standardless analysis procedures, particularly with EDS, very attractive (Pouchou and Pichoir 1992). The achievable accuracy should be comparable to that of conventional procedures. The main danger in standardless analysis is the possibility of slow changes in the detector and window characteristics, such as contamination and deposition of ice.

---

[1] Communications of this Congress are noted by their number in square brackets.

The application of EPMA to soft biological tissue, an important goal, was never fully achieved with the conventional instrument. The Analytical Electron Microscope (i.e., Duncumb's EMMA instrument with a solid-state EDS), promises a radical change in that regard (Somlyo 1992). This instrument has a significantly increased spatial resolution, which is also very useful in metallurgical and meteorite analysis (Williams and Goldstein 1991, [1.5.1]).The correction procedures which apply in this case are much simpler than those for thick specimens.

Most discussions about the analysis of biological tissue center on the problem of preparing the specimen without migration or removal of elements. Special SEM instruments are also developed in which wet specimens can be investigated in an atmosphere of reduced pressure, at electron energies typically below 5 keV.

DATA REDUCTION PROCEDURES

For many years, there were essentially two procedures: the 'theoretical' or ZAF procedure (Philibert and Tixier 1968) and the empirical procedure (Ziebold and Ogilvie 1946) based on a hyperbolic approximation to the analytical function of relative intensity as a function of concentration. Because detection of low-energy x-rays has become available, the time-honoured procedures are now being replaced by a variety of new ones of greater accuracy, particularly in the absorption correction [1.2.1]. Some of them (Scott and Love 1991, [1.2.11], Duncumb 1991) are ZAF procedures, i.e., they use a model for energy loss and x-ray production for the stopping power calculation, another calculation for backscatter losses, an absorption function for characteristic lines derived from a generalization of the experimental $\phi(\rho z)$ curves first developed by Castaing et al (Castaing and Descamps 1955, Castaing and Henoc 1966), and a correction for fluorescence, usually that developed by Reed and Long (1963) and recently improved by Reed (1990). Pouchou and Pichoir (1991) have proposed a model along different lines, which, however, can be described as a modified ZAF model since the Z, A, and F factors are derived separately, although their assembly is presented in a different manner.

The $\phi(\rho z)$ method (Brown 1991, Packwood 1991) is based solely on the experimental $\phi(\rho z)$ curves. Many new tracer experiments have been performed in recent years, particularly at low operating energies (Brown and Parobek 1972, Karduck and Rehbach 1991). Bastin et al have used this method in their extensive and important work on analysis of elements of atomic number below ten (Bastin and Hejligers 1991a). They concluded that it was more accurate for the absorption correction than for the atomic-number correction, and substituted the PAP method of Pouchou and Pichoir for the latter (see also [1.2.1]). Bastin et al have also studied the effects of low electrical conductivity in specimens (Bastin and Hejligers 1991b), a problem that for many years had been neglected or treated superficially.

The PAP method was developed with the aim to extend EPMA to thin layers, coated specimens, and even to the characterisation of distributions in depth of diffused specimens. There is an increasing interest in such specimens, with numerous publications (Bastin et al 1992, [1.2.7], [1.2.14]). Special techniques are also being developed for the analysis of submicron particles and fibers (Armstrong,1991).

The most detailed description of the electron-target interaction is the Monte-Carlo method (MC, Henoc and Maurice 1991, [1.2.14]). The relevant events are applied randomly but with the appropriate frequency to simulations of the paths of the electrons penetrating into the specimen. The method is very flexible, since it is easy to adapt it to any specimen shape which determines the end of the trajectories of those electrons which emerge from the specimen.

Although MC was used very early by Bishop (1966) to predict the energy distribution of backscattered electrons, it cannot be applied yet in on-line calculations, because of the large number of trajectories and interactions required to obtain statistically meaningful results. It has, however, been used in various ways: several authors have calculated with it parameters such as the backscatter loss and approximated the result by algebraic formulae; usually, the models were simplified (multiple scattering models, assumption of continuous energy loss), and certain interactions such as collective scattering were neglected.The scattering cross-sections were recently subjected to critical studies, and recent calculations check well with experimental data. Duncumb (1991) showed that for the conventional specimen configuration good results can be obtained by approximating algebraically the results obtained with a simple MC.

In the application of several correction programs to a large number of data on binary specimens we experienced occasional failures. These could be eliminated when the formal integrations used in the models of reference were replaced by numerical integration, without any penalty regarding the speed of execution [1.2.1]. Abandoning the use of the Laplace transformation and the logarithmic integral also allows a wider choice of models for $\phi(\rho z)$ and for the electron deceleration.

PHYSICAL MODELS AND PARAMETERS

The accuracy of procedures based on the description of the physical processes in the target is limited by the accuracy of models and parameters. Perhaps the most conspicuous of these is the x-ray mass absorption coefficient (MAC). A new set of equations for the calculation of MAC's (Heinrich 1987a) has been reported to provide improvement and is widely used. However, for radiation of energy below 0.8 keV this set provides rough approximations only, not intended for accurate quantitative work. In this energy range most authors use Henke's publications (see Bastin and Hejligers 1991a and 1991b), and some critical values only provide acceptable results after being empirically adjusted. The poor accuracy of MACs, of fluorescent yields, weights of lines, chemical effects and Coster-Kronig factors in certain areas complicate quantitative analysis with the use of low-energy x-ray bands, which should be approached by the casual user with great caution.

The stopping power equation has also been the subject of various proposals. The equation of Bethe for energy loss was not intended for use at the low energies typical for EPMA, and leads to inaccuracies in the atomic number correction. PAP (Pouchou 1991) and Scott and Love (1991) use empirical equations which approach the Bethe equation at higher energies and diverge from it at lower energies. Joy (1989) instead proposes an equation which modifies the value of the mean energy loss (J) at low voltages; this proposal is used by Duncumb in the aforementioned model. The choice of expression for J seems to be somewhat arbitrary and imbedded in the empirical adjustment of various data reduction routines.

The value at zero depth of the $\phi(\rho z)$ curve is of great importance in the calculation of absorption of primary x-rays which are strongly absorbed. This subject is treated in communications [1.2.6] and [2.1.2].

In spite of the work of Henoc (1973), the effects of fluorescence caused by the continuum are still not given the attention which they deserve. One suspects that some of these effects are hidden in the current models for atomic number effects. The binary data set used by Heinrich [1.2.1] does not contain a sufficient number of significant cases to be of great help. The effect is expected to be large when high-energy radiation is measured in matrices of low atomic numbers. While the formal calculation of this correction is quite cumbersome, much simpler procedures can be followed if numerical integrations are performed over the energy range of interest (Heinrich 1987b).

STANDARD REFERENCE MATERIALS

Standard reference materials (SRM's) can be used during analysis for obtaining relative x-ray intensities or for testing quantitative accuracy of methods. With regard to the second purpose, in spite of the number of binary data pairs now available, many situations are still poorly covered. One of the advantages of MC is the possible use of standard materials of a variety of shapes (foils, fibers, particles) to test its accuracy. The production of new SRM's is a very time-consuming and expensive task, particularly for layered or diffused specimens. International cooperation would be very useful in preventing duplications or production of materials of limited usefulness.

WHAT CAN WE EXPECT IN THE FUTURE?

The advances in spatial resolution of the ATM and other microanalytical techniques hold the promise that resolution on the atomic scale, which has been demonstrated already, will be a practical tool of immense importance in the future. But even the conventional EPMA techniques require further improvements in order to reach their full potential. Several physical parameters are in need of better definition. The x-ray mass absorption coefficients are poorly defined at lower energies (Bastin and Hejliger 1991a, Pouchou and Pichoir1991); fluorescent yields, Coster Kronig parameters (Lábár 1991) and relative line intensities are still poorly known, at least for emissions from shells other than the K shell. The controversy concerning stopping power is still alive, but the proposed solutions converge rapidly. The same cannot be said of the handling of fluorescence due to the continuum.

Once an agreement has been reached concerning usable schemes for quantitative data reduction, our instruments will be equipped with automated programs for both ED and WD spectrometers which will not only provide the results but also issue warnings when the working conditions are less than adequate. This, in turn, will make a simple inexpensive SEM with an ED detector and automated quantitation the preferred tool for most routine EPMA. The treatment of energy-dispersive spectra by Fiori et al (1991) indicates what we can expect in that regard.

There exist already schemes for standardless quantitative analysis applied to routine ED work. When the detector efficiency across the x-ray spectrum is well known, the accuracy of this technique is similar to that of conventional EPMA. What is needed to increase the confidence in the procedure is a single-standard calibration method which will correct automatically for grad-ual changes in the detector efficiency due to contamination, and produce a warning if major changes have occurred.

With the improvements in speed and accuracy of the MC techniques and the development of al-gebraic formulae derived from MC we can expect that rapid simulations for special specimen configurations, and for the construction of $\phi(\rho z)$ curves at various conditions will soon be gen-erally available. A wider use of this method will also encourage us to consider secondary target events which are not yet handled efficiently.

Quantitative scanning techniques have been developed (Newbury 1992) and will be used exten-sively to investigate materials problems. It is possible to store all significant signals obtained in an area scan, and after initial observation of the scanning images of the entire area, to select re-gions, lines or points of particular interest for further quantitative evaluation. There will also be interesting developments in the presentation of the analytical results: the stunning images often seen in modern computer work, experimentally or for entertainment ('virtual reality') could be applied to produce instantaneously extraordinary visual images of three-dimensional element distributions on a microscopic scale.

The experience at the National Institute of Standards and Technology (formerly NBS) as well as in other laboratories has demonstrated the great power of combining EPMA information with that obtained from other techniques such as Auger microscopy (Cazaux 1992), secondary ion, electron energy loss, and laser Raman spectroscopies. This list is widening: we foresee that serious material research will rely even more strongly on the coordination of techniques in the future. A further step in that direction is the combination of specimen preparation and anal-ysis by various means in situ in the chamber of a single instrument. This idea has been realized in some cases, but it will be developed to a much larger extent. By the same token it is obvious that in-depth investigations require teams of investigators with knowledge in diverse fields of science, such as microanalysis, statistics, chemistry, crystallography, solid-state physics, and, in specific cases, of metallurgy, geology or biology (Somlyo, 1992). We will also have access to large networks for reception and transmission of data, programs and publications, on an un-precedented scale. Our libraries are already being rendered obsolete by storage of their contents in computer memory.

Unfortunately, the same process will probably also affect the organization of large conferences and congresses, which could be replaced by the transmission of presentations through information networks. I trust, though, that there will still be a place for small informal gatherings of experts armed with pencils and paper, preferably in a whitewashed room on a Mediterranean island (recorders and electronic notebooks might be admitted). In the mean time, let us enjoy the present state of affairs while it lasts!

REFERENCES

Armstrong JT (1991) Electron Probe Quantitation, eds Heinrich KFJ and Newbury DE, (New York: Plenum Press) 261

Bastin GF and Hejligers HJM (1991a) Electron Probe Quantitation, eds: Heinrich KFJ and Newbury DE (New York: Plenum Press) 145

Bastin GF and Hejligers HJM (1991b) Electron Probe Quantitation, eds: Heinrich KFJ and Newbury DE (New York: Plenum Press) 163

Bastin et al (1992) Mikrochim. Acta 12 93

Bishop HE (1966) Proc. 4th ICXOM, eds: Castaing R, Deschamps P, Philibert J, (Paris: Hermann) 153

Brown JD (1991) Electron Probe Quantitation, eds Heinrich KFJ and Newbury DE (New York: Plenum Press) 77

Brown JD and Parobek L (1972) proc. 6th Conf.X-ray Optics and Microanalysis. eds Shinoda G et al (Tokio: U of Tokio Press) 163

Castaing R and Descamps J (1955) J Phys Radium 16 304

Castaing R and Henoc J (1966) Proc 4th ICXOM, eds: Castaing R, Deschamps P, Philibert J (Paris: Hermann) 120

Cazaux J (1992) Mikrochim Acta suppl 12 37

Duncumb P (1991) Proc. Electron Microscope Society of America, Microbeam Analysis Society and Microscopical Society of Canada (San Francisco: San Francisco Press) 1674

Fiori CD et al (1992),NIST/NIH DeskTop Spectrum Analyzer, Standard Reference Data NIST Gaithersburg MD 20899

Heinrich KFJ (1987a) Proc.11th ICXOM, eds Brown J D and Packwood R, U Western Ontario, Canada 67

Heinrich KFJ (1987b) Microbeam Analysis (San Francisco: San Francisco Press) 24

Henoc J et al (1973) NBS Tech. Note 769, U.S. National Bureau of Standards, Gaithersburg, MD

Henoc J and Maurice F (1991) Electron Probe Quantitation, eds: Heinrich KFJ and Newbury DE (New York: Plenum Press) 105

Joy DC and Luo S (1989) Scanning 11 176

Karduck P and Rehbach W Electron Probe Quantitation, eds: Heinrich KFJ and Newbury DE, (New York: Plenum Press) 191

Lábár JL (1991) Electron Probe Quantitation, eds: Heinrich KFJ and Newbury DE (New York: Plenum Press) 219

Newbury DE (1992) Microscopy 22 1 11

Packwood R (1991) Electron Probe Quantitation, eds: Heinrich KFJ and Newbury DE (New York: Plenum Press) 83

Pouchou JL (1992) Proc. Electron Microscope Society of America, Microbeam Analysis Society and Microscopical Society of Canada (San Francisco: San Francisco Press) 1650

Pouchou JL and Pichoir F (1991) Electron Probe Quantitation, eds Heinrich KFJ and Newbury DE (New York: Plenum Press) 31

Pouchou JL and Pichoir F (1992) Proc. Electron Microscope Society of America , Microbeam Analysis Society and Microscopical Society of Canada (San Francisco: San Francisco Press)

Philibert J and Tixier R (1968) NBS Spec Publ 298 13

Reed SJB (1990) Proc. Microbeam Analysis Society, eds: Michael JR and Ingram P (San Francisco: San Francisco Press) 109

Reed SJB and Long JVBI (1963) Proc. 3d CXOM eds: Pattee HH et al (New York: Academic Press) 317

Scott VD and Love G (1991) Electron Probe Quantitation, eds: Heinrich KFJ and Newbury DE (New York: Plenum Press) 19

Somlyo AP et al (1992) Proc. Electron Microscope Society of America, Microbeam Analysis Society and Microscopical Society of Canada (San Francisco: San Francisco Press) 1472

Williams DB and Goldstein JI, (1991) Electron Probe Quantitation, eds: Heinrich KFJ and Newbury DE, (New York: Plenum Press) 371

Ziebold TD and Ogilvie RE (1964) Anal. Chem. 46 322.

Author Index

Subject Index

Milton Keynes UK
Ingram Content Group UK Ltd.
UKHW031123141024
449569UK00006B/468